Salters Horners Advanced Physics for Edexcel A2 Physics

STUDENT BOOK

A PEARSON COMPANY

Pearson Education
Edinburgh Gate
Harlow
Essex
CM20 2JE
United Kingdom

and Associated Companies throughout the world

www.pearson.com

First published 2000
Second edition published 2008
This edition 2009

ISBN 978-1-40820-5860

Designed and illustrated by Pantek Arts, Maidstone, Kent
Indexer John Holmes
Printed and bound by Graficas Estella, Bilboa, Spain

The publisher's policy is to use paper manufactured from sustainable forests.

'Please cite as: Salters Horners Advanced Physics Project, A2 Student Book, Pearson Edexcel, London, 2009.'

We are grateful to the following for permission to reproduce copyright material:

Driving Standards Agency for Highway Code stopping distances, Crown © copyright, Crown Copyright material is reproduced with permission under the terms of the Click-Use License; Edexcel Limited for the Edexcel exam questions: Q30 from Edexcel paper PSA1 June 1999 (q5); and Q31 from Edexcel paper PSA1 Jan 2004 (q2) copyright © Edexcel Limited; Faber and Faber Ltd for extracts from the poem 'The Hollow Men' by T.S. Eliot and 'In the Beginning' from Shema by Primo Levi translated by Ruth Feldman and Brian Swann, Menard Press, 1976, reproduced with permission of Faber and Faber Ltd; John Fox and Norman Palmer, SHAP physics teachers at De Aston School, Market Rasen for their eyewitness accounts of February 2008 Market Rasen earthquake; Popular Science Magazine for an extract adapted from Window Solar Collector: Venetian blinds with a new slant by Rochelle Chadakoff originally printed in the November 1978 Issue of Popular Science Magazine, reproduced with permission; Rail News Ltd for an extract from Rail News, June 1992 copyright © Rail News Ltd; and The Society of Authors for an extract from the poem "The Torch Bearers" by Alfred Noyes, reproduced with permission from The Society of Authors as the Literary representative of the Estate of Alfred Noyes.

In some instances we have been unable to trace the owners of copyright material and we would appreciate any information that would enable us to do so.

Picture Credits

The publisher would like to thank the following for their kind permission to reproduce their photographs:

(Key: b-bottom; c-centre; l-left; r-right; t-top)

airbus.com: 74, 75c, 75t, 88; Alamy Images: Colin Palmer Photography 202; Eddie Gerald 73; Harriet Cummings 70; Heiner Heine/imagebroker 102; Image Source Pink 90tr; Paul Rapson 58b; qaphotos.com 5b, 13, 19; Robert Bird 291; Robert Clayton 2b; sciencephotos 164; Tony Harrington/StockShot 122tc; alveyandtowers.com: 5t, 16, 29b, aviation images.com: 103b; BAE Systems Electronics Ltd: 89, 90b; Bancroft School: 204b; Gareth Boden: 40, 76, 79, 99t, 219c; Bridgeman Art Library Ltd: Archaeological Museum of Heraklion, Crete, Greece /Bildarchiv Steffens 184; CERN Geneva: 154, 156, 171b, 171c, 172; Corbis: Bettmann 72, 259, 285; Digital Art 116b; ISSEI KATO/ReutersWorld 219; Jim Sugar 270t; Joachim Pfaff/zefa 336; Jonathan Blair 331tr; Jonathan Selkowitz/NewSport 122tl; KF012578 120; Macduff Everton 259t; Reuters 322; Roger Ressmeyer 282, 333; Skyscan 275t; Courtesy of Museum of New Zealand Te Papa Tongarewa: 225; DK Images: Sarah Ashun 97b, 97c; EFDA-JET: 274; Getty Images: 83b; Sean Gallup 71; Grand Central Railway Company Ltd: 51; Guralap Systems Limited: 193; iStockphoto: Matej Michelizza 122bc; Kobal Collection Ltd: Paramount television 129; Milepost 92 1/2: 31; NASA: 132, 249tl, 251, 319; MSFC-0102088 122bl; SOHO 257bl, 257br; NERC 2008: Paul Witney BGS Photographer 195b, 195c; No Trace: 230, 296t, 352; PA Photos: 50br, 163; AP Photo/Eugene Hoshiko 182; AP/Julie Jacobson 162; Associated Press 200; Xinhua/Chen Xiaowei/wll /Landov 122tr; photographersdirect.com: Dennis Jackson Photography 202br, Tom Thorpe; 296b; Railnews: 43, 47, 122cl; Science & Society Picture Library: 3, 108, 271br, 289; National Railway Museum 50bl; Science Photo Library Ltd: A. Barrington Brown 331tl; 133, 144, 255t, 265, 271bl, 287t, 290b, 302tl, 302tr, 324, 325t, 336cl; Adam Block 325b; Andrew Lambert 84b, 90tl, 148; Cern 114, 115; David A. Hardy 260; David Parker 84t, 117, 124; Dr Rudolph Schild 84c; Dr Seth Shostak 320; Eckhard Slawik 281; Edward Kinsman 48; Emilio Segre Visual Archives / American Institute of Physics 287c, 328; European Southern Observatory 309, 313; European Space Agency 280b; Giphotostock 78; John Chumack 270b; Larry Landolfi 279; Michael Abbey 262; Mount Stromlo and Siding Spring Observatories 316; NASA 129b; NASA / ESA / STSCI / A. Fruchter, Ero Team 351t; NASA / ESA / STSCI / J. Hester & P. Scowen, ASU 248; NASA / ESA / STSCI / R.Williams, HDF Team 249tr; Physics dept., Imperial College 271c; Royal Astronomical Society 296c; Royal Observatory, Edinburgh / AATB 298; Russell Croman 287; Stanford Linear Accelerator Center 161; Tony Mcconnell 183, 233; US Department of Energy 275c; Stable Micro Systems: 122c; University of Canterbury New Zealand: 151; View Pictures Ltd: Ronald Halbe/Artur 227

Cover images: Front: Science Photo Library Ltd: Mehau Kulyk

Picture Research by: Charlotte Lippmann

Every effort has been made to trace the copyright holders and we apologise in advance for any unintentional omissions. We would be pleased to insert the appropriate acknowledgement in any subsequent edition of this publication.

Contributors

Many people from schools, colleges, universities industries and the professions have contributed to the Salters Horners Advanced Physics (SHAP) project and the preparation of SHAP course materials.

Authors of this A2 edition

Steven Chapman (Institute of Education, University of London)
Bryan Berry (Science Learning Centre South West)
Steven Chapman (Institute of Education, University of London)
Frances Green (formerly of Watford Girls' Grammar School)
Nick Fisher (Rugby School)
Greg Hughes (de Ferrers College, Staffs)
Paul Lee (Franklin College, Grimsby)
Chris Pambou (City & Islington College, London)
Sandy Stephens (formerly of The Lady Eleanor Holles School, Middlesex)
Wendy Swarbrick (Lancing College, Sussex)
Elizabeth Swinbank (University of York)
David Swinscoe (City & Islington College, London)
Carol Tear (York)
Clare Thomson (Institute of Physics)

Project director and general editor

Elizabeth Swinbank (University of York)

Acknowledgements

We would like to thank the following for their advice and contributions to this edition.

Paul Denton (British Geological Survey)
Gary Hambling (Virgin Trains)
Graham Meredith (Gloucester)
Robin Millar (University of York)
Steve Pickersgill (Wycombe High School, Bucks)
Somak Raychaudhury (University of Birmingham School of Physics and Astronomy)
Graham P Smith (University of Birmingham School of Physics and Astronomy
Sandra Wilmott (University of York)
Engineers at Alstom and Angel Trains Ltd
Japanese Embassy, London
National Railway Museum, York

Authors of previous edition

This edition is based on the original SHAP course materials and incorporates the work of the following authors.

Jonathan Allday
Chris Butlin
Steve Cobb
Tony Connell
Howard Darwin
Nick Fisher
Alasdair Kennedy
Bob Kibble
Maureen Maybank
Averil Macdonald
David Neal
Kerry Parker
David Sang
Tony Sherborne
Richard Skelding
Elizabeth Swinbank
Carol Tear
Nigel Wallis

Sponsors

We are grateful to the following for sponsorship that has continued to support the Salters Horners Advanced Physics project after its initial development and has enabled the production of this edition.

The Worshipful Company of Horners
The Worshipful Company of Salters
Corus UK Ltd

Advisory Committee for the initial development

Prof. Frank Close
Prof. Cyril Hilsum FRS
Prof. Robin Millar
Prof. Sir Derek Roberts FRS

Contents

How to use this book

Context-led study

Welcome to the A2 part of the Salters Horners Advanced Physics course.

Each teaching unit in the course starts by looking at particular situations in which physics is used or studied, and then develops the physics you need to learn to explore this 'context'.

We have tried to select contexts to give you some idea of how physics can help improve people's lives, how physics is used in engineering and technology, and how physics research extends our understanding of the physical world at a fundamental level. These will show you just some of the many physics-related careers and further study that might be open to you in the future.

Within each teaching unit, you will develop your knowledge and understanding on one or more areas of physics. In later units, you will meet many of these ideas again – in a completely different context – and develop them further. In this way, you will gradually build up your knowledge and understanding of physics and learn to apply key principles of physics to a variety of contexts.

About this book

Each chapter includes the following features:

Main text

This presents the context of each teaching unit and explains the relevant physics as you need it.

Within the main text, some words are printed in **bold**. These are key terms relating to the physics. We suggest that you make your own summary of the these terms (and others if you wish) as you go along. Then you can refer back to it when you revisit a similar area of physics later in the course and when you revise for exams.

Activities

The text refers to many Activities. These include practical work, the use of information technology (e.g. CD-ROMs and the Internet), reading, writing, data handling and discussion. Some activities are best carried out with one or more other students, others are intended for you to do on your own. For some activities, there are handout sheets giving further information, details about apparatus and so on.

There is an introductory activity you can do to help you decide how you will make your summary of key terms. Ask your teacher for the activity sheet.

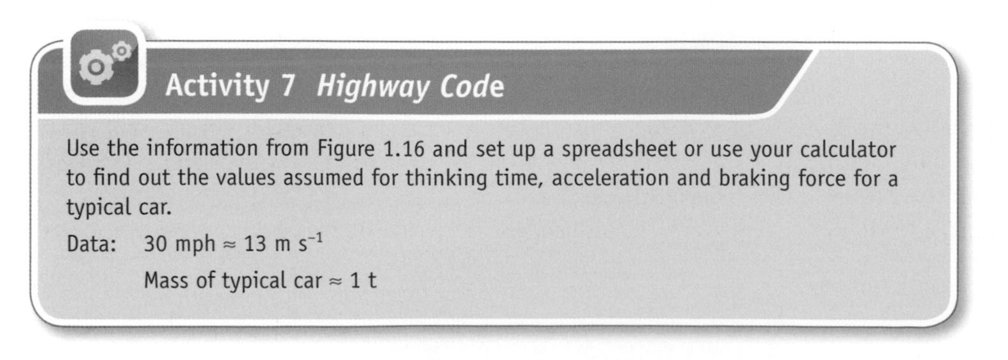

Questions

You will find plenty of Questions in this book. Some are to do as you go along and at the end of each main section. The answers to these questions are given at the end of each teaching unit.

Once you have had a go at a question, check your answer.

If you have gone wrong, use the answer (and the relevant part of the book chapter) to help you sort out your ideas. Working in this way is not cheating! Rather, it helps you to learn.

Maths notes

Maths references in the main text will direct you to the Maths notes, which are to help you with the maths needed in physics. This may involve calculations, rearranging equations, plotting graphs, and so on. You will probably have covered most of what's needed at GCSE, but you may not be used to using it in physics.

The Maths notes at the end of the book summarise the key maths ideas that you need in the AS course, and show how to apply them to situations in physics.

Maths reference

Using log graphs
See Maths note 8.7

Study note

You met Newton's third law and force vector diagrams in the AS chapter *Higher, Faster, Stronger*.

Study notes

These notes in the margin are intended to help you to get to grips with the physics – for example, they indicate links with other parts of the course.

Further investigations

If you continue into the second year of this course, you will spend two weeks on a practical project exploring a topic of your own choice as part of your coursework. You will be asked to research some background information on your chosen topic, plan and carry out your laboratory work, and write a report. As you proceed through the first-year teaching units, keep a note of any areas you might like to pursue further. We have included some suggestions under the heading Further investigations, but any unanswered question that intrigues you could form the basis of a future investigation.

Further investigations

If violently shaken, a jelly can be made to split apart. Use an earthquake table to explore the response of jellies to various vibrations.

Achievements

At the end of each teaching unit you will find a list of Achievements. This is a summary of the key points that you have covered in that unit, and shows what you can expect to be tested on in the exams. (It is copied from the Exam Specification.) Look through the Achievements when you check back over your work after finishing a unit. If there is anything that looks unfamiliar, or that you think you have not properly understood, consult your teacher and the explanations in this book.

CHAPTER 1

Transport on Track

Why a chapter called *Transport on Track*?

In 1825 George Stephenson's steam-powered *Locomotion* (developed from his famous *Rocket*) drew the world's first passenger/goods train from Stockton to Darlington, a 37 mile journey, at an average speed of only 12 miles per hour. Trains can now travel from Britain to France in roughly 20 minutes (See Figure 1.1).

The desire to improve and extend the rail transport system has constantly tested scientists and engineers. The rails themselves needed to be strong enough to cope with the weight of fully loaded trains, while incorporating designs to limit the effect of expansion. Engine and coach production has had to make use of new techniques of propulsion and new materials of construction to become faster and cleaner. Finally, most importantly, increasing train speeds, passenger numbers and service density (number of trains per interval) have necessitated continuous advances in the area of railway safety, using the foremost technologies of the day. As Clive Avery, a research engineer at AEA Technology Rail in Derby, puts it, 'People see the running of trains being a very simple process, yet the science and technology involved in propelling them is staggering.'

Rail enthusiasts claim that this century will see the birth of a new railway age, as environmental worries and road congestion force trains back under the spotlight. The extraordinary technical success of the Channel Tunnel, and the trains that thread their way at speed through its 50 confined kilometres, is said to have marked the beginning of a new era.

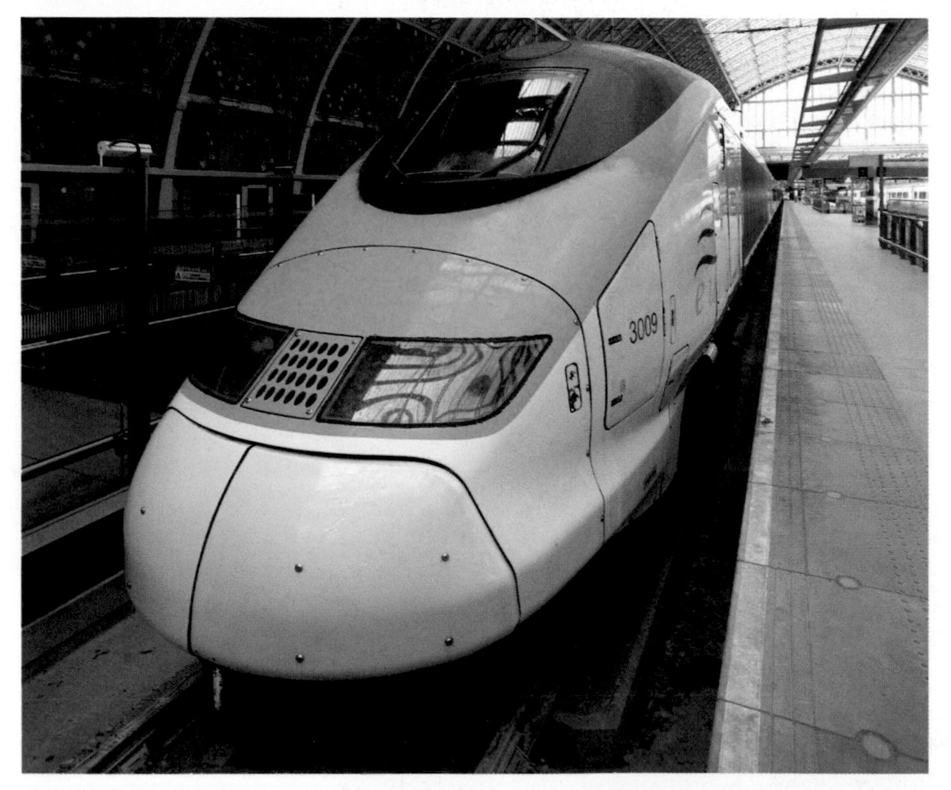

Figure 1.1 (a) *Eurostar*

Figure 1.1 (b) Stephenson's *Rocket*

Overview of physics principles and techniques

This chapter builds on your understanding of basic electrical circuits and mechanics gained from AS chapters. After a review of DC electricity, ideas about magnetism will be extended. This will lead into electromagnetism, electric motors and the laws of electromagnetic induction. You will also investigate circuits containing capacitors and resistors used to time events. By considering what happens when objects start, stop and collide, you will meet the concept of momentum and will apply energy conservation. Finally, you will consider how science works in relation to some of the issues involved in running a rail network.

In the course of this chapter you will be involved in a wide range of practical activities, some of which will have an electrical and electronic basis. You will also use spreadsheets to model situations involving the motion of trains, and a circuit-modelling application to test circuit designs.

In this chapter you will extend your knowledge of:

- DC circuits from the AS chapters *Technology in Space* and *Digging Up the Past*
- energy conservation from the AS chapters *Higher, Faster, Stronger, Spare Part Surgery* and *Technology in Space*.

In other chapters you will do more work on:

- momentum and magnetic fields in *Probing the Heart of Matter*
- exponential changes in *The Medium is the Message* and *Reach for the Stars*
- capacitance in *The Medium is the Message*
- forces and motion in *Build or Bust?*
- kinetic energy and work in *Reach for the Stars*.

1 Getting on track

At the end of this chapter, you will be asked to write either a newspaper article about safety and rail transport or a technical advertisement selling some of the technology that is used to make trains safe. The following short article, compiled from items published in *Snippets* and *Electronics Education*, provides you with some initial information and sets the scene for the chapter.

1.1 Tunnel trains

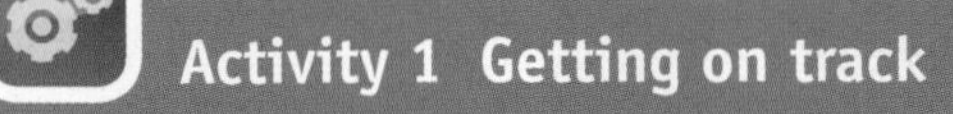

Activity 1 Getting on track

Read this article on tunnel trains, and think which aspects of physics you might expect to find being used in each of the four areas mentioned. Use the internet to find out more information about Channel Tunnel trains and other modern rail systems.

The Channel Tunnel was designed as a quick and reliable route for passengers and freight between Britain and continental Europe. Le Shuttle *trains, specially designed and constructed for the tunnel, run between purpose-built terminals at the end of it.* Eurostar *trains provide a fast link to the continent – London to Paris in 2 hours 15 minutes. To meet the demand for cross-channel transport, the Eurotunnel company needs to send 24 trains through the tunnel per hour, possibly even rising to 30 per hour in the future. These trains travel at up to 160 km per hour, forcing air pressure pulses in front of them that generate high winds, making it impossible to work in the tunnels. In the entry and exit sections of the tunnel, trains travel over gradients as steep as 1 in 90.*

The Eurostar *train in Figure 1.1(b) carries hundreds of people on board, all of whom hope to arrive safely at their destination. It is almost half a kilometre long and has roughly the mass of a thousand small cars; and yet it can travel at 300 km per hour (83 m s^{-1}). When moving at speed, it has enough energy to raise itself and its contents to the top of the Eiffel Tower. Each of its electrically powered locomotives needs to be equivalent to 20 Formula One racing cars.*

Figure 1.2 shows Le Shuttle. *This train is three-quarters of a kilometre long and it can travel at 140 km per hour (39 m s^{-1}). At this speed, it needs 1.5 km to stop. It contains hundreds of vehicles: cars, lorries, coaches and their passengers. To cope with passenger demand, there is generally a similar train not far behind it, and one in front.*

Both the Eurostar *and* Le Shuttle *travel through a narrow tunnel 7.6 m wide, only slightly wider than the train itself (Figure 1.3). The Channel Tunnel actually consists of three separate tunnels, each 50 km long. There are two rail tunnels, one for traffic in each direction, and a central service tunnel. They were bored, largely through chalk, at an average depth of 45 m below the seabed; for most of its length the Channel Tunnel runs through the virtually impermeable stratum of chalk marl (Figure 1.4).*

The facts and statistics presented above pose obvious safety problems including:

- *How can the vast energies of these trains be dissipated safely as the train is slowed?*
- *How can we ensure that the trains will not collide with each other or anything else?*

Figure 1.2 *Le Shuttle* locomotives are the most powerful electric railway locomotives in the world; each train has one at each end

Figure 1.3 Channel Tunnel

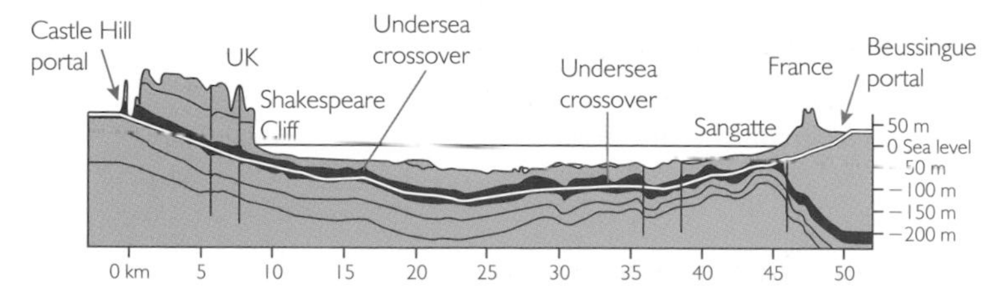

Figure 1.4 Geological cross-section showing the Channel Tunnel

- *How can we ensure that if a train breaks down in the dangerous environment of the Channel Tunnel, where access is difficult, it can be started again and moved to safety?*

To solve these problems, some of which affect all modern railway systems, sophisticated engineering techniques are used. For example, in the Channel Tunnel, an automated electronic signalling system is employed to allow very small gaps between trains. This involves a technically advanced sensing system to ensure the position and speed of each train is known at all times. Despite the complexity of modern rail safety systems, you will find their principles of operation are based in fundamental physics, and often straightforward to understand.

This chapter looks at some of the physics behind making high-speed, high-momentum trains acceptably safe. It focuses on four areas:

- *Signalling* The technique of detecting a train and thereby providing signals to indicate its presence.
- *Stopping and starting* Bringing trains to a halt in an acceptably safe time and distance interval; and starting them moving again, if necessary on a gradient.
- *Sensing speed* Automatically detecting the speed of a train to enable safe intervals to be computed and applied.
- *Structure and safety* How to make trains 'crashworthy'– that is, able to undergo collisions, with as little damage and danger to passengers as possible.

2 Signalling

2.1 Short circuits and train safety

Early days

Reliable signalling is important for the safety of passengers on today's railways, but in the early days of train travel, safety could be ensured only by having large gaps between trains. Getting information about hazards, breakdowns etc. back to other trains on the line was very difficult. The earliest solution was to use railway policemen who gave hand signals in a similar way to road traffic police. At night time, they held up coloured lights instead. As traffic increased, more policemen were needed; this put up the costs of running a railway, so a system was needed that allowed one person to operate several signals. An early hand-operated signal on the Great Western Railway (GWR; affectionately known as God's Wonderful Railway) was the 'disc and crossbar'. With its red disc facing the driver, the indication was all clear, but if the crossbar showed, the train had to stop. At night time, the disc and crossbar had lights attached red for 'stop' and white for 'all clear'. The idea of using coloured lights became the norm for both day- and night-time signalling and is still the main method used today.

Track circuit signalling – how it works

The route followed by a train is set from a control centre or signal box. Its exact position along that route needs to be communicated to other train drivers on the same track. This needs to be done by signals, set by the train itself, providing a **short**

circuit across the rails through its wheels and axles – this is called 'track circuiting'. The track is divided up into electrically insulated 'blocks', and wherever a train is in a block, it activates signals behind it to inform other trains that the block is occupied. In 'three-aspect' signalling, three colours of light are used. The idea is that the train automatically puts the signal closest behind it at red ('stop'), and the previous signal at yellow ('caution – start braking'). A green light indicates all-clear, meaning that the next two sections, at least, are free of trains (see Figure 1.5).

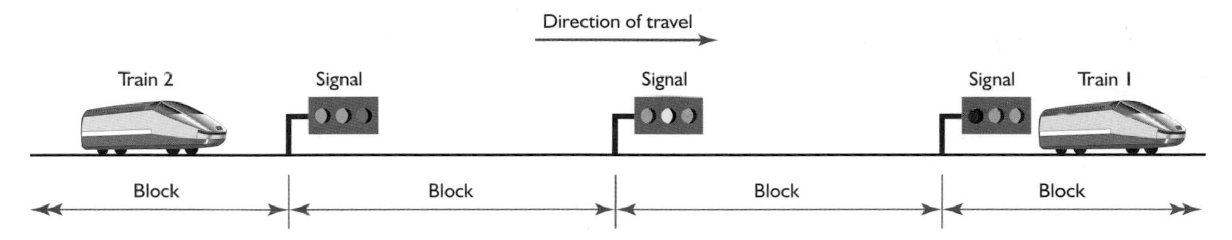

Figure 1.5 Three-aspect signalling

When dealing with electricity, you may well have been told that it is essential to avoid short circuits. These circuits (having negligible resistance and excessive current) can flatten batteries, blow fuses, cause fires and generally stop things from working properly, so it may be a surprise to find them as a method of setting signals.

Activity 2 Short circuits on the line

Set up a model railway line from the circuit diagram in Figure 1.6 and note the ammeter reading before adding a short circuit. Note how a brief short circuit between the two rails affects the current. The granite bed is called the ballast. See what happens to the current if the ballast is slightly salty and damp. Write a statement about short circuits and the effect of ballast on the signalling current.

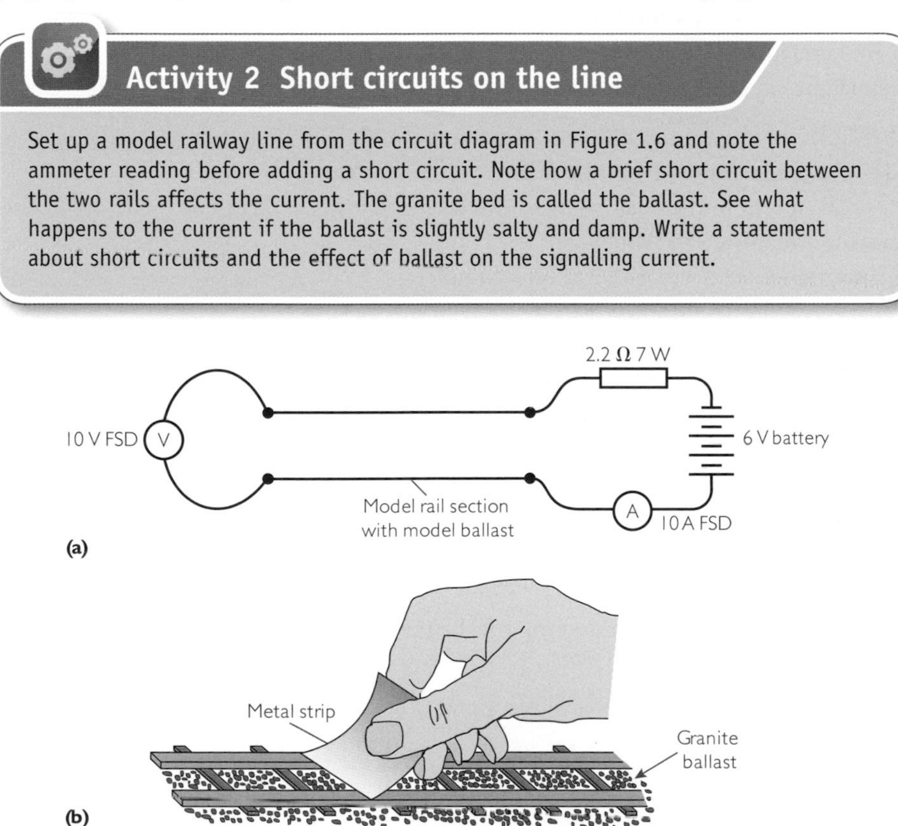

Figure 1.6 Model track circuit: (a) circuit diagram, (b) one way to short-circuit it

The **electromagnetic relay** is one of the main devices used to control a signal in a track circuit system. It is an enclosed mechanical switch that uses one electrical circuit to operate another, electrically isolated, circuit. The relays shown in Figure 1.7 are only for two-aspect signalling (red and green lights only), but interconnection between additional relays would enable three-, four- and five-aspect signals to operate.

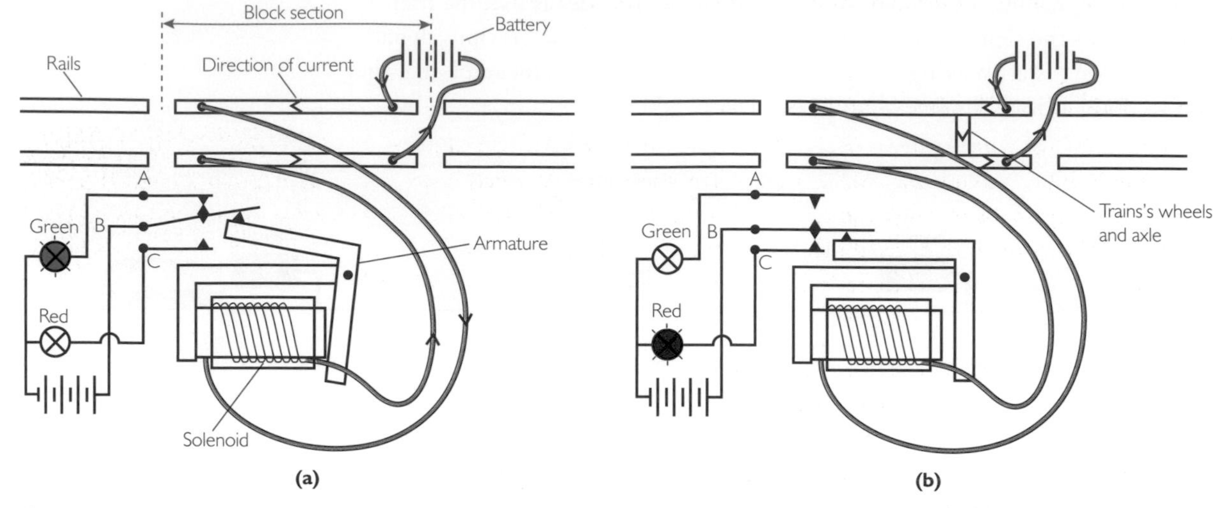

Figure 1.7 Signals (a) with relay current high and (b) with relay current low

With no train on a section of track electric current flows along the rails into the solenoid of the relay (Figure 1.7(a)). This makes the solenoid into an **electromagnet**, which causes it to attract the soft-iron armature. The armatures movement closes the switch contacts A and B. The green lamp's circuit is now complete and it lights.

With a train on the track (Figure 1.7(b)), the situation is different. The short circuit through the train's wheels and axles causes a fall in the potential difference across them and the relay. There is now only a small current in the relay, so its solenoid's magnetic field is not strong enough to attract the armature. The switch contacts B and C now spring back together and the red lamp comes on as the circuit is now complete.

Notice that the system is fail-safe; if there was a power failure to the relay solenoid, the signal would set itself to the safest position, ie red, in order to stop all traffic. All signalling systems have to be fail-safe.

What can go wrong?

If, for any reason, the train can't short-circuit the lines, it can't be detected. Alternatively, if the lines are short-circuited by something other than the train, signals will falsely be set to stop. Examples of each of these conditions occurring might include:

- Torrential rain making the ground and sleepers more conductive (low ballast resistance) so that the two rails short circuit.
- Leaves on a track, whose waxy surface coatings are good insulators, preventing a short circuit occurring.
- Very rusty rolling stock or track. (Rust is iron oxide, which conducts badly.) In normal use this is not a problem as friction between the wheels and the rails soon moves the corrosion.
- Where the track is adjacent to the sea or in undersea tunnels. In clean, dry, well-insulated conditions track-to-track or 'ballast' resistance should be around 100 ohms per km. Moist saline conditions can reduce this to 0.5 ohms per kilometre.

- A train coming off the rails, but blocking the track. There would be no short circuit and so the track would appear unoccupied. (To deal with this eventuality the guard's van contains a length of thick copper cable with a large clip at each end. If such an accident occured, the guard would clip the cable across the track to short-circuit it.)

Even safer systems for the future

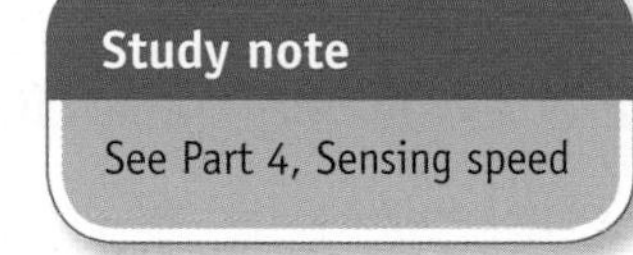

More modern, and even safer, systems of signalling have now been devised, and are gradually being installed around the world. Some control the train remotely. The system for controlling trains through the Channel Tunnel is very sophisticated, although it still follows the idea of fixed block sections. The difference is that the speed and position of a train are automatically sensed and the driver receives an in-cab display of a required speed for the end of the next block.

Another modern approach, which uses radio communications between train and control centre, can still be thought of as involving blocks, but with the block moving as a sort of 'safety envelope' (Figure 1.8). The control centre allocates each train its own particular safety envelope and no envelopes are allowed to overlap. The big advantage of individual trains 'carrying' their own safety envelope is that slower trains can have smaller envelopes, thus increasing the line capacity – ie the number of trains occupying a particular length of track at any one time.

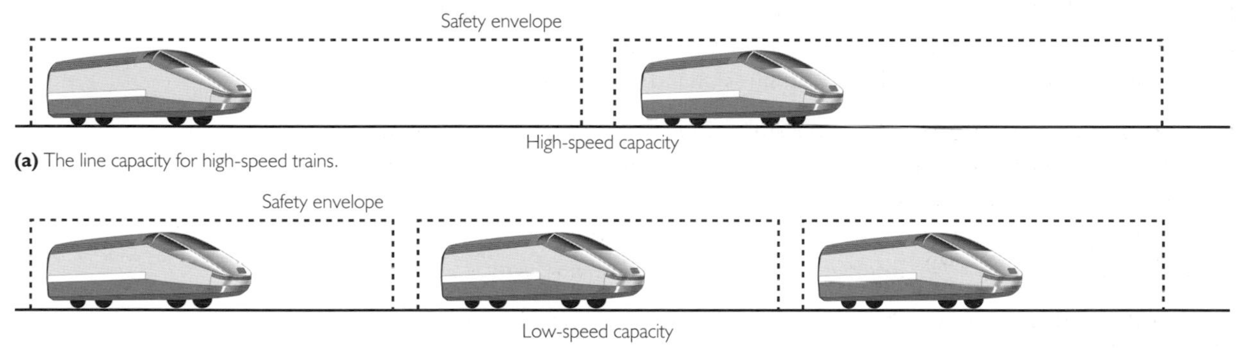

(a) The line capacity for high-speed trains.

(b) The line capacity for slower freight trains. It is higher than for high-speed trains because of the smaller braking distances

Figure 1.8 Safety envelopes

Use Questions 1 to 5 to check your understanding of Activity 2 and the passage you have just read.

Questions

1 What do you understand by the term, 'electrically insulated blocks'?

2 A solenoid is a coil of wire along a straight axis. How does passing electric current through a solenoid alter its properties?

3 Track circuit signalling is said to be a 'fail-safe.' Explain what you think this means with reference to the two-aspect signalling shown in Figure 1.7.

4 Explain why 'leaves on the line' or a rusty coating to the rail can be a serious problem for train safety.

5 In new and future train signalling systems, each train will be assigned a safety envelope based on a calculation of its theoretical deceleration curve.

(a) The deceleration curve calculated does not use the shortest practically possible time to stop the train. Suggest reason for this.

(b) Explain why freight trains can operate in smaller envelopes.

(c) What is the advantage of having smaller envelopes?

2.2 Exploring track circuits

Why should a short circuit in parallel with the relay cause a drop in potential difference? If there were no other resistance in the circuit, the potential difference (pd) across the combination would be the same as the supply voltage. The key point is that there is another resistance in series with the relay, so the circuit is really a potential divider (Figure 1.9). Adding very low resistance(s) (a short circuit) in parallel with the relay alters the way in which the supply voltage is 'shared'. The pd across the relay-plus-short-circuit is small, and so there is only a small current in the relay. Activity 3 and Questions 6 and 7 illustrate what happens.

Study note

You might wish to look back at work you did in the AS chapters *Technology in Space and Digging Up the Past.*

Maths reference

Reciprocals
See Maths note 3.3.

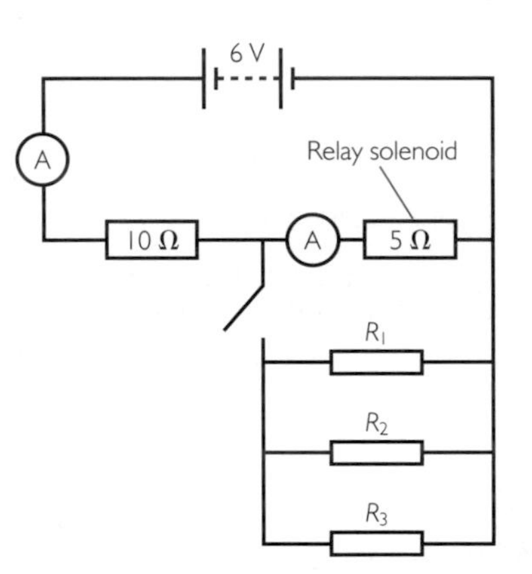

Figure 1.9 Track circuit

Activity 3 Modelling a track circuit

Use a computer simulation or work with components to model and explore a simple track circuit for two-aspect signalling.

Questions

6 A 6 V supply provides the current to operate a relay solenoid of resistance 5 Ω as shown in Figure 1.9. Unfortunately other resistances R_1, R_2 and R_3 may occur in parallel with the relay.

(a) Without these extra resistances, calculate the current drawn from the supply and the pd across the relay.

(b) If the resistors are all present and $R_1 = R_2 = R_3 = 5\ \Omega$:

(i) What is the new circuit resistance?

(ii) How much current is now drawn from the supply?
(iii) What is the new pd across the 5 Ω relay?
(iv) What is now the current in the relay?
(v) Comment on any problem that might arise as a result of additional resistors in parallel with the relay.

7 Figure 1.10 shows a model track circuit. *R* is a series variable resistor used to regulate the current and voltage in the parallel part of the circuit.

(a) If the relay is to have a pd of 6V across it, what resistance value of *R* is required?

(b) If the variable resistor is set as suggested in part (a), what is the electric current through the relay?

(c) What is the current through the variable resistor in this situation?

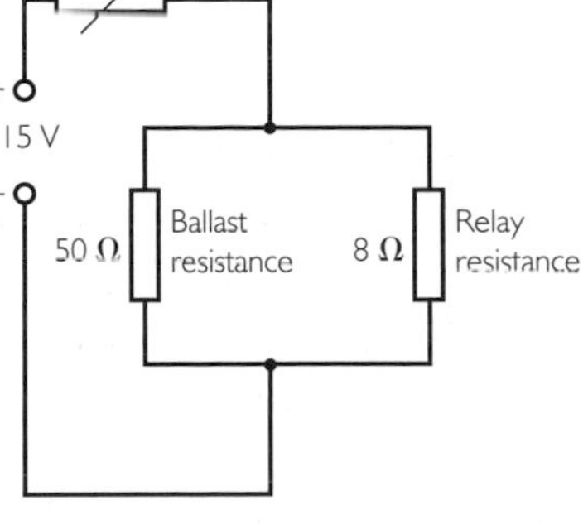

Figure 1.10 Circuit for Question 7

Near the sea, or within the Channel Tunnel, damp, saline (salty) conditions between the rails can be a problem because salt water is a good electrical conductor. If there is enough of it around, it can lower the ballast resistance to a level sufficient to deactivate the track circuit relay and indicate the presence of a train erroneously. The more complex track signals used in the Channel Tunnel can also be disrupted.

Activity 4 Trouble between the tracks

In Activity 3, you found the minimum ballast resistance required to deactivate a track circuit relay and give a false signal. Determine the resistivity of damp, salty kitchen towel and hence make an estimate for the effect of salt water on railway ballast.

Study note

You might need to look back at the AS chapter *Digging Up the Past* for a reminder of resistivity.

Further investigations

Measure the electrical conductivity of leaves, or rust, and see how that might affect track circuit signalling.

Explore how the contact resistance between track and train might be affected by the weight of the train.

2.3 Summing up Part 2

This part of the chapter has mainly involved revision of work from AS about DC electric circuits.

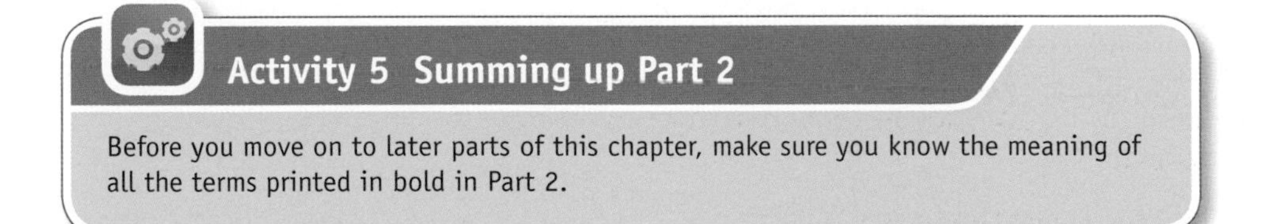

Activity 5 Summing up Part 2

Before you move on to later parts of this chapter, make sure you know the meaning of all the terms printed in bold in Part 2.

Questions

8 Figure 1.11 shows a record of a test done on a track circuit in 1909, when track circuit signalling was first being investigated by railway engineers. The 'lorry' being tested is an open wagon on the rails. Figure 1.12 represents the test on a diagram.

E 6bbc/09

Derby August 12th/09

Mr Holt.

Automatic signalling Keighley - Skipton.
Platelayers lorry on track circuit

In accordance with Superintendent's verbal instructions, a platelayer's lorry was tested on the Utley - Skipton Down Line track circuit, on Friday August 6th.
The voltmeter was connected from rail to rail, and fluctuations noted each time the lorry was run on to track circuit.
The lorry was first tested light, and weight gradually increased.

Test.

Weight on lorry.	Voltage rail to rail with lorry on track. Minimum.	Maximum.
Light	·5	·68
100 lbs	·5	·69
200 lbs	·5	·7
400 lbs	·5	·7
1000 lbs	·4	·7
2000 lbs	·25	·65
2500 lbs	·16	·6
3000 lbs	·16	·6
3500 lbs.	·15	·5

Test commenced at 8.0 a.m. Track very dry Rails bright and clean, with frequent traffic. Weather Fine. (Brilliant sunshine)

Figure 1.11 Extract from engineer's records

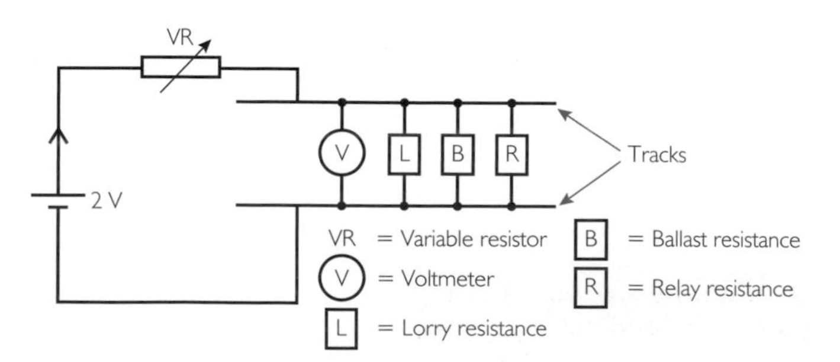

Figure 1.12 Circuit diagram for the test

(a) If the voltmeter reads 0.5 V, what is the voltage across the relay?

(b) If the voltmeter reads 0.5 V, what is the voltage across the variable resistor?

(c) Why is there a maximum and minimum voltage for each load?

(d) What happens, in general, to the voltage measured from track to track as the load is increased?

(e) From your answer to (d), what must happen to the resistance *L* from track to track across the lorry?

(f) Suggest a reason for your answer to (e).

(g) What problems may arise from trains with very light loads?

3 Stopping and starting

3.1 The technical challenge of *Eurostar*

Moving and stopping the massive *Eurostar* trains (Figure 1.13) has proved an enormous technical challenge. It has required the application of intriguing uses of electromagnetism plus an understanding of dynamics. The following paragraphs set the scene for this part of the chapter.

Figure 1.13 *Eurostar* locomotive

Supplying electricity

The *Eurostar* service runs through three European countries and each one has a different electrical power supply. This means that the trains have to be able to collect their supply either from the 'third rail' on the ground or via a pantograph from an overhead cable. The demands of imposed restrictions on axle loads means that the power developed in the new motors has to be 50% higher than in any previous locomotive for less weight.

Brakes

In the 19th century mechanical braking was used for wagons and trains. While a mechanical braking lever coupled to a brake shoe could stop a wagon fairly quickly, it required great force and long levers when used on a train. Each coach had a separate mechanical brake, so that when a train went downhill the brakemen had to run along the top of the train turning a cranking wheel to set the brakes on each coach, and repeat the process to turn the brakes off. A single mechanical system was impossible because trains were so long and made of separate freight wagons or passenger coaches. The invention of the air brake eventually solved this problem for the railways and is a system still in use. A simple high-pressure air hose can easily be connected or disconnected between coaches and the system can be supplied from a single high-pressure air pump in the locomotive.

Even when *Eurostar* is slowing down quite gradually, it needs a braking power of many megawatts. To achieve this braking power, the train uses three braking systems. Air lines operate the conventional friction action of disc brakes (just as in a car) on the passenger coaches, while on the locomotive two types of electric braking are available. Knowledge of the stopping distances required by these new trains is essential if the signals are to be put at suitable intervals along the line.

Motors

Eurostar has AC traction motors rated at 1400 kW with two motors per locomotive. Each drive system includes a power conversion unit to change a DC or AC fixed-voltage supply into an AC supply of variable voltage and variable frequency. Domestically we run hairdryers, electric drills etc. using motors of a type called commutator motors, and they will run very effectively on 50 Hz AC. But these applications need only a few hundred watts. *Eurostar* needs much, much more. The more powerful AC motors are generally of a type called induction motors. The winding on a static part generates a rotating magnetic flux that induces current in copper bars in a rotating part. These currents react with the changing flux to produce rotation. The copper bars look like the sort of exercise wheel given to hamsters or pet mice and, presumably for this reason, AC induction motors are often called 'squirrel-cage motors' (see Figure 1.14).

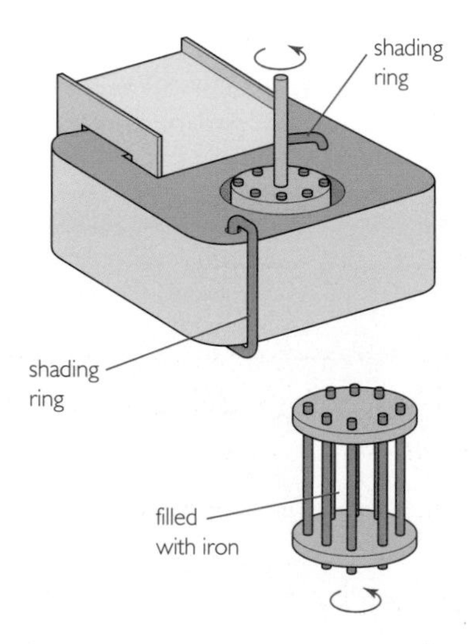

Figure 1.14 Induction motor

Forces

A high force of friction between the wheels and the track enables a large driving force or tractive effort to be developed. For *Eurostar* up to 5.6 MW of power can be provided using two locomotives. This gives around 400 kN of tractive effort to start the train off.

3.2 Moving theory

In this section we will consider the motion of trains as they start and stop. This will involve building on the basic theories and equations of motion that you met in the AS chapter *Higher, Faster, Stronger*.

Question

9 (a) If the *Eurostar*, mass 750 t, is subject to a net tractive (pulling) force of 400 kN, what is its acceleration?

(b) What power is required to produce this acceleration when the train reaches a speed of 10 m s^{-1}?

Study note

1 tonne (1 t) is 1000 kg

Momentum

In *Higher, Faster, Stronger* you learned that the relationship between force, mass and acceleration can be expressed as:

$$F = ma$$

and that this can also be written as:

$$F = \frac{m\,\Delta v}{\Delta t} = \frac{m(v-u)}{\Delta t}$$

These equations are one way of expressing Newton's second law of motion, but in this chapter we look at another way.

If the velocity changes from u to v in time t, then:

$$\frac{\Delta v}{\Delta t} = \frac{(v-u)}{\Delta t}$$

so:

$$\Delta F = \frac{(mv - mu)}{\Delta t}$$

$$\Delta F = \frac{\Delta(mv)}{\Delta t} \qquad (1)$$

The quantity *mass* × *velocity* is the **momentum** of the object. Momentum is a vector quantity and is usually given the symbol p. It has SI units kg m s^{-1} (or Ns, as these are identical).

Study note

Here we are using symbols *F, u, v* and *p* to represent the magnitudes of vector quantities.

Equation 1 can be written as:

$$F = \frac{\Delta p}{\Delta t} \qquad (1a)$$

This equation is more in line with Newton's original meaning. It shows that force on an object is equal to its rate of change of momentum.

If Δp and Δt are small this can give an instantaneous value for force, written:

$$F = \frac{dp}{dt} \qquad (1b)$$

When Newton's second law is expressed in terms of momentum change, we do not have to assume that the mass remains constant. If the mass changes from an initial value of m_1 to a final one of m_2:

$$\Delta mv = (m_2 v - m_1 u)$$

This widens the application of Newton's second law to cover, for example, a steam train carrying its own coal for fuel (not a complication needed for *Eurostar* trains as they run on electricity).

Another rearrangement gives us:

$$F\Delta t = \Delta(mv) = \Delta p \qquad (2)$$

This relationship is used so frequently that the product of force and time has gained its own name. It is called the **impulse** of a force and is particularly useful when analysing impacts since the force and time of an impact must be the same for both participants. It can be expressed in words as:

impulse = change in momentum

Note that impulse and momentum are both vector quantities, so the direction as well as magnitude should be included.

We are now able to address such important questions as 'How much force is needed to stop a train?'. Use Equations 1 and 2, and other equations of motion, to answer Questions 10 to 15.

Questions

10 What are two possible SI units for impulse and momentum?

11 A railway coach of mass 50 t is in motion at + 20 m s^{-1}. Calculate the change in momentum needed to produce each of the following final velocities (1 t = 1000 kg):

(a) v = +40 m s^{-1}

(b) v = –20 m s^{-1}

(c) v = 0 m s^{-1}

12 A 1.2 t car accelerates from 0 to 25 m s^{-1} in 10 s. Calculate (a) the change in momentum, (b) the tractive force applied (ignore any drag due to air resistance etc.), (c) the distance travelled, (d) the power of the engine.

13 A *Shuttle* train (Figure 1.15) has mass of 2400 t, ie 2000 times greater than the car in Question 12. To achieve the same performance it would require a power 2000 times greater, ie 75 MW, but it is pulled by power units rated at only 5.6 MW. Why is this sufficient?

Figure 1.15 *Le Shuttle*

14 A fully laden *Eurostar* of 750 t leaves the tunnel at 44 m s^{-1} and takes 14 minutes to get to its top speed of 83 m s^{-1}.

(a) What tractive force was required?

(b) How many times smaller is the tractive force required to give the same acceleration to a train of mass 250 t?

(c) How does your value for the tractive force compare with the braking force required to slow down the same train as it enters the tunnel on the way back?

(d) For both *Eurostar* and the lighter train, drag at high speeds is about 40 kN. What difference does that fact make to your answer to (a) and (b)?

15 Suppose a train must come to a halt in 60 s.

(a) Find the braking force required to do this inside the tunnel, where the maximum speed is 44 m s^{-1}, when the train is (i) a fully loaded *Shuttle*, mass 2400 t and (ii) a *Eurostar*, mass 750 t.

(b) Above ground the top speed of *Eurostar* is 83 m s^{-1}. If it stops in 3 minutes, what net braking force must be applied?

(c) In Question 14 you calculated tractive forces. Compare the sizes of the forces required from the engine to those needed by the brakes.

Stopping distances

Once the brakes are applied to a fast-moving train it still takes some time to come to a stop, during which the train can move a considerable distance. Information about the probable stopping distance is critical for normal service in order to ensure that the train stops exactly alongside its platform. It is also important in deciding on the positions of signals. This type of question can be tackled using force and momentum and other equations of motion, but another approach, using energy, is more direct.

Imagine that a train of mass m is moving at an initial speed u, before it is brought to a standstill by a force F acting in the opposite direction to its motion. All the kinetic energy E_k of the train is transferred by the work of the resultant braking force. We can use:

work done = energy transferred

and:

work done = force × distance moved in direction of force

Let Δs be the distance covered while the brakes are applied. Since the force acts along the direction of motion, we have:

$$-F\Delta s = \Delta E_k = \frac{1}{2}mv^2 - \frac{1}{2}mu^2$$

Since the final kinetic energy of this train is zero:

$$-F\Delta s = -\frac{1}{2}mu^2 \qquad (3)$$

Rearranging Equation 3, the stopping distance for a known brake system can be calculated from:

$$\Delta s = \frac{mu^2}{2F}$$

Study note

You met this idea in the AS chapter *Higher, Faster, Stronger*. You might wish to refer back to that earlier work now.

Questions

16 (a) Calculate the stopping distances for a *Shuttle* train which is partly loaded and has a mass 2100 t; it is stopped from a speed of 40 m s^{-1} by a braking force of 1.2 MN.

(b) A *Eurostar*, mass 750 t, is travelling at 80 m s^{-1} and must stop within 1500 m. What force is required?

(c) If there is a drag force of 40 kN at a speed of 80 m s^{-1}, suggest what difference that would make to your answers to (a) and (b).

17 (a) How long would it take the *Eurostar* in Question 16 to come to rest?

(b) What would be its acceleration?

18 (a) The speed of a train is doubled. What is the change in (i) its momentum (ii) its kinetic energy?

(b) Suppose two trains are travelling at the same speed, but one has twice the mass of the other. What is the ratio of (i) their momenta (ii) their kinetic energies?

We could go on with all the possible combinations of initial speed, mass of train and distance to station, and calculate the braking force required to stop in time. It would require far too many hours at your calculator. In Activity 6 you will let a computer take the strain.

Activity 6 Stop the train

Set up spreadsheets to calculate the braking forces required to stop a *Eurostar* train within a specified distance or time. The initial speed could be anything up to 300 km h^{-1}. Analyse any patterns that you find by plotting graphs from your spreadsheets.

So far, we have dealt only with situations involving trains. However, the same approaches can be used to analyse the motion of any object – as is the case in Activity 7, which uses data from the *Highway Code*.

In the *Highway Code* there is a table of stopping distances for cars travelling at different speeds (Figure 1.16). It is there to inform drivers on the correct distance to allow between themselves and the vehicle in front in case they have to make an emergency stop. The figures include an allowance for a delay between recognising danger and applying the brakes – the thinking time.

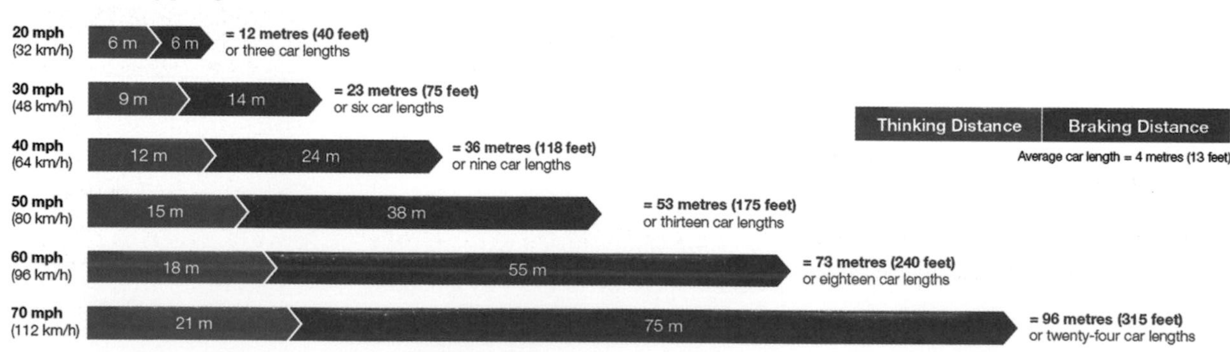

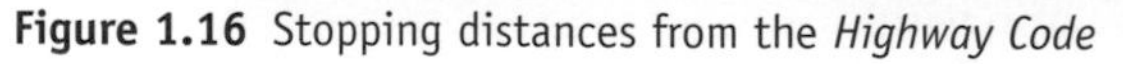
Figure 1.16 Stopping distances from the *Highway Code*

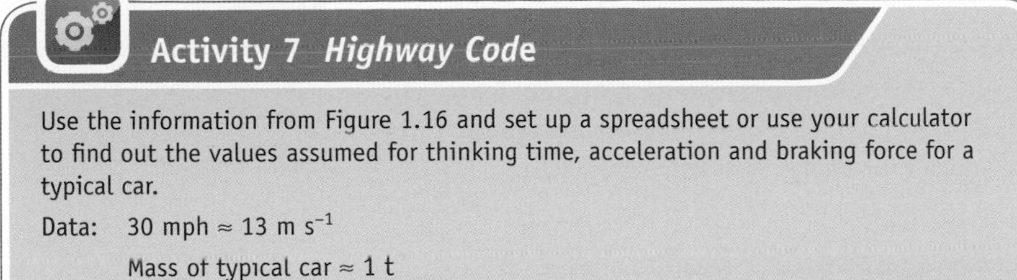

Activity 7 *Highway Code*

Use the information from Figure 1.16 and set up a spreadsheet or use your calculator to find out the values assumed for thinking time, acceleration and braking force for a typical car.

Data: 30 mph ≈ 13 m s^{-1}

Mass of typical car ≈ 1 t

An uphill struggle

It is important that Channel Tunnel trains do not get stuck – especially not underground. The engine specifications for Channel Tunnel trains particularly mention that the two locomotives must be able to restart the train from rest with an acceleration of 0.13 m s^{-2} up the steepest slope on the track (Figure 1.17). This section has a gradient of 1 in 90. If one locomotive fails the other must still be able to restart the train, albeit at a reduced speed.

In such a situation, the forces do not all act along the same direction. In order to analyse what's going on, you need to resolve force vectors into components.

Figure 1.17 Hauling a train uphill

Questions

19 (a) How much tractive force is required to give the specified acceleration to a fully loaded *Shuttle* on the flat against a constant drag force of 40 kN? (*Shuttle* mass: 2400 t.)

(b) If the train stops on a slope of 1 in 90 how much forward tractive force does it need to exert to stay still; ie just stop rolling back? (Use g = 9.81 N kg^{-1}.)

(c) Combine your answers to determine the total force needed to accelerate the train from rest (with the required acceleration) on this slope.

(d) How much time does it take for the train in (c) to get up to a speed of 44 m s^{-1}?

(e) (i) If a single locomotive has a tractive force of 400 kN what acceleration can it produce on the slope?

(ii) How long is it before the *Shuttle* speed reaches 44 m s^{-1}?

(iii) How far has it travelled?

20 A *Shuttle* loses all engine power at the lowest point in the Channel Tunnel when it is travelling at 44 m s^{-1} and has 25 km to travel along a 1 in 90 slope to get up and out.

(a) Assuming no friction, would the train be able to free-wheel to the surface?

(b) If the drag forces average 30 kN, how far will the free-wheeling train travel?

Study note

You met components of vectors in the AS chapter *Higher, Faster, Stronger*. You might wish to look back at the work before you tackle questions 19 and 20.

3.3 Motors

In Section 3.1 we mentioned that *Eurostar* depended on electrical motors and brakes to operate at its extremely high speeds. The motors and braking systems of Channel Tunnel trains exploit electricity and magnetism to generate an **electromagnetic force** that drives motors and to recover some of the energy 'lost' during braking.

Study note

Brakes are the subject of Section 3.4.

Activity 8 Brush up on magnetism

Use some permanent magnets and a current-carrying coil to review your knowledge of the nature and behaviour of magnetic fields.

In an electric motor, the magnetic effect of a current is used to drive and maintain the rotation of the wheels. Activities 9 and 10 build up the basic principles behind the working of an electric motor.

Activity 9 An electromagnetic force

Use the arrangement shown in Figure 1.18 to explore the force on a current-carrying conductor when it is in a magnetic field. Use a top-pan balance to measure the size of the force between an electric current in a wire and a magnetic field (Figure 1.19). Vary the conditions to show how the size of the force varies with the length of wire, the strength of field and the size of the current.

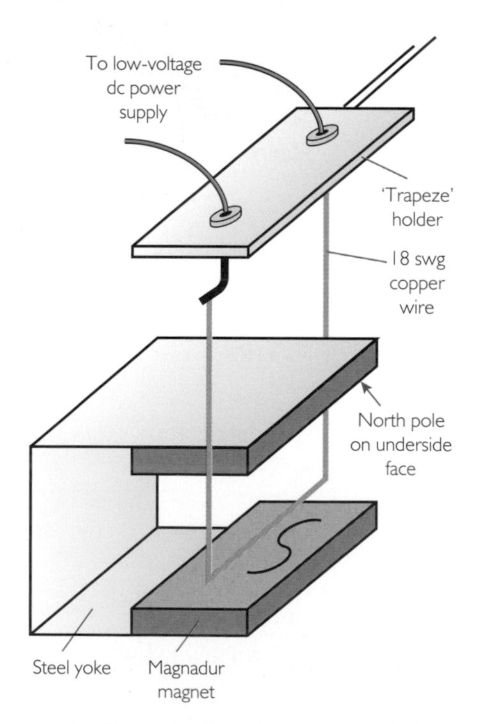

Figure 1.18 Demonstrating the direction of an electromagnetic force

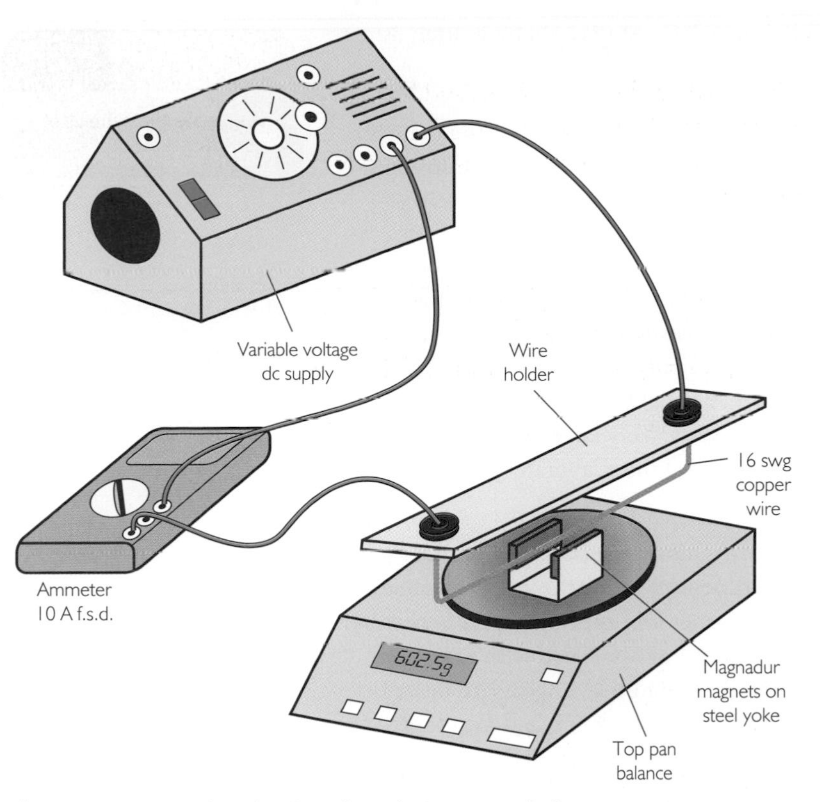

Figure 1.19 Measuring the size of an electromagnetic force

The results of the first part of Activity 9 are usually expressed in a rule known as **Fleming's left-hand rule**. The thumb and first two fingers of the left hand are held so that they are all at right angles to each other, like the corner of a box (see Figure 1.20). If the **f**irst finger points in the direction of the magnetic **f**ield and the se**c**ond finger points in the direction of the **c**urrent, then the thu**m**b shows the direction of the **m**otion and therefore of the force.

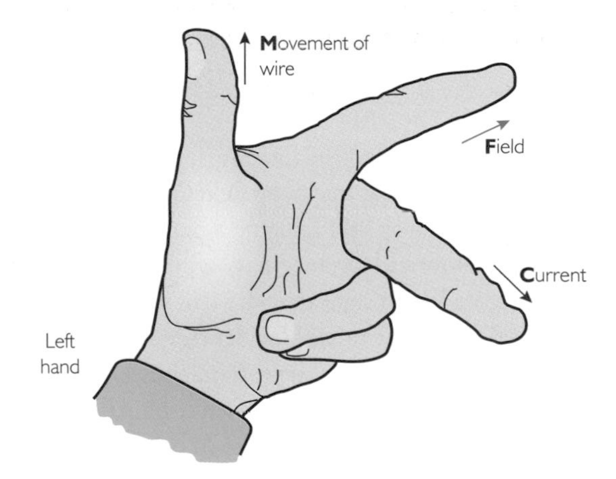

Figure 1.20 Fleming's left-hand rule

The size of the force

Several factors determine the size of the force: the strength of the magnet, the size of the current and the length of the wire. Experiments such as the second part of Activity 9 lead to the conclusion that the size of the force, F, on a wire is proportional to:

- the current, I, in the wire
- the length of the wire, l, that lies within the magnetic field
- the strength of the magnetic field

and that F is a maximum when the current is at right angles to the field and zero when the current and field are parallel. The force still acts in a direction that is at right angles to both the current and the field, but its magnitude decreases as the angle between the field and current is reduced.

When the field and current are at right angles, the relationship could be written as:

$$F = kIlm$$

where k is a constant of proportionality and m represents the strength of the magnets. In practice, though, this relationship is used to define what we mean by the strength of the magnetic field and, rather than introducing an extra constant, we express it as:

$$F = BIl \qquad (4)$$

Maths reference

Constant of proportionality
See Maths note 5.1

where the symbol B represents the **magnetic flux density** (loosely speaking, the 'strength' of the field). The SI unit of magnetic flux density is the tesla, T (named after Nikola Tesla, 1856–1943, a Croatian–American physicist). $1\ \text{T} = 1\ \text{N A}^{-1}\ \text{m}^{-1}$. A field of 1 T containing a wire of length 1 m carrying 1 A at right angles to the field will produce a force of 1 N on the wire. How large is a tesla? Very large as it happens. The magnetic field of the Earth is typically about 50 μT and most laboratory solenoids produce a few mT at their centres.

We use the term **magnetic field** to describe the three-dimensional region of space where the magnet has some influence, but the term 'flux density' requires some explanation, as 'flux' is a word generally associated with flow – you met it in the AS chapter *Technology in Space*, where it was used to describe the 'flow' of energy associated with radiation. The same word is used to describe the flow of fluids. The formal language to describe electromagnetism was developed by Michael Faraday (1791–1867), and he adopted the term 'flux' because there are some similarities between the mathematical descriptions of magnetic field patterns and fluid flow patterns. If you think of a magnetic field plotted out by field lines, then B represents the density of the lines – where the lines are closer together the magnetic flux density is higher. Loosely speaking, the magnetic flux density is related to the number of lines crossing a given area.

Study note

To review your knowledge of components, refer to the AS chapter *Higher, Faster, Stronger.*

Angles

To see how the size of the force varies with angle between field and current, we need to note that magnetic field and current are both vectors – they have direction as well as size. Imagine a magnetic field acting vertically down into this page. Now add an imaginary wire running along the foot of the page. If the current goes left to right the wire will want to move up the page (as described by Fleming's left-hand rule). Now let the side of the page lift up so your imaginary 'wire' slopes uphill at an angle, θ, to the vertical. Only the component of the magnetic field at right angles to wire, $B\sin\theta$, produces a force (see Figure 1.21). The component $B\cos\theta$ is parallel to the wire and ineffective. (Alternatively you can think of the current being resolved into components, $I\sin\theta$ and $I\cos\theta$, perpendicular and parallel to the field – it comes to the same thing.) The full version of Equation 4 now becomes:

$$F = BIl\sin\theta \qquad (4a)$$

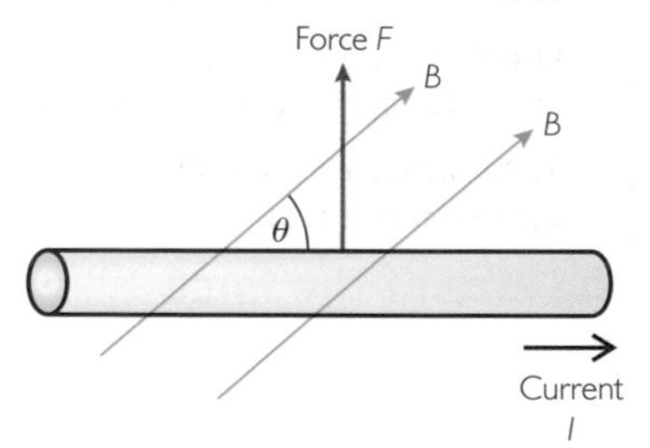

Figure 1.21 Angles and the electromagnetic force

Questions

21 There is a current in each of the wires shown in Figure 1.22. State the direction of the force on the wire in each case.

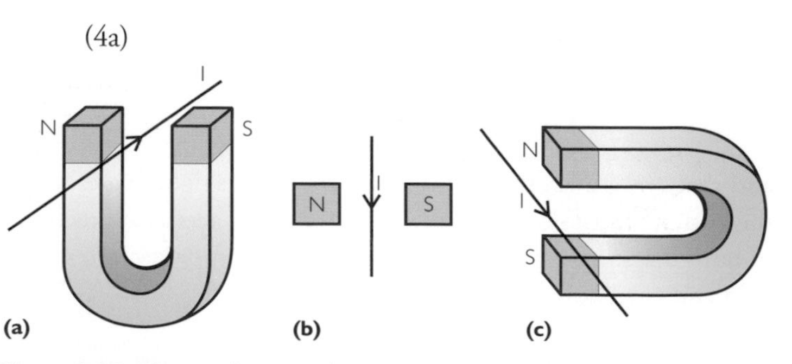

Figure 1.22 Diagram for Question 21

22 The flux density between the poles of a powerful electromagnet is 2 T. What is the force exerted on 10 mm of wire carrying 4 A when the wire is (a) at right angles to the field, (b) parallel to the field and (c) at an angle of 30° to the field?

23 A horizontal wire of length 100 mm lies perpendicular to the field between the poles of a magnet. When a current of 5.0A is passed through the wire, it is pushed vertically upwards. A rider (ie a small mass that rests on the wire) of mass 15 g placed on the wire just brings it back to its original position. Calculate the mean magnetic flux density between the poles of the magnet ($g = 9.81\ \text{N kg}^{-1}$).

A force that just kicks a wire into or out of a horseshoe magnet is not sufficient to move a train from London to Brussels. The wheels have to keep turning for hours, driven by forces acting to produce a turning effect – a **torque**. Figure 1.23 and Activity 10 show how a force on a wire can produce a torque that will maintain the rotation of an axle. This is the basis of a simple electric motor.

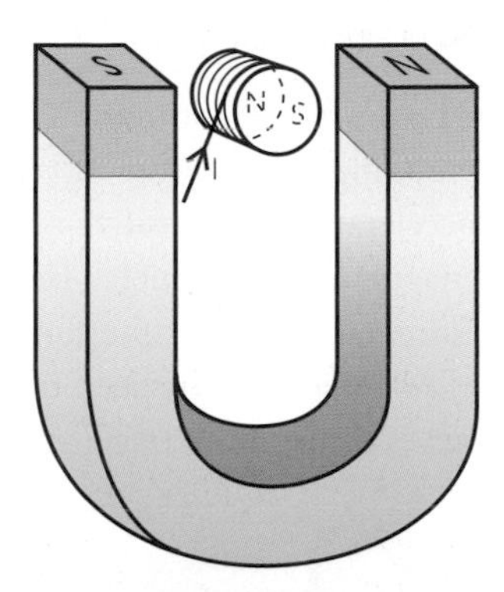

Figure 1.23 Current-carrying coil in a magnetic field

When a wire is wound into a coil and the current switched on, the two faces have magnetic poles. Freely suspend such a coil and place it at an angle to a magnetic field, and the pull of the attraction between opposite poles will turn the coil. That is the first stage of a continuously moving motor. Unfortunately the turning stops once the coil faces its opposite poles. This problem goes away if the current then reverses, since the poles are then on the opposite faces of the coil and the coil has to move on. In the following activity you will look at motors. The paper clip version maintains rotation by inertia – the coil turns past the place where it is facing the poles because it is moving rapidly and there is nothing to stop it. The more elaborate split ring version is more similar to the construction of working motors.

Activity 10 A simple electric motor

Construct a simple motor such as one of those shown in Figure 1.24. If you are working with other groups of students, compete to see whose motor can rotate for more than 30s.

Motors have their own vocabulary: armature, split ring commutator, brushes etc. Use this activity to make sure that you know and recognise them.

As an optional extension, use the motor demonstration program available via **www.shaplinks.co.uk**.

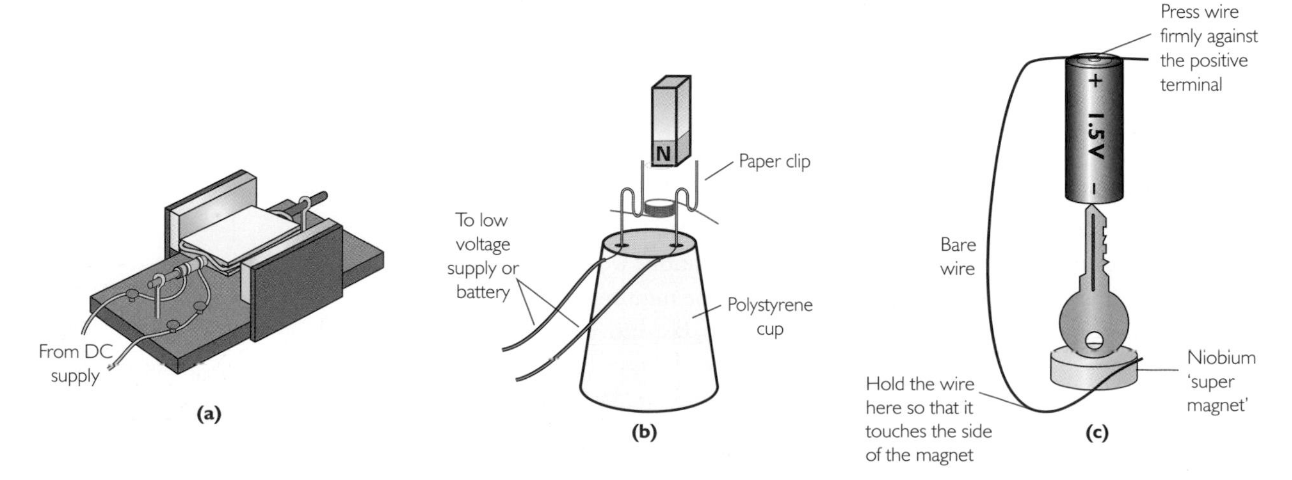

Figure 1.24 Simple electric motors: (a) Westminster model, (b) paper-clip model, (c) key model

Question

24 The *Eurostar* demands very large forces from its motors (around 400 kN). Assume the engineers have designed a modified motor of the sort you have met in Activity 10. Using Equation 4, make estimates for the necessary sizes of the quantities involved if the forces are to be anything approaching the right order of magnitude. (There is one extra quantity you need to include – think back to the construction of the coils.)

3.4 Brakes

The motor effect can get *Eurostar* going, but doesn't look promising as a means of stopping a train. And, as you have seen, it takes just as much force to give a negative acceleration as a positive one. Activity 11 shows you how it is possible to make magnetic brakes. Such a braking system was considered for the *Advanced Passenger Train* (APT) but at the time was not found to be effective. (The APT was an experimental high-speed train developed in the UK in the 1970s and 1980s, which never came into service.)

Activity 11 Eddy current braking

Use the apparatus shown in Figure 1.25 to explore the strange effects of magnetic brakes.

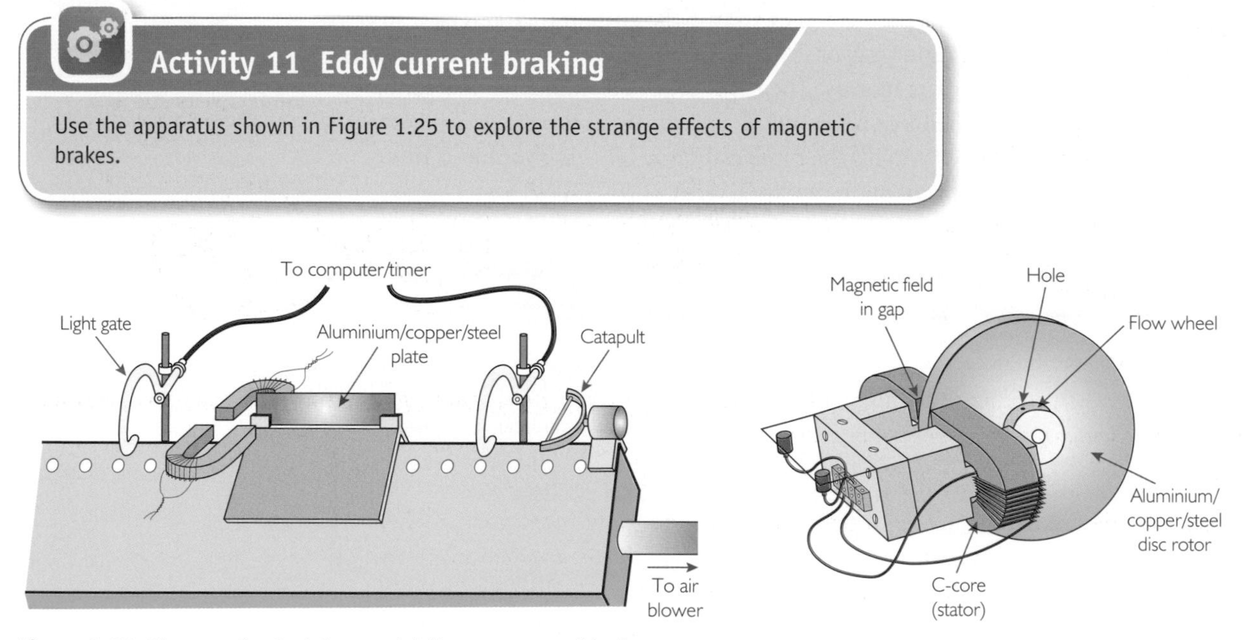

Figure 1.25 Diagrams for Activity 11: (a) linear system, (b) disc system

In Activity 11 you saw examples of **eddy-current** braking. Many lorries and coaches use this method for their initial braking at high speed. In a wire circuit the path of the electrons is clearly round the wire but, if the conducting material is a whole thick sheet of metal, the possible loops for a flow of electrons are not so clear. Indeed, in the changing situations we are talking about here, the eddies of current change path from one minute to the next (Figure 1.26), just like the swirls of water (eddies) in a fast-moving river.

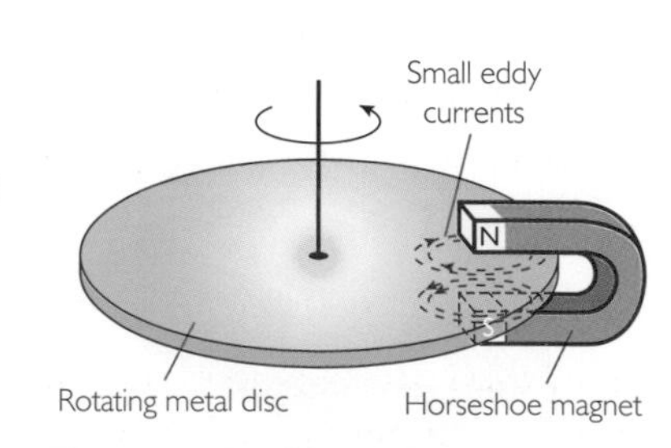

Figure 1.26 Eddy currents

Each temporary current loop produces its own magnetic field and hence gives rise to forces of attraction and repulsion between it and the external magnet. But how do the eddy currents arise in the first place? In order to produce a current, an emf is required. One way to provide an emf is to use a battery or a solar panel, but as you will see in Activity 12, an emf can also be produced using magnetism – this effect is called **electromagnetic induction**, and the emf and current thus produced are referred to as an **induced emf** and an **induced current**.

Study note

You met the term 'emf' in the AS chapter *Technology in Space*.

Activity 12 Electromagnetic induction

Use the apparatus shown in Figure 1.27 to show how a current can be induced in a wire using magnetism. Explore the factors that affect the size of the induced emf and current.

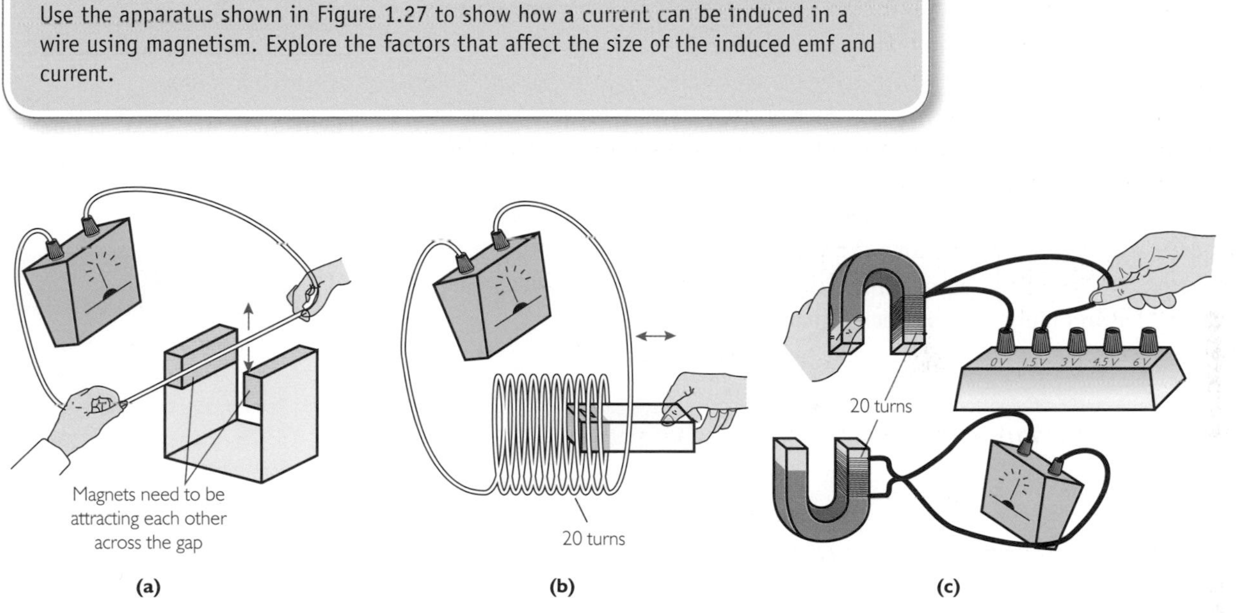

Figure 1.27 Demonstrations of electromagnetic induction: (a) moving a wire in a magnetic field, (b) moving a magnet relative to a coil, (c) changing a magnetic field in a coil

We can now describe some of the properties of electromagnetic induction: ie when there is relative motion between a conductor and a magnetic field, an electromotive force is induced in the circuit. In each case there is an induced emf when there is a change in the magnetic field around the wire and the size of the induced emf $\mathcal{E}$ is proportional to:

- the strength of the magnetic field
- the rate at which the wire or coil is moved, or the rate at which the magnet is moved
- the number of wire loops or turns on the coil.

The emf changes direction if the magnet is reversed or the direction of movement is reversed.

In the third part of Activity 12 you saw that an emf can be induced in one solenoid when the current in another, nearby, solenoid is changed. Again, the key point is that an emf is induced in the first solenoid because it is in a changing magnetic field – only now the change is made electrically rather than by moving a magnet or the coil.

Laws of electromagnetic induction

We can make the above description more formal using Faraday's way of describing magnetic fields. Think first of the situation when an emf is induced by moving a magnet into a single loop of wire (Figure 1.28), and picture the magnet's field lines. When the magnet is some distance from the loop only a few lines thread through the loop, but as the magnet approaches the coil, the field within the loop becomes stronger – more lines thread through the loop. Faraday expressed the 'amount of field' within a loop in terms of the **magnetic flux** enclosed by (or 'linking') the loop. For a (flat) loop square-on to the field direction,

magnetic flux = magnetic flux density × area of loop

Flux is usually given the symbol Φ, so in symbols:

$$\Phi = BA \qquad (5)$$

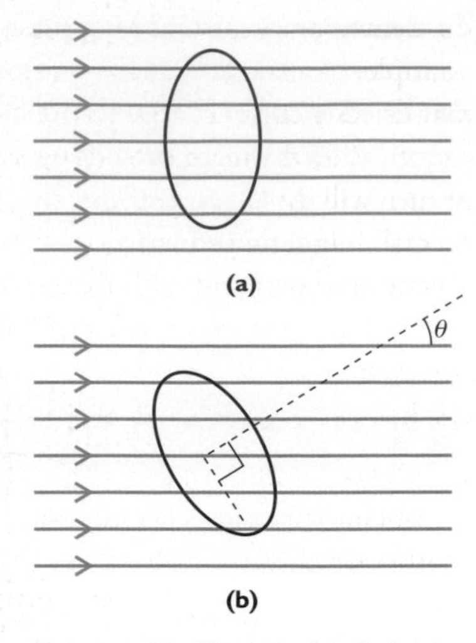

Figure 1.28 Single loop of wire in a magnetic field (a) square-on (b) at an angle

where A is the area of the loop. If you think of B as the number of magnetic field lines per square metre, then Φ is the total number of lines enclosed by the loop. The SI unit of magnetic flux is the weber, Wb. 1 Wb = 1 T m^2 = 1 N A^{-1} m. Magnetic flux density is sometimes expressed in units of Wb m^{-2} rather than tesla – the two are exactly equivalent. If the normal to the loop makes an angle of θ to the field as in Figure 1.28(b), then Equation 5 becomes:

$$\Phi = BA\cos\theta \qquad (5a)$$

Study note

The normal is a line at 90° to the plane of the loop. You met the term in the AS chapter *The Sound of Music* in the context of reflection and refraction at a surface.

The size of the induced emf

Faraday noted that the size of the emf, $\mathscr{E}$, induced in a loop is proportional to the rate at which the flux within the loop is changed; this relationship is known as **Faraday's law of electromagnetic induction** and is written mathematically as:

$$\mathscr{E} \propto \frac{\Delta\Phi}{\Delta t} \qquad (6)$$

If a coil contains N turns (N loops) of wire, then an emf is induced in each loop; as the loops are connected in series, the net emf from the whole coil is multiplied by N, and Equation 6 becomes:

$$\mathscr{E} \propto \frac{\Delta(N\Phi)}{\Delta t} \qquad (6a)$$

$N\Phi$ is called the **magnetic flux linkage**. For a coil with more than one loop, Faraday's law states that the size of the induced emf is proportional to the rate of change of flux linkage. Flux linkage is sometimes given the unit weber-turns to distinguish it from the flux through a single loop – but N is just a number so 1 weber-turn has the same dimensions as 1 weber.

Maths reference

Dimensions
See Maths note 2.5

In SI units, ie with Φ in webers, t in seconds and $\mathscr{E}$ in volts, the size of the emf is *equal* to the rate of change of flux linkage – not merely directly proportional. However, there is one important point that we have not yet included, and that is the direction of the induced emf (and hence of the induced current). Here, an energy argument is useful.

The direction of the induced emf

People have long searched for a perpetual motion machine, ie one that needs no input of energy to keep on turning for all eternity. It's a pipe dream, of course. At first sight the demonstration of induction in Activity 12 appears to offer untold riches. For example, in a bicycle dynamo, movement of a coil inside a magnet produces an emf that drives a current through the coil which, from Activity 10, you know will make it rotate inside the magnet. As the coil moves faster, it will generate a larger emf, which in turn will drive a larger current making the coil rotate faster still… and so on… A small initial movement would produce runaway motion, apparently generating kinetic energy out of nothing. So what is the catch? It is that emf is, in practice, always induced in a direction such that any forces will oppose the motion that produces it. A wire moving into the magnet is slowed down by repelling action. A coil rotating clockwise is opposed by a torque that is trying to rotate it anti-clockwise.

Lenz's law sums up the situation. 'The direction of the current induced in a conductor by moving it relative to a magnetic field is such that its own field opposes the motion.' Mathematically, Lenz's law is expressed by inserting a minus sign into Equation 6:

$$\mathscr{E} = -\frac{\Delta(N\Phi)}{\Delta t} \qquad (6b)$$

thus producing an expression that encapsulates both Faraday's and Lenz's laws. In calculus notation:

$$\mathscr{E} = -\frac{\mathrm{d}(N\Phi)}{\mathrm{d}t} \qquad (6c)$$

The change in flux linkage can be brought about by a change in any or all of N, B or A. If only one of these is changed, then we can rewrite Equation 6 in yet more ways. Changing B only:

$$\mathscr{E} = -NA\frac{\mathrm{d}B}{\mathrm{d}t} \qquad (6d)$$

changing A only:

$$\mathscr{E} = -NB\frac{\mathrm{d}A}{\mathrm{d}t} \qquad (6e)$$

or changing N only:

$$\mathscr{E} = -\Phi\frac{\mathrm{d}N}{\mathrm{d}t} \qquad (6f)$$

Questions

25 Show that the SI units on the left-hand side of Equation 6 (any version) are the same as those on the right.

26 A magnetic flux of 10 Wb passes through a coil of 50 turns. The flux decreased to zero over 2.0 s. What would be the size of the emf induced in the coil?

27 A 10-turn coil has a cross-sectional area of 0.03 m^2. It is placed so that the plane of the coil is perpendicular to a field of magnetic flux density 2 T. It is then rotated through 90° in 0.2 s so that it lies parallel to the field. What size emf would be induced in the coil?

28 If large currents are turned on or off near your radio you may hear a crack or a click. Suggest an explanation for this.

Another look at electromagnetic induction

We started by considering a single wire loop that can be extended to make a coil with many turns, but Equation 6 also applies to all other ways of inducing an emf. In Figure 1.27(c), changing the current on one solenoid changes the magnetic flux linking the other solenoid; in Figure 1.27(a), the movement of the wire between the poles of magnet changes the flux linking the circuit made by the wire and the meter. It is sometimes useful, too, to consider an isolated section of wire as shown in Figure 1.29, moving in a magnetic field. The wire is straight, has a length l, and moves at a velocity v in a direction that is at right angles to its length and to the direction of a uniform field B. Notice the symbol that represents the field (or any other vector) directed into the page; think of an arrow travelling away from you – you see its feathers, which are represented by a cross. A field (or other vector) directed towards you out of the page would be represented by a dot in a circle (the tip of the arrow).

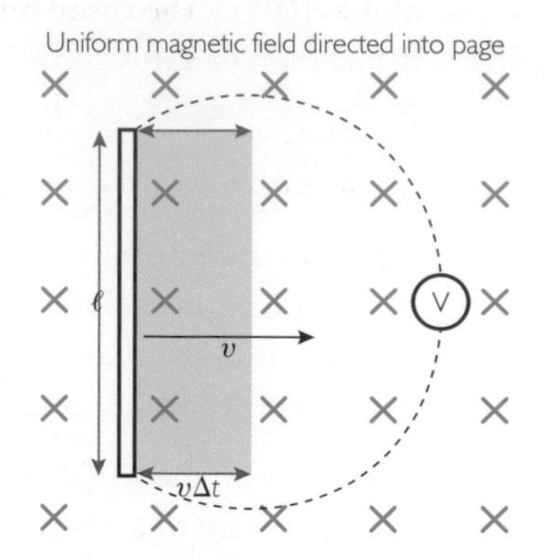

Figure 1.29 Moving an isolated wire in a magnetic field

Imagine that the wire is connected in a circuit. In a time Δt, the wire moves through a distance $v\Delta t$ and so the area of the circuit changes by an amount:

$$\Delta A = lv\Delta t$$

The motion will thus give rise to an emf which, from Equation 6e with $N = 1$, is given by:

$$\mathscr{E} = -\frac{Blv\Delta t}{\Delta t} = -Blv \qquad (7)$$

The emf could be detected as a potential difference by a voltmeter connected between the ends of the wire. If the wire is connected into a complete circuit, the induced emf will give rise to an induced current whose size will depend on the emf and on the resistance of the circuit.

Inspection of Figure 1.29 and consideration of Lenz's law provides us with another useful way of predicting the direction of the induced emf and current. We know that the electromagnetic force due to the induced current must oppose the motion, ie it must act to the left of Figure 1.29. Using Fleming's left-hand rule with the first finger pointing into the page and the thumb pointing to the left shows that the current must be flowing towards the top of the page.

The direction of the induced current can be deduced directly using **Fleming's right-hand rule** (see Figure 1.30). Now the thumb points in the direction of the motion giving rise to the induced current, and the first and second fingers have the same meaning as in Fleming's left-hand rule. But it is easy to confuse left and right, and forget which hand to use in which situation, so it might be wise just to remember the left-hand rule for the motor effect, and use Lenz's law to deduce the direction of an induced current.

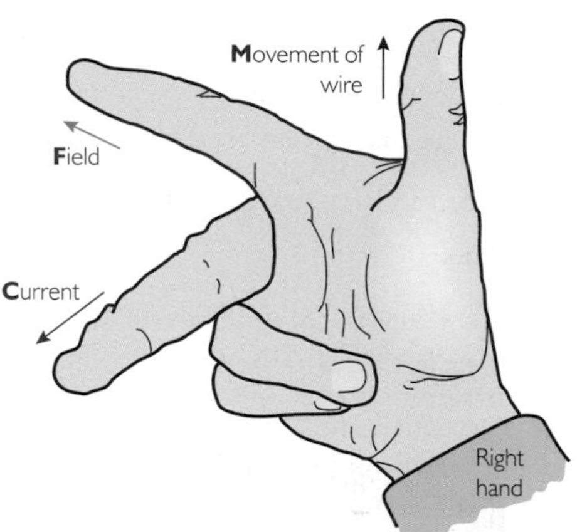

Figure 1.30 Fleming's right-hand rule

Questions

29 Imagine an aeroplane as two metal conductors (the fuselage and wing tip to wing tip) connected together in the shape of a cross. What are the electromagnetic consequences as it flies towards the Earth's magnetic north pole?

30 At a certain location, the Earth's magnetic flux density has a vertical component of 4×10^{-5} T. The metal body of a train is 4 m wide and 5 m high, and the train is moving at 60 m s^{-1} (see Figure 1.31).

(a) State the direction of any induced emf and current resulting from moving across the vertical component of the Earth's magnetic field.

(b) Calculate the size of the induced emf that is likely to develop.

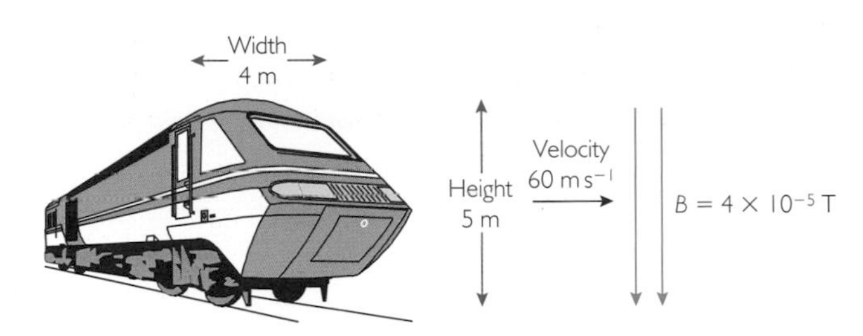

Figure 1.31 Moving train in Question 30

Back to brakes

We are now in a position to explain eddy-current braking such as you saw in Activity 11 when a metal disc rotated in a magnetic field. The motion of the metal in the field induces emfs in the disc that vary according to the local flux density and the speed of the motion. The net effect is to drive eddy currents in the disc. The currents, plus the external magnetic field, give rise to electromagnetic forces that, by Lenz's law, must oppose the motion, ie they retard the disc.

Questions

31 (a) A bicycle operates by a block of rubber being pressed against the rim of the wheel. It is a contact brake. Is an eddy-current brake a contact brake?

(b) Is an eddy-current brake any use as a parking brake? Explain.

32 In designing an eddy-current brake, how would you design the disc so as to ensure that the induced currents were large?

33 From a given speed, any braking system must allow strong or gentle braking. Suggest one way to get this kind of control from eddy-current braking.

Electromagnetism is used to good effect in designing braking systems in trains. The following short article from the June 1992 issue of *Rail News* describes the braking systems on a class of trains known as *465* (Figure 1.32).

Figure 1.32 Class 465 train

Brakes save wear and tear

Trains have three braking systems: three-step air braking, rheostatic braking and regenerative braking.

Air brakes are standard three-step automatic electrically-controlled 'energise to release' air brakes with enhanced emergency braking.

Rheostatic braking is an electric braking system. When braking is selected the traction motors are made to act as generators, producing current which is fed to braking resistances (rheostats) similar to electric-fire elements. They glow and produce thermal energy when current is passed through them. Generating current in this way produces a retarding effect on the traction-motor armatures, which acts through the traction-motor gearing to individual wheels and slows them down. Thus the kinetic energy of the train is converted into thermal energy. The rheostatic circuits ensure that the braking effect produced in the motors matches the braking rate selected by the driver.

As speed falls, however, the retarding effect diminishes. Special circuitry – electronic on the Class 465 *– detects this and makes up the shortfall in braking effect by progressively blending in the air brakes.*

Regenerative braking operates on a similar principle to rheostatic braking, but with the difference that the current produced by the motors is fed back into the third rail and used to power other trains. If there are no trains to accept the current, the braking circuitry detects this and switches to rheostatic braking.

Regenerative braking uses a dynamo to 'recover lost energy' by generating a current while braking. Figure 1.33 shows a circuit in which a motor, when connected to a power supply, turns a flywheel. If no power is connected to it, the motor runs as a dynamo as long as it is rotated by something. Activity 13 demonstrates this.

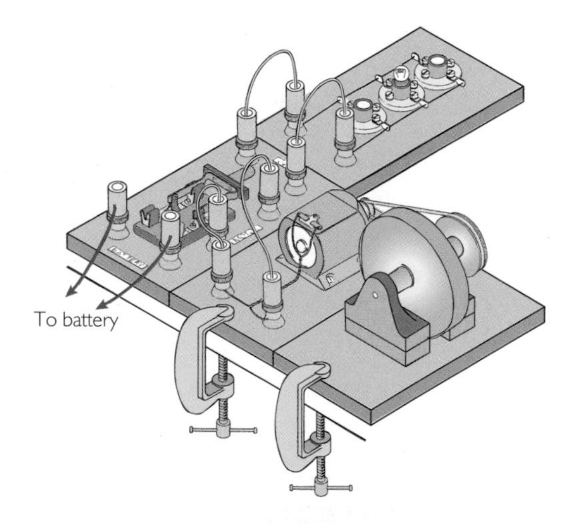

Figure 1.33 Circuit for demonstrating regenerative braking

Activity 13 Electrical Braking

Use the apparatus shown in Figure 1.33 to demonstrate regenerative braking. First allow the flywheel to reach a steady speed, then switch off the power and allow it to freewheel to a stop. Repeat, but this time connect the circuit through the light bulb when the power is switched off. Note any differences, and explain them in terms of energy transfer.

Question

34 In the article 'Brakes save wear and tear',

(a) What do you think is done to change the traction motors to generators?

(b) What path might the induced current follow?

3.5 Getting the voltage right

The European challenge

The Channel Tunnel trains from London to France and Belgium have had to cope with the vagaries of an electrical power supply system that is radically different in the three countries. The original line from London Waterloo to the tunnel was not a high-speed line so its third rail provided 750 V DC, not the main-line higher voltages. In France and in the tunnel the supply is 25 000 V AC. In Belgium, it is taken from an overhead line – this time at 3000 V DC. Add to these problems the fact that the motors themselves run at 1500 V, and we have another problem for physics to solve.

Figure 1.34 Transformer used in a rail power supply

Flexible power-conversion equipment is needed to provide a suitable supply of 1500 V. First the input voltage size is adjusted, then the AC to DC bit is addressed. When regenerative braking is used, the whole problem goes into reverse and the generated power has to be returned to the power lines at a suitable voltage.

This is done with the help of **transformers** (Figure 1.34). A transformer is a device consisting of two electrically insulated coils, of differing numbers of turns, both wound on the same common laminated (layered) iron core that passes through the centre of each coil, often forming a complete loop. In Activity 12, a changing magnetic field in one solenoid induced a current in a nearby solenoid. It was a case of sudden surge and that was that. But if the magnetic field is produced by an alternating current in one solenoid, the magnetic field also changes continuously, and so produces a continuous but alternating current in the second solenoid. To get the largest effect possible the solenoid coils are wound one on top of the other. This means that all the field from the first solenoid threads through the second, thus increasing the flux linkage between the two solenoids, which is further improved by using a core of so-called soft iron (see Figure 1.35).

Study note

The term 'soft' here refers to magnetic properties. Soft iron can easily be magnetised or demagnetised. (It is not soft like putty.)

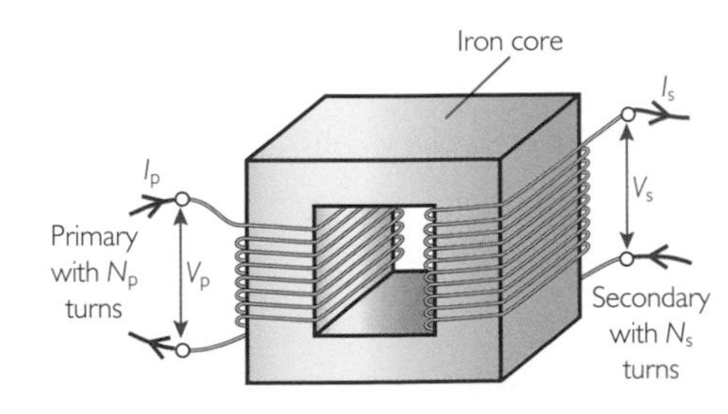

Figure 1.35 Transformer circuit

The primary solenoid has N_p turns and is connected to an input voltage V_p. An output voltage of V_s is obtained from the secondary coil, where the number of turns is N_s. Activity 14 illustrates the relationship between voltages, currents, powers and the **turns ratio** $\frac{N_p}{N_s}$.

Activity 14 Transformers

Build circuits with C-cores and coils or with the software *Crocodile Clips* to produce transformers and investigate the relationships between input and output voltage, current and power.

In Activity 14 you probably found that the ratio of voltages equals the ratio of the numbers of turns:

$$\frac{V_p}{V_s} = \frac{N_p}{N_s} \qquad (8)$$

If N_s is greater than N_p this gives a **step-up transformer**. A **step-down transformer** has N_s less than N_p.

In an ideal transformer, the transfer of energy from primary to secondary will be 100% efficient, so the output power is the same as the input power:

$$V_pI_p = V_sI_s \qquad (9)$$

Comparing Equations 8 and 9 shows that:

$$\frac{V_p}{V_s} = \frac{I_s}{I_p} \qquad (10)$$

so a step-up transformer steps up the voltage but steps down the current. However, as you probably saw in Activity 14, a real transformer has an output power less than the input power and so the current ratio is less than predicted.

One reason for this is the core design. For 100% efficiency, all the magnetic flux from the primary coil should link the secondary coil; in practice this may not quite be the case. Also, not all the input power will be 'transformed' into electrical output power.

Some energy is wasted in heating the coils, and some in producing motion as parts of the transformer are repeatedly attracted to one another and released – this produces the characteristic low-frequency hum that you may have noticed in Activity 14. Question 37 identifies another way in which energy can be wasted in a transformer.

Questions

35 In France the electricity supply is at 25 kV, but the motors require 1500 V. If the primary coil has 1000 turns, how many turns are required in the secondary coil?

36 The UK supply of 750V is converted to AC and then stepped up for the 1500 V motors. The motor power is 1200 kW. (a) What is the turns ratio? (b) What is the primary current?

37 (a) When investigating the magnetic field of a coil carrying alternating current, a student notices that an iron nail on which the coil is wound becomes very hot. Explain the source of this heating. (Hint: look back to your work in Section 3.4 of this chapter.)

(b) The iron cores used for solenoids and transformers are normally laminated, ie made from thin layers of iron glued together, rather than from solid iron. Use your answer to (a) to suggest a reason for this.

3.6 Summing up Part 3

This has been a long part of the chapter. After some revision work on mechanics in which you were introduced to the idea of momentum and reviewed your knowledge of vectors, you met two important areas of the physics of electricity and magnetism. You saw how a current-carrying wire can experience an electromagnetic force in a magnetic field, and how a changing magnetic field can give rise to electromagnetic induction.

Activities 15 and 16 are designed to take you back through some of the ideas you met earlier and at the same time they give you an opportunity to extend your knowledge of electromagnetism.

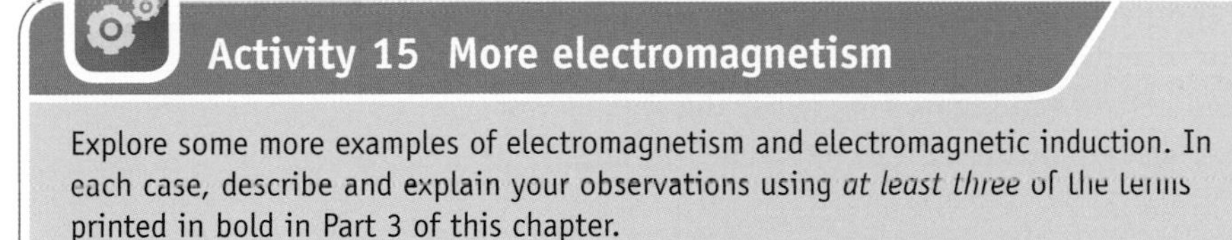

Activity 15 More electromagnetism

Explore some more examples of electromagnetism and electromagnetic induction. In each case, describe and explain your observations using *at least three* of the terms printed in bold in Part 3 of this chapter.

Activity 16 Induction motor

Construct and investigate an induction motor. Compare its operation with that of the motor that you used in Activity 10.

Questions

38 Imagine a situation where engineers were ignorant of the effects of electromagnetic induction and the Channel Tunnel was allowed to be lined with hoops of metal connected together. A *Eurostar* motor, as you have seen, has magnets as a vital element in its motor construction and is going to rush towards the tunnel at 44 ms^{-1}. Why would this cause problems?

39 An aeroplane is travelling horizontally with a velocity of 268 m s^{-1} in a region where the vertical component of the Earth's field is 4.1×10^{-5} T and the horizontal component is 1.8×10^{-5} T. If the wing span is 47 m, find the size of the emf induced between the wing tips.

40 A rectangular coil with 100 turns, 5.0 cm by 8.0 cm, hangs in a vertical plane with a north-west to south-east orientation. A horizontal magnetic field of size 3.5 μT is aligned due north.

(a) What flux passes through the coil's area?

(b) What is the flux linkage through the coil?

(c) If the magnetic field through the coil collapses to zero in 0.10 ms, find the size of the emf induced in the coil.

4 Sensing speed

In Part 2 you saw how signals could be set by the train. It is no use having signals unless they are seen and acted upon by the drivers. The problem is to how get information to a fast-moving object reliably under all conditions. The key elements in the solution are the design of sensors and the electronics to decipher the outputs.

The driver of a train has access to computer-generated information about the journey. This is shown on a screen and includes a three-digit number, which is either a maximum allowable speed (if all is clear) or a target speed for the end of the block section (if slowing down is required). If a signal is flashing, it means that there will be a new signal at the beginning of the next block section, thus giving the driver some forewarning of a required change. For each block a new target speed is set. A graph modelling the projected journey ahead is shown in Figure 1.36 above the sequence of signals. This shows a *Shuttle* train that needs to be brought to rest from 140 km h^{-1}.

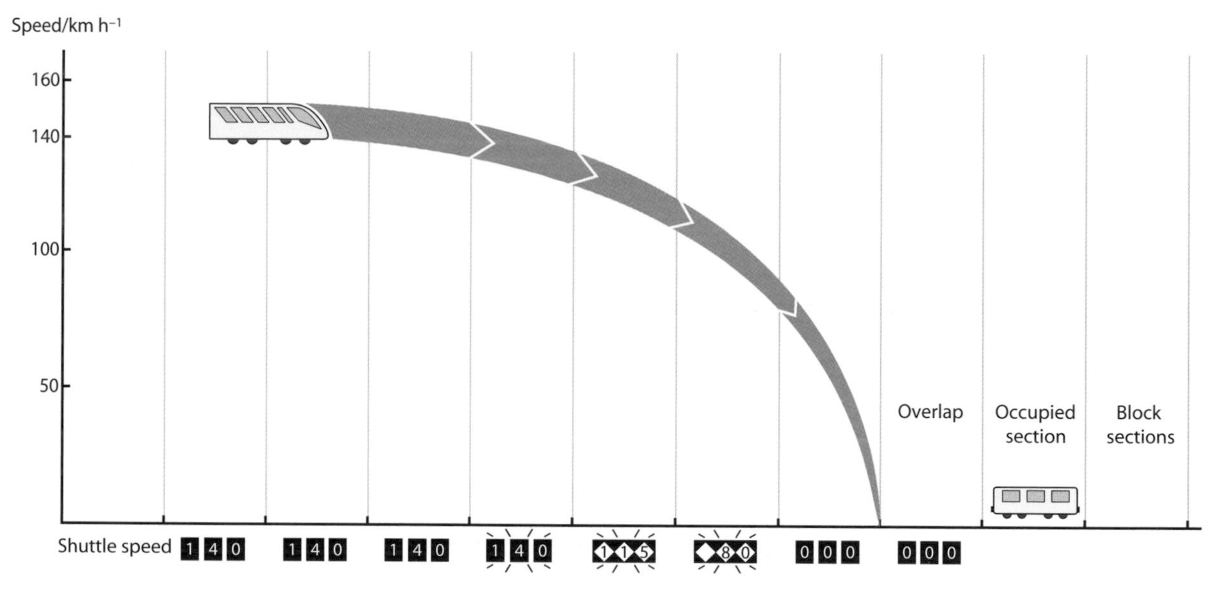

Figure 1.36 Slowing down *Le Shuttle*

The most important task of a driver is to control the train's speed; for example in anticipating a steep bend, the approach to a station or the speed limits of the Channel Tunnel. Speed is continuously monitored directly on the train and if the driver exceeds a maximum speed or ignores a target speed, automatic braking comes into effect operated by the on-board computer. Measuring speed is the subject of this part of the chapter.

4.1 Inductive speed sensors

You will probably have had some experience at GCSE using light gates to measure the linear speed of vehicles. Activity 17 introduces a similar method using magnetism.

Activity 17 Sensing train speed

Use a system of magnets and coils, such as the one shown in Figure 1.37, to show how electromagnetic induction can be used to sense the speed and acceleration of a model train. Figure 1.38 shows some typical results for you to analyse if you are unable to obtain your own.

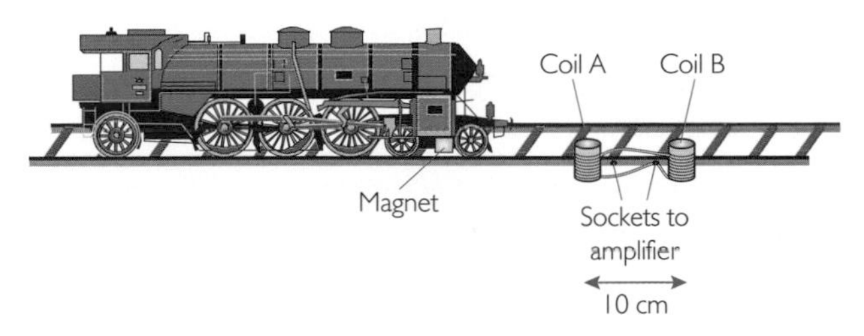

Figure 1.37 Arrangement for sensing the speed of a model train

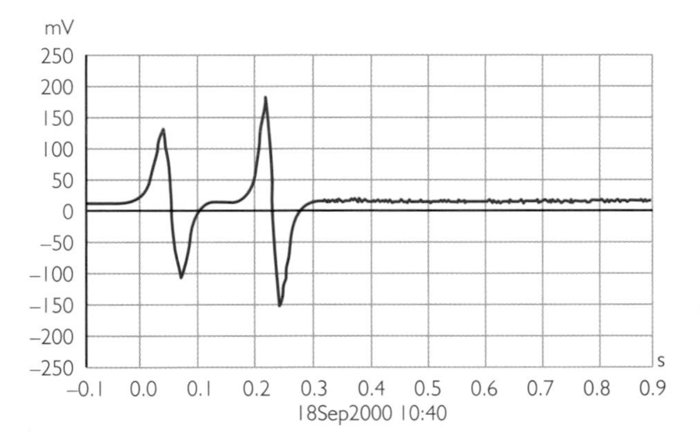

Figure 1.38 Output from coils in Activity 17 displayed on Pico Technology Picoscope

The arrangement in Activity 17 illustrates some basic principles of the inductive speed sensors used on trains, although a practical speed sensor must have all its parts on the train.

4.2 Timing

In measuring speed electronically, one very important requirement is the ability to generate a voltage pulse that is exactly one second long. To do this, we can use devices called **capacitors**, which can store electric charge and release it at a precisely known and predictable rate. Such devices are the subject of this section.

Introducing capacitors

Capacitors are components that can store electrical charge. The simplest capacitor design consists of two parallel metal plates separated by an air gap, so capacitors are symbolised as two parallel lines, one of which may be shown as an open plate or with a positive sign to indicate which terminal of the capacitor should be connected nearest to the positive terminal of a power supply (see Figure 1.39).

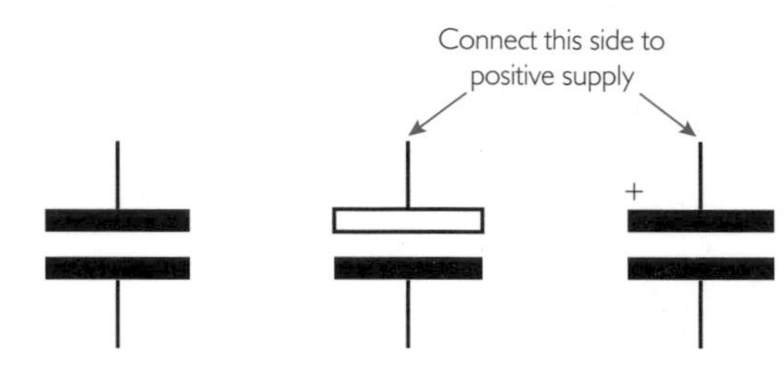

Figure 1.39 Capacitor symbols

Activity 18 Capacitors

Explore the way the current in a circuit changes as various capacitors are charged from a battery and then discharged using the circuit shown in Figure 1.40.

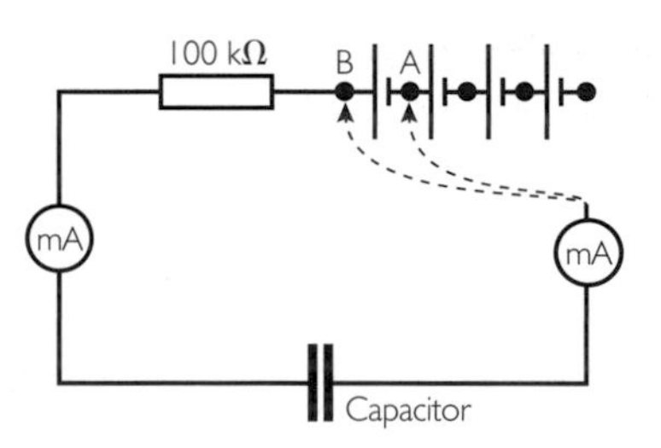

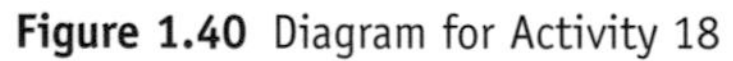
Figure 1.40 Diagram for Activity 18

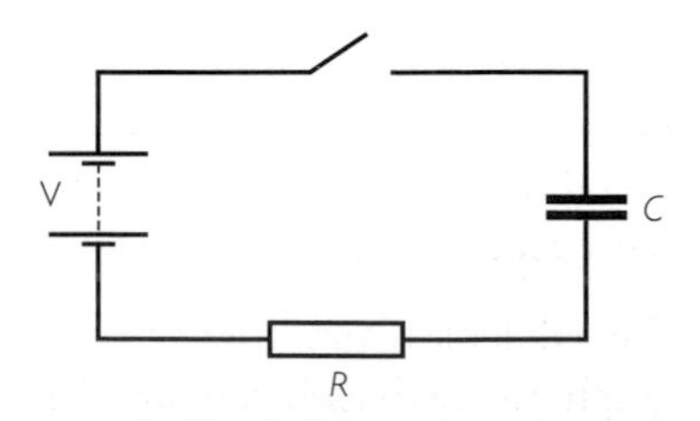

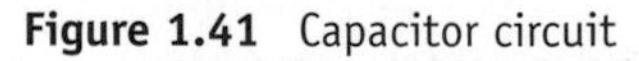
Figure 1.41 Capacitor circuit

For a circuit as shown in Figure 1.41, when the switch is closed charge flows on to one side of the capacitor from the battery; an equal amount flows away from the other side of the capacitor into the other battery terminal. Charge flows until the potential difference across the plates becomes the same as that across the battery. The capacity then has a charge $+Q$ on one side and $-Q$ on the other. In a sense it 'stores' a charge Q, though 'separates' would be a better word, and it is actually more accurate to say that a capacitor stores energy by separating charge. The charge that a capacitor can store is directly proportional to the battery voltage:

$$Q \propto V$$

the ratio $\frac{Q}{V}$ is constant, and is defined to be the **capacitance**, C, of the capacitor:

$$Q = CV \qquad (11)$$

Capacitance is the amount of charge stored per unit potential difference across the plates. The SI unit of capacitance is the farad, F (named after Michael Faraday).

Study note

You will learn about capacitors as energy storage devices in the chapter *The Medium is the Message*.

$1\ F = 1\ C\ V^{-1}$. One farad is a very large capacitance. Capacitances generally range between 1 picofarad (pF) and 10 000 microfarads (μF).

If the potential difference or voltage across the capacitor plates is too high, the insulation (called dielectric) between the plates may start to fail and start to conduct. For this reason capacitors are marked with a maximum working voltage above which the capacitor should not be operated.

Maths reference

SI prefixes
See Maths note 2.4

Charging and discharging

In order to see how a capacitor can be used in a timing circuit, it is important to know what happens to voltage and current in a capacitor–resistor circuit as the capacitor accumulates charge and releases it. In Activity 19 you will investigate charging and discharging experimentally using a large-valued capacitor and resistor, which will slow down the changes in the circuit sufficiently so that they can be easily observed.

Activity 19 Slow charge and discharge

Observe the way voltages and currents change as a capacitor charges and discharges.

Activity 20 Modelling charge and discharge

Use *Crocodile Clips* to investigate more closely how the charging and discharging of a 100 μF capacitor through a 100 kΩ resistor affects the current and voltage in the circuit, and to see how capacitance and resistance values determine the time of charging and discharging.

In Activities 19 and 20, you have seen graphs representing the charge and discharge of a capacitor. They belong to a family of graphs known as **exponentials**. Many naturally occurring changes are exponential, as you will see when you study the chapters *The Medium is the Message* and *Reach for the Stars*. One important characteristic of an exponential discharge (or decay) graph is that *the changing quantity changes by equal fractions in equal times.* You should have seen this when you found the times for the voltage to halve, halve and halve again.

We will look at various ways in which the shape of a capacitor discharge graph can be described mathematically; they are all equivalent to one another.

Figure 1.42 shows a capacitor, C, discharging through a resistor, R. The capacitor behaves rather like a cell with a changing terminal potential difference. This pd, V, is related to the charge, Q, stored on the capacitor:

$$V = \frac{Q}{C} \qquad \text{(from Equation 11)}$$

Figure 1.42 Capacitor discharging through a resistor

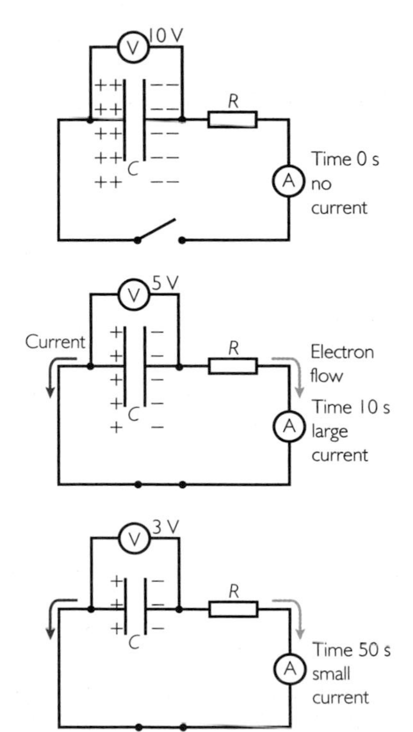

As the capacitor discharges, Q decreases and hence V also decreases. Since we also know that the current, I, in the resistor is related to the pd, we can write:

$$I = \frac{V}{R} = \frac{Q}{RC} \qquad (12)$$

Since Q is decreasing, I must also be decreasing. Remembering that current is the rate of flow of charge:

$$I = \frac{\Delta Q}{\Delta t}$$

we can write:

$$\frac{\Delta Q}{\Delta t} = -\frac{Q}{RC} \qquad (13)$$

The negative sign indicates that the charge is decreasing with time.

If we are dealing with very small time intervals, we can use calculus notation:

$$\frac{\mathrm{d}Q}{\mathrm{d}t} = -\frac{Q}{RC} \qquad (13a)$$

Equation 13 says that *the rate of flow of charge is proportional to the charge itself.* This is another important characteristic of exponential discharge. Equation 13 is actually equivalent to the 'equal fractions in equal times' pattern as you can see by rearranging it:

$$\frac{\Delta Q}{Q} = -\frac{\Delta t}{RC} \qquad (13b)$$

This version tells us that the fraction $\frac{\Delta Q}{Q}$ is directly proportional to the time interval Δt.

Activity 21 Exponential discharge

Examine the discharge curve shown in Figure 1.43 and verify that it is exponential.

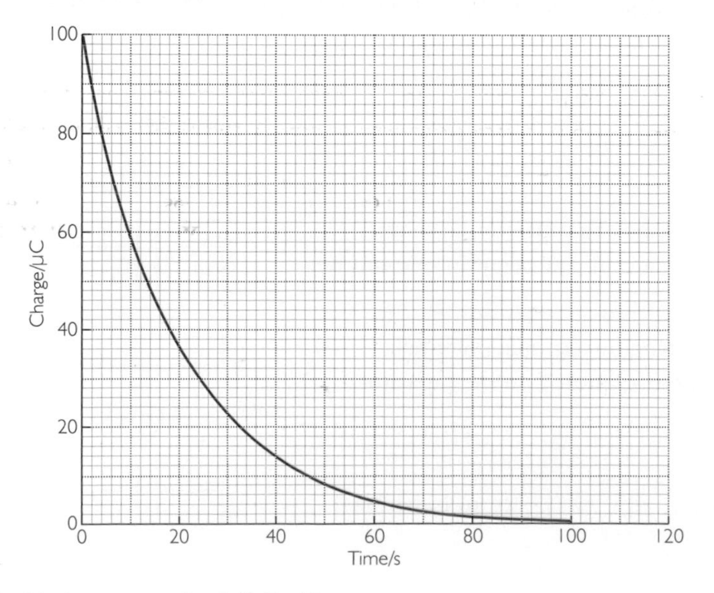

Figure 1.43 Discharge curve for Activity 21

Equation 13 describes a capacitor discharge curve by telling us how the gradient at any one time is related to the stored charge at that time. But it does not directly tell us how much charge is stored at any one time, or answer the most commonly asked questions like: 'how much charge is left after 0.5 s?' For this, we need another equation, which can be derived from Equation 13 using calculus:

$$Q = Q_0 e^{-t/RC} \qquad (14)$$

Here Q_0 is the charge stored when $t = 0$, and e is a number that arises from the maths: e ≈ 2.718.

We can use Equation 14 to tell us how the voltage and current change as well. Substituting $Q = CV$ (from Equation 11) and cancelling C, we get:

$$V = V_0 e^{-t/RC} \qquad (15)$$

Then similarly, using $I = V/R$ and cancelling R, we get:

$$I = I_0 e^{-t/RC} \qquad (16)$$

In other words, the voltage and current also decrease exponentially.

Maths reference

Exponential changes
See Maths note 9.1

Exponential functions
See Maths note 9.2

Study note

If you are familiar with calculus you can show that the 'equal fractions in equal times' pattern will always be described by an expression such as Equation 14.

Questions

41 (a) A 100 μF capacitor is discharged through a 250 kΩ resistor for 20 s. What fraction of its original charge will remain stored at this time?

(b) Initially the pd across the plates was 10 V. How much charge remained stored after 20 s?

42 (a) By inspecting Equation 13, argue convincingly that RC must have dimensions of time (and hence SI units of seconds).

(b) A capacitor, C, discharges through a resistor, R. What fraction of the starting voltage (or charge) will remain across the capacitor after a time equal to RC?

If you answered Question 42 correctly, you should have seen that approximately 0.37 (or 37%) of the initial voltage or charge will remain on a capacitor after a period of time equal to RC. The time RC is called the **time constant** of the circuit and is used as a measure of how fast the resistor–capacitor combination discharges. It is sometimes represented by the Greek letter τ 'tau':

$$\tau = RC \qquad (17)$$

As the charge and voltage drop by 37% in each interval of τ, we can argue that they never actually reach zero. In practice, provided the time interval is several times τ, we can say the capacitor is fully discharged. Question 43 and Activity 22 illustrate this.

Activity 22 One step at a time

Use a spreadsheet to examine the discharge of a capacitor over a succession of small time intervals.

Questions

43 (a) What percentage of a capacitor's initial charge will remain after three time constants?

(b) A student claims that a discharging capacitor is 'effectively fully discharged' after five time constants. Do a calculation and comment on this viewpoint.

44 Figure 1.44 shows a timing circuit. When the push switch is pressed, the capacitor charges almost instantly. When the switch is released, it discharges more slowly through the resistor. As it does so, the voltage at the output falls and pd across the capacitor falls. (a) What is the voltage at the output after the push switch is pressed? (b) What is the time constant for the resistor–capacitor combination? (c) What will be the output voltage after 5 s?

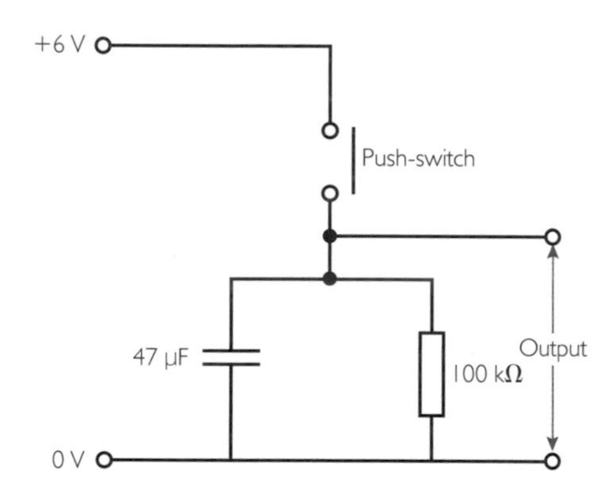

Figure 1.44 Timing circuit for question 44

An electronic timer

In Question 44 you saw how a simple capacitor circuit might be used to produce a set time delay. In this section, we look at a small, electronic integrated circuit (chip), the 555 timer, which uses capacitor charge/discharge in a more versatile timing device.

Figure 1.45 555 chip

An integrated circuit (IC) is simply a package that includes an array of electronic components miniaturised to fit on a single silicon 'chip'. The 555 IC is a very cheap and versatile chip that works on any DC supply from 5 to 15 V and can give out or take in a current up to 200 mA (see Figure 1.45). It is particularly useful as a simple timing device since it can easily be adapted to form part of either an **astable** or a **monostable** circuit.

An astable circuit has no stable output states. It continually switches from its first state to its second state, and back again. The two states are usually a high voltage and 0 V. (An example of an astable circuit would be that operating the flashing lights warning of a train or on a pedestrian crossing.) The 555 chip is used to control the switching period.

A monostable circuit has one stable output state. This means the output will never leave that state unless it is switched, and when changed to the other state it is unstable and will return eventually to the first, stable, state after a pre-determined time delay. (An example of monostable circuit might be the timing circuit to set a video recorder to switch on to record a chosen programme.) The 555 chip is used to set the length of the time delay.

It is not necessary to know the details of how the 555 IC works; it is sufficient to know that the eight pins are organised as shown in Figure 1.46. The 555 IC contains the equivalent of 40 resistors and transistors, and needs just an external resistor or two and a capacitor to make up the final circuit. It is these external components that are used to set the required time delays (monostable) or time periods (astable).

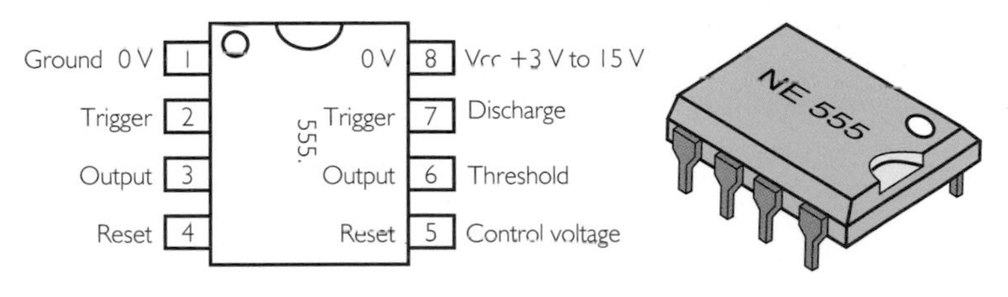

Figure 1.46 Pins on a 555 chip

Figure 1.47 shows how to connect a 555 chip to form a monostable circuit. When push switch S is momentarily pressed, the stable 0 V output of the monostable goes 'high' – close to the supply voltage, V_s. It then stays high for a certain time period, T, before falling back to 0 V. The time period is determined by the size of resistor R and capacitor C; it is controlled by how fast C charges through R. The time period, T, can be approximated to a reasonable degree of accuracy by the formula:

$$T = 1.1RC$$

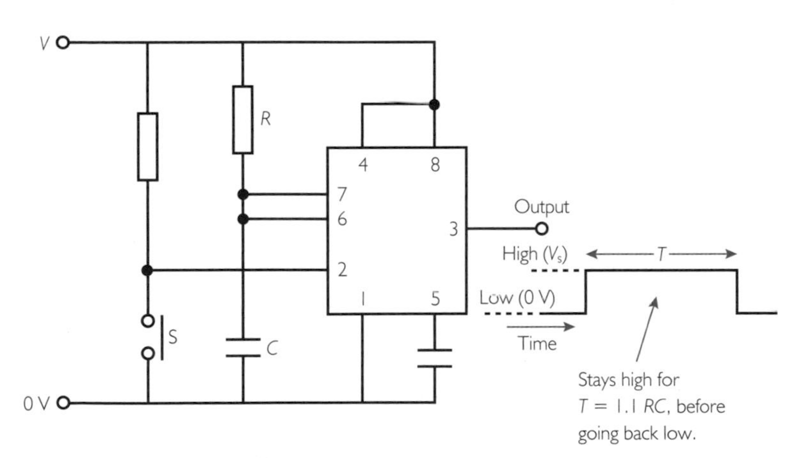

Figure 1.47 Monostable circuit with a 555 chip

Activity 23 Modelling a 555 monostable circuit

Use *Crocodile Clips* or other circuit design software to model the operation of a monostable circuit incorporating a 555 chip.

Near the start of this part of the chapter we considered a method of counting pulses to determine the speed of a train and said that it was necessary to create a voltage pulse exactly one second long. Now that we have the means to do this accurately, we are in a position to go back to our model with greater understanding.

Activity 24 Counting revolutions

Model a counting circuit with *Crocodile Clips* or other circuit design software.

4.3 Summing up Part 4

In this part of the chapter you have learned about capacitors and about some properties of exponential graphs, and seen how the discharge of a capacitor can be used to control a timing circuit.

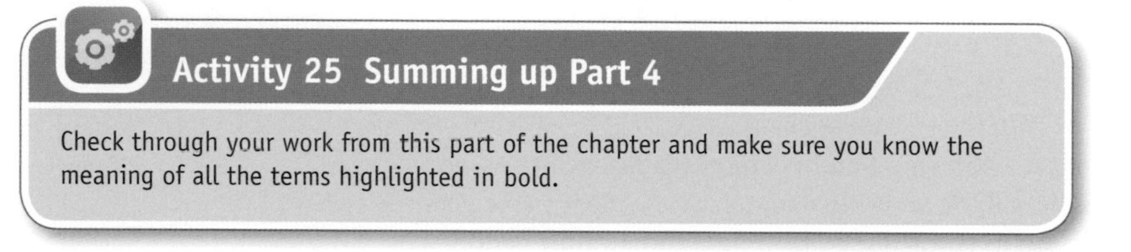

Activity 25 Summing up Part 4

Check through your work from this part of the chapter and make sure you know the meaning of all the terms highlighted in bold.

Question

45 A model-rail enthusiast wants to create a block signalling system that gives a red light when a train is occupying one section of track, and gives a green light 10 s after the train has left the block. The circuit is shown in Figure 1.48.

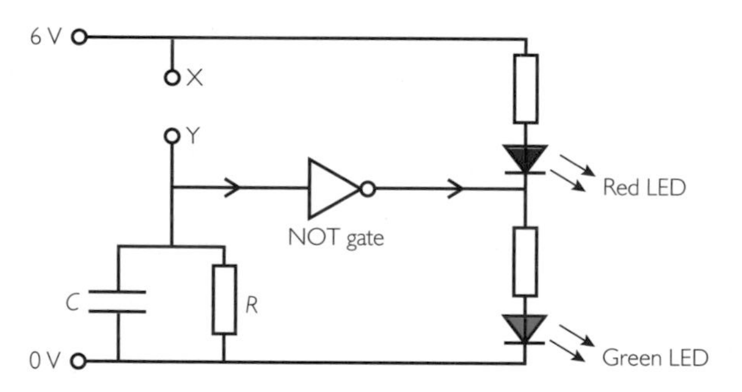

Figure 1.48 Model railway circuit for Question 45

X and Y are connected to each of the metal tracks in the 'block' to be monitored.

The NOT gate turns an analogue (continuously varying) input voltage into a digital signal. If the input voltage to the NOT gate is above 1.5 V, the output is low (close to 0 V), and if the input is below 1.5 V the output is high (around 6 V).

The LEDs (light-emitting diodes) emit light only when the potential decreases in the direction of the arrow that forms part of their symbol.

(a) Explain how the circuit works.

(b) Show that, for a time delay of 10 s, $RC \approx 7.2$ s.

(c) Suppose the circuit was made with $R = 1000$ kΩ and $C = 10$ μF. Comment on whether this would produce a time delay that was too short or too long.

5 Structure and safety

5.1 How safe is the Channel Tunnel?

Train passengers want to get to their destination on time and in safety. As trains move ever faster and travel in more extreme conditions the potential for danger increases, but so does the skill of the engineers applying physics in the modern world.

Despite what some people think, flooding is not a major concern as the amount of sea water actually seeping into the tunnel is very slight; some of it evaporates and pumps remove the rest. Fire is the biggest worry and there is a complex system of fire detection, suppression and smoke control in place. In the case of a fire, the passengers would be sealed off and would breathe recycled air until the fire was under control.

General wear and tear is also a concern. High speeds and heavy loads deform and move the track under a train. Tiny kinks and dips develop and become exaggerated with time – like potholes in roads. Eventually they can lead to the train swaying about (which disturbs the passengers) and, if left for a very long time, they could cause a derailment. *Eurostar* reduces the problem of wear by being lighter than traditional trains. Axle loads are limited to 17 tonnes and the line is maintained every single night so that the high-speed service can continue.

The lighter weight may be ideal for track maintenance but seems to offer poor protection in the event of hitting something or being hit. In fact, you have already seen one advantage of reducing the mass of a train in Part 3. It makes it easier to stop a train before it collides with anything dangerous; less force is required.

The manufacturers use materials that may be light but are also strong and so resist deformation. But, most important of all, the locomotives and coaches incorporate sophisticated impact-absorbing structures. These designs were the result of the British Rail Research's Crashworthy Development Programme. To protect passengers in a collision, this team constructed trains with collapsible zones in the ends of the coaches. Hydraulic couplers will now sheer off at high-speed impact and slide back into the undercarriage. Each cab and coach end has special anti-climber bars fitted to prevent riding up in a crash. Winston Rasaiah, project engineer, said, 'With crash damage restricted to collapsible zones the passenger compartment remains intact' (see Figure 1.49).

Figure 1.49 Modified coach after collision

We need to be able to predict the likely damage caused by moving objects and for that we need to know the size of any forces involved, and also the energy transferred. This part of the chapter deals with both these aspects of collisions.

5.2 Forces in collisions

A train crash can be caused by many different events. Here we focus on collisions between two trains that are unfortunately on the same track. A theory of collisions will enable you to calculate the likely outcome of such a meeting or to deduce the speeds just prior to impact.

Activity 26 Observing collisions

Observe some collisions between trolleys on a runway or between air-track vehicles. Include situations where one vehicle is initially at rest and when both are initially moving towards one another. Include some collisions in which the two vehicles become coupled together and some in which they bounce off one another. Notice what happens to the velocities after impact. Try altering the mass of one or both vehicles.

In Activity 26 you probably noticed that, if the moving mass is increased (as when a moving vehicle couples to a stationary one and they both move off together), the velocity decreases. Also, if two vehicles of equal mass approach one another with equal speeds, they come to rest when they couple together. Mass and velocity are both important when it comes to determining the outcome of a collision, and so are the material properties of the colliding vehicles (do they bounce, stick together or crumple?). To gain more insight into what happens, we need to use Newton's laws of motion and the concept of momentum that you met in Part 3 of this chapter. If one train meets another train we may have situations like the one shown in Figure 1.50.

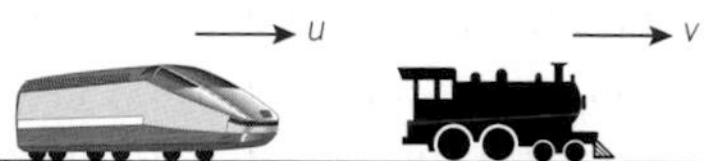

Figure 1.50 Two trains about to collide – the rear train is travelling faster

For all collisions, the impact forces on the two bodies during the collision are equal and opposite and last for equal times. This is a direct application of Newton's third law and is always true regardless of whether the two bodies bounce apart or stick together. In Part 3 you met the impulse of a force ($F\Delta t$). Both objects in a collision experience the same impulse during the impact. In Figure 1.50, unequal amounts of damage may be caused by the force of the collision but the force will be of equal size on each train. The *Eurostar* will experience a force of magnitude F to the left and the steam train a force of the same magnitude to the right; both forces will act for exactly the same time interval Δt.

As you saw in Part 3, the change in momentum of a moving object (*mass* × *velocity*) is related to the impulse:

$$F\Delta t = \Delta(mv) \qquad \text{(Equation 2)}$$

Since the forces at impact act in opposite directions the changes in momentum produced are also in opposite directions (remember that both force and momentum are vector quantities). Let the initial velocities be u_1 and u_2, and the final velocities v_1 and v_2, as shown in Figure 1.51 for trains 1 and 2 with masses m_1 and m_2.

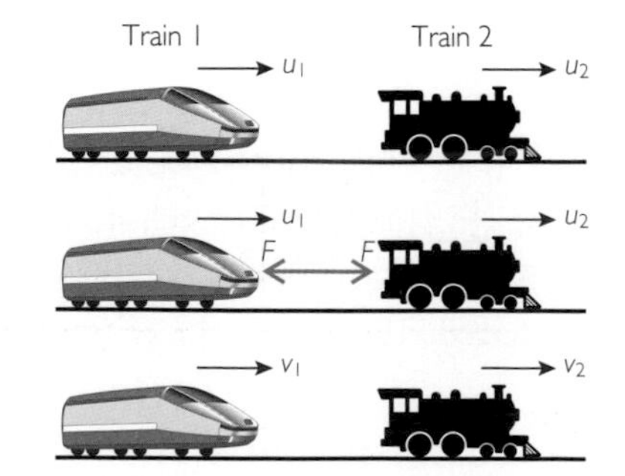

Figure 1.51 Two trains colliding

For train 1:

$$-F\Delta t = m_1v_1 - m_1u_1$$

For train 2:

$$F\Delta t = m_2v_2 - m_2u_2$$

So:

$$-(m_1v_1 - m_1u_1) = m_2v_2 - m_2u_2$$

which can be rearranged to give:

$$m_1u_1 + m_2u_2 = m_1v_1 + m_2v_2 \qquad (18)$$

Equation 18 can also be written using p to represent momentum:

$$p_1 + p_2 \text{ before collision} = p_1 + p_2 \text{ after collision} \qquad (18a)$$

and also generalised to a system consisting of more than two interacting objects:

$$\Sigma p = \text{constant} \qquad (18b)$$

(in words: the sum of all the momenta remains constant).

The terms on the left of Equation 18 represent the total momentum of the system before the impact, and those on the right the total momentum of the system afterwards. This is very important! Expressed in words, *the total momentum of a system remains constant provided no external forces act on the system*; this is the **principle of conservation of linear momentum**. ('Linear' means it applies to objects moving in a straight line.) It applies to *all* interactions without exception. The idea of a system is important. For the duration of the impact the only forces that count must be between the two trains. In Activity 26, you should have found that the velocities before and after impact were (approximately) as described by Equation 18.

Maths reference

The summation sign
See Maths note 0.3

Explosions

At first sight an explosion is very different from a collision, but this is another event where the principle of conservation of momentum can be applied. When a gun fires a shell forward the gun recoils backward. The two forces produced by the explosion have to be equal but opposite in direction. This means that impulses are equal for gun and shell, and so the total momentum of the gun-shell system is unchanged (see Figure 1.52).

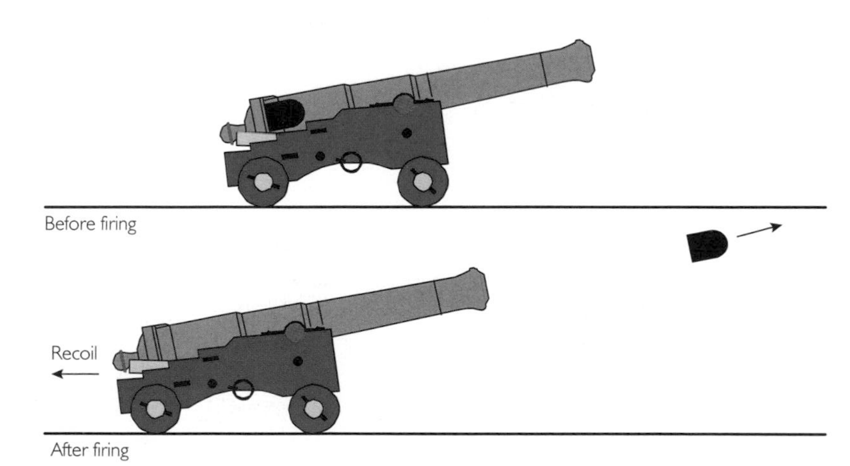

Figure 1.52 An explosion

Questions

46 A 5 t wagon runs into the back of a stationary 41 t *Eurostar* locomotive and gets jammed there. They move off together at 5 m s^{-1}. How fast was the wagon travelling before the collision?

47 As shown in Figure 1.53, two railway coaches are about to collide head on. After the collision the 90 t mass moves on in the same direction at a velocity of 30 m s^{-1}. What is the velocity of the 20 t mass after the collision?

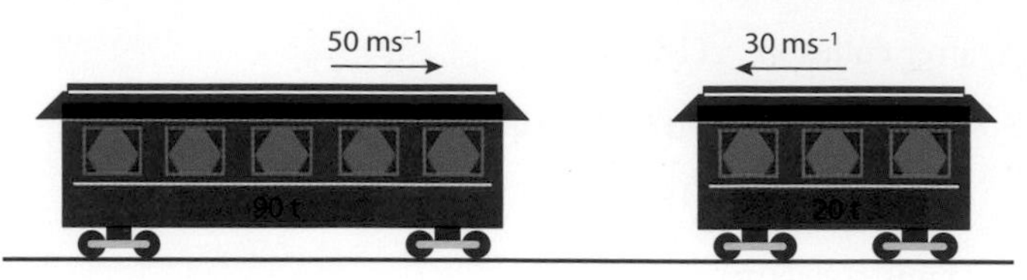

Figure 1.53 Diagram for Question 47

48 A party trick is to let the air out of a balloon and watch it fly all over the room. How is this an example of an 'explosive' collision?

Impulse and momentum change

The principle of conservation of momentum is useful for analysing the situation immediately before and after a collision. But if we want to know about possible safety implications, we need to know something about the forces that act during the collision. You have already met useful expressions for this:

$$\text{impulse} = F\Delta t$$

and:

$$F = \frac{\Delta p}{\Delta t} \qquad \text{(Equation 2)}$$

If the impact time is known, then the (average) force acting can be found. Activity 27 and Questions 49 to 52 illustrate this.

Activity 27 Forces in collisions

Use *Logger Pro*, *Multimedia Motion* or other motion analysis software that you met in the AS chapter *Higher, Faster, Stronger* to explore the forces acting in various impacts. For example *Multimedia Motion* has a soccer sequence where you can step through the frames to find how long a foot is in contact with a ball. There are also sequences showing collisions between air-track vehicles. The *Data* button gives you access to values for subsequent speeds, and the mass of the ball can be found using the *Text* button. Additional help is available through the *Help* screen.

Questions

49 A snooker ball is struck by a cue. Figure 1.54 shows the force–time graph for the shot from the point of view of the ball.

(a) Draw the force–time graph for the same event for the cue.

(b) What information can you obtain from the area under such a graph?

(c) What shape graph would you get for uniform acceleration?

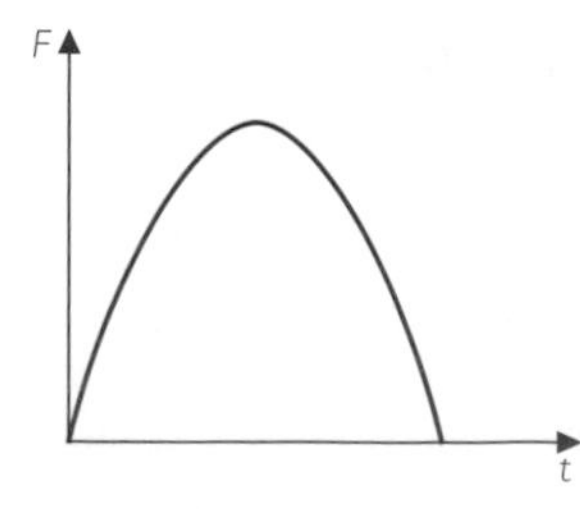

Figure 1.54 Diagram for Question 49

50 When you catch a hard cricket ball it hurts less if you pull your cupped hands back at impact. Why does this work?

51 A hole in a water pipe leading from a tanker in the Channel Tunnel allows 70 kg of water to escape every second in a horizontal jet moving at 30 ms^{-1}. The jet strikes the tunnel wall and then falls to the ground. (a) What horizontal force does the wall exert on the water? (b) What is the horizontal force on the wall?

52 Suppose the arrival of a hurricane is predicted for the Channel area with wind speeds of 120 mph. The forces on the side of a *Eurostar* train are known for a 60 mph gale. Will the hurricane forces on the same train be

A: the same

B: twice as large

C: three times as large

D: four times as large?

5.3 Energy in collisions

There is another useful way to look at what happens in a collision. Figure 1.55 shows test coaches that have been damaged in a collision. Each coach experiences a force, and the force does work producing some plastic deformation of structure, so in an **inelastic collision** there is a net loss of kinetic energy (although energy overall is conserved). On the other hand, an **elastic collision** is one in which there is no net loss of kinetic energy – any energy that is temporarily stored due to elastic deformation is recovered as kinetic energy as the colliding objects spring apart.

Study note

This might remind you of some ideas you met in the AS chapters *Good Enough to Eat* and *Spare Part Surgery*, where you considered elastic and plastic deformation of materials.

Figure 1.55 Result of collision between two test coaches

Mathematically, an elastic collision can be described as follows:

$$\frac{1}{2}m_1 u_1^2 + \frac{1}{2}m_1 u_2^2 = \frac{1}{2}m_1 v_1^2 + \frac{1}{2}m_1 v_1^2 \qquad (19)$$

or:

$$E_{k1} + E_{k2} \text{ before} = E_{k1} + E_{k2} \text{ after} \qquad (19a)$$

where E_k represents kinetic energy and the other symbols have the same meanings as in Equation 18. For a system consisting of more than two interacting objects, Equation 19 can be written in a more general form:

$$\Sigma E_k = \text{constant} \qquad (19b)$$

From the viewpoint of a safety engineer, which type of collision is more desirable – elastic or inelastic? How can trains or cars be designed to be 'crashworthy'?

From the perspective of train crashes the chances of an elastic collision are remote and positively undesirable. Bits rebounding all over the place would be a nightmare in an accident. The whole point of crashworthiness is that kinetic energy must be absorbed in ways that do as little damage as possible. One way to achieve this is to make the outer parts of a vehicle from a material that readily undergoes plastic deformation, such as is done in the 'crumple zones' of cars, while surrounding the passengers with a strong and fairly rigid structure.

In some situations, though, elastic collisions are desirable, as in some items of sports equipment; a tennis racket that can return a demon serve with little loss of kinetic energy is very desirable (Figure 1.56). You may have seen a desk toy called Newton's cradle, which also relies on collisions being elastic in order that the ball bearings can continue to bound back and forth for a long time.

Figure 1.56 Tennis racket at full stretch

Activity 28 Impact forces and crumple zones

Using the apparatus shown in Figure 1.57, explore the forces that arise during collisions.

The sharpened dowel rod provides a means of estimating the force involved in the impact. Impacts will throw the dummy onto the spike, which penetrates the plasticine. Note the depth of penetration. To obtain a value for the force, pull the dummy onto the spike using a thread fastened to a force meter. Obtain enough readings to plot a graph of force against depth of penetration. Explore the effects of colliding at different speeds, changing the mass of the trolley and having a crushable front to the trolley.

Study note

In order to do Activity 28 and some of Questions 53 to 55, you will probably need to look back at Part 3 of this chapter, and possibly at the work you did in the AS chapter *Higher, Faster, Stronger*.

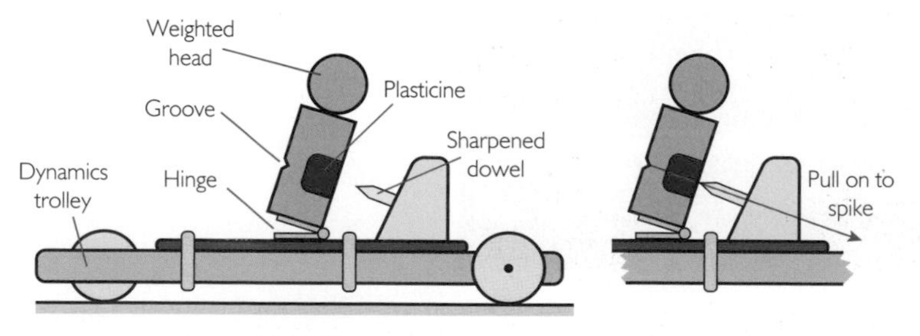

Figure 1.57 Apparatus for Activity 28

Questions

53 The new collapsible front on trains mentioned in Section 5.1 provides for a total of 1 MJ energy absorption in a deformation length of 1 m at each end of the locomotive, and further energy absorption at the interfaces between coaches. A multiple unit of 24 t is built to this specification.

(a) What is the approximate acceleration as the front of the locomotive deforms? Comment on the size of your answer and the likely effect of this acceleration on people within the vehicle.

(b) If a 750 tonne *Eurostar* has the same features, what is its acceleration on deforming?

54 In Part 1 of this chapter it was stated that the kinetic energy of *Eurostar* at full speed (83 m s^{-1}) is almost exactly that required to raise its 750 t from the ground level to the top of the TV aerial at the top of the Eiffel Tower, which is 318 m high.

(a) Calculate the energies involved to show how close 'almost' is to having enough energy to do this (g = 9.81 N kg^{-1}).

(b) In practice, could the *Eurostar* be made to rear straight up in the air to such a great height? (Imagine this taking place outside the tunnel!)

55 Examine some of the collisions from Questions 46 and 47 and Activities 26 and 27 to see whether they were elastic.

5.4 Coming together

In this part of the chapter, you have seen how collisions can be analysed first in terms of force, momentum and impulse and then in terms of energy transfer. It is important to be clear that momentum and energy are *different* quantities. Momentum is always conserved in any event where no external forces act. Energy, too, is always conserved, but *kinetic* energy is conserved only in *elastic* collisions. So, when analysing a collision or explosion, it is usually best to deal with momentum first, and then see what happens to the kinetic energy.

The questions and activities in this section involve both momentum and energy and thus provide you with an opportunity to review your knowledge and understanding.

Activity 29 Summing up Part 5

Check through your work on this part of the chapter and make sure you know the meaning of all the terms printed in bold.

Activity 30 Crashworthy?

At a press conference, after a nasty railway accident, a railway engineer was asked why his trains weren't built to be as strong as it is physically possible to build. He replied 'our trains are as strong as is necessary to withstand normal service loads but we use clever engineering at the ends of the vehicles to disperse energy by structural collapse while maintaining the integrity of the occupied zone'.

Explain in your own words what he meant, using *at least three* of the terms printed in bold in this part of the chapter.

Question

56 A man of mass 70 kg stands on a trolley of mass 330 kg that rolls smoothly on frictionless rails. The man fires a rivet gun in a direction parallel to the rails. The gun contains 50 g rivets that leave the gun at a speed of 20 m s^{-1} relative to the trolley.

(a) How does firing a single rivet affect the speed of the man-plus-trolley?

(b) If the gun fires continuously at a rate of two rivets per second, calculate the speed of the trolley after the first minute.

(c) Which acquires more kinetic energy per minute, rivets or man-plus-trolley?

(d) The man gets a warning of an on-coming train, 5 km away, approaching at a constant speed of 10 m s^{-1}. Can he get up enough speed to escape?

6 Running a railway

In Victorian times tycoons including Isambard Kingdom Brunel (Figure 1.58) set up railway companies all over the UK. Since then there have been a succession of entrepreneurs such as Richard Branson (Figure 1.59) who want to run their own railway. Most recently, in 2007 Giles Fearnley set up Grand Central Trains (Figure 1.60) to run trains from Sunderland to London. So what does it take to run a railway?

Figure 1.58 Isambard Kingdom Brunel and George Stephenson, two early tycoons

Figure 1.59 Richard Branson and a Pendolino train

Figure 1.60 A Grand Central Train

6.1 How safe is safe?

The majority of modern trains are very expensive, with most costing millions of pounds. What sort should you use? What are the limitations? In Europe fast trains run on dedicated high-speed track, but most UK companies operate on standard track. Grand Central went for refurbished 30-year-old High Speed Trains, in the hope that they would be reliable but relatively straightforward. Virgin went for high-tech tilting trains, which tilt to go around corners but have had teething troubles and cost many millions. (You can read about this on the BBC website via **www.shaplinks.co.uk**.)

What are the benefits and risks involved with having a dedicated high-speed line (like the HS1 from London)? What sort of rolling stock should one use? (HS1 is the UK's first, and at the moment only, high-speed rail link. It takes *Eurostar* trains from St Pancras to Kent.)

Automatic train protection safety systems have been installed on trains to apply the brakes automatically to prevent signals being passed at danger (SPAD). Many UK fast passenger trains, such as *Eurostar* and the West-Coast Main Line have it. But the colossal cost (around £6 billion in 2003) prevents it being fitted everywhere.

In Parts 2 and 5 of this chapter, the emphasis has been on rail safety. But making railways safe comes at a cost, and this must also be considered when designing a rail system.

Acceptable risk?

The Rail Safety & Standards Board (RSSB) is in charge of ensuring the safety of UK railways. The railway companies have a legal duty to make the risk of accidents and injury on the railways 'as low as reasonably practicable' (ALARP). You can read a paper about this on their website: *Taking safe decisions – how Britain's railways take decisions that affect safety*. The website also provides a summary of the report. Both are available via **www.shaplinks.co.uk**.

The more money that is spent on reducing a risk, the lower the chance of an accident. If this principle is followed to a logical conclusion, a company could spend millions of pounds preventing an accident that is extremely unlikely to happen, but the cost of travel on such a railway would be so high that no-one could afford it, and so the company would go bankrupt. To decide whether to spend money on a safety improvement, over and above the statutory legal requirements to ensure safety, companies use several ways of analysing risk and benefits.

Figure 1.61 shows the tolerability of risk (TOR) framework for reducing risks. The Health & Safety Executive (HSE) suggests guidelines of a 1:1 000 000 fatality risk per year for the boundary between broadly acceptable and tolerable, and 1:10 000 per year (for the public) and 1:1000 per year (for the workforce) fatality risk for the boundary between tolerable and unacceptable.

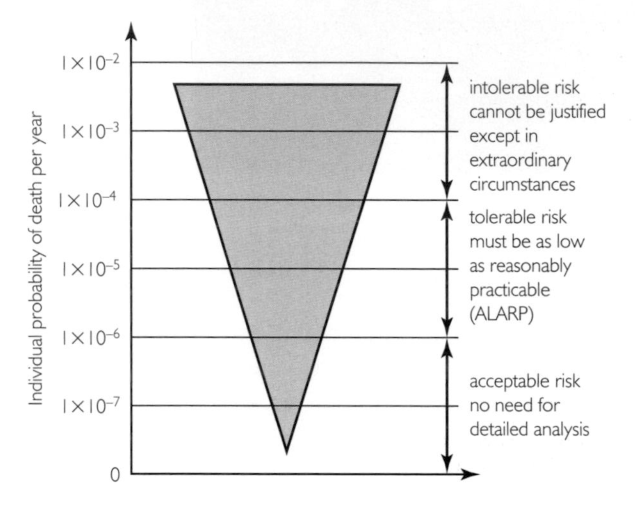

Figure 1.61 Tolerability of risk (TOR) framework

Cost–benefit analysis

Another method that companies use to consider risk is to do a cost–benefit analysis (CBA). This involves calculating the value of preventing a fatality (VPF) and the costs of an accident, and comparing them to the cost of implementing the safety measure. Safety measures are then put in place if the value of the benefits are greater than the cost. All the costs and benefits to everyone, not just the company, must be included in the calculations. You can see immediately that it is important to fix the VPF at the right level.

Costs such as medical costs, damage to trains, track and buildings, and the service disruption can be calculated directly. The cost of loss of life and permanent injury is more difficult, and the method used to value the prevention of a fatality is to ask members of the public how much they would be willing to pay for safety measures that gave a small reduction in risk. These surveys are then used to estimate a mean value. The official 2008 value of preventing a fatality was set at £1.652 million. It is important to understand that this fatality is a statistical fatality – the actual number may be higher or lower in any year. The value put on a life does not represent the life of a real person, as is shown by the fact that if an accident occurs every effort is made to save everyone; CBA would be unethical once real people were involved.

Other indirect costs of accidents include the fact that some people may stop using the train because they think it is not safe. These are very difficult to judge in advance, so the results of a CBA are used with caution.

The main objection to CBA is that it balances the cost of safety against other financial considerations and that this is unethical. However, when you think about it, this always happens – when people consider buying a car, they don't usually find out which is the safest model on the market, and then buy it regardless of cost. What CBA does is to make the choice clear and objective, and make it more likely that the interests of the general public will be given a high priority. Another objection is that railways can put 'profit before safety', but provided firstly, that all the costs and benefits are taken into account, not just those of the railway companies, and secondly, that the value of the VPF is not set by the railway company but by the general public, a proper CBA should prevent this.

Figure 1.62 shows how the CBA and TOR are used together to decide what safety measures should be implemented.

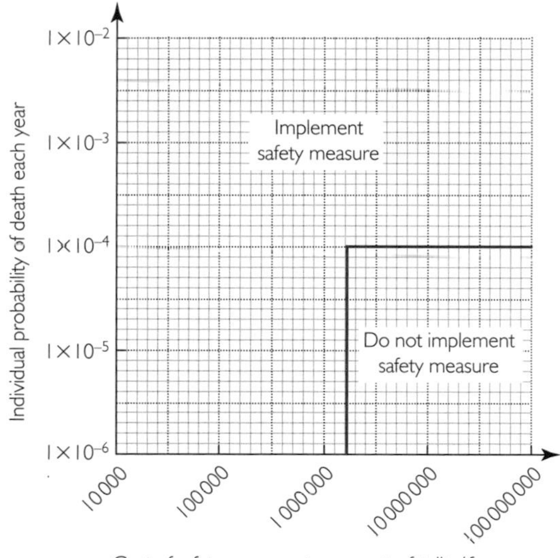

Figure 1.62 How individual risk and cost–benefit analysis is used to decide whether to implement safety procedures

Question

57 Make a rough sketch of the CBA diagram shown in Figure 1.62.

(a) Divide your diagram into four regions and label them A, B, C, D so that:

A: Low risk, high cost of implementing safety measure (the area labelled 'Do not implement...')

B: Low risk, low cost of implementing safety measure

C: High risk, low cost of implementing safety measure

D: High risk, high cost of implementing safety measure.

(b) Mark on your diagram a point representing a risk of 1 in 100 000 of a fatality which could be prevented by a safety measure costing £10 million. Say whether this would be implemented.

(c) Use your diagram to give an example of a risk and safety measure cost that (i) would be implemented, (ii) would not be implemented.

6.2 Where to build the line?

The UK railway network was mostly constructed between 1840 and the First World War. Some routes and stations have closed due to movement of population and industry, and some were closed in the 1960s when road travel became more popular and affordable. The current routes are often overcrowded and there are new towns and businesses that are a long way from existing lines. Passenger numbers are now increasing so there are good reasons for installing additional modern high-speed rail lines, but the UK has only one, the HS1, between London and the Channel Tunnel. It is just 130 km long. In Activity 31 you use the internet to find out which countries have invested in high-speed lines and compare their situation to that in the UK.

Activity 31 Drawing the line

Use the internet to look at a map of high-speed rail lines in Europe and other countries such as Japan. Compare the land mass and populations of countries with high-speed lines. Suggest reasons for the lack of high-speed lines in the UK.

Some years ago, the government in Spain authorised the construction of new high-speed rail lines and by 2008 a few were already operating. Spain will soon be the country with the most high-speed rail lines in the world; Figure 1.63 shows a map of the planned high-speed lines in 2012. The distance from Madrid to Barcelona is 660 km and this now takes just over two and half-hours. When the link with the French TGV is opened in 2012 it will be possible to travel by high-speed train from London to Malaga.

Figure 1.63 High-speed lines in Spain 2012

Looking at the map, it is easy to imagine a similar scheme in the UK, linking major cities, or maybe a route linking the new Ebbsfleet station (on the HS1), to Stansted Airport and then to Cambridge, Lincoln, Hull, Newcastle and Edinburgh. If you live near one of these places, no doubt you immediately feel concern about how it would affect you, and that is the problem. On the one hand we have the advantages of new rail links:

- fewer cars and airplanes (and so less CO_2 and other pollutants released)
- less land required per traveller (a six-lane motorway is required for the same number of travellers as a two-track railway line)
- lower number of injuries and deaths (about 3000–3500 people are killed on the roads each year and 250 000 injured; on the railways, including staff (but excluding trespassers and suicides), about 20 are killed and 2000–3000 are injured).

On the other hand, land has to be acquired for the track and stations, and this land already belongs to someone. Even if a fair price is paid for the land, the owner may not want to move, or the land may have special significance. A railway track through an area of outstanding natural beauty would spoil it for us all.

The route of the HS1 (Figure 1.64) had to be negotiated with national and local government, with businesses and pressure groups, and with individuals. In this part of the UK there are heavily inhabited areas, and areas of ancient woodland and wetland that are ecologically important (Figure 1.65). It was not possible to choose a route that avoids all these sensitive areas. People were very concerned about the impact of the project; during the two years that Parliament considered the Bill half of the petitions were to do with the environment. Extensive consultation before the route was finalised helped people to realise that their views were being taken into account. An environmental statement was prepared with hundreds of commitments to environmental groups, agreeing to minimise damage to sensitive habitats and to compensate for unavoidable damage.

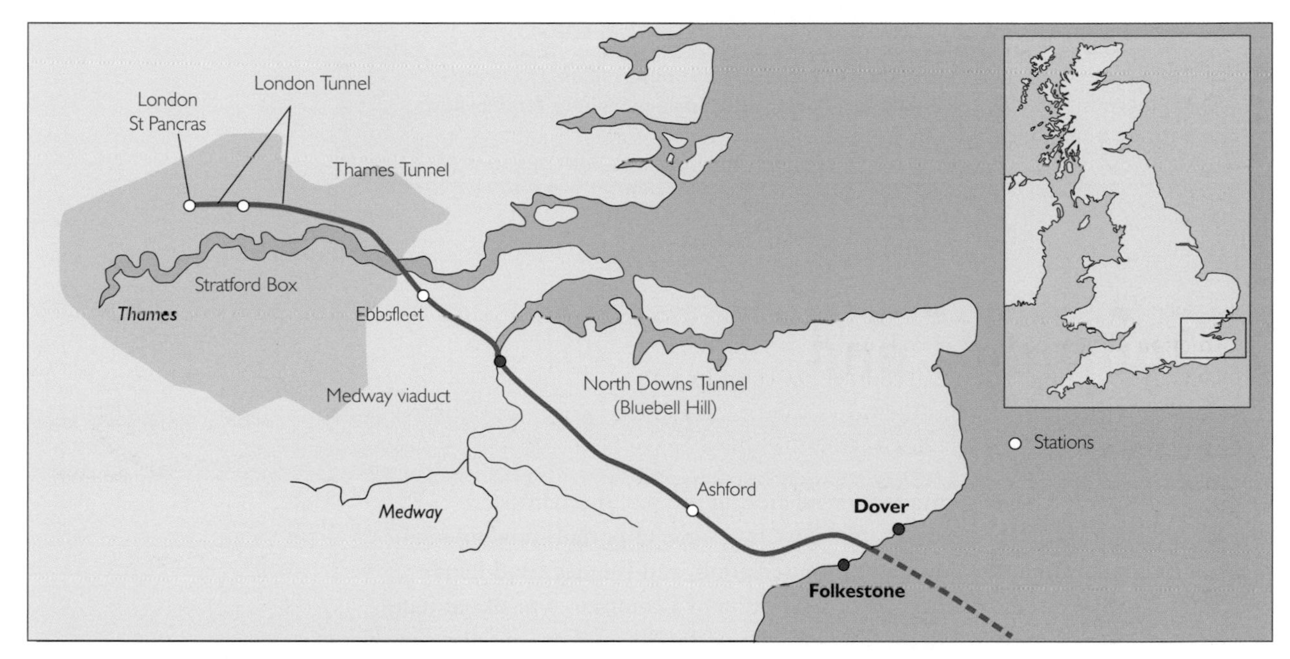

Figure 1.64 Map of the route of the HS1

Figure 1.65 HS1 passes through some environmentally sensitive areas

In the early 1990s the favoured route for the second part, from Rochester to London, was one that passed south of London, through Dulwich and Peckham, finishing at Kings Cross. This passed through some beautiful countryside and there were strong objections from locals. As a result, a route that had been suggested earlier, through East London and finishing at St Pancras, was chosen because of the opportunities for regeneration.

In Activity 32 you consider the environmental and other ethical implications of choosing a route for a new railway, using the HS1 between London and the Channel Tunnel as an example.

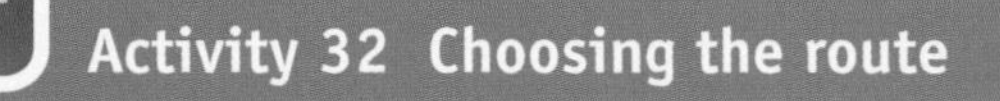

Activity 32 Choosing the route

Use the internet to research the route chosen for the HS1 and read an article about similar problems from 1840. Working in a group, write some guidelines for deciding where to site a new railway track.

7 Journey's end

7.1 On track

In this chapter, you have encountered several areas of physics: mechanics, electromagnetism and DC electricity. You have met some important fundamental physical laws and principles: momentum conservation, and Faraday's and Lenz's laws. And you have used a mathematical description of a common type of naturally occurring change: exponential decay. The following activities, and the questions in the next section, are designed to help you look back over your work from this chapter.

Activity 33 Tracking the physics

You learned something about each of the following in this chapter. Look back through your work and make brief notes under each heading, listing the examples that were used in the chapter and noting any ways in which the ideas were developed or refined. Add cross-references to earlier chapters where you met the same areas of physics. You could add references to later chapters when you have studied them.

- *DC circuits* Potential divider, resistivity.
- *Capacitors* Behaviour in DC circuits, charge and pd, exponential discharge, time constant.
- *Electromagnetism* Magnetic flux and flux density, electromagnetic force, electromagnetic induction, transformers.
- *Mechanics* Momentum and impulse, kinetic energy, elastic and inelastic collisions.

Activity 34 Hold the front page

Watch a train-crash video. Either act as reporter for a non-specialist publication and write an article on train safety or the nature of the collision you have observed or act as a member of a company that makes and sells devices that help make trains safe, and produce a technical advertisement for one of your products.

Activity 35 Down the tubes

Observe a small cylindrical magnet being dropped down three different vertically mounted tubes: a plastic tube much wider that the magnet; a plastic tube just wider than the magnet; and a copper or aluminium tube just wider than the magnet. Explain your observations.

Activity 36 Runaway train

In this activity, a model train is allowed to run, out of control, down an inclined track and is brought to rest by an energy-absorbing buffer. Hold a competition to see who can construct the safest buffer. Use your knowledge of dynamics and capacitor circuits to analyse your results and hence to test your structures as fully and quantitatively as possible.

Further investigations

Explore ways in which cracks in rails might be detected and located.

Investigate the effectiveness of seatbelts or airbags in cars.

7.2 Questions on the whole chapter

58 Figure 1.66 is taken from the notes of an accident investigator.

The driver of car A said he stopped at the junction and looked. When he continued he was struck by car B. He thinks that car B must have been speeding and he had not seen him because of the bend in the road

The driver of car B says he was driving within the speed limit and car A did not stop, but drove out in front of him.

Car A has a mass of 1450 kg and travelled 26.7 m after the collision in the direction shown on the map.

Car B has a mass of 2020 kg and travelled 22.8 m after the collision in the direction shown on the map.

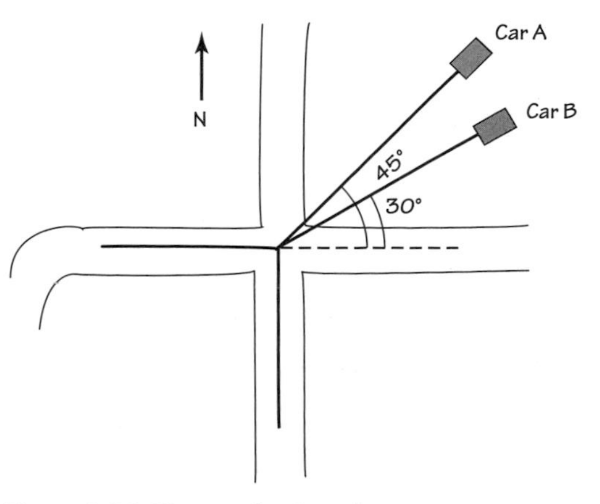

Figure 1.66 Diagram for Question 58

Investigators measured a value for the deceleration produced by the road surface after the collision of 3.10 m s^{-2}.

(a) Show that: (i) the speed of car A immediately after the collision was about 13 m s^{-1} (ii) the speed of car B was about 12 m s^{-1}.

(b) Calculate the momentum of (i) car A and (ii) car B immediately after the collision.

(c) Calculate the component of momentum to the east, immediately after the collision of (i) car A (ii) car B.

(d) Use your answer to (c) to calculate the speed of car B immediately before the collision.

(e) The speed limit was 40 mph (17.8 m s^{-1}). Was car B speeding before the collision?

(f) Use a similar analysis to determine the speed of car A immediately before the collision.

(g) Comment on your answer to (f).

59 A student makes some measurements using the plastic toy shown in Figure 1.67. Table 1.1 shows the data for one collision.

Mass of sphere A, m_A	63 g
Mass of sphere B, m_B	31 g
Speed of sphere A just before collision, u_A	0
Speed of sphere B just before collision, u_B	2.45 m s^{-1}
Speed of sphere A just after collision, v_A	0.90 m s^{-1}

Table 1.1 Data for Question 59

Figure 1.67 Diagram of toy for Question 59

(a) Show that the speed, v_B, of sphere B just after the collision is about 0.6 m s^{-1}.

(b) Determine whether this is an elastic or inelastic collision.

(c) When experimenting with another toy in which both spheres have m = 30 g, the student finds that sphere B is always stationary after the collision.

Show that for this toy (i) the speed v_A of sphere A after the collision is always the same as the speed, u_B, of B immediately before and (ii) the collision is always elastic.

60 Figure 1.68 shows a so-called 'everlasting torch' that operates without batteries. When the torch is shaken, a strong permanent magnet moves through a copper coil and generates an electric current.

(a) By referring to Faraday's law, explain (i) how the current is generated and (ii) how the speed of shaking affects the size of the current.

Figure 1.68 Everlasting torch

The current is used to charge a capacitor, which provides a power source for an LED. Figure 1.69 shows how the capacitor voltage varies with charge.

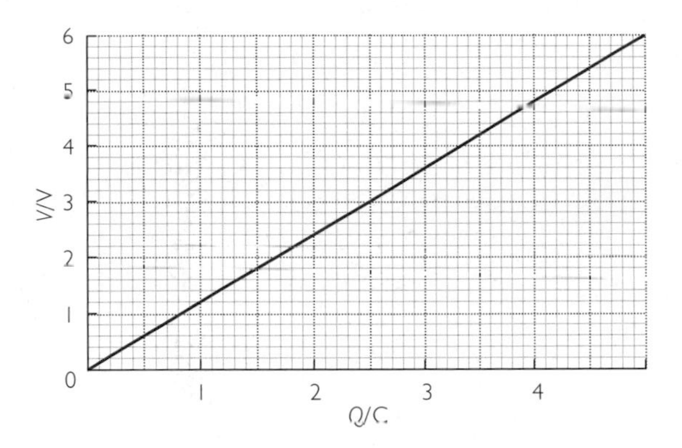

Figure 1.69 Capacitor voltage when the torch is shaken

(b) Calculate the capacitance of the capacitor.

Figure 1.70 shows how the capacitor voltage changes with time when the torch is in use after the shaking has stopped.

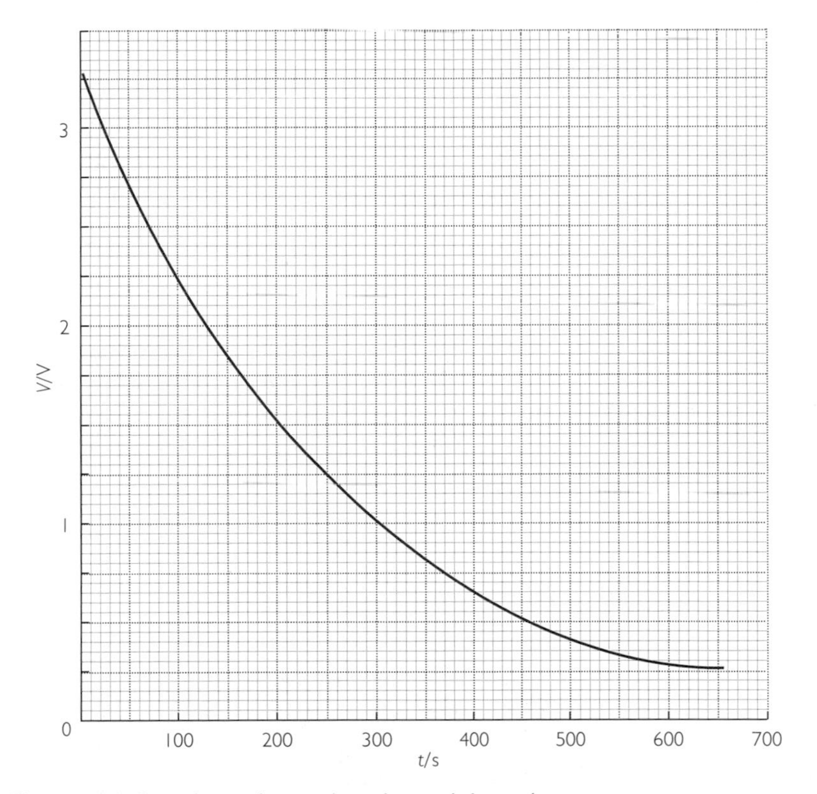

Figure 1.70 Capacitor voltage when the torch is used

(c) Explain why the brightness of the LED fades over a few minutes.

(d) Use the graph in Figure 1.70 to calculate (i) the time constant for the capacitor–LED circuit and (ii) the circuit's resistance.

61 There are two possible routes for a new railway track, as shown in Figure 1.71.

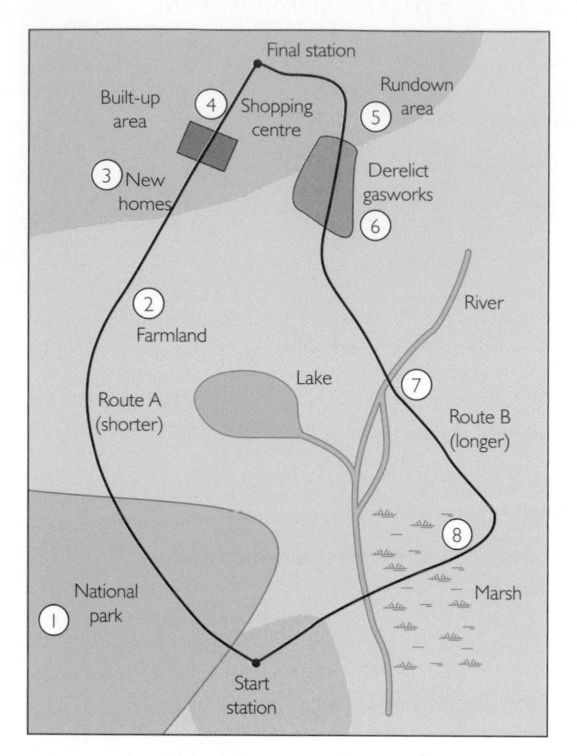

Figure 1.71 Route map for Question 61

Discuss the advantages and disadvantages of each route and say what science and technology solutions might be used to solve some of the problems involved.

7.3 Achievements

Now you have studied this chapter you should be able to achieve the outcomes listed in Table 1.2.

Table 1.2 Achievements for the chapter *Transport on Track*

	Statement from examination specification	*Section(s) in this chapter*
73	use the expression $p = mv$	3.2
74	investigate and apply the *principle of conservation of linear momentum* to problems in one dimension	3.2, 5.2
75	investigate and relate net force to rate of change of momentum in situations where mass is constant (Newton's second law of motion)	3.2, 5.2
78	explain and apply the principle of *conservation of energy*, and determine whether a collision is *elastic* or *inelastic*	5.3
87	investigate and use the expression $C = \frac{Q}{V}$	4.2 (and see MDM)
89	investigate and recall that the growth and decay curves for resistor–capacitor circuits are exponential, and know the significance of the *time constant RC*	4.2
90	recognise and use the expression $Q = Q_0 e^{-t/RC}$ and derive and use related expressions for *exponential discharge* in *RC* circuits eg $I = I_0 e^{-t/RC}$	4.2
91	explore and use the terms *magnetic flux density B*, *flux* Φ and *flux linkage* $N\Phi$	3.3, 3.4
92	investigate, recognise and use the expression $F = BIl\sin\theta$ and apply *Fleming's left-hand rule* to currents	3.3
94	investigate and explain qualitatively the factors affecting the emf induced in a coil when there is relative motion between the coil and a permanent magnet and when there is a change of current in a primary coil linked with it	3.4, 3.5, 4.1
95	investigate, recognise and use the expression $\mathscr{E} = \frac{d(N\Phi)}{dt}$ and explain how it is a consequence of Faraday's and Lenz's laws	3.4

Answers

1 Electrically insulated blocks are twin lengths of rail separated from the rest of the rail by an electrically insulating material.

2 A solenoid with current in it becomes an electromagnet whose magnetic field is very similar to that produced by a simple bar magnet.

3 Track circuiting is said to be fail-safe because if something short-circuits the track, or if the relay power supply fails, the signal is automatically set to red.

4 Leaves on the line or a coating of rust may prevent proper contact between a train's wheels and the track. If this occurs, the low-resistance path required to cause a short-circuit and activate the 'danger' lights may not be present; ie a train could go undetected. (Rusty rails are a particular problem because as trains get lighter due to modern materials and engines get smoother, coatings of rust may not be removed by the passage of trains. To combat this, a device called a track circuit actuator has been developed to break down the rust electrically.)

5 (a) Theoretically, we have the technology to bring trains to a halt extremely rapidly. If we did this, however, the inertia of unrestrained passengers would keep them moving forward (think about Newton's first law) and they would undergo dangerous collisions with structures within the train. (See Part 5 *Structure and Safety*.)

(b) Freight trains can operate in smaller envelopes as their speeds are lower and they therefore have shorter braking distances.

(c) Smaller envelopes mean more trains can fit on a particular section of track, and so more trains can pass along the track in a particular time interval.

6 (a) Without the additional resistors in parallel with the relay, circuit resistance, $R = 15\,\Omega$:

$$\mathrm{I} = \frac{V}{R} = \frac{6\mathrm{V}}{15\,\Omega} = 0.4\ \mathrm{A}$$

$$\text{pd across relay} = \left(\frac{5\,\Omega}{15\,\Omega}\right) \times 6\mathrm{V} = 2\mathrm{V}$$

(the circuit can be treated as a potential divider)

(b) (i) Now there are four $5\,\Omega$ resistors in parallel, which have a net resistance R_{par} given by:

$$\frac{1}{R_{\text{par}}} = \frac{1}{5\Omega} + \frac{1}{5\Omega} + \frac{1}{5\Omega} + \frac{1}{5\Omega}$$

$$\text{so } R_{\text{par}} = \frac{5\Omega}{4} = 1.25\,\Omega$$

This resistance is in series with the $10\,\Omega$ resistor so the circuit resistance is now reduced to $11.25\,\Omega$. (The circuit resistance decreases whatever the value of R_1, R_2 and R_3.)

(ii) $I = \frac{V}{R} = \frac{6\mathrm{V}}{11.25\Omega} = 0.53\ \mathrm{A}$ (to 2 sig. fig.)

(iii) As you can treat the circuit like a potential divider, the pd across the relay (and the additional resistors) is given by

$$V = \left(\frac{1.25\,\Omega}{11.25\,\Omega}\right) \times 6\,\mathrm{V} = 0.67\,\mathrm{V}$$

(to 2 sig. fig.)

(Whatever the additional resistors, the pd will always be reduced.)

(iv) Current in relay,

$$I = \frac{V}{R} = \frac{0.67\,\mathrm{V}}{5\Omega} = 0.13\ \mathrm{A}\ (2\ \text{sig. fig.})$$

(v) The relay current will be reduced (whatever the size of the additional resistors in parallel) and the magnetic field in the solenoid might not be large enough to operate the switch.

7 (a) First, find the combined resistance, R_c, of ballast and relay – they are in parallel:

$$\frac{1}{R_c} = \frac{1}{50\,\Omega} + \frac{1}{8\,\Omega} = 0.145\,\Omega^{-1}$$

$$R_c = \frac{1}{0.145\,\Omega^{-1}} = 6.9\,\Omega$$

Now use the potential divider formula to get resistance, R:

$$\left(\frac{R_c}{(R + R_c)}\right) \times 15\,\mathrm{V} = 6\,\mathrm{V}$$

$$R_c \times 15 = 6 \times (R + R_c) = 6R + 6R_c$$

$$6R = 9R_c$$

$$R = \frac{9R_c}{6} = 9 \times \frac{6.9\,\Omega}{6} = 10.35\,\Omega$$

(b) Through relay,

$$I_{relay} = \frac{V}{R} = \frac{6\text{V}}{8\,\Omega} = 0.75\text{ A}$$

(c) Current through variable resistor = $I_{ballast} + I_{relay}$

Through ballast,

$$I_{ballast} = \frac{V}{R} = \frac{6\text{V}}{50\,\Omega} = 0.12\text{A}$$

So through variable resistor,

$I = 0.12\text{A} + 0.75\text{ A} = 0.87\text{A}$.

8 (a) 0.5 V as they are in parallel.

(b) Cell voltage – track voltage
$= 2\text{V} - 0.5\text{V} = 1.5\text{V}$

(c) The voltmeter was recording as the 'lorry' rolled along the track. At different points, different conditions of contact between wheels and track gave different readings.

(d) The voltage decreases with increasing load.

(e) The resistance from track to track must decrease (to make the combined resistance of *L*, *B* and *R* smaller).

(f) The resistance decreases as the contact between the lorry's wheels and the track gets better. This 'contact resistance' is the most important factor in the resistance of a vehicle between two tracks.

(g) If a train is too light, it may not make a good enough contact with the rails. This could prevent it from causing the pd across the relay to fall below the 'drop-off' voltage and the train may go undetected.

9 (a) Using $F = ma$, $a = \frac{F}{m}$:

$$a = \frac{400 \times 10^3\text{ N}}{750 \times 10^3\text{ kg}} = 0.53\text{ m s}^{-2}$$

(b) Power = work done per second, $P = Fv$:

$P = 400 \times 10^3\text{ N} \times 10\text{ m s}^{-1}$
$= 4 \times 10^6\text{ W} = 4\text{ MW}$

10 N s or kg m s^{-1}. The units are interchangeable (see Equation 2).

11 (a) $\Delta p = 50\,000\text{ kg} \times +40\text{ m s}^{-1} - 50\,000\text{ kg} \times +20\text{ m s}^{-1}$

$= 1.00 \times 10^6\text{ kg m s}^{-1}$

(b) $\Delta p = 50\,000\text{ kg} \times -20\text{ m s}^{-1} - 50\,000\text{ kg} \times +20\text{ m s}^{-1}$

$= -2.0 \times 10^6\text{ kg m s}^{-1}$

(c) $\Delta p = 50\,000\text{ kg} \times 0\text{ m s}^{-1} - 50\,000\text{ kg} \times +20\text{ m s}^{-1}$

$= -1.00 \times 10^6\text{ kg m s}^{-1}$

12 (a) $\Delta p = (1200\text{ kg} \times 25\text{ m s}^{-1} - 1200 \times 0\text{ m s}^{-1})$

$= 30 \times 10^3\text{ kg m s}^{-1}$

(b) $F = \frac{\Delta p}{\Delta t}$

$$F = \frac{30 \times 10^3\text{ kg m s}^{-1}}{10\text{ s}}$$

$F = 3$ kN in direction of the motion

(c) $\Delta s = \frac{1}{2}(u + v)\,\Delta t$

$s = 0.5(0\text{ m s}^{-1} + 25\text{ m s}^{-1}) \times 10\text{ s} = 125\text{ m}$

(d) $P = \frac{\text{work}}{\text{time}} = \frac{\text{force} \times \text{distance}}{\text{time}}$

$$P = \frac{(3000\text{ N} \times 125\text{ m})}{10\text{ s}} = 3.75 \times 10^4\text{ W}$$

$= 37.5$ kW

13 Although the mass for the power unit to pull is so much greater it would be very uncomfortable to leave a station at the acceleration quoted for a car. Trains contain passengers who expect to be able to walk about.

14 (a) $F = \frac{\Delta(mv)}{\Delta t}$

$F = (750 \times 10^3\text{ kg} \times 83\text{ m s}^{-1} - 750 \times 10^3\text{ kg} \times 44\text{ m s}^{-1})$

$\div (14 \times 60)\text{ s}$

$= 35$ kN

(b) To produce the same acceleration with only one third the mass requires a tractive force one third of that found in (a).

(c) It takes the same size force to change speed from 83 m s^{-1} to 44 m s^{-1} over the same length of time, but it now acts in the opposite direction to the direction of travel.

(d) To produce the same resultant force, the tractive force provided by the engines must increase by 40 kN to 75 kN in the case of *Eurostar* and 52 kN for the lighter train.

15 (a) Using $F = \frac{\Delta p}{\Delta t}$:

(i) $F = 2\,400\,000 \text{ kg} \times \frac{44 \text{ m s}^{-1}}{60 \text{ s}} = 1\,760\,000 \text{ N}$

$= 1.8$ MN opposed to the direction of travel.

(ii) The same method gives $F = 550$ kN for *Eurostar*.

(b) $F = (750\,000 \text{ kg} \times 83 \text{ m s}^{-1} - 750\,000 \times 0 \text{ m s}^{-1}) \div 180 \text{ s}$

$= 350\,000 \text{ N} = 350$ kN opposite to the direction of travel.

(c) Far larger forces are needed to stop these trains quickly than it does to get them going, because the time interval is shorter.

16 (a) Equation 3: $\Delta s = \frac{mu^2}{2F}$

$\Delta s = 2100 \times 10^3 \text{ kg} \times (40 \text{ m s}^{-1})^2 \div (2 \times 1.2 \times 10^6 \text{ N})$

$= 1400$ m

(b) $F = \frac{mu^2}{2\Delta s}$

$= 750 \times 10^3 \text{ kg} \times (80 \text{ m s}^{-1})^2 \div (2 \times 1500 \text{ m})$

$= 1.6$ MN

(c) At high speeds the drag will help to slow down the trains but 40 kN is a small compared to the mega-newton braking forces. As speed decreases, so will the drag force. Both answers will thus be smaller than calculated in (a) and (b), but only slightly.

17 (a) Now Equation 1 is again the best one to use:

$\Delta t = \frac{\Delta p}{F} = 750 \times 10^3 \text{ kg} \times 80 \text{ m s}^{-1} \div 1.6 \times 10^6 \text{ N} = 37.5 \text{ s}$

(b) $a = (0 - 80) \text{ m s}^{-1} \div 37.5 \text{ s} = -2.1 \text{ m s}^{-2}$

18 (a) (i) Momentum doubles (mv to $2mv$).

(ii) Kinetic energy is four times greater: $0.5\, mv^2 \rightarrow 0.5\, m(2v)^2$.

(b) Both quantities are in the ratio 2:1. For the more massive train, (i) momentum doubles (mv to $2mv$) and (ii) kinetic energy is also doubled $0.5\, m\, v^2 \rightarrow 0.5\; 2\, m\, (v)^2$.

19 (a) *Net* force is found using:

$F = ma = 2.40 \times 10^6 \text{ kg} \times 0.13 \text{ m s}^{-2} = 312 \text{ kN}$

Force = 40 kN, so to produce the required resultant, we need a tractive force $T = 352$ kN.

(b) Resolving along the slope, with T representing tractive force and W the weight of the train:

$T = W\sin\theta = mg\sin\theta$

$= \frac{2.40 \times 10^6 \text{ kg} \times 9.81 \text{ N kg}^{-1} \times 1}{90} = 262 \text{ kN}$

(c) Total tractive force required $= 352 \text{ kN} + 262 \text{ kN} = 614 \text{ kN}$

(d) $\Delta t = \frac{\Delta p}{F}$

where F is the *net* accelerating force; ie $F = 312$ kN (from (a))

$\Delta t = \frac{2.40 \times 10^6 \text{ kg} \times 44 \text{ m s}^{-1}}{312\,000 \text{ N}}$

$= 338 \text{ s} = 5 \text{ min } 38 \text{ s}$

(e) (i) From (a) and (b), net force along slope is:

$F = 400 \text{ kN} - 262 \text{ kN} - 40 \text{ kN} = 98 \text{ kN}$

$a = \frac{F}{m} = \frac{98 \times 10^3 \text{ N}}{2.40 \times 10^6 \text{ kg}} = 0.041 \text{ m s}^{-2}$

(ii) $\Delta t = \frac{\Delta v}{a} = \frac{(v - u)}{a}$

$= \frac{44 \text{ m s}^{-1}}{0.041 \text{ m s}^{-2}}$

$= 1073 \text{ s} = 17 \text{ min } 53 \text{ s}$

(iii) $\Delta s = \frac{\Delta E_k}{F}$

$= \frac{0.5 \times 2.40 \times 10^6 \text{ kg} \times (44 \text{ m s}^{-1})^2}{98 \times 10^3 \text{ N}}$

$= 2.37 \times 10^4 \text{ m} = 23.7 \text{ km}.$

20 (a) One possible approach is to note that acceleration due to gravity is a vector, and so can be resolved as shown in Figure 1.72.

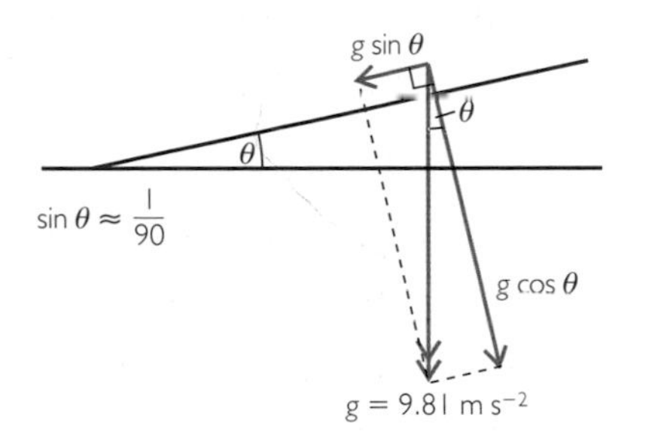

Figure 1.72 Diagram for the answer to Question 20

Along the slope, $u = 44 \text{ m s}^{-1}$, $v = 0 \text{ m s}^{-1}$,

$a = 9.81 \times \frac{1}{90} \text{ m s}^{-2}$:

$v^2 = u^2 + 2as$

$s = \frac{(v^2 - u^2)}{2a}$

$= \frac{(-44 \text{ m s}^{-1})^2}{2 \times -9.81 \times \frac{1}{90} \text{ m s}^{-2}}$

$= 8.89 \times 10^3 \text{ m} \approx 9 \text{ km}$;
ie not far enough to get out.

(We have chosen the positive direction to be the direction of travel, ie up the slope.)

(b) The net force along the slope = drag + component of weight acting along slope. This has magnitude:

$F = D + W\sin\theta$

which is related to the acceleration by $F = ma$.

So $a = \frac{F}{m} = \frac{D}{m} + g\sin\theta$

$= \frac{30 \times 10^3 \text{ N}}{2400 \times 10^3 \text{ kg}} + \frac{9.8 \text{ m s}^{-2}}{90}$

$= 0.121 \text{ m s}^{-2}$

Taking the positive direction to be up the slope:

$a = -0.121 \text{ m s}^{-2}$

Using the same method as in Question 19:

$s = \frac{(v^2 - u^2)}{2a}$

$= \frac{(-44 \text{ m s}^{-1})^2}{2 \times -0.121 \text{ m s}^{-2}}$

$= 8.0 \text{ km}$

21 (a) downwards, (b) out of the page, (c) to the right.

22 (a) $F = 2 \text{ T} \times 4 \text{ A} \times 10 \times 10^{-3} \text{ m} = 0.08 \text{ N}$

(b) $F = 0 \text{ N}$

(c) $F = 2 \text{ T} \times 4 \text{ A} \times 10 \times 10^{-3} \text{ m} \times \sin 30° = 0.04 \text{ N}$

23 The electromagnetic force must be equal in size (but opposite in direction) to the weight of the rider:

$BIl = mg$

$B = \frac{mg}{Il}$

$= \frac{15 \times 10^{-3} \text{ kg} \times 9.81 \text{ N kg}^{-1}}{5.0 \text{ A} \times 100 \times 10^{-3} \text{ m}}$

$= 0.3 \text{ N A}^{-1} \text{ m}^{-1} = 0.3 \text{ T}$

24 The missing ingredient is the number of turns, n. The force is increased for every extra turn in the coil. This gives $F = nBIl$ for the force on each side of the coil.

These are some plausible estimates: suppose we have current between 10 and 100 A, B between 0.1 and 1 T, length of coil is limited by space, say 2 m, n depends on thickness of wire (too thin and the resistance is too high for required high currents we need but thicker means fewer turns fit on coil) – let's say 1000 turns.

These values give:

$F \approx 1000 \times 1 \text{ T} \times 100 \text{ A} \times 2 \text{ m} = 2 \times 10^5 \text{ N}$

This is half our target, which is close. We can increase each quantity slightly, or we can always use two motors per locomotive and have two or more locomotives per train.

25 The SI units of the right-hand side are Wb s^{-1}:

$1 \text{ Wb s}^{-1} = 1 \text{ N A}^{-1} \text{ m s}^{-1}$

But 1 N m = 1 J, and 1 A s = 1 C (because $1 \text{ A} = 1 \text{ C s}^{-1}$), so:

$1 \text{ Wb s}^{-1} = 1 \text{ J C}^{-1}$ and $1 \text{ J C}^{-1} = 1 \text{V}$ – which is the SI unit for emf as required for the left-hand side.

26 Using Equation 6b, with $\Delta\Phi = 50 \times 10 \text{ Wb}$, $\Delta t = 2.0 \text{ s}$:

$\mathcal{E} = \frac{50 \times 10 \text{ Wb}}{2.0 \text{ s}} = 250 \text{ V}.$

27 Using Equations 6e, with $\Delta A = 0.03\ \text{m}^2$, $\Delta t = 0.2\ \text{s}$, $B = 2\ \text{T}$ and $N = 10$:

$$\mathcal{E} = \frac{10 \times 2\ \text{T} \times 0.03\ \text{m}^2}{0.2\ \text{s}} = 3\ \text{V}.$$

28 The changing current produces a changing magnetic field. A radio receiver is sensitive to the changing fields that make up electromagnetic waves, so will also respond to the changing field produced by a current and hence may produce an audible signal.

29 As the field has a vertical component, any horizontal conducting path on the aeroplane will have an induced emf across it during flight. If flying due north, this will be from right to left (east to west) across wings and fuselage. (Check this with the right-hand rule looking down on the plane.) There will be no emf along the length of the fuselage as this is not 'cutting' across the field.

The induced emf is a potential danger as it could interfere with control systems within the plane.

30 (a) The induced emf will be away from you – into the page.

(b) Use Equation 7 with $B = 4 \times 10^{-5}\ \text{T}$, $v = 60\ \text{m s}^{-1}$, $l = 4\ \text{m}$ (the height is irrelevant):

$$\mathcal{E} = 4 \times 10^{-5}\ \text{T} \times 4\ \text{m} \times 60\ \text{m s}^{-1} = 9.6\ \text{mV}$$

31 (a) Magnetic fields extend across space so no contact is required.

(b) No, it depends on motion for its effect so would not work when at rest.

32 The disc should have low electrical resistance; ie it should be thick and made of good conducting material.

33 Vary the distance of the magnet from the disc. Alter the strength of the magnet (only possible if the system uses an electromagnet).

34 (a) The electricity supply to the coils is switched off. The motor no longer drives the wheels round. The reverse happens – the momentum in the wheels keeps the coils rotating in the magnetic field and so inducing a current; ie acting as a generator.

(b) Either the current is switched to pass through resistances which heat up readily or the current is passed back into the supply system available along the rail track. The third rail is the British method. On the continent it would be to the overhead supply line.

35 From Equation 8, $N_s = N_p \times \left(\frac{V_s}{V_p}\right) = \frac{N_p V_s}{V_p}$.

The motors always require $V_s = 1500\ \text{V}$, and we always have $N_p = 1000$, so $N_p V_s = 1.5 \times 10^6\ \text{V}$:

$$\text{France: } N_s = \frac{1.5 \times 10^6\ \text{V}}{25 \times 10^3\ \text{V}} = 60$$

36 (a) Equation 8: $\frac{N_p}{N_s} = \frac{V_p}{V_s} = \frac{750\ \text{V}}{1500\ \text{V}} = 0.5$

(b) Assuming that input power = output power = 1200 kW,

$$I_p = \frac{P}{V_p} = \frac{1200 \times 10^3\ \text{W}}{750\ \text{V}} = 1600\ \text{A}.$$

37 (a) The alternating magnetic field induces eddy currents within the iron nail. These currents heat the iron.

(b) The layers of glue have high electrical resistance and so inhibit the circulation of eddy currents. This reduces the heating of the core. In a transformer, eddy-current heating of the core is one factor that would reduce the efficiency.

38 The tunnel lining would act like a giant solenoid. Any electric train, with very strong motor magnets, would set up an induced current in it, which would be directed to push back on the train like a brake. More energy is then required to drive the train into the tunnel; ie increased fuel costs – a potentially expensive mistake.

39 Only the vertical component of B contributes to the emf. Using Equation 7:

$$\mathcal{E} = 4.1 \times 10^{-5}\ \text{T} \times 47\ \text{m} \times 268\ \text{m s}^{-1} = 0.516\ \text{V}$$

40 (a) $\Phi = BA\cos\theta$

$B = 3.5 \times 10^{-6}\ \text{T}$, $\theta = 45°$,

$A = 5.0 \times 10^{-2}\ \text{m} \times 8.0 \times 10^{-2}\ \text{m} = 4.0 \times 10^{-3}\ \text{m}^2$

$\Phi = 3.5 \times 10^{-6}\ \text{T} \times 4.0 \times 10^{-3}\ \text{m}^2 \times \cos 45° = 9.9 \times 10^{-9}\ \text{Wb}$

(b) Flux linkage = $N\Phi$

$= 100 \times 9.9 \times 10^{-9}\ \text{Wb} = 9.9 \times 10^{-7}\ \text{Wb}$

(c) Use Equation 6, with $\Delta\Phi = 9.9 \times 10^{-7}\ \text{Wb}$, $\Delta t = 1.0 \times 10^{-4}\ \text{s}$

$$\mathcal{E} = \frac{9.9 \times 10^{-7}\ \text{Wb}}{1.0 \times 10^{-4}\ \text{s}} = 9.9 \times 10^{-3}\ \text{V} = 9.9\ \text{mV}$$

41 (a) From Equation 14, fraction of charge left =

$\frac{Q}{Q_0} = e^{-t/RC}$:

$RC = 250 \times 10^3\ \Omega \times 100 \times 10^{-6}\ F = 25\ \Omega F$

$\frac{t}{RC} = \frac{20}{25} = 0.8$

$e^{-t/RC} = 0.45$ (2 sig. fig.)

(b) From Equation 11, initial charge

$Q_0 = CV_0 = 100 \times 10^{-6}\ F \times 10\ V = 10^{-3}\ C$

After 20 s, $Q = 0.45\ Q_0 = 4.5 \times 10^{-4}\ C$

42 (a) In Equation 13b, the left-hand side has units of charge divided by charge; ie it is dimensionless. The right-hand side, $\frac{\Delta t}{RC}$, must also be dimensionless. As Δt has dimensions of time (SI units of second) then so must *RC*.

(b) If $t = RC$, then $\frac{V}{V_0} = e^{-RC/RC} = e^{-1}$

$= \frac{1}{e} = \frac{1}{2.718} = 0.37$ to 2 dp

43 (a) From Equation 13, $\frac{Q}{Q_0} = e^{-3} \approx 0.05 = 5\%$.

(b) Similarly, $\frac{Q}{Q_0} = e^{-5} \approx 0.0067 \approx 0.7\%$.

Strictly mathematically speaking, the charge *never* actually reaches zero, but more than 99% of the initial charge will have been released from a discharging capacitor after five time constants. Depending on how precisely one can detect this small amount of remaining charge, one can say that the capacitor is 'fully discharged' for all practical purposes.

44 (a) 6 V

(b) $\tau = RC = 1 \times 10^5\ \Omega \times 47 \times 10^{-6}\ F = 4.7\ s$

(c) From Equation 15, $V = V_0 e^{-t/RC}$

$\frac{t}{RC} = \frac{5\ s}{4.7\ s} = 1.06,\ V_0 = 6\ V$

$V = 6 \times e^{-1.06} = 2.08\ V\ (\approx 2\ V)$

45 (a) When a train is occupying the block, there is a connection across XY. This instantly charges the capacitor and also makes the NOT gate input high. The NOT gate output is therefore low, or close to zero volts, allowing the red LED to light. When the train leaves the block, the capacitor discharges and the pd across it gradually falls. Hence, the input voltage to the NOT gate also falls. When this voltage drops below 1.5 V, it is read as low and the NOT gate output goes high, close to 6 V. This allows the green LED to light, thus indicating all clear.

(b) We need to show that the capacitor voltage falls to 1.5 V when $t = 10$ s. At this time, $\frac{t}{RC} = \frac{10\ s}{7.2\ s} = 1.39$.

From Equation 15, $V = 6V \times e^{-1.39} = 1.5\ V$ as required.

(c) $RC = 1000 \times 10^3\ \Omega \times 10 \times 10^{-6}\ F = 10\ s$. This is a little longer than the required τ of 7.2 s, so discharge will be slower which gives a delay that is longer than the required 10 s.

46 Using Equation 18 with $m_1 = 5$ t (wagon), $m_2 = 41$ t (*Eurostar* loco), $u_2 = 0\ m\ s^{-1}$ and $v_1 = v_2 = v = 5\ m\ s^{-1}$, we have:

$m_2u_2 = 0$ and so

$m_1u_1 = (m_1 + m_2)v$

$u_1 = \frac{(m_1 + m_2)v}{m_1}$

$= \frac{(46\ t \times 5\ m\ s^{-1})}{5\ t} = 46\ m\ s^{-1}$ (this is over 100 mph!)

(Notice that we have kept all masses in tonnes. You could convert to kg, but the factors of 1000 would cancel at the final stage.)

47 Using Equation 18 and taking left-to-right as positive:

$m_1 = 90$ t, $m_2 = 20$ t (as in Question 46, we have kept masses in t.)

$u_1 = +50\ m\ s^{-1}$, $u_2 = -30\ m\ s^{-1}$, $v_1 = +30\ m\ s^{-1}$

Momentum beforehand = $m_1u_1 + m_2u_2$

$= 90\ t \times 50\ m\ s^{-1} - 20\ t \times 30\ m\ s^{-1} = 3900\ t\ m\ s^{-1}$

Momentum afterwards = $m_1v_1 + m_2v_2$

$= 90\ t \times 30\ m\ s^{-1} + m_2v_2$

$= 2700\ t\ m\ s^{-1} + m_2v_2$

Since momentum must be conserved:

$m_2v_2 = 3900\ \text{t m s}^{-1} - 2700\ \text{t m s}^{-1} = 1200\ \text{t m s}^{-1}$

$v_2 = \dfrac{1200\ \text{t m s}^{-1}}{20\ \text{t}} = 60\ \text{m s}^{-1}$.

Notice that the final answer gives v_2 in the correct units (the tonnes cancel) and because the number is positive we know the coach must be moving from left to right.

48 Air plus balloon initially have no velocity and so no momentum. The air is held inside at high pressure and so once released the air moves out at high speed. To conserve momentum the balloon must move in the opposite direction with a momentum equal to that of the air.

49 (a) See Figurc 1.73.

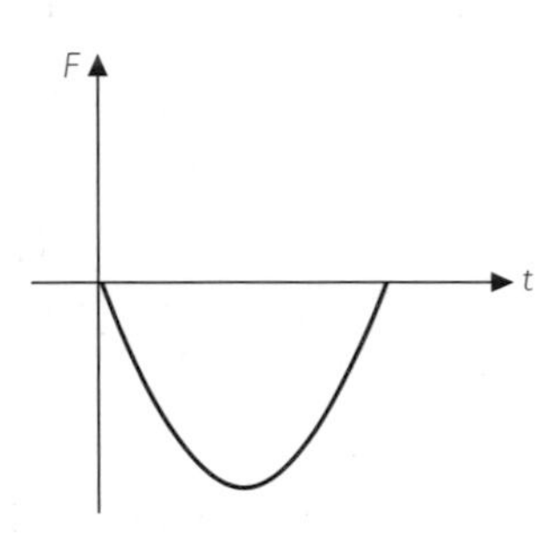

Figure 1.73 Answer to Question 49

(b) The area gives the impulse and hence the change in momentum.

(c) A straight line parallel to the time axis (since the force has to be constant)

50 It lengthens the time of the impact. The required change in momentum to stop the ball can be achieved by either a large force for a short time (painful) or by a smaller force for a longer time (comfortable).

51 (a) Using Equation 2: $F = \dfrac{\Delta(mv)}{\Delta t}$. In a time interval $\Delta t = 1\text{s}$, 70 kg strikes the wall and its horizontal velocity decreases from 30 m s^{-1} to zero so:

$\Delta(mv) = 70\ \text{kg} \times 30\ \text{m s}^{-1} = 2100\ \text{kg m s}^{-1}$

and so $F = 2100$ N in the opposite direction to the water's motion.

(b) The force exerted on the water is also 2100 N, but in the opposite direction, ie outwards.

52 D

Force is equal to rate of momentum change (see Question 51). The air mass is moving twice as quickly so twice the mass hits the train per second. Twice the mass moving at twice the velocity has four times as much momentum, so the force will increase fourfold.

53 (a) Using $\Delta W = F\Delta s$ and $F = ma$:

$F = \dfrac{\Delta W}{\Delta s},\ a = \dfrac{F}{m} = \dfrac{\Delta W}{m\Delta s}$

$= \dfrac{1 \times 10^6\ \text{J}}{(1\ \text{m} \times 24 \times 10^3\ \text{kg})} = 41.7\ \text{m s}^{-2} \approx 4 \times g.$

This is a very large acceleration, so would be very uncomfortable to experience.

(b) Using the same method as in (a), $a = 1.3$ m s^{-2} (which is much more comfortable but still a real jolt).

54 (a) $E_k = \frac{1}{2}mv^2$

$= 0.5 \times 750 \times 10^3\ \text{kg} \times (83\ \text{m s}^{-1})^2$

$= 2.58 \times 10^9\ \text{J}$

$\Delta E_{grav} = mg\Delta h$

$= 750 \times 10^3\ \text{kg} \times 9.81\ \text{N kg}^{-1} \times 318\ \text{m}$

$= 2.34 \times 10^9\ \text{J}$

so there is *more* than enough energy to reach the top of the Eiffel Tower.

(b) The track would need to curve upwards, enabling the train to 'coast' to a great height – and it would need to be engineered so as to reduce frictional energy losses almost to zero.

55 From Question 46, with $m_1 = 5$ t (wagon), $m_2 = 41$ t (*Eurostar* loco), $u_1 = 46$ m s^{-1}, $u_2 = 0$ m s^{-1} and $v_1 = v_2 = v = 5$ m s^{-1}:

before collision, total

$E_k = \frac{1}{2}m_1u_1^2 + \frac{1}{2}m_2u_2^2$

$= 0.5 \times 5 \times 10^3\ \text{kg} \times (46\ \text{m s}^{-1})^2 = 5.3 \times 10^6\ \text{J}$

after collision, total

$E_k = \frac{1}{2}(m_1 + m_2)v^2$

$= 0.5 \times 46 \times 10^3\ \text{kg} \times (5\ \text{m s}^{-1})^2 = 5.8 \times 10^5\ \text{J}.$

This collision is *not* elastic, since kinetic energy is not conserved.

From Question 47, with $m_1 = 90$ t, $m_2 = 20$ t:

$u_1 = +50 \text{ m s}^{-1}$, $u_2 = -30 \text{ m s}^{-1}$,

$v_1 = +30 \text{ m s}^{-1}$, $v_2 = 60 \text{ m s}^{-1}$

before, $E_k = 0.5 \times 90 \times 10^3 \text{ kg} \times (50 \text{ m s}^{-1})^2 + 0.5 \times 20 \times 10^3 \text{ kg} \times (30 \text{ m s}^{-1})^2$

$= 1.215 \times 10^8$ J

after, $E_k = 0.5 \times 90 \times 10^3 \text{ kg} \times (30 \text{ m s}^{-1})^2 + 0.5 \times 20 \times 10^3 \text{ kg} \times (60 \text{ m s}^{-1})^2$

$= 7.65 \times 10^7$ J

Again, this is not an elastic collision since there is a net loss of kinetic energy.

56 (a) Suppose the system is initially at rest; ie momentum is zero. If a rivet, mass $m_1 = 50$ g, is fired with a velocity $v_1 = 20 \text{ m s}^{-1}$, then the trolley-plus-man, total mass $m_2 = 400$ kg, must acquire a velocity v_2 where $m_1 v_1 + m_2 v_2 = 0$ (because the total momentum must still be zero):

$$v_2 = \frac{-m_1 v_1}{m_2} = \frac{0.050 \text{ kg} \times 20 \text{ m s}^{-1}}{400 \text{ kg}}$$

$= -2.5 \times 10^{-3} \text{ m s}^{-1}$.

ie the trolley-plus-man moves at 2.5 mm s^{-1} in the opposite direction to the rivet.

(b) One way to approach this problem is to find the force in the trolley-plus-man, using

$$F = \frac{\Delta p}{\Delta t}.$$

With $\Delta t = 1$ s, $\frac{\Delta p}{\Delta t}$

$$= \frac{2 \times 400 \text{ kg} \times 2.5 \times 10^{-3} \text{ m s}^{-1}}{1 \text{ s}}$$

$= 2 \text{ kg m s}^{-2} = 2$ N.

(Or use the momentum change of two rivets to get the same result.)

Acceleration of trolley, $a = \frac{2 \text{ N}}{400 \text{ kg}}$

$= 0.0050 \text{ m s}^{-2}$.

$v = u + at$, with $t = 60$ s

$u = 0 \text{ m s}^{-1}$,

$v = 0.0050 \text{ m s}^{-2} \times 60 \text{ s}$

$= 0.30 \text{ m s}^{-1}$.

(Or use the same method as in (a), with $m_1 = 2 \times 60 \times 0.050$ kg.)

(c) For rivets, in 1 minute

$\Delta E_k = 0.5 \times 2 \times 60 \times 0.050 \text{ kg} \times (20 \text{ m s}^{-1})^2$

$= 1200$ J

For trolley-plus-man, in 1 minute:

$\Delta E_k = 0.5 \times 400 \text{ kg} \times (0.30 \text{ m s}^{-1})^2 = 18$ J

(d) No chance! The train takes only 500 s to reach the man's original position. By this time, his speed is only 2.5 m s^{-1} (using the same method as in (b)) and he will still be very close to the on-rushing train.

(By plotting *s*–*t* graphs for the train and for the trolley, perhaps using a spreadsheet, you could find the time for the train to catch up with the trolley.)

57 (a)

C	D
B	A

(b) No

(c) (Answers will vary.)

CHAPTER 2

The Medium is the Message

Why a chapter called *The Medium is the Message*?

The world today depends on communications – speaking to someone, sending a letter, e-mailing, texting, hearing the news on the radio, television or internet, or transferring data from and to the Stock Exchange or bank (Figure 2.1a). Recently there have been major advances in display technology (Figure 2.1b), bringing us liquid-crystal display (LCD), light-emitting diode (LED) and plasma screens that can provide bigger, clearer and brighter displays with lower power demands than their traditional predecessors.

Electricity and magnetism first played their part in communication when the idea of telegraphy was put forward in 1820 by the French physicist André Ampère (1775–1836). Early versions involved switching electromagnets so that needles pointed to particular characters. By 1843 Samuel Morse (1791–1872) had invented his famous code, enabling the much faster transmission of messages. In 1878 Alexander Bell (1847–1922), the Scottish-born American scientist, invented the telephone. Radio followed a little later with Marconi sending a Morse-code signal across the Atlantic Ocean in 1901 and with the first speech transmission by Reginald Fessenden in 1906. Television was established in 1925.

Figure 2.1a Communications

Since then electronic developments have come thick and fast with the sending of signals through fibre-optic cables, faster broadband speeds, greater storage capacity of computers, digital television, satellite communications and our increasing reliance on mobile phones, where a wide range of information can be displayed on liquid-crystal displays. The world seems to want to communicate faster and more often, and progress in physics has helped this come about; the telecommunications industry is one of the major employers of physicists.

While we tend to think of communications in terms of messages and images sent between people, communication can involve any transmission, reception and display of information. In this chapter, we will look mainly at how information is sent and displayed.

Figure 2.1b Display

Overview of physics principles and techniques

In this chapter you will learn how sensors based on potentiometers are used to gather information electronically. The output from sensors is often sent along a fibre-optic cable, you will also explore how a signal is attenuated as it travels along such a cable.

Next you will explore the operation of a charge-coupled device (CCD), which uses the physics of charge, voltage and capacitance to produce and display two-dimensional images. Finally, you will look at the physics behind other displays: cathode-ray tubes, light-emitting diodes, liquid crystal and plasma displays, where electric and magnetic fields play a key part.

In this chapter you will extend your knowledge of:

- using graphs from *Higher, Faster, Stronger* and *Digging Up the Past*
- charge and potential difference from *Technology in Space* and *Digging Up the Past*
- the behaviour of light from *The Sound of Music* and *Technology in Space*
- capacitors and exponential change from *Transport on Track.*

In other chapters you will do more work on

- electric and magnetic fields in *Probing the Heart of Matter*
- electromagnetic radiation in *Reach for the Stars.*

1 Sensing and sending information

Over the past century, the ways information is gathered and transmitted have changed dramatically, and systems involving electrical (or electronic) control have become commonplace. We can illustrate these changes with a particular example: powered flight.

Figure 2.2 Wilbur Wright at the controls of a Model A

Control of powered flight is attributed to the two bicycle manufacturers from Dayton, Ohio, USA – the brothers Orville and Wilbur Wright. In the first flights of their biplane *The Flyer*, Orville was able to exert control on movable wing tips, and move the rudder and wing flaps, all by movements of his body. The first flight, in 1903, lasted 12 seconds. In 1911 the Wright Brothers' Model A (Figure 2.2) had wing warping, rudder and elevator control wires operated by the pilot.

Until the 1970s, with the exception of experimental aircraft, control of an aircraft's flying surfaces (Figure 2.3) was through mechanical linkages; though with connections generally made with rods, gears and hydraulics rather than with wires. Most motor vehicles today still use mechanical linkages. For example, in most cars a foot on the brake pedal produces a force on a piston that pushes against incompressible brake fluid in a pipe; this force is transmitted by the fluid to push another piston at the far end of the pipe, which in turn pushes the brake pads into contact with the wheels. Similarly, a turn of the steering produces movement of a rack and pinion gear system through which the wheels are changed in direction, and information about the speed of a car may be transferred via a rotating cable linking the wheels to the speedometer.

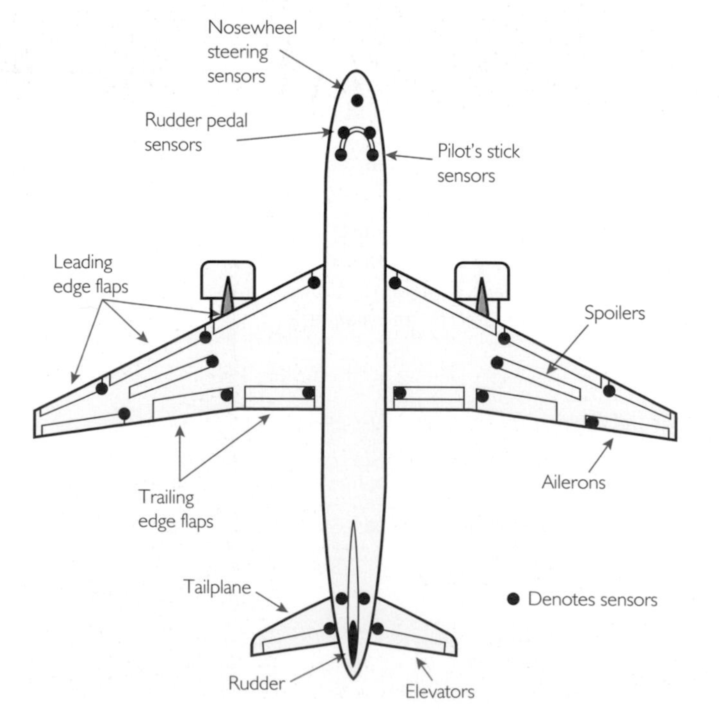

Figure 2.3 Flying surfaces of an aircraft

These days both motor vehicles and aircraft are becoming more computer controlled, with on-board sensors relaying electronic information to computers. Flying an aircraft, be it a civil airliner or a military fighter, requires the collection of lots of data, often at very frequent intervals. The position of the aircraft, its heading (direction of travel), its altitude, orientation (pitch, yaw and roll), its airspeed, the position of the various flight control surfaces, the engine thrust, coolant temperatures, cabin pressure ... and so much more – all need to be sensed, transmitted, processed and, where necessary, displayed in order that the pilot can take the required actions.

1.1 Avionics

Avionics – an aircraft's electronic control, sensing and display systems – nowadays accounts for at least 10% of the cost of new aircraft. The high initial expenditure reduces running costs, allowing just two pilots to operate the flight deck where previously a flight engineer was also needed to check the vast array of instruments. Computer-based instrument displays have brought many instruments together into a smaller space and any malfunction can be instantly highlighted and brought to the attention of the pilot (see Figure 2.4).

In modern avionics systems, information is sensed electronically and signals are sent along metal or fibre-optic cables – know as fly-by-wire and fly-by-light, respectively. The control mechanisms and linkages are less bulky, and mechanical devices are needed only to move the control surfaces, lower the wheels and so on.

Figure 2.4 An aircraft instrument panel

Such systems produce great savings in weight and volume compared with mechanical systems, 5:1 at least. Using a technique known as multiplexing, several signals can be combined and sent down the same cable, thus reducing the weight still further.

There are other advantages as, being linked through computers, the actions of the pilot can be moderated within safe and comfortable boundaries. For example, passengers do not appreciate the nose being tipped up at angles greater than 25° or tipped down through more than 10°, or being banked over at more than 45°.

Lots of redundancy is built into the system, with sensors usually being connected to computers in parallel by four independent channels. The computers compare signals with one another and act on the 'majority vote' if there is a discrepancy.

You might be concerned about computers running aircraft. In 1996, the Ariane 5 rocket failed on its maiden flight. The subsequent Inquiry Board suggested that one of the main problems was in transferring software from Ariane 4 without changing all the anticipated sensor readings in line with the new and more powerful rocket. Because of this human error, the system was unable to interpret the readings it sensed correctly, and acted as though something was wrong. The control software from aircraft has to satisfy certain standards, much as you saw for the HACCP (Hazard and Critical Control Points) system for food safety in the AS chapter *Good Enough to Eat*. Aircraft software is subjected to many checks and has 60-70% redundancy built in so that failures can be accommodated.

Fly-by-wire or fly-by-light?

Communications along screened electrical and fibre-optic cables, respectively, with computer control, are now commonly known as fly-by-wire (FBW) and fly-by-light (FBL) technology.

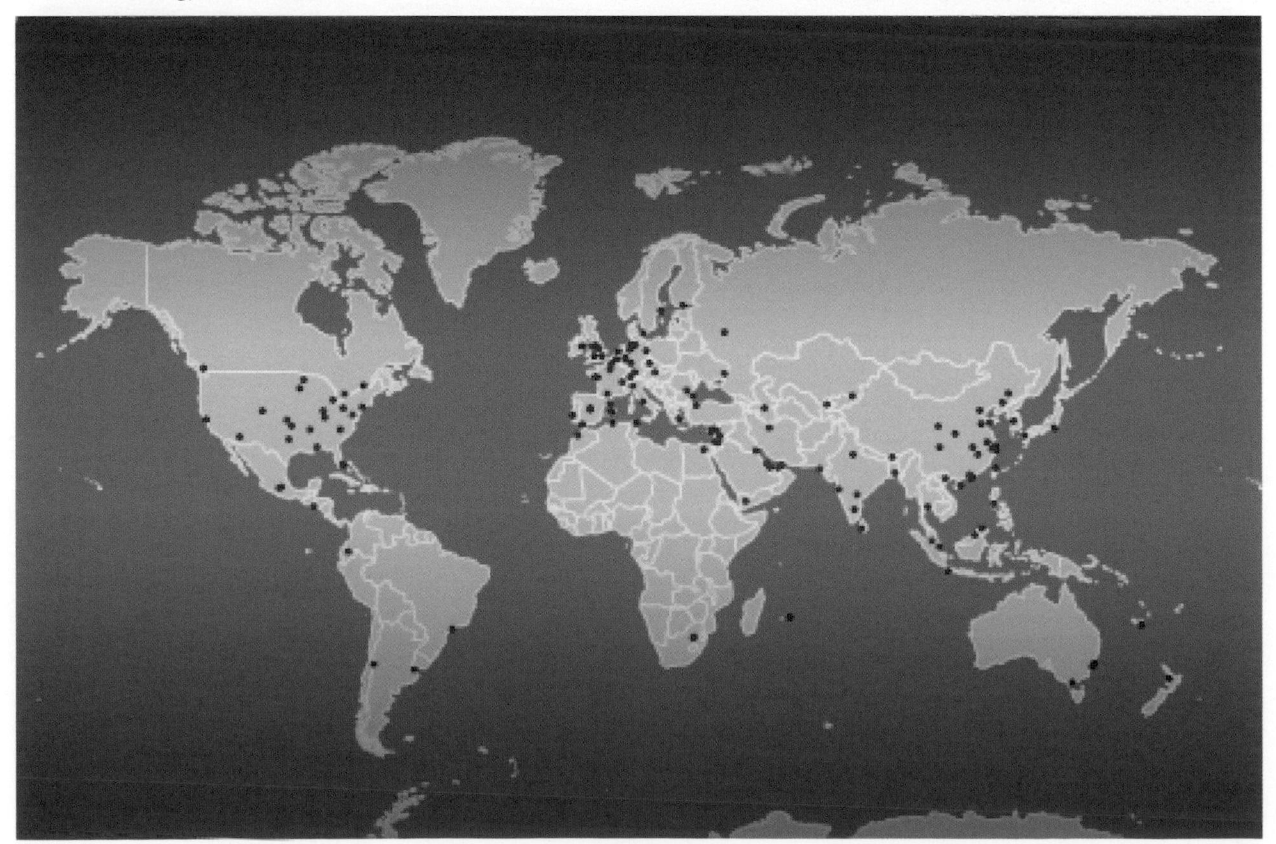

Figure 2.5 Airbus activities worldwide

FBW was first introduced on an airliner by Airbus Industrie, a European multinational consortium owned by Aerospatiale of France, British Aerospace of Great Britain, CASA of Spain and Daimler-Benz Aerospace of Germany, with Italy's Alenia, Fokker in the Netherlands and Belairbus in Belgium acting as associate risk-takers on special projects. In 2001, the consortium became a single fully integrated company, Airbus, registered in France. Manufacturing, production and sub-assembly of parts for Airbus aircraft are distributed around 16 sites in Europe, with final assembly in Toulouse and Hamburg (see Figure 2.5). There are also centres for engineering design, sales and support in North America; and sales and customer support centres in Japan and China. Airbus has a joint engineering centre in Russia with Kaskol. By September 2008 Airbus had taken over 9000 orders for aircraft and delivered over 5000 to world markets.

Figure 2.6 Airbus A319, A320 and A321 aircraft

In the first ten years after the A320 (Figure 2.6) went into service as the first FBW airliner in April 1988, some 700 Airbus aircraft or similar types covered seven million flying hours without major mishap caused by the system. The catastrophe rate for a fleet of 3000 such aircraft averaging 3000 hours per year is estimated at one in a hundred years. Lightning strikes on an Airbus 320 and an A330-300 (Figure 2.7) have resulted in visible damage to the aircraft skin but have left the avionics system unaffected.

Figure 2.7 Lightning damage on and A330

By 2007, Airbus had developed the A380, which was the world's first twin deck, twin aisle airliner. Increasingly, fly-by-light (FBL) or fly-by-optic systems are being used rather than fly-by-wire (FBW).

Fly-by-light (FBL) relies on exactly the same type of system except that now all the sensors are optically rather than electronically or electrically based, with fibre-optic cable providing the communications links. Such a system will further reduce weight and provide links that are immune to electromagnetic interference that might occur in lightning storms or overflying high-power radio and radar transmitters. With a military aircraft it would also provide immunity from the electromagnetic pulse (EMP) that is emitted in a nuclear explosion and which would otherwise interfere with or even destroy any on-board electronics. Such systems are being developed by NASA and they have performed tests on an F-18 fighter with FBL control and are planning tests on a Boeing 757. The Skyship 600 airship has had FBL controls since 1988.

Both systems use digital communication, but fibre optics has the advantage of being lighter, so often these are being used to replace the traditional wiring. You will find out more about fibre optics in Part 2. The digital control system means that the pilot can fly using computers. The computer monitors the position and force inputs from the pilot's controls and aircraft sensors.

In the military context both FBL and FBW enable an aircraft to be far more manoeuvrable than those with conventional controls. The airframe can be turned and twisted at high speed and yet can be safely controlled through the fast responses of its on-board computer system. The Eurofighter Typhoon is a fly-by-wire aircraft.

A further advance is power-by-wire (PBW) where the heavy and bulky hydraulic circuits have been replaced by electrical power circuits. Again, this has significant weight saving and helps with efficiency.

Activity 1 Building an Airbus

Look at the Airbus website (see www.shaplinks.co.uk) to get an overview of the operation of such a large company. Identify some of the problems that such an organisation has to overcome in designing and making aircraft co-operatively. Factors might include fitting it all together, languages, finance, and harmony and agreement.

Drawing on what you have read in this chapter so far, make brief notes on the following:

- In mechanical control systems for an aircraft, what problems would be associated with long pipes, lots of fluid and quite a large mass and volume of equipment?
- How does the reduction of weight and staff increase an airline's profit?
- How does PBW help with this?
- What are some of the safety advantages of FBW and FBL technology?

1.2 Sensing the situation

On an aircraft, there are many different transducers or sensors; these are devices that produce an electrical signal that varies in response to changes in some other physical factor. Some are used to sense pressure, which, indirectly, provides a measure of airspeed. More complex are the gyros and accelerometers that sense the attitude and position of the aircraft, and optical pyrometers that provide information on the engine temperature. Somewhat simpler transducers sense the position of control surfaces (see Figure 2.3), the pilot's stick or control column, or the angle of the nosewheel. How might this be achieved?

The Potentiometer

A simple but commonly-used position sensor depends on the rotary **potentiometer** or **potential divider**. Figure 2.8 shows a rotary potentiometer with a resistive element of uniform thickness, width and resistivity. As you saw in the AS chapter *Digging Up the Past*, the potential difference between the sliding contract and a fixed end depends on the position of the sliding contact.

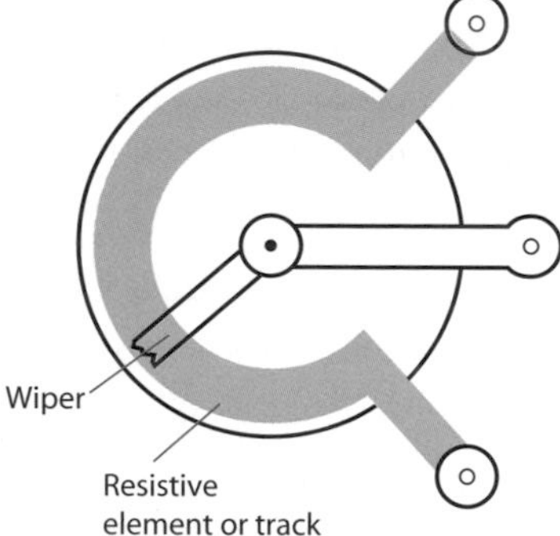

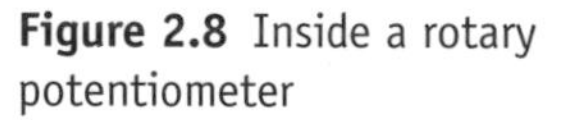
Figure 2.8 Inside a rotary potentiometer

Activity 2 Sensing position

Discuss how a potentiometer, attached to a model flap, aileron or nosewheel, could be arranged to give a voltage output that depends on their position.

Then use a potentiometer attached to a model aircraft part to show how the voltage output varies with angular position (see Figure 2.9).

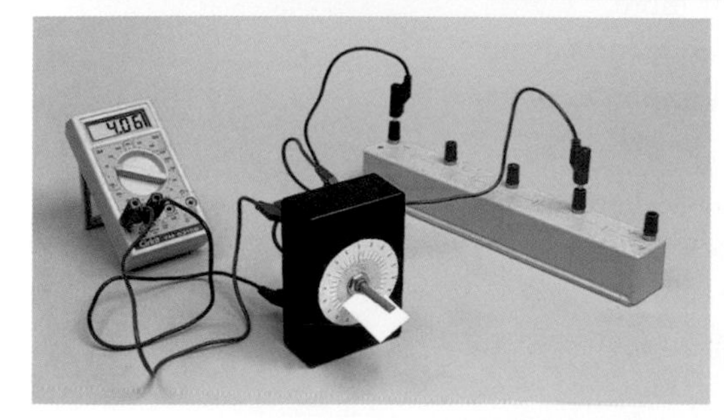

Figure 2.9 Model flap attached to potentiometer

In Activity 2 you should have found that the variation of a voltage output with angle was fairly linear. In practice this variation will be affected by the resistance of the device that is measuring the voltage output. If there is time, you might be able to explore this further in Activity 3.

Activity 3 What if the voltmeter's resistance is low?

Investigate the effect of taking the potentiometer output to a low-resistance voltmeter. See how a voltage follower or buffer amplifier can help deal with the problems that arise.

The voltage output from the position sensor needs to be transmitted to the computers that are managing the aircraft's flight. This could be achieved by simply having long wires connected, via an appropriate interface or link, to the computers. However, long wires produce large voltage drops and so what started at 5V might end up as only 2V, with a consequent faulty interpretation of position. So other means of communicating information are needed that do not get degraded en route – these lie outside the scope of this chapter.

This section has involved some revision of work from the AS chapters *Technology in Space* and *Digging Up the Past*. Use Questions 1 and 2 to check your progress and understanding.

Questions

1 Why is a large voltage drop associated with a current in a long cable?

2 Assume that the voltmeter in Figure 2.10 has an infinite resistance. State the reading in each case. (In (c), the contact is at the midpoint on the potentiometer.)

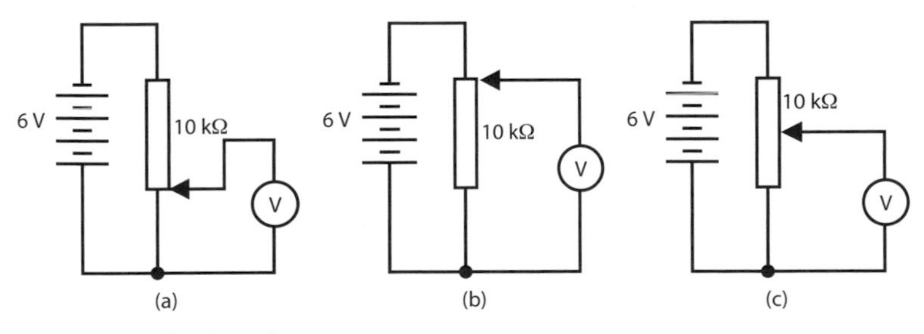

Figure 2.10 Diagrams for Question 2

1.3 Fibre-optic cables

In this section, we consider a signal in transit along a cable. In a fly-by-wire system, signals are transmitted down coaxial cables. A number of the Airbus civil aircraft – A319, A320, A321, A330, A340 and the Eurofighter Typhoon – have such systems. The air balloon Skyship 600 and other prototype aircraft have fly-by-light systems that use fibre optic cables to carry infrared or visible light signals. More generally, fibre-optic cables are used in many ground-based communications systems such as telephone systems and computer networks.

Figures 2.11 and 2.12 show the structure and relative sizes of the cables. We will concentrate on fibre-optic cables.

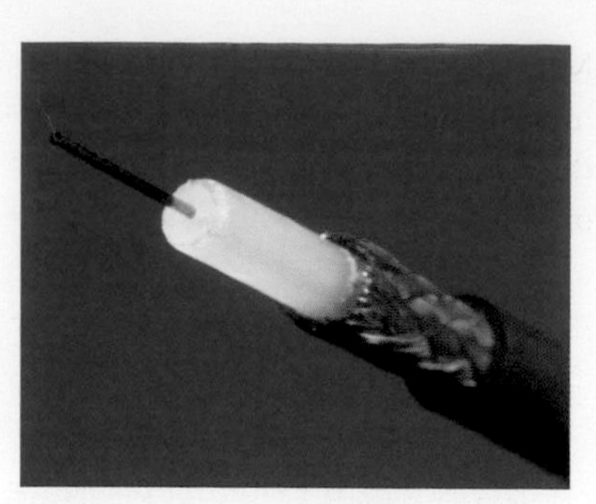

Figure 2.11 Coaxial cable

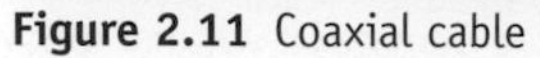

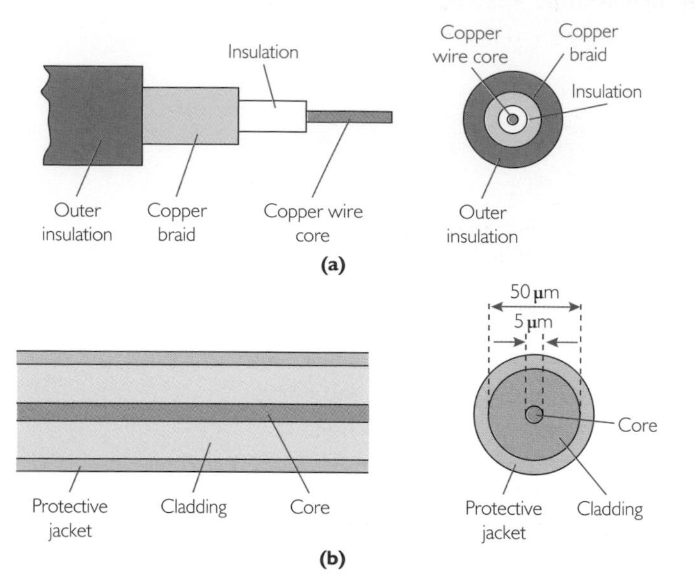

Figure 2.12 Structure of (a) coaxial cable and (b) fibre-optic cable

Study note

You might wish to refer back to your work in the AS chapter *The Sound of Music* for a reminder of these ideas.

In fibre-optic cables the signal travels inside a thin transparent fibre, meeting the surface at an angle of incidence greater than the critical angle. It therefore undergoes total internal reflection and is confined to the interior of the fibre. Figure 2.13 shows this for a fibre made of a single material. The radiation is reflected at the surface, ie $i = r$ in Figure 2.13, and if the angle of incidence, i, is greater than the critical angle, C, the total internal reflection takes place. C depends on the refractive index, μ, between the fibre and its surroundings:

$$\mu = \frac{1}{\sin C} \qquad (1)$$

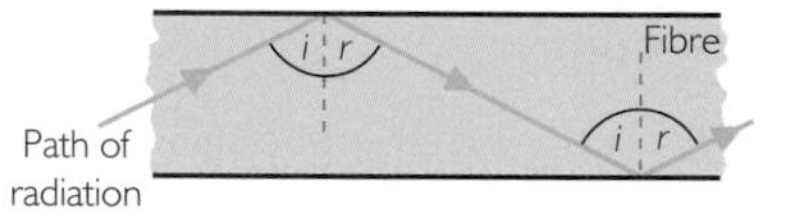

Figure 2.13 Radiation undergoes total internal reflection within an optical fibre

Unfortunately a pulse of radiation (light or infrared) entering a simple fibre is not just directed one way. It spreads sideways and results in many rays travelling by different routes down the fibre (Figure 2.14). The **dispersion** (spreading) means that a signal that starts as a sharp pulse is smeared out after travelling along the fibre. Because the refractive index is the same throughout the fibre, the speed of propagation of the light through it is the same regardless of route. Hence ray C in Figure 2.14 takes the least time and A the most. The result is that a sharp pulse gets spread out. The longer the fibre, the worse things get. This effect is known as **multipath dispersion**. The net effect is shown in Figure 2.15.

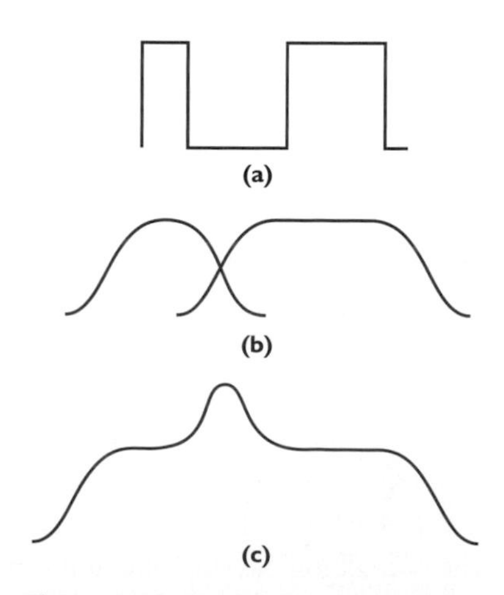

Figure 2.15 (a) Original pulses (b) after dispersion and (c) combined effect

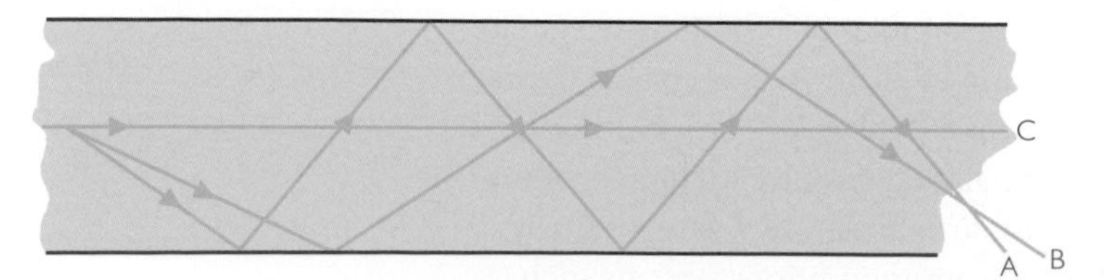

Figure 2.14 Rays travelling different routes in a fibre

Multipath dispersion can be reduced somewhat by limiting the range of angles of incidence that result in total internal reflection. This can be achieved by surrounding the core fibre with a cladding material whose refractive index is slightly lower than that of the core to produce a stepped-index fibre. Activity 4 illustrates how this works.

Activity 4 Multipath dispersion

Draw accurate diagrams to show how rays of light travel along (a) an optical fibre of refractive index $\mu_1 = 1.6$ surrounded by air and (b) $\mu_2 = 1.4$. Compare the ranges of angles of incidence that result in total internal reflection in each case.

The most recent development has been what is called single or **monomode** fibres in which the rays can pass down only the centre of the fibre. You might then think that no dispersion would occur as all the rays would have travelled the same path in the same time. This would be so if the light was of a single wavelength, but even laser light has a small spread of wavelengths: 1 to 2 nm or less. The refractive index of a material varies with wavelength, which causes the small spread of wavelengths to arrive at different times due to their different speeds. The result is again a slight spreading of the pulse.

Questions

3 On aircraft, cable lengths are typically less than 100 m. If the dispersion is 1 ns km^{-1}, and the pulse rate is 10^9 s^{-1}, explain whether dispersion is likely to be a major problem.

1.4 Attenuation

Both coaxial and fibre-optic systems have their advantages and drawbacks, and **attenuation** (loss of intensity) occurs in both. Attenuation in metal cables occurs where there is a transfer of energy from the signal to the cable due in part to its resistance and leakage through the insulation. Optical fibres cannot in practice be completely transparent, and some of the signal is absorbed by the material through which it passes. The attenuation in an optical fibre is modelled in Activity 5 using a jelly 'fibre' that absorbs strongly in the infrared region (see Figure 2.16). This activity also demonstrates a mathematical way of describing the attenuation.

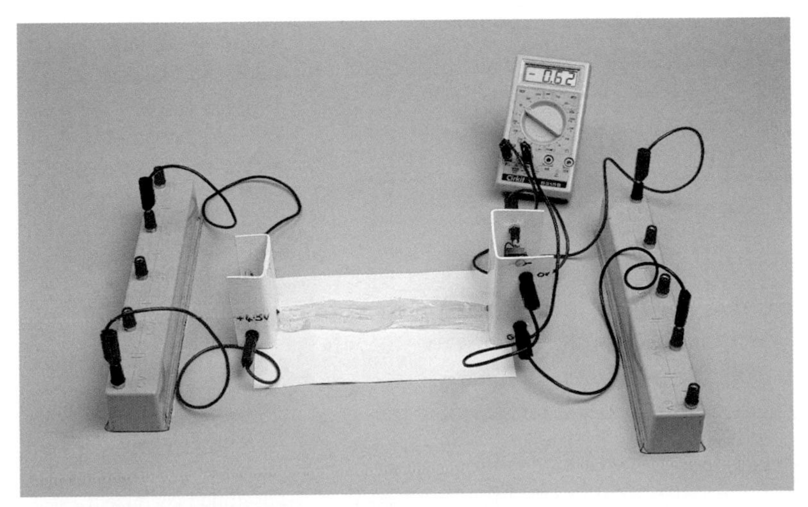

Figure 2.16 Investigating attenuation in jelly fibre

Activity 5 The jelly fibre

Using an infrared-emitting diode as the source and an infrared-sensitive phototransistor as the detector, investigate attenuation in a glass fibre modelled with jelly.

Figure 2.17 show some typical results from Activity 5. As you should have found, the intensity (as measured by the voltage) always changes by the same fraction when the length is changed in equal steps. For example, in Figure 2.17, a length of $x = 7$ cm reduces the voltage to half its initial value (V falls from 1.0 V to 0.5 V), after a further 7 cm the voltage has halved again ($x = 14$ cm, V = 0.25 V), and after another 7 cm it has halved yet again ($x = 21$ cm, $V = 0.125$ V).

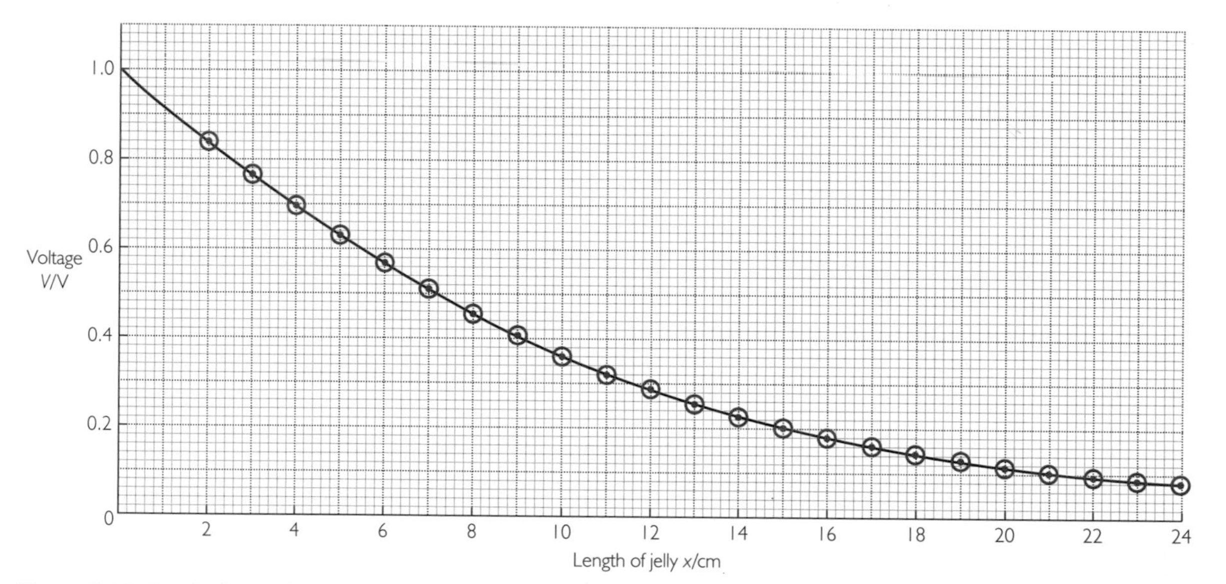

Figure 2.17 Graph of V against x for a jelly fibre

Exponential change

The attenuation of radiation intensity in a fibre is an example of an **exponential** change. Exponential changes are characterised by the 'equal steps resulting in equal fractional changes' pattern. In the case of capacitor discharge, the steps are time intervals, whereas here they are equal changes in length.

Mathematically the exponential change in intensity with distance can be described by the equation:

$$I = I_0 e^{-\mu x} \qquad (2)$$

where I_0 is the intensity of radiation that enters the jelly, I the intensity after it has travelled a distance x, and e is the exponential number (e ≈ 2.718). μ is the absorption coefficient which describes how rapidly the intensity is attenuated with distance – if μ is large, then $e^{-\mu x}$ rapidly becomes small as x is increased; ie the radiation is strongly attenuated.

Study note

If you have already studied the chapter *Transport on Track*, you will have met another example of such a change when looking at the way a capacitor discharges.

Study note

The symbol μ is, unfortunately, used both for refractive index and absorption coefficient.

After a distance $x = \frac{1}{\mu}$, the intensity will fall to $\frac{1}{e}$ of its initial value ($\mu x = 1$, and so $I = I_0 e^{-1}$). The value of μ can therefore be found from a graph such as that in Figure 2.17 by finding the distance over which I falls to $\frac{1}{e}$ of its initial value, and then find the reciprocal of that distance. This reasoning also tells us the units of μ – if x is expressed in m, then μ has units m^{-1}, and if x is in cm, then μ is in cm^{-1}.

Maths reference

Reciprocals
See Maths note 3.3

Exponential changes
See Maths note 9.1

Exponential functions
See Maths note 9.2

Many, but by no means all, naturally occurring changes are exponential. How can you tell whether a change is exponential? One way is to see whether it fits the 'equal steps result in equal fractional changes' pattern, but this can be time-consuming, and if there are large experimental uncertainties it is hard to decide where to draw a best-fit curve. A neater way is to plot a **log–linear graph**. Taking the log of both sides of Equation 2 gives:

$$\log(I) = \log(I_0) - \mu x \log(e) \qquad (3)$$

Equation 3 describes a relationship of the form $y = mx + c$; $\log(I_0)$, μ and $\log(e)$ are all constants. If $\log(I)$ is plotted on the y-axis and x on the x-axis, then the graph is a straight line as shown in Figure 2.18. This gives another way to test whether a change is exponential – a log–linear graph produces a straight line. This is usually quicker than looking for the equal-steps equal-fractions pattern, and it is often easier to tell whether points lie close to a straight line than it is to draw a best-fit curve through points with error bars.

Study note

If you are familiar with calculus, you can show that the pattern of equal steps, resulting in equal fractional changes will always be described by an equation such as Equation 2.

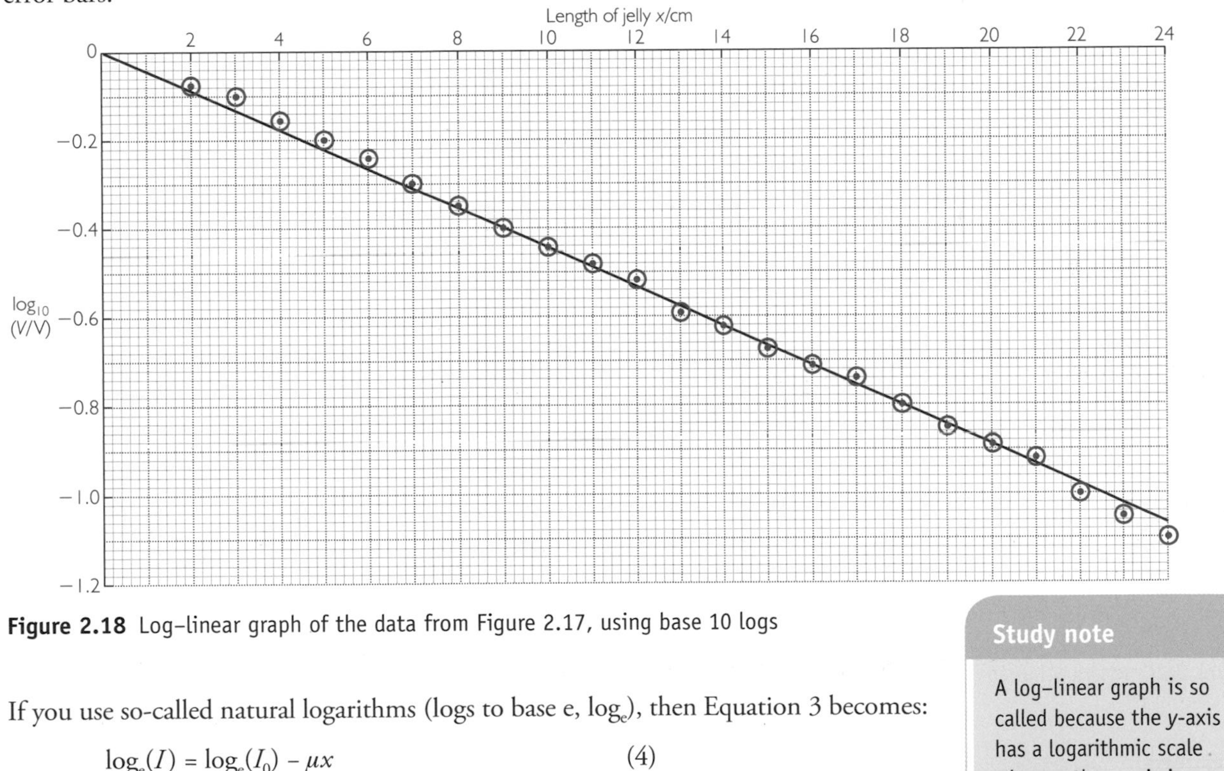

Figure 2.18 Log–linear graph of the data from Figure 2.17, using base 10 logs

Study note

A log–linear graph is so called because the y-axis has a logarithmic scale whereas the x-axis is linear.

If you use so-called natural logarithms (logs to base e, $\log_e$), then Equation 3 becomes:

$$\log_e(I) = \log_e(I_0) - \mu x \qquad (4)$$

because $\log_e(e) = 1$. A graph of $\log_e(I)$ against x is a straight line with gradient $-\mu$. Figure 2.19 shows a plot of the same data in Figures 2.17 and 2.18, using logs to base e. From the triangle drawn, the gradient of this graph is approximately –0.1, so the absorption coefficient in this case is approximately $0.1 cm^{-1}$.

Maths reference

Logs to base e
See Maths note 8.5

Maths reference

Using log graphs
See Maths note 8.7

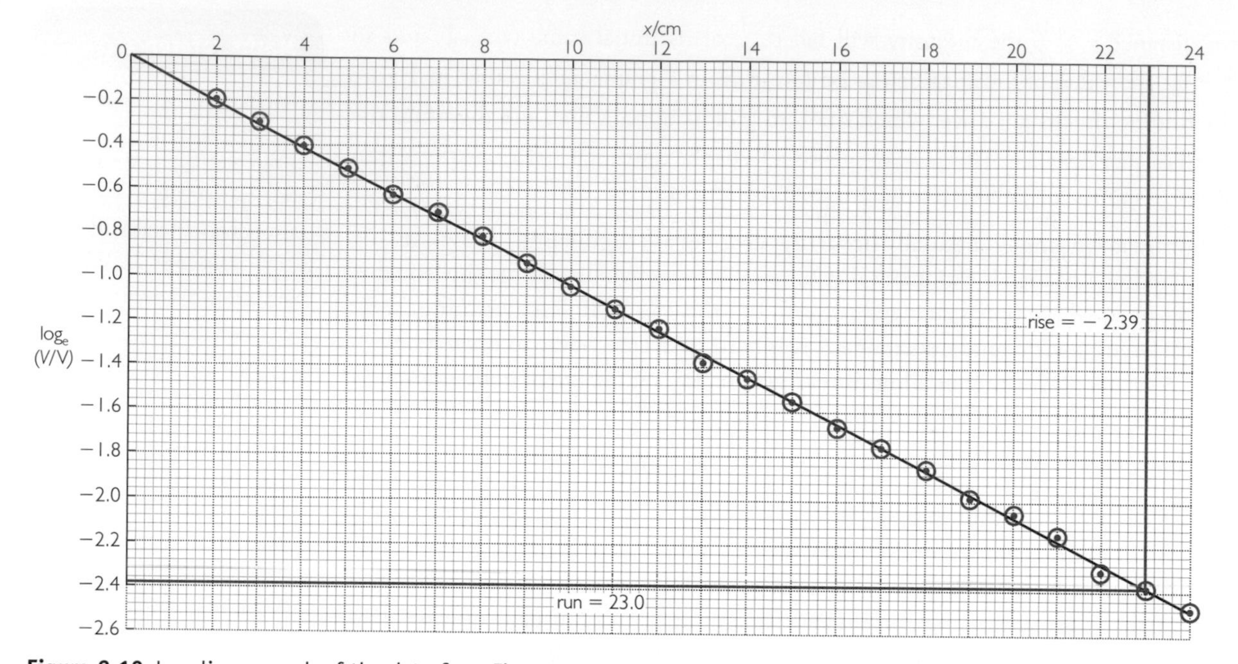

Figure 2.19 Log-linear graph of the data from Figure 2.17, using natural logs (logs to base e)

The fibre-optic cables used in communications have *much* smaller absorption coefficients than a jelly fibre – absorption coefficients are typically about 10^{-5} m^{-1} or less. On board an aircraft any attenuation is negligible. However, for long-distance communications attenuation can still be a problem, and so it is important to reduce it as much as possible. One way in which attenuation can be reduced is by careful choice of fibre material and wavelength of radiation. As shown in Figure 2.20, attenuation can depend markedly on wavelength.

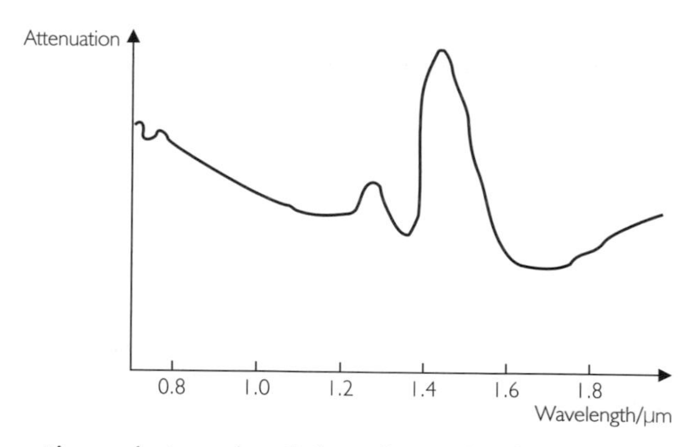

Figure 2.20 Attenuation against wavelength for a glass optical fibre

Questions

4 An optical fibre has an absorption coefficient $\mu = 2.0 \times 10^{-6}$ m^{-1}.

(a) Over what distance will the intensity of a signal fall to $\frac{1}{e}$ of its initial value?

(b) What will be the intensity of the signal, expressed as a fraction of its initial value, after it has travelled 100 km along the cable?

1.5 Summing up Part 1

Part 1 began by explaining why FBW and FBL are useful technologies for aerospace, how they have increased safety, and how a potentiometer can be used as a position sensor. You then looked in detail at the degradation or attenuation of signals due to dispersion and absorption. Attenuation in a jelly fibre illustrated an exponential change, and showed how logarithmic graphs can help analyse such changes. Use Activities 6 and 7 and Questions 5 and 6 to help you check your progress.

Activity 6 Summing up Part 1

Look back through your work and check that you know the meaning of all the terms highlighted in bold.

Activity 7 Fly-by-wire/light handbook

Many passengers, or indeed pilots, may be concerned about the technology used on modern aircraft. Produce an illustrated pamphlet that introduces people to the avionic systems used in fly-by-wire or fly-by-light, and outline how safety is enhanced by such systems.

Questions

5 In a certain optical fibre, the intensity of the signal falls to half its initial value after travelling 50 km. What is the absorption coefficient of the material expressed in units of m^{-1}?

6 To minimise attenuation, what would be the best wavelength to use for sending signals along the fibre in Figure 2.20?

2 Making an image

In Part 1 of this chapter, we were concerned with measurements (eg of position) that could each be expressed as a single number or as a sequence of numbers that changed with time. Sometimes, however, a two-dimensional picture is required, which can be encoded electronically and transmitted in exactly the same way as a single measurement. Several examples of communicating two-dimensional images will be familiar to you – TV pictures, for example. A less familiar example might be the device shown in Figure 2.21 which is used to give a pilot or a soldier 'night vision'. It works by first producing an infrared image of the surroundings and then displaying the image in a helmet-mounted display. The image is projected into the user's field of vision, to give a realistic picture of the view ahead.

Figure 2.21 Helmet-mounted display

2.1 Charge-coupled devices

This section of the chapter is about one particular device, the CCD (charge-coupled device), which is used to obtain two-dimensional images. CCDs are used to obtain the images for helmet-mounted displays, and are also common in many mobile phones. They have also largely replaced photographic plates for recording astronomical images obtained by telescopes.

Figure 2.22 CCD 'chip'

A CCD is a small 'chip' made of layers of semiconductor material. It is typically about 2 cm by 2 cm, made up of about 1000 by 1000 tiny light- or infrared-sensitive pixels (pixels comes from 'picture element') (see Figure 2.22). When a photon is absorbed by a pixel, a photoelectron is released from the semiconductor – this is rather similar to what happens in the photovoltaic cells that you used in the AS chapter *Technology in Space*. However, in this case, the released electron remains trapped within the pixel. As more photons are absorbed, more electrons are released. If an image is projected onto a CCD (eg through a system of lenses) then charge builds up in each pixel according to the number of photons that have reached it, thus recording the image.

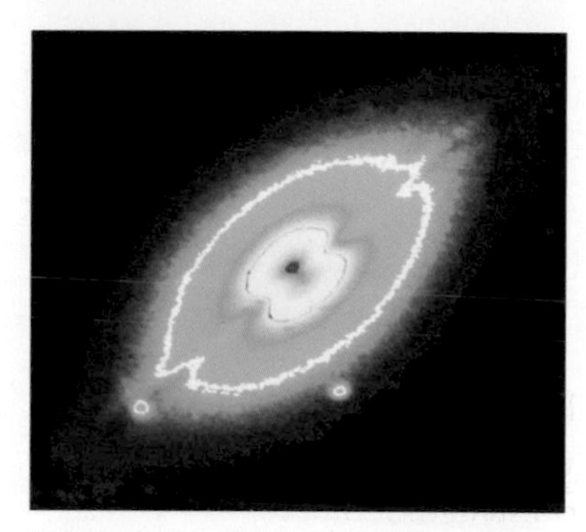

Figure 2.23 CCD image taken through a telescope

In order to 'read' a CCD image, the electrons are electrically shunted along the CCD step by step in a sort of 'bucket brigade', producing a small pulse of current as each 'package' reaches the edge. The size of each pulse can be detected and encoded to produce a digital signal, and the signals can then be processed and transmitted in just the same way as those that represent single measurements.

If the image is bright, then only a very short exposure time is needed – the CCD images that give a pilot 'night vision' are produced so rapidly that there is no noticeable time delay. However, if a CCD is being used in an astronomical telescope to build up an image of a very faint galaxy then this takes much longer. Figure 2.23 shows some CCD images taken with various exposure times. You can see that the image gradually builds up as more and more photons arrive (if your school or college has a telescope and CCD imager, you might be able to see this for yourself).

> **Study note**
>
> You will have met capacitors if you have studied the chapter *Transport on Track*.

Each pixel in a CCD behaves rather like a tiny solar cell attached to a **capacitor** – a device that stores energy by separating electric charge. The simplest type of capacitor consists of two parallel conducting plates separated by an air gap. When a capacitor becomes charged, electrons simply redistribute themselves between the plates, making one positive and one negative. There is no net charge added to the device (see Figure 2.24). Most practical capacitors (Figure 2.25) consist of thin layers of conducting material separated by thin layers of electrical insulation and fitted into a compact space, thus concealing the layered structure.

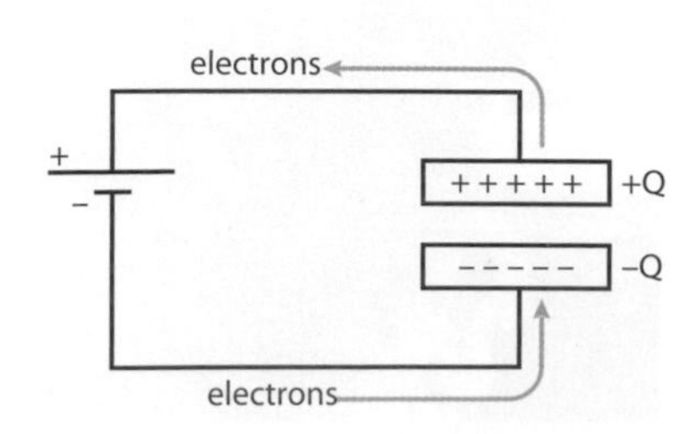

Figure 2.24 Charging a capacitor

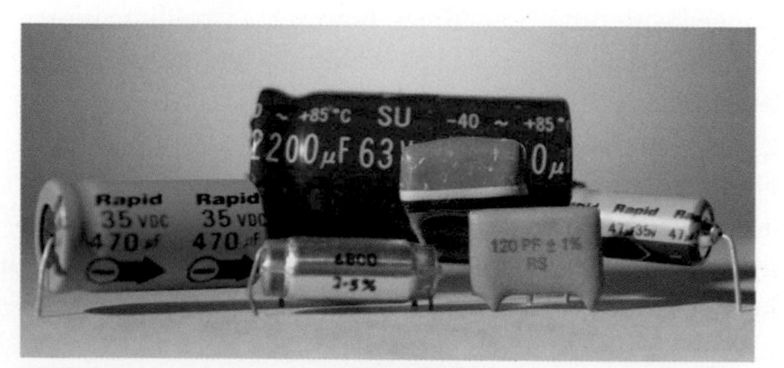

Figure 2.25 Various types of capacitor

Activity 8 Model CCD imager

Use an array of capacitors and solar cells, each connected as in Figure 2.26, to model the behaviour of a CCD imager.

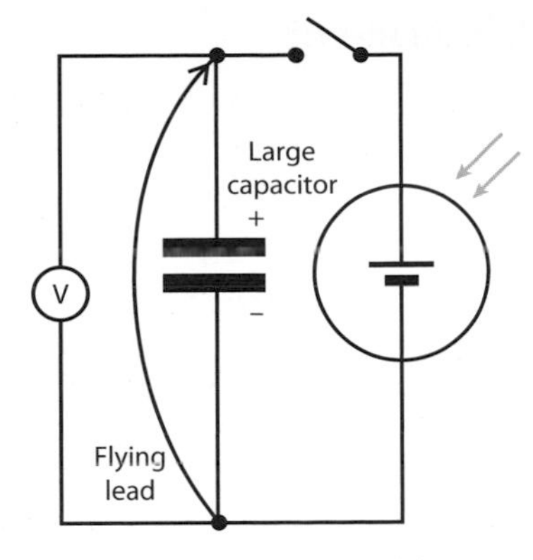

Figure 2.26 Circuit of model CCD pixel

In Activity 8 you probably took it on trust that the voltage across a capacitor is proportional to the charge it stores. You can explore this in more detail in Activity 9.

Activity 9 Spooning charge onto a capacitor

Transfer measured amounts of charge onto a capacitor and observe how the voltage across the capacitor changes.

In Activity 9 you will have found that equal additions of charge raised the voltage across the capacitor by equal amounts. In other words, $Q \propto V$. The constant of proportionality, C, is the **capacitance** of the device.

$$Q = CV \quad \text{or} \quad C = \frac{Q}{V} \qquad (5)$$

The SI unit of capacitance is the farad, F, where $1\text{ F} = 1\text{ C V}^{-1}$. Most practical capacitors are rated in microfarads, μF ($1\ \mu\text{F} = 1 \times 10^{-6}$ F) or picofarads, pF ($1\text{ pF} = 1 \times 10^{-12}$ F); the ones you used in Activity 8 were unusual in being rated at about 1 F – they were 'memory back-up' capacitors for use in computers.

Questions

7 A typical MOS (metal oxide semiconductor) capacitor in a CCD imager has a capacitance of 1 pF and can store a charge of $Q \approx 0.16$ pC.

(a) How many electrons would have to be collected by the MOS capacitor to give it a charge of 0.160 pC? (Electron charge $e = 1.60 \times 10^{-19}$ C.)

(b) What will be the voltage across such a capacitor when it is charged to this extent?

8 The capacitors used in the model CCD imager were rated at about 1 F. If the highest voltage the solar cells can provide is 0.45 V, what is the maximum charge that one of these capacitors could have stored?

9 If a solar cell provides an average current of 50 mA at 0.45 V to a 1 F capacitor, calculate how long it is before it reaches its maximum charge for that voltage. (The current actually varies as the capacitor charges, as you may recall if you have studied the chapter *Transport on Track*.)

CCD read-out

In order to extract information from the array, each packet of charge has to be moved to an output system (like the voltmeter in the model in Activity 8) and measured. Each package of charge is moved to a separate read-out register. To enable this to be done, the CCD is designed so that each tiny light-sensitive capacitor is separated from its neighbour by another capacitor that remains unaffected by light. One complete step in the read-out for one pixel is shown in Figure 2.27. First a temporary voltage is applied to the 'empty' capacitor so that charge spreads into it. The first capacitor's voltage is then reduced to zero leaving all the charge on the adjacent one.

This process is repeated until each capacitor's initial charge gets to an output sensor where the voltage across the capacitor is finally measured. All of this is done with the aid of a computer, which controls the application of the voltages, and a device (an analogue to digital converter) which encodes each measured voltage in order to determine how bright an image it should display at each point on a screen.

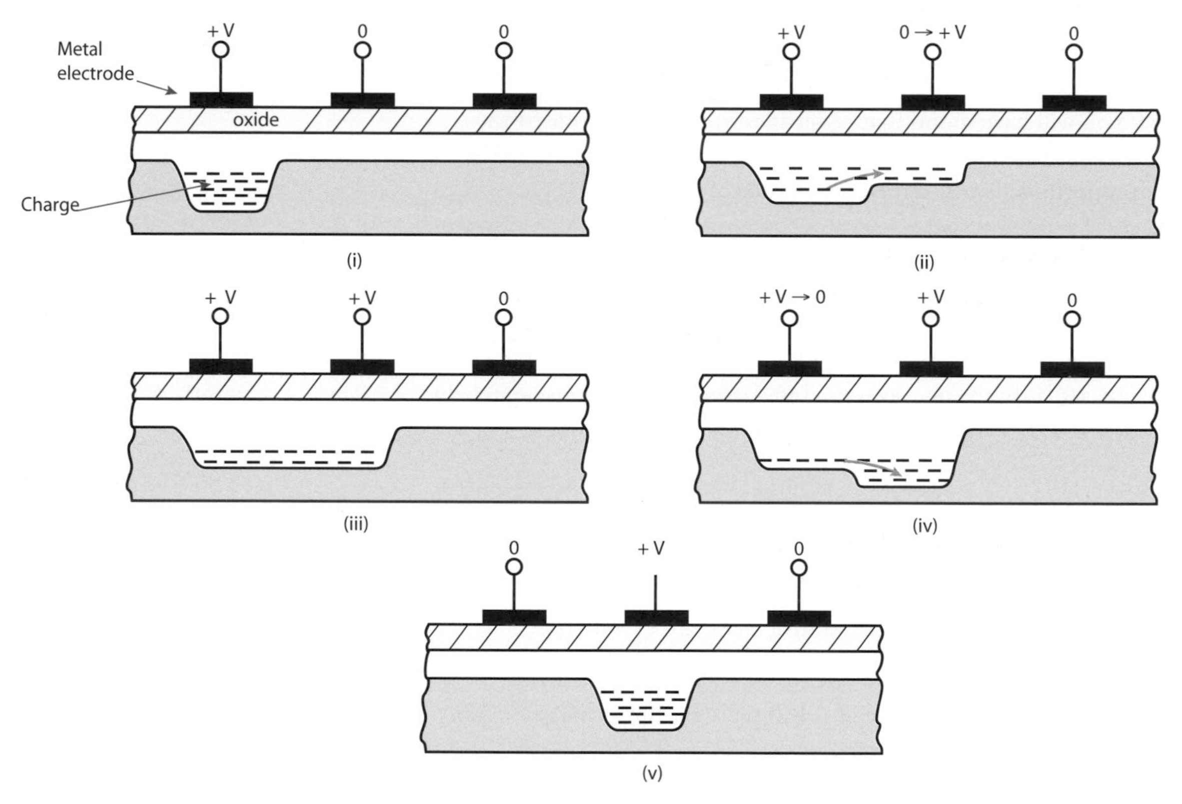

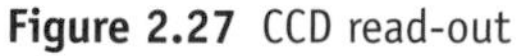

Figure 2.27 CCD read-out

2.2 Energy and capacitors

A capacitor is a device that stores energy by separating charge. In Activity 10 you will see that the stored energy is related to the capacitor voltage.

Activity 10 How many bulbs will it light?

By charging a 10 000 μF capacitor to different voltages and discharging it into various numbers of small light bulbs, suggest a relationship between capacitor voltage and the energy it stores.

In Activity 10, you should have found that 3 V provided enough energy to light one bulb with a moderately bright flash; doubling the voltage to 6 V provided enough energy to light four bulbs, and tripling the voltage enabled nine times as many bulbs to be lit. This suggests that the stored energy is proportional to the square of the capacitor voltage. The following argument shows why this indeed the case.

We already know that charge stored is proportional to the voltage to which the capacitor has been charged. A graph of voltage, V, against charge, Q, will therefore be a straight line through the origin, as in Figure 2.28.

Think now of a very small amount of extra charge (δQ) that has been placed on the capacitor, so small that the voltage across the capacitor remains at V. The small amount of work done, (δW), is given by:

$$\delta W = V \times \delta Q \qquad (6)$$

which is the shaded area in Figure 2.28. (Remember in the AS chapter *Technology in Space* you saw that $W = QV$.)

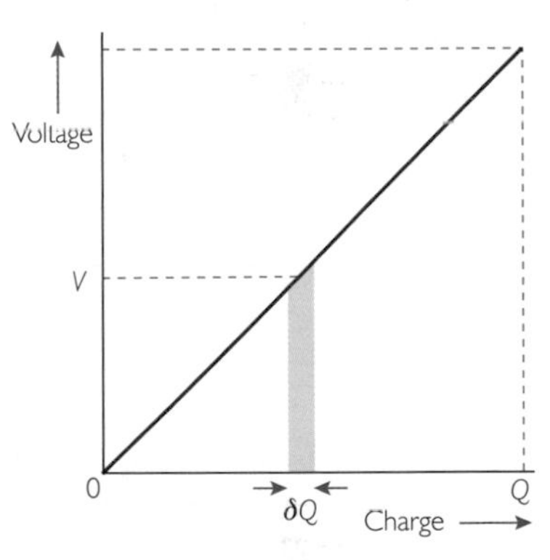

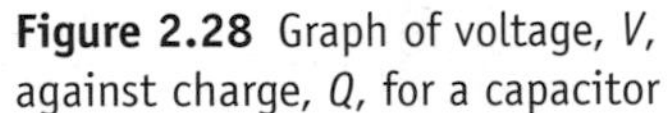
Figure 2.28 Graph of voltage, V, against charge, Q, for a capacitor

To charge the capacitor from zero to V would require the addition of lots of extra tiny amounts of charge δQ, each of which would need work done equal to the area of another tiny strip on the graph. The total energy transferred would be given by the areas of all the tiny strips added together, ie. the area under the graph. Since the graph is a straight line, we can write:

$$W = \frac{1}{2}QV \qquad (7)$$

where W is the total work done when the capacitor is charged to a potential difference V. Since energy is conserved, this must also be the energy stored in the capacitor and which is recovered when it discharges.

Combining Equations 5 and 7 we get:

$$W = \frac{1}{2}CV^2 \qquad (8)$$

and

$$W = \frac{1}{2}\frac{Q^2}{C} \qquad (9)$$

Question

10 How much energy is stored in a 10 000 μF capacitor when it is charged to: (a) 3 V, (b) 6 V, (c) 9 V?

2.3 Summing up Part 2

Part 2 has shown you how a CCD responds to incident radiation, building up a pattern of charge that represents the brightness of an image, and has involved some aspects of the physics of capacitors.

Activity 11 Summing up Part 2

Look back through your work and make sure you know the meaning of each term printed in bold.

3 On display

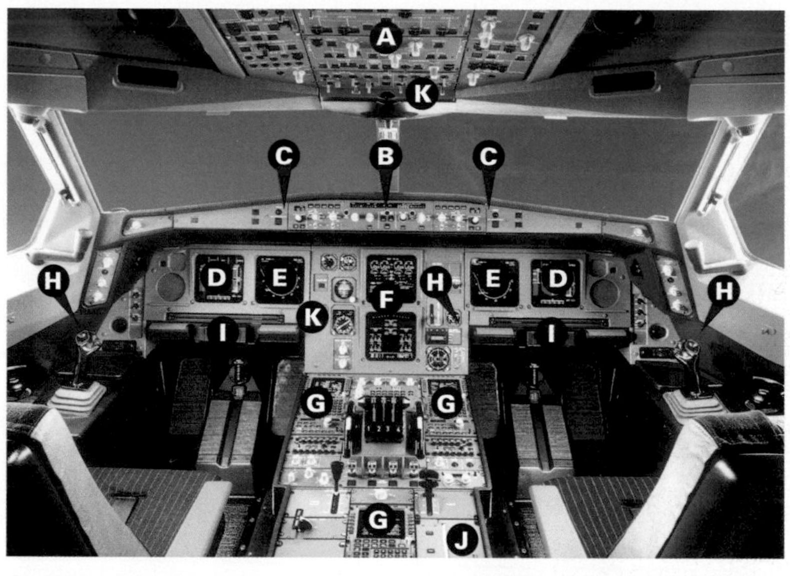

A OVERHEAD PANEL
System panels used more frequently are in lower part, centre row for engine related systems, flow scheme from bottom to top. Push-button controls, dark cockpit philosophy

B FLIGHT CONTROL UNIT (FCU)
Engages autopilot and autothrust. Selection of modes HEADING, SPEED, MACH, ALTITUDE, VERTICAL SPEED, LOC, APPROACH

C EFIS CONTROL PANEL
Select modes, ranges and options of Electronic Flight Instruments System, BARO:STD selection, master warning, master caution, autoland warning and sidestick priority lights

D PRIMARY FLIGHT DISPLAY (PFD)
Engage status of Flight Director, autopilot and autothrust. Flight Mode Annunciation. Indication of ATTITUDE, AIRSPEED, ALTITUDE, VERTICAL SPEED, HEADING, ILS-DEVIATION, MARKER, RADIO ALTITUDE

E NAVIGATION DISPLAY (ND)

F ELECTRONIC CENTRALISED AIRCRAFT MONITORING SYSTEM - ECAM
UPPER DISPLAY UNIT (DU)
Engine primary indication, fuel quantity, slats/flaps position, warning/caution/memo message
LOWER DISPLAY UNIT (DU)
Aircraft system synoptics, status of systems

G MULTI-PURPOSE CONTROL DISPLAY UNIT (MCDU)
Controls the Flight Management System (FMS) and the Central Fault and Display System (CFDS)

H SIDESTICK

I PULL-OUT WORKING TABLE
In stowed position - Footrest pedals right and left

J FULL-SIZE PRINTER

K STAND-BY INSTRUMENTS
Attitude, altitude, speed, DDRMI, compass

Figure 2.29 Instrument displays in the A380

Over the past decade or so, there have been many developments in display technology that have involved cutting-edge research and major investment from industry. These provide much of the material for this chapter. But we will also look a bit further back and examine some devices that were developed during the 20th century. Returning to the example we used in Part 1, a brief history of aircraft instrumentation and display illustrates the rapid and fundamental changes that have taken place.

A pilot needs to be able to see clearly how an aircraft is functioning, and so the display of information is a crucial aspect of avionics. Various techniques have been developed to improve the quality of such displays so that they can be easily seen and interpreted. Many of these techniques are not specific to avionics, and can also be found in many other types of communication system.

Modern aircraft, such as the A380 and the Eurofighter Typhoon, have arrays of instruments giving details of speed, heading, altitude, position … and a myriad of other information (see Figures 2.29 and 2.30). Some displays appear on a screen only when required.

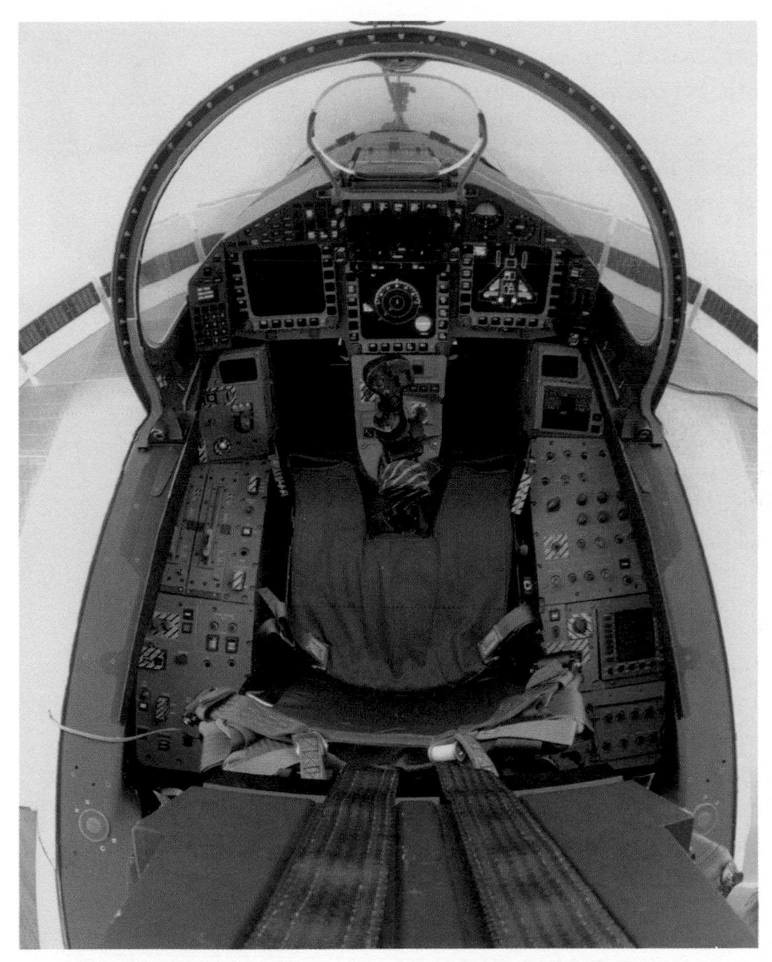

Figure 2.30 Instrument displays in the Eurofighter Typhoon

Originally, virtually all aircraft instruments used needle dial displays, much like the analogue electrical meters and pressure gauges (see Figure 2.31) you probably met earlier in your science and technology education.

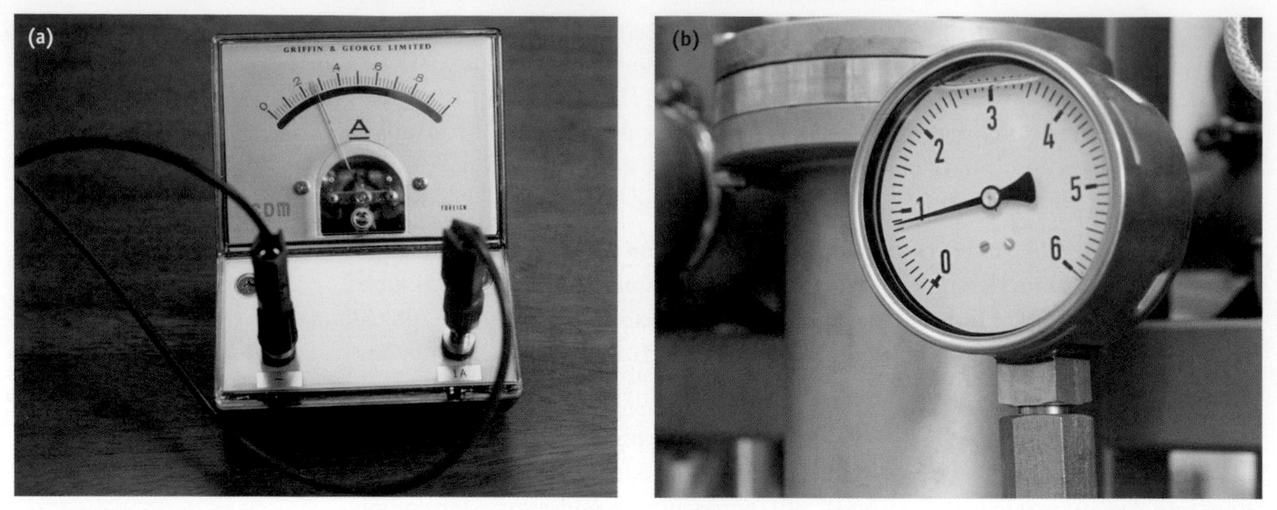

Figure 2.31 Needle dial instrument displays (a) electrical meter and (b) pressure gauge

Displays in modern aircraft are of three types.

- In head-down displays (HDDs) the viewer has to look directly at the instrument – it might be a case of looking down, but could also be upwards or sideways.
- Head-up displays (HUDs) project an image directly in front of the viewer to combine with the view of the environment (see Figure 2.32).
- Helmet-mounted displays (HMDs) are essentially HUDs provided direct to an image-forming system attached to the pilot's helmet (see Figure 2.21). If you have studied Part 2 of this chapter, you will already have come across one of the key elements of a helmet-mounted display – the CCD imager. The display moves with the head.

Figure 2.32 Head-up display

Here we will concentrate on head-down displays: making these into head-up or helmet-mounted displays involves optical projection of the image, which lies outside the scope of this chapter. We will deal in turn with three types of display – the light-emitting diode (LED), the liquid-crystal display (LCD) and the cathode-ray tube (CRT).

3.1 Light-emitting diode displays

You will be familiar with visible LEDs as 'power on' indicators on a hi-fi or video, but they have much wider usage. On aircraft single LEDs tend to be used for warnings, engine monitoring, as on–off indicators in programmable switches and in small matrix displays. Much larger screen displays can also be made using a million or so small LEDs. By grouping the LEDs in sets of red, green and blue mounted close together a full-colour image can be produced by adjusting the brightness of each LED appropriately. To make a moving image, each LED is switched on or off by a computer at the correct time to produce the picture. Switching times can be quite short – in the nanosecond (10^{-9} s) region.

Activity 12 involves an exploration of LEDs, and includes some revision of ideas about charge and voltage that you met in the AS chapter *Technology in Space*, and about photons and energy levels that you met in the AS chapters *The Sound of Music* and *Technology in Space*.

Activity 12 Illuminating LEDs

Using LEDs of at least two different colours, find the voltage that will just light each one, and suggest an explanation for your results.

Measure the power required to light each LED.

In Activity 12 you should have found that differently coloured LEDs required different voltages to make them light. For an explanation, think back to what you have already learned about electronic energies. In a semiconductor light-emitting diode, charge carriers are accelerated across a gap between energy bands, which results in a photon being emitted whose energy is equal to the size of the energy gap. The voltage that is just big enough to light the diode corresponds to charge carriers just crossing the gap. If a particle with charge q is accelerated through a potential difference ΔV, then it acquires energy ΔE where:

$$\Delta E = q\Delta V \qquad (10)$$

If this energy is then radiated as a photon, then the frequency, f, and wavelength λ of the radiation are related to the energy via the expressions

$$\Delta E = hf = \frac{hc}{\lambda} \qquad (11)$$

where c is the speed of light in a vacuum and h is Planck's constant.

Questions

The following data will be needed when you tackle Questions 11–13:

electron charge, $e = 1.60 \times 10^{-19}$ C
1 eV = 1.60×10^{-19} J
Planck constant, $h = 6.63 \times 10^{-34}$ J s
speed of light in vacuum, $c = 3.00 \times 10^{8}$ m s^{-1}.

11 The bandgaps for compounds of indium and antimony can be as small as 0.18 eV and those of gallium and nitrogen up to 3.4 eV. Calculate the wavelength of the radiations given off by the LEDS with each of those bandgaps.

12 What bandgap energies would be needed to produce visible LEDs that emit light over the range 400 nm to 600 nm? Express your answer in joules and in eV.

13 (a) A standard red LED was found to be passing a current of 15 mA when there was a pd of 2.15 V across it. What power was it drawing from the supply?

(b) An infrared-emitting diode, operating at a voltage of 1.3 V, passed a current of 100 mA. What power was it drawing from the supply?

(c) If you need 10^6 LEDs to produce a screen display, estimate the power required.

3.2 Liquid-crystal displays

Liquid-crystal displays (LCDs) are used in mobile phones, DVD players, hand-held games, calculators, on cookers, video-recorders, televisions, computer monitors and so on – and in aircraft cockpits. The production of LCDs is a multi-billion pound industry.

A **liquid crystal** is a liquid in which the molecules arrange themselves in some sort of ordered pattern. Although we tend to think of molecules in liquids being completely randomly distributed, many liquids are in fact liquid crystals – these include DNA, RNA, soaps and detergents, and even the brain is about 70% liquid crystal. Liquid-crystal display technology has its origins in the 1970s, when George Gray of Hull University made the first stable synthetic liquid crystals known as nematic alkylcyanobiphenyls, which have the important characteristic that their optical properties can be controlled electrically.

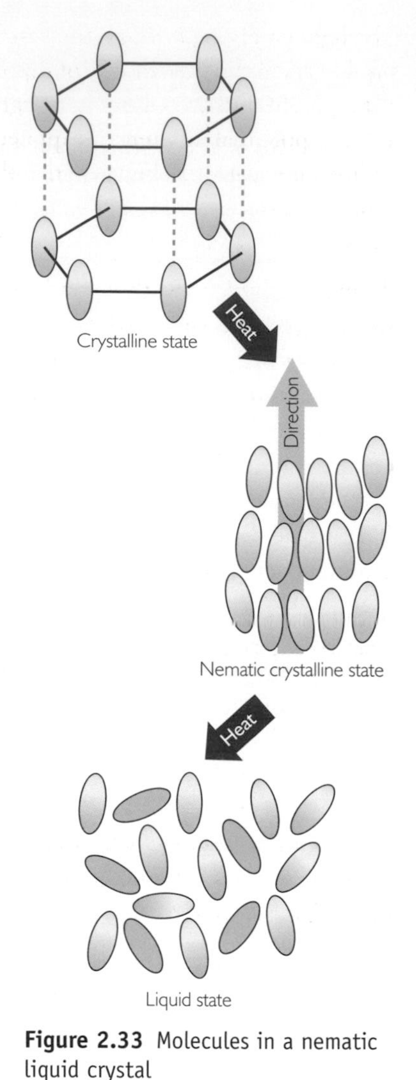

Figure 2.33 Molecules in a nematic liquid crystal

A so-called nematic liquid crystal consists of long thin molecules (the Greek word *nematos* means 'thread-like') that tend to arrange themselves with their long axes parallel to each other, as shown in Figure 2.33. This comes about because the molecules are **polar** – that is, their charge is neutral overall but not evenly distributed, so one end is slightly positive and the other slightly negative, forming an **electric dipole** (Figure 2.34). They line up because the positive end of one molecule attracts the negative end of another.

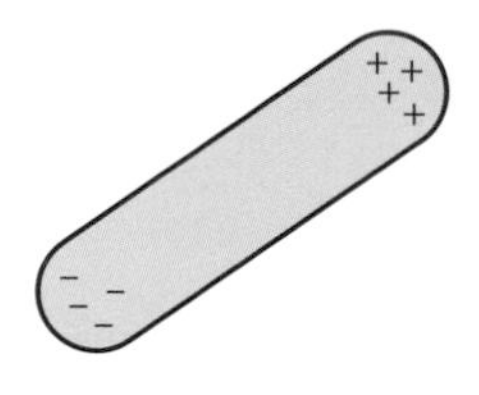

Figure 2.34 An electric dipole

The LCDs in use today are of a type known as twisted nematic (TN) or super twisted nematic (STN). A TN display consists of a thin layer of nematic liquid crystal between two glass plates whose surfaces have been coated with a very thin conducting layer and treated with a rubbed polymer layer that has the effect of creating small parallel grooves. The molecules prefer to lie along, rather than across, the grooves. The glass plates are oriented with one set of grooves running at right angles to the other, which introduces a 90° twist in the molecular arrangement as shown in Figure 2.35(a). (In an STN display, there is a twist of 270°.) This 'sandwich' is mounted between two pieces of polarising filter set in the 'crossed' position.

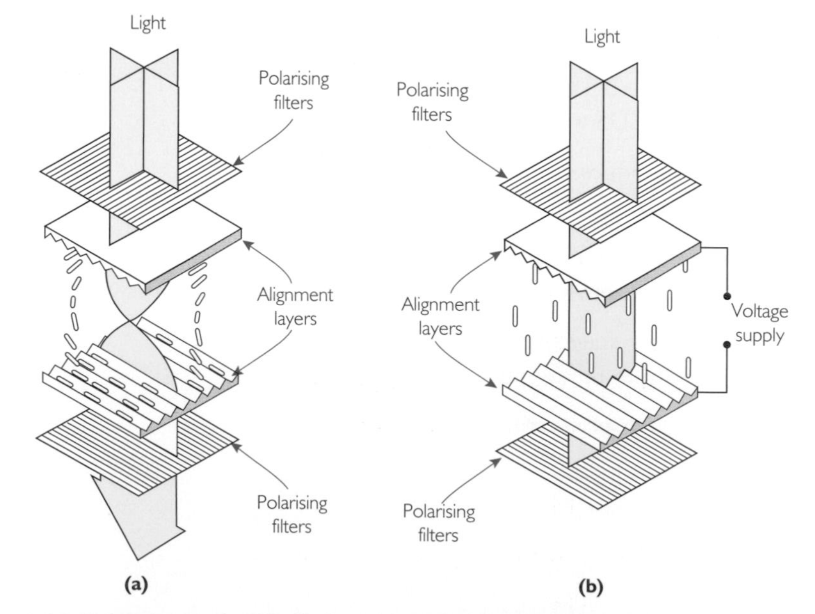

Figure 2.35 Twisted nematic LCD display: (a) voltage off, transparent and (b) voltage on, opaque

Study note

You met polarised light and polarising filters in the AS chapter *The Sound of Music*.

The light emerging through the first filter is polarised, and as it travels through the liquid crystal the twisted line of dipoles guides its plane of polarisation so that it rotates through 90° and can emerge through the second filter – the 'sandwich' is transparent. But if a potential difference is applied across the liquid crystal, the molecules are forced to line up end-to-end between the plates as shown in Figure 2.35(b). There is no longer any rotation of the plane of polarised light, so the light cannot emerge through the second filter and the 'sandwich' is opaque.

To make a display, the thin conducting coating is applied in a suitable pattern (eg line segments to make up numbers), each connected to a separate small electrode that can be controlled by a digital on–off signal. Colours can be obtained by placing coloured filters in front of the display and shining white light through it. TV screens, computer monitors and various instrument displays can be made from arrays of many thousands of individual LCDs.

Electric field

To explain how the molecules are aligned by the applied potential difference, it is useful to introduce the idea of an **electric field**, which is defined as a region in which a charged object experiences a force. The **electric field strength**, E, is defined as the force experienced by a charge of 1 coulomb, in other words, the force per unit charge. If a charge q experiences a force of magnitude F, then the magnitude of field strength is

$$E = \frac{F}{q} \qquad (12)$$

The SI units of electric field strength are N C^{-1}. As force is a vector, electric field is also a vector; ie it has direction as well as magnitude. The **direction of an electric field** is defined as that of the force experienced by a positive charge.

It is quite difficult to measure electric field strength by directly measuring the force on a charged object; mainly because in most practical cases the force is quite small. However, there is an alternative way of thinking about electric fields, which both provides an easier way to measure E and also gives some insight into the type of field that is present in an LCD.

Figure 2.36 shows an arrangement for producing an electric field between two parallel conducting plates, separated by a distance Δx and with a potential difference ΔV applied between them.

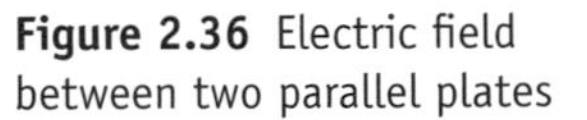

Figure 2.36 Electric field between two parallel plates

Now think of placing a single small positive charge, q, on plate A in Figure 2.36. As A is connected to the positive terminal of the power supply, the charged particle will move straight across the gap to plate B – the field direction is straight across from A to B as shown in Figure 2.36. As the particle moves through the potential difference ΔV between the plates, it will gain energy ΔW, where:

$$\Delta W = q\Delta V \qquad (13)$$

Another way to look at the same situation is to say that there is a force of magnitude F acting on the particle. As the particle crosses the gap, width Δx, this force does work ΔW, where:

$$\Delta W = F\Delta x \qquad (14)$$

As Equations 13 and 14 are just two ways of describing the same thing, we can write:

$$F\Delta x = q\Delta V \qquad (15)$$

As the electric field strength is defined as the force per unit charge, we can rearrange Equation 15 to get its magnitude:

$$E = \frac{F}{q} = \frac{\Delta V}{\Delta x} \qquad (16)$$

In other words, the field strength is equal to the potential gradient – the rate at which the potential changes with distance across the gap. As field strength is a vector, we should really be careful with signs. The field is directed from A to B, which is the direction in which the potential is decreasing from positive to zero – that is, the potential gradient is negative. We can therefore write Equation 16 as:

$$E = -\frac{\Delta V}{\Delta x} \qquad (16a)$$

Equation 16a provides an alternative way to define and measure electric field strength, and also gives its alternative SI unit: V m^{-1}.

An electric field can be represented diagrammatically by drawing **electric-field lines** which show the path and direction a free positive charge would take, as shown in Figure 2.37. The spacing between the electric-field lines shows the strength of the field; for a strong electric field the electric field lines are close together. If voltage across the plates is increased, and the plates remain the same distance apart, the field strength will increase and the field lines will be drawn closer together as in Figure 2.37(b). If the plates are moved closer together and the voltage across the plates remains the same the field strength will increase and the field lines will be drawn closer together as in Figure 2.37(c). The electric field between two parallel plates is called a **uniform electric field** because the electric field strength is the same at every point between the plates, and so all of the field lines are the same distance apart between the plates.

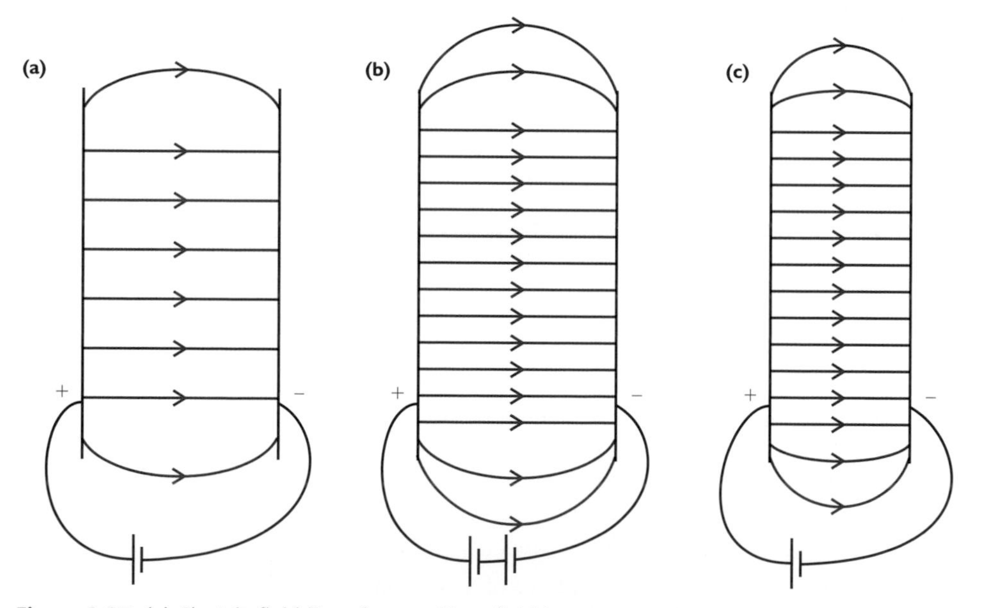

Figure 2.37 (a) Electric field lines for a uniform field between parallel plates (b) when the pd between the plates is increased (c) when the plates are moved closer together

If an electric dipole (such as a polar molecule) is placed in such a field, then the positive and negative end will experience forces in opposite directions, ie there will be a couple and the dipole will twist around until it is aligned with the field direction as shown in Figure 2.38.

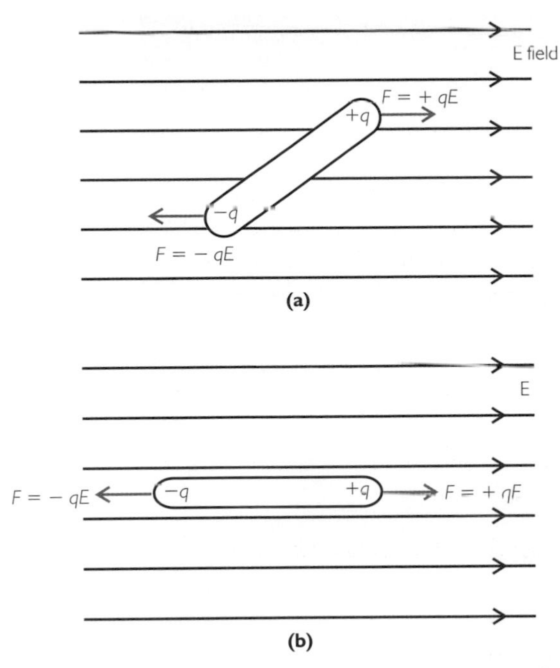

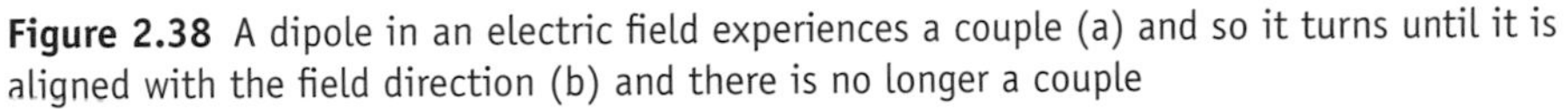

Figure 2.38 A dipole in an electric field experiences a couple (a) and so it turns until it is aligned with the field direction (b) and there is no longer a couple

The field around a point charge is a **radial electric field**. The field lines point towards or away from the centre of the charge because they show the direction a small positive charge would take. The field lines of a positive charge point outwards as they show the direction of the force on another small positive charge. The field lines of a negative charge (eg an electron) point inwards; see Figure 2.39.

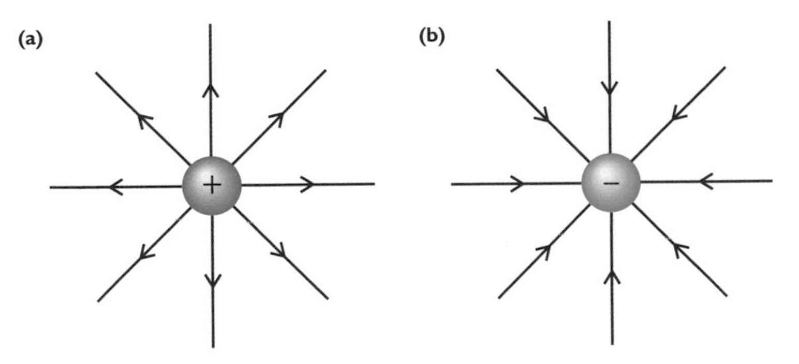

Figure 2.39 Electric field lines for (a) a positive and (b) a negative point charge

The field of a point charge is not uniform. It is strong near the point charge where the field lines are close together, and weaker further away where there are bigger spaces between the field lines.

Figure 2.40 shows the apparatus that can be used to measure field strength – a double-flame probe. There are two tiny flames at the tips of two fine metal tubes (which are connected to a gas supply). A flame provides some ionisation in the air, and the ions move until there is no potential difference between the flame and its immediate surroundings. The two fine tubes are connected to the cap and case of an electroscope, where the deflection of the leaf depends on the potential difference between the cap

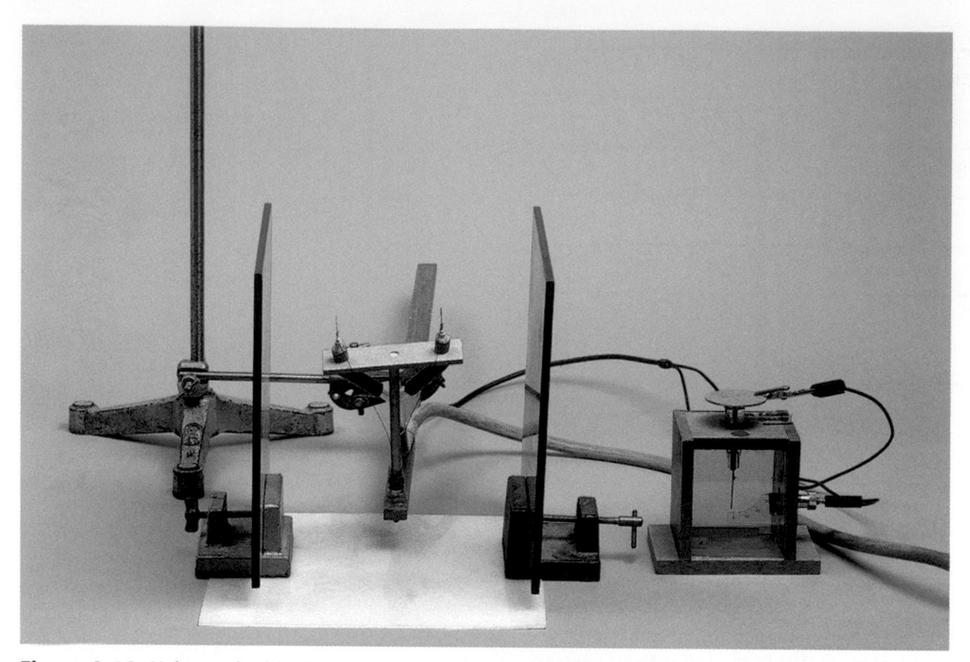

Figure 2.40 Using a double flame-probe to explore an electric field

and the case. The electric field strength close to the tips of the fine tubes can then be found from a measurement of the potential difference indicated by the electroscope and the physical separation of the tips.

Activity 13 Probing the field

Use a double-flame probe to investigate the size and direction of the electric field between parallel plates and in some other arrangements.

Further investigations

Measure the capacitance of various arrangements of metal plates, including some where they are not parallel.

Questions

14 Show that the SI units N C^{-1} and V m^{-1} are exactly equivalent.

15 (a) If the voltage between two parallel plates is 1kV and they are separated by 0.1 m, what is the magnitude of the electric field strength between them? Give your answer in both possible SI units.

(b) How large a force would this electric field exert on a charge of 3 μC?

16 An LCD operating at a voltage of 2.5 V passed a current of 140 pA (1 pA = 1×10^{-12} A). What power was it drawing from the supply?

17 Figure 2.41 shows the electric field lines between two parallel plates. The potential difference between the plates is 200 V and the distance between the plates is 4.0 cm.

(a) What is the strength of the electric field between the plates?

(b) For each of the following, calculate the field strength and draw a diagram showing the electric field lines.

(i) $V = 400$ V and $x = 4.0$ cm

(ii) $V = 200$ V and $x = 8.0$ cm

(iii) $V = 100$ V and $x = 2.0$ cm

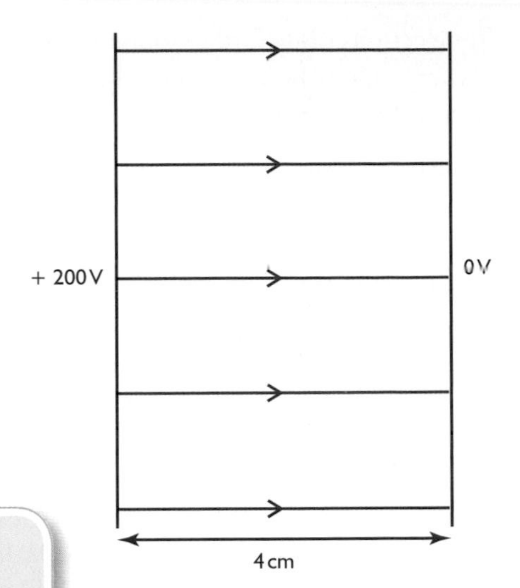

Figure 2.41 Diagram for Question 17

Activity 14 Comparing LCD and LED displays

Compare the voltages and powers required to operate an LCD and an LED seven-segment display.

Further investigations

Measure and compare the response times of LEDs and LCDs.

3.3 Shine a light

In September 2007, the UK Environment Minister, Hilary Benn, announced that the traditional tungsten filament light bulb should be phased out within two years. This followed on from similar plans in other countries such as Australia and has been supported by Greenpeace, with the aim of reducing CO_2 emissions within the UK. The alternatives to these incandescent bulbs (ie those in which there is a heated wire filament) are either LED lighting or energy-efficient ones called compact fluorescent light bulbs (CFL) (see Figure 2.42). Generally a CFL lasts 8–12 times longer than an incandescent bulb. It also uses less energy; a 15 W compact fluorescent could replace a 60 W incandescent bulb because it has greater efficiency.

Figure 2.42 Energy-efficient light bulbs (a) LED (b) CFL

A CFL contains mercury vapour. At room temperature, a few of the atoms are ionised, so as well as neutral atoms there are some free electrons and positive ions. A potential difference is applied across the tube containing the vapour, and the resulting field accelerates the electrons and ions. Provided they can acquire enough kinetic energy between collisions, the free electrons cause further ionisation when they collide with the neutral atoms. The gas very rapidly becomes ionised (an ionised gas is called a **plasma**) and is then a good electrical conductor.

Collisions between the electrons and the mercury atoms and ions excite the bound electrons to higher energy levels. The excited electrons lose energy by emitting ultraviolet photons as they fall to lower energy levels. The inside of the CFL is coated with phosphor which absorbs the ultraviolet and emits visible radiation.

Although CFLs are more energy efficient, there is a downside: concerns have been raised about disposing of the CFLs due to the mercury vapour contained within them.

Question

18 The excited mercury vapour in a CFL emits ultraviolet radiation where the dominant photon energy is 4.9 eV. In order to excite the mercury ions so that they give out this ultraviolet radiation, the free electrons in the plasma must acquire energy between collisions that is at least 4.9 eV.

If the average distance between collisions is 8.0×10^{-3} m, what must be the magnitude of the uniform electric field within the lamp?

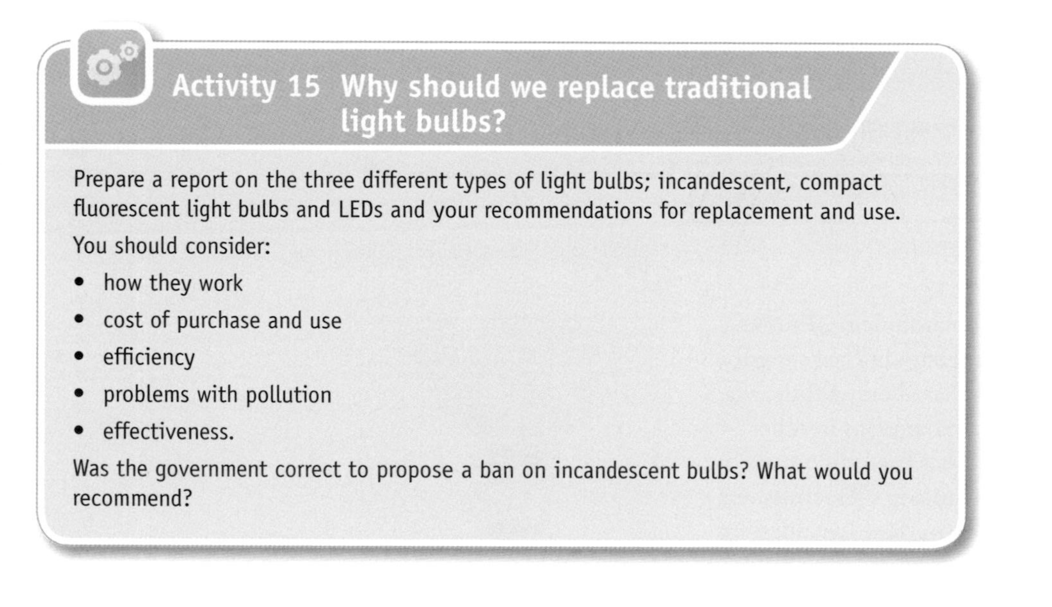

Activity 15 Why should we replace traditional light bulbs?

Prepare a report on the three different types of light bulbs; incandescent, compact fluorescent light bulbs and LEDs and your recommendations for replacement and use.

You should consider:

- how they work
- cost of purchase and use
- efficiency
- problems with pollution
- effectiveness.

Was the government correct to propose a ban on incandescent bulbs? What would you recommend?

3.4 Cathode-ray tube

The oldest type of electronic display is the cathode-ray tube (CRT), which was the most common type of display in use until LCD sales overtook CRT sales in 2004. CRTs have been very widely used in TVs, oscilloscopes and computer monitors, and still form the basis for many of the displays used in avionics.

Figure 2.43 shows one type of cathode-ray tube, known as a Perrin tube after its designer Jean Baptiste Perrin (1870–1942), and Figure 2.44 shows a CRT from a television. Inside each tube there is an **electron gun** in which electrons are released from a negative electrode (a **cathode**), and accelerated in an electric field between the cathode and **anode** (positive electrode). The resulting stream of high-energy electrons hits a phosphor screen, causing it to glow. (Originally the electron beam was called a 'cathode ray' before people knew what it consisted of.) There must be a vacuum inside

Study note

Electron guns are used in electron microscopes, which you studied in the AS chapter *Digging Up the Past*.

the tube, otherwise air molecules would scatter the electron beam. Around the outside of the tube are deflection coils – current in these coils produces a magnetic field that deflects the electron beam.

Figure 2.43 Perrin tube

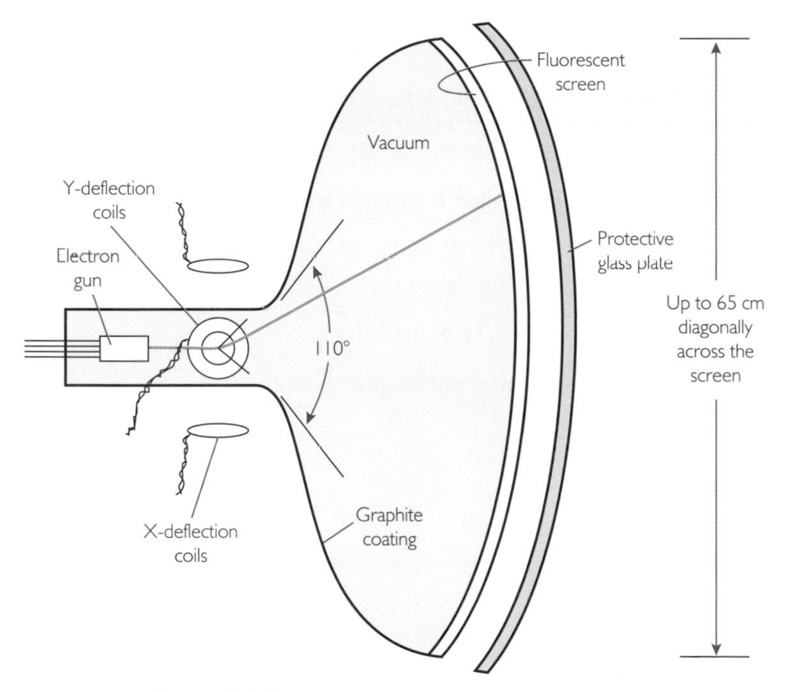

Figure 2.44 CRT used in a television

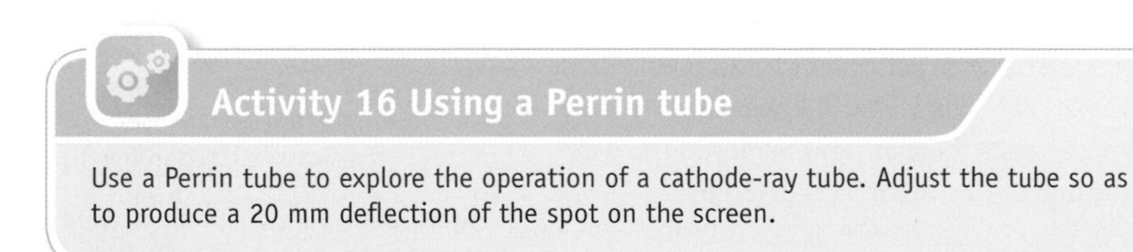

Activity 16 Using a Perrin tube

Use a Perrin tube to explore the operation of a cathode-ray tube. Adjust the tube so as to produce a 20 mm deflection of the spot on the screen.

We will now look in more detail at the operation of each part of a CRT and see how it contributes to the formation of an image.

Electron gun

Figure 2.45 shows details of an electron gun. The filament, F, is heated by an electric current and, by a process known as **thermionic emission**, electrons that have enough energy escape from the filament's surface much as water evaporates from a puddle of rain. These electrodes and the radiant energy from the filament heat up a nickel cathode (C) coated with a mixture of barium and strontium oxides, which results in the thermionic emission of a large number of electrons from the cathode. A potential difference (often called the gun voltage) between the cathode C and anodes A_1 and A_2 produces an electric field that accelerates the electrons. A_1 is either a hollow cylinder or a disc with a hole in it and focuses the electrons into a narrow beam. A_2 is usually a cylinder with a disc with a hole in at the screen end through which the accelerated electrons pass on their way to the screen. Most guns also incorporate a grid (G), which is negative with respect to the cathode and so controls the overall rate at which electrons pass through the whole system.

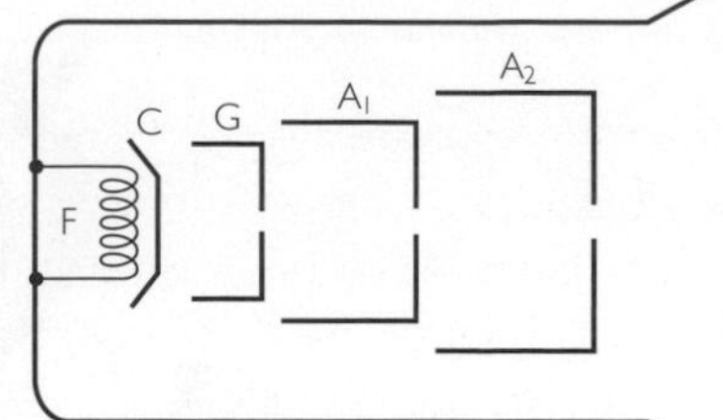

Figure 2.45 Electron gun

The electrons hit the phosphor screen and are brought to rest. Their kinetic energy is transferred to the atoms and molecules in the phosphor screen coating, which become excited and then lose energy by emitting photons.

Questions

19 (a) If an electron is accelerated by a gun voltage of 5000 V, how much energy does it acquire?

(b) What is the electron's (i) kinetic energy and (ii) speed when it emerges from the gun?

(c) How would the presence of air in the CRT affect your answer to (b)?

(Electron charge, $e = 1.60 \times 10^{-19}$ C. Electron mass, $m_e = 9.11 \times 10^{-31}$ kg.)

20 Explain how each of the following would affect the brightness of the spot on the screen:

(a) increasing the potential of anode A_2

(b) increasing the voltage across the filament

(c) making the grid more negative with respect to the cathode.

Deflecting the beam

To produce an image, the beam is swept rapidly across the screen in a scanning pattern (Figure 2.46). As you saw in Activity 16, the deflection is produced by current-carrying coils.

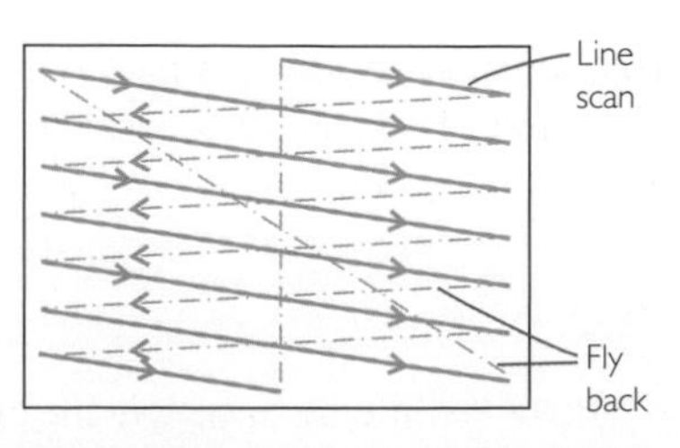

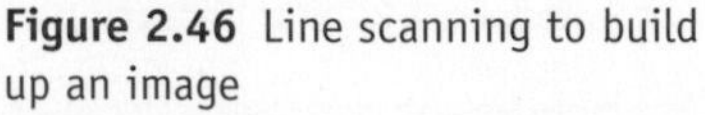
Figure 2.46 Line scanning to build up an image

You may have wondered why an electron beam is deflected in a magnetic field. If you have studied the chapter *Transport on Track* you will know that a current-carrying wire in a magnetic field experiences a force at right angles to the wire and to the field direction. The magnitude of this force, F, is given by:

$$F = BIl \sin\theta \qquad (17)$$

where B is the magnetic flux density, I the current, l the length of the wire and θ the angle between the wire and the field direction (see Figure 2.47). The direction of the force is given by Fleming's left-hand rule (Figure 2.48): where the first finger points along the direction of the magnetic field, the second finger along the direction of the (conventional) current and the thumb points in the direction of the resulting motion, ie the direction of the force.

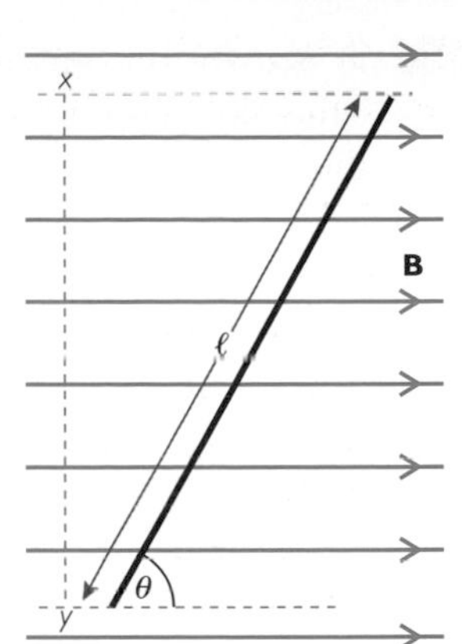

Figure 2.47 Current-carrying wire in a magnetic field

This force arises because of the motion of the charged particles in the magnetic field – so electrons that are not confined to a wire will still experience a force when they move in a magnetic field. The size of the force can be related directly to the electrons' speed. So what about the force on an electron, or indeed any other charge carrier? To deal with this we need to peer inside our current-carrying conductor (as in Figure 2.49) and consider the individual charge carriers.

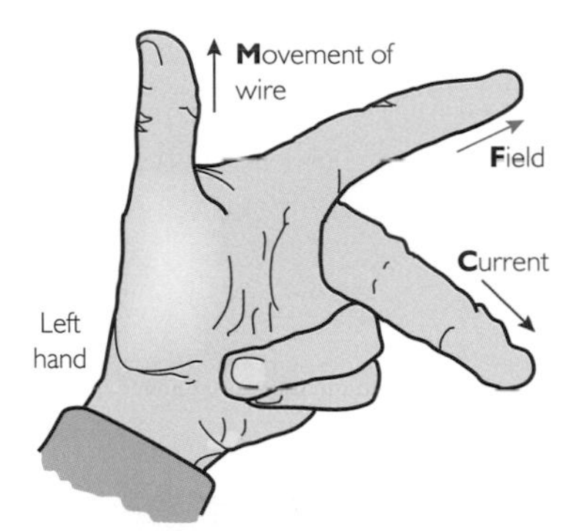

Figure 2.48 Fleming's left-hand rule

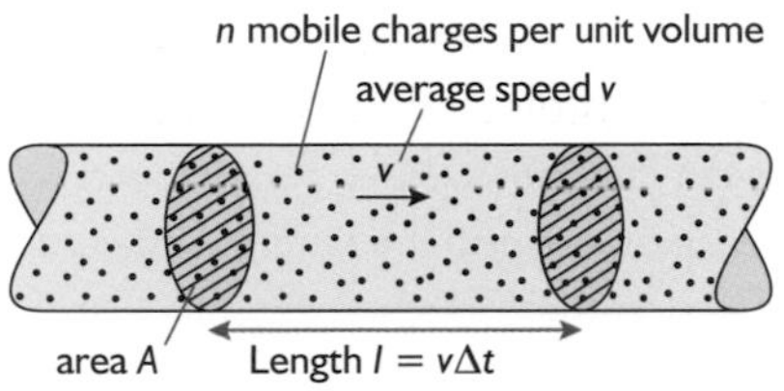

Figure 2.49 Charged particles in a conductor

There are n charge carriers per unit volume. The volume of a piece of wire with length l and area A is Al, so the number of particles, N, in this length of wire is:

$$N = nAl \qquad (18)$$

In the AS chapter *Technology in Space* you used an expression that related the current, I, to the charge carriers' motion:

$$I = nAqv \qquad (19)$$

Substituting this into Equation 17 gives:

$$F = BnAqvl \sin\theta \qquad (20)$$

Dividing this force by the number of particles N (Equation 18) gives an expression for the force on a *single* mobile charge carrier:

$$F = Bqv \sin\theta \qquad (21)$$

The direction of the force is still given by Fleming's left-hand rule. But beware – the direction of conventional current is that of the apparent movement of *positive* charge, ie in the opposite direction to the movement of electrons.

Study note

The direction of conventional current is that in which positive charge appears to move, ie from the positive to the negative terminal of a power supply. Electrons, having negative charge, move in the opposite direction.

Questions

21 What is the magnitude of the force that acts on an electron moving at a speed of 8.0×10 m s^{-1} at right angles to a field whose flux density is 0.10 T? (Electron charge $e = 1.60 \times 10^{-19}$ C.)

22 In Activity 16, you saw that a pair of coils, known as Helmholtz coils, placed either side of a CRT deflected the beam vertically. What can you deduce from this about the direction of the magnetic fields produced by the coils?

Helmholtz coils are named after their inventor, the Geman scientist Hermann Helmholtz (1821–1894). They are rather special, being designed to produce a fairly uniform field in the region between them. This is achieved by having their centres separated by a distance equal to their radius (Figure 2.50). The magnitude, B, of the magnetic flux density thus produced can be calculated using the following expression:

$$B = \frac{8\mu_0 nI}{5r\sqrt{5}} \qquad (22)$$

where I is the coil current, r the radius and separation of the coils, n the number of turns on each coil, and μ_0 is the permeability of a vacuum. $\mu_0 = 4\pi \times 10^{-7}$ N A^{-2}.

In a CRT used to produce a display (eg on a TV screen, Figure 2.51), the coils are mounted close to the electron gun, though they are not in fact Helmholtz coils. As you can see in Figure 2.44, two sets of coils are used – this is in order to scan the beam in two dimensions and build up an image. The coils and the current they carry must be designed to produce a deflection that matches the screen size. In Activity 16, you found the field that would deflect a particular electron beam through 20 mm, as would be required for a screen 40 mm from top to bottom. To produce a larger image, eg on a wide-screen TV, a larger field would be needed.

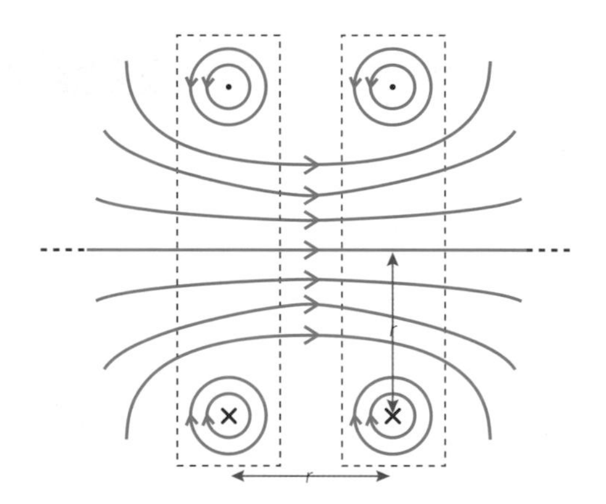

Figure 2.50 Magnetic fields produced by Helmholtz coils (x indicates current flowing into page; • indicates current flowing out of page)

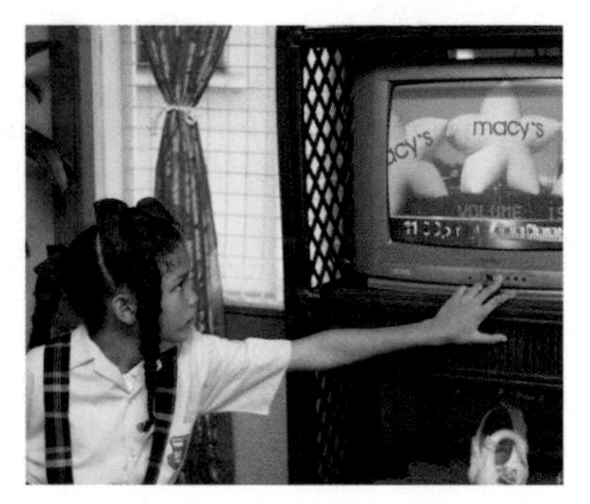

Figure 2.51 TV screen

Activity 17 Scanning the screen

Explain how the magnetic field must change in both direction and magnitude in order to produce the line scanning shown in Figure 2.46.

The image on the screen

To produce a monochrome (single-coloured) image, all that is required is that signals fed to the gun control the brightness of the spot as it is swept across the screen. There are various ways to produce a coloured image. The system most commonly used in aircraft display is the shadow-mask type shown in Figure 2.52.

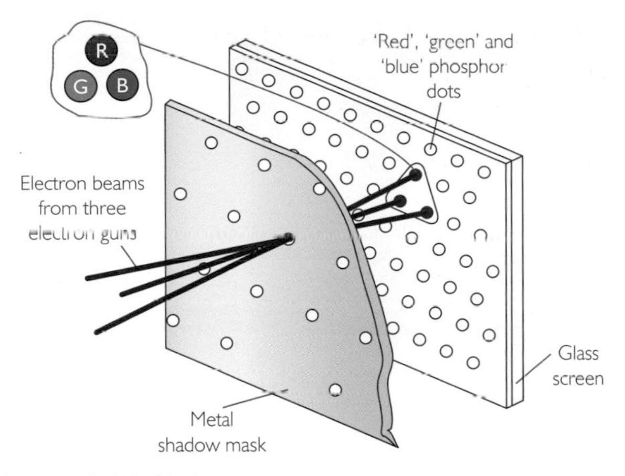

Figure 2.52 Shadow-mask CRT display

This system has three electron guns, each accelerating electrons through voltages of around 25 kV. The three electron beams are moved across and down the screen by the magnetic fields provided by the deflection coils. In doing so each beam passes through the shadow mask (a metal plate with holes in it) to hit the red, green and blue phosphor dots, respectively, on the screen. Equal exciting of each phosphor would cause white light to be emitted from the screen. Exciting just the blue and green phosphors equally would give cyan, the red and green equally would give yellow, and the red and blue equally would give magenta. Any colour, and its intensity, can be controlled by the speed and number of electrons hitting the combination of phosphors. These, and the position on the screen, are controlled by the signals fed to the electron gun and the deflecting coils by a computer. Television often uses the same type of tube, but the images are controlled by signals put out by the broadcasters.

Various other versions of CRT exist, such as the beam index tube, which has just a single electron gun that scans over tiny vertical red, green and blue phosphor stripes. Panasonic have designed what is known as a field emission display, which has over 10 000 electron guns, each targeting its electrons on a tiny portion of the screen.

Cathode-ray tubes produce good colour, brightness and definition. Their problem is power ratings, which are in the range 500 W and above, which would be undesirable on a military fighter aircraft such as the Eurofighter Typhoon (Figure 2.53).

Figure 2.53 Eurofighter Typhoon

Question

23 Why might having a lot of CRT displays prove a disadvantage on a military fighter aircraft?

3.5 Plasma screens

For many purposes, CRT displays have been replaced by LCDs, LEDs or plasma screens. Plasma displays were invented in 1964 at the University of Illinois in the USA. A plasma screen has about a million small **cells** which can emit spots of coloured light. Each cell is about around 1 mm across and is filled with gas (usually a mixture of neon and xenon). Each cell has electrodes on the top and bottom (Figure 2.54) and acts as a minute fluorescent light, creating a very bright display.

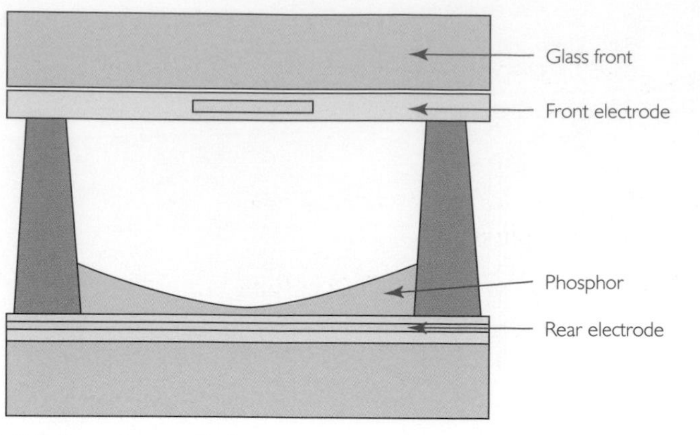

Figure 2.54 Cell for a plasma display

When a voltage is applied to the electrodes across the cell it produces a strong electric field in the cell that causes ionisation as in a CFL (Section 3.3). The cell then contains a plasma (ionised gas) that gives the display its name.

When a free electron combines with a positive ion of neon or xenon it gives off a photon of ultraviolet radiation. The photon may be absorbed by other atoms causing them to ionise, which quickly leads to a very large number of ions in the cell. Many of the photons will hit the back of the cell which has a fluorescent phosphor coating which gives off visible light when it absorbs UV radiation. The light comes out of the front of the cell which is made of glass and is seen by the person viewing the screen. The screens are produced so that in each pixel there is a red cell, a green cell and a blue cell using different fluorescent phosphors so that coloured images can be produced. Plasma screens are used for large displays to be viewed from some distance away as each cell is relatively large.

Some displays use cells with two electrodes at the front (Figure 2.55). The field in the cell is not uniform; the electric field strength is not the same at all points in the cell. The gas in the cell is ionised where the electric field is strong. The electrodes can be close together if they are placed at the front of the cell, which means that a lower voltage can be used to get the same electric field strength.

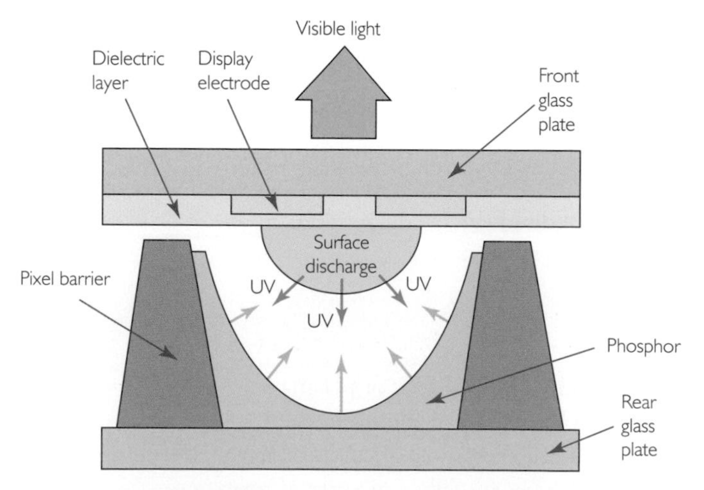

Figure 2.55 Cell with front electrodes

Questions

24 The cells in a plasma screen are 150 μm high and have 200 V applied to them when they are turned on. What is the electric field strength in a plasma cell when it is turned on?

25 When an electron combines with a xenon ion it produces an ultraviolet photon of wavelength of 147 nm. What is the energy of this photon?

26 Sketch a copy of Figure 2.55 and on it draw a diagram of the electric field lines. Write a few sentences to explain where in the cell the gas is ionised.

3.6 Summing up Part 3

In Part 3 you have looked at instrument displays and light sources: cathode-ray tubes, light-emitting diodes, liquid-crystal displays, compact flourescent light bulbs, plasma displays and the physics behind them. Magnetic and electric fields played their part and you saw how to calculate and measure these. Use Activity 18 to check your progress and understanding.

Activity 18 Summing up Part 3

Look back through your work and check that you know the meaning of all the terms printed in bold.

Activity 19 Display power

Small LCD displays require much less electrical power to operate than CRTs, but many consumers are buying large displays which have high power consumption. Compare the power consumption for televisions of similar display size that use different display technologies to find the most efficient displays.

Suppose a member of your family wants to buy a new TV set, but has heard that some TVs have high power demands and is concerned about the environmental effects of electricity usage. What advice would you give them?

Question

27 An LCD in a pocket calculator is operated by a 1.5 V power supply.

(a) If the conducting layers in the display are separated by 10 μm, what is the strength of the electric field within the display?

(b) What would be the magnitude of the force experienced by a free electron within this field? (Electron charge $e = 1.60 \times 10^{-19}$ C.)

4 Message received

4.1 Delivering the message

In this chapter you have covered many aspects of modern communication technology. A substantial part of this chapter has been about displays.

Cathode-ray tubes used to be used for most domestic televisions. They produce good colour, brightness and definition and have a quick response time. However, their power demands are large and they need a lot of space for the three electron guns. As customers began to want larger televisions, the alternatives became more popular. LCD or plasma televisions are much thinner and allow for both wall mounting and greater screen size. The cost of these was initially high and did not meet the quality of CRTs. LCD television suffered from a slow refresh rate – this is the time it takes for the pixels to switch on and off to cope with a moving image. However, with improvements in technology and quality, CRTs are now rarely used as computer monitors, and LCDs are frequently used instead.

Plasma televisions have also increased in popularity. The size of screen (ie its diagonal) can be over a metre, but some of the cheaper models do not have the same contrast or brightness of CRTs. There are also concerns about their lifespan and they can be susceptible to screen 'burn-in'. CRTs still offer better quality, so with more HD television available, the unfashionable CRT may still have a future. A number of new display technologies may also revolutionise the future of the humble TV. One of these is the organic light-emitting diode (OLED). This offers all of the benefits of a traditional LCD display, but each pixel radiates its own light. The main advantage is that you no longer need a backlight as in traditional LCD TVs, massively reducing power consumption and making the screen considerably thinner.

In Activity 20, you will apply what you have learnt about display technology and find out about some of the most recent developments.

Activity 20 What else is going on?

Displays are very important to the electronics industry and many different types of display are being developed including:

- OLED (organic light-emitting diode display)
- SED (surface-conduction electron-emitter display)
- FED (field emission display)
- ELV (electrowetting light valve)
- EL (electroluminescent).

Prepare a brief report on one type of display technology now available or currently being researched. Your report should outline how the display works, the advantages and disadvantages of the display technology and the main uses of the type of display.

Try typing 'display technology' into a search engine. Possible websites, accessible via **www.shaplinks.co.uk**, include Cambridge Display Technology.

4.2 Questions on the whole chapter

Questions

28 (a) Plasmas have been studied using the fast discharge of banks of capacitors to deliver a burst of energy. If a bank of capacitors totalling 100 μF was to be charged to 50 kV and then discharged into the plasma in 1 ms, what mean power would be delivered?

(b) A 100 μF capacitor is charged up to just 10 V, and the energy stored is then discharged into a thermally insulated coil of wire of very low resistance. The temperature rise of the coil is recorded by a very sensitive thermometer as 0.005°C. What rise in temperature would you expect if this same capacitor was charged to 20 V and then discharged into the same coil?

29 In 1982 the first pocket-sized television went on sale. It was the Sinclair Microvision flat-screen TV from Sinclair Research Ltd (Figure 2.56). It had a cathode-ray tube measuring only 10 cm × 5 cm × 2 cm and, compared with a normal size tube of that time, was much brighter and drew far less power from the supply.

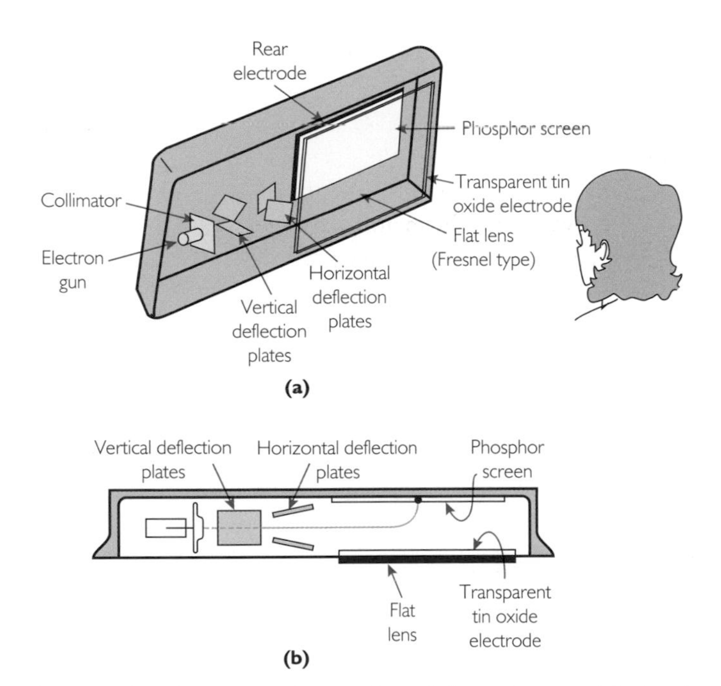

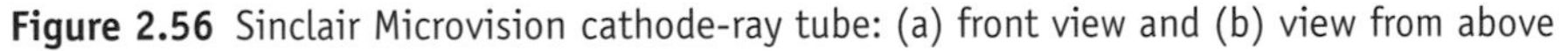

Figure 2.56 Sinclair Microvision cathode-ray tube: (a) front view and (b) view from above

The tube consisted of an electron gun, collimator (to produce a narrow electron beam), vertical and horizontal deflection plates, a phosphor screen mounted on the rear electrode, a transparent front electrode and a flat lens (a so-called Fresnel lens like those used on overhead projectors). The electron beam was fired in from the side of the screen.

The electric field between the deflection plates and another between the front tin oxide electrode and the rear electrode caused the electrons to be deflected towards the phosphor screen. The viewer looked through the flat lens at the screen at the back of the tube.

(a) (i) What must have been the direction of the electric field between the front tin oxide and rear electrodes?

(ii) The potential difference between these electrodes was 800 V and the separation between them 14 mm. What was the electric field strength between these electrodes?

(iii) Calculate the force on an electron in this electric field.

(Electron charge $e = 1.60 \times 10^{-19}$ C)

(b) The electrons were produced in a conventional electron gun by the process of thermionic emission.

(i) If fewer electrons were required to be given off each second, what would have to be done to the filament?

(ii) What effect would this have on the screen image?

(iii) If the accelerating voltage of the electron gun was increased, what effect would this now have on the image of the screen?

(c) The phosphor screen was mounted on a metal plate. What was the advantage of this over the more usual glass?

(d) This television operated from a 6 V lithium battery with a current of approximately 0.08 A.

(i) What was the power drawn from the supply?

(ii) The battery was rated at approximately 1.2 Ah (ampere-hours). For how long could you have watched this television before the battery was discharged?

(e) Why do modern pocket televisions and portable computers have LED or LCD displays rather than a CRT?

30 In 1897 Sir Joseph John Thomson (1856–1940), who believed that cathode rays consisted of a stream of charged particles, devised an experiment with which to measure their specific charge, ie the charge per unit mass. Thomson's result confirmed that the 'rays' were indeed streams of the particles we now call electrons; their specific charge is usually written as $\frac{e}{m}$.

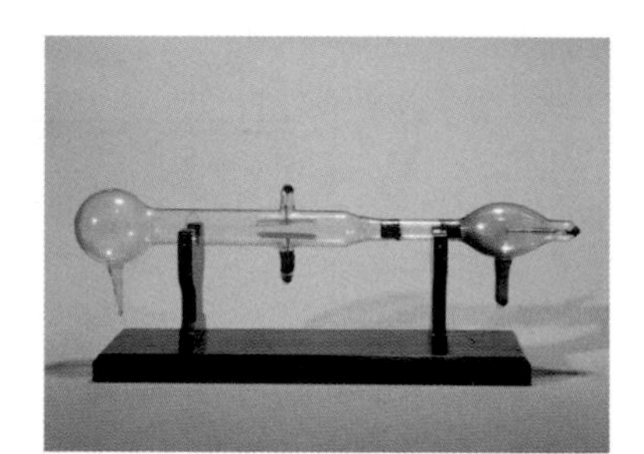

Figure 2.57 J. J. Thomson's apparatus

Thomson's apparatus is shown in Figure 2.57. Electrons were accelerated into the electric field between the parallel metal plates and deflected downwards from the initial straight-through path. A magnetic field was then applied across the tube deflecting the electron beam back upwards, so nullifying the original deflection.

(a) (i) If the separation between the plates was 2.0 cm and the potential difference across the plates 2000 V, calculate the electric field, E, produced.

(ii) If the charge on the electron is given the symbol e, write an expression for the magnitude of the force, F_{el}, on an electron in this electric field.

(b) A magnetic field of flux density 3.125×10^{-3} T is applied that brings the electron beam back up to the straight-through line. If v is the speed of these electrons, write an expression for the magnitude of the force, F_{mag}, on these electrons due to this magnetic field.

(c) (i) From your answers to (a) and (b), obtain an expression for an electron's speed, v, in terms of E and B.

(ii) Use your previous answers to obtain a numerical value for v.

(d) (i) If V is the accelerating potential difference, write an expression for the energy transferred to each electron as it is accelerated by the electron gun.

(ii) Assuming that all the energy transferred by the gun is used to increase the electron's kinetic energy, write an expression relating an electron's kinetic energy to the accelerating voltage.

(e) (i) Use your previous answers to obtain an expression for $\frac{e}{m}$ in terms of E, B and V.

(ii) The electrons in this apparatus had been accelerated through a potential difference of 3000 V. Calculate a value for $\frac{e}{m}$.

4.3 Achievements

Now you have studied this chapter you should be able to achieve the outcomes listed in Table 2.1.

Table 2.1 Achievements for the chapter *The Medium is the Message*

	Statement from examination specification	*Section(s) in this chapter*
83	draw and interpret diagrams using lines of force to describe radial and uniform electric fields qualitatively	3.2
84	explain what is meant by an electric field and recognise and use the expression electric field strength $E = \frac{F}{q}$	3.2, 3.3
86	investigate and recall that applying a potential difference to two parallel plates produces a uniform electric field in the central region between them, and recognise and use the expression $E = \frac{V}{d}$	3.2
87	investigate and use the expression $C = \frac{Q}{V}$	2.1 (and see TRA)
88	recognise and use the expression $W = \frac{1}{2}QV$ for the energy stored by a capacitor, derive the expression from the area under a graph of charge stored against potential difference, and derive and use related expressions, for example, $W = \frac{1}{2}CV^2$	2.2
91	explore and use the terms magnetic flux density B, flux Φ and flux linkage $N\Phi$	3.4 (and see TRA)
93	recognise and use the expression $F = Bqv \sin\theta$ and apply Fleming's left-hand rule to charges	3.4
98	recall that electrons are released in the process of thermionic emission and explain how they can be accelerated by electric and magnetic fields	3.1, 3.4, 3.5 (and see PRO)

Answers

1 Think of the expression $V = IR$. If the cables have any resistance (R), then any current (I) must also be associated with a potential difference (V) between the ends.

2 (i) 0 V, (ii) 6 V, (iii) 3 V

3 Over a path of 100 m, a pulse will be spread by 0.1 ns. Each pulse has a length of 1 ns, so a spread of 0.1 ns would only make a slight difference to the pulse length and so dispersion will not be a major problem.

4 (a) $I_x = \frac{I_0}{e}$ when

$$x = \frac{1}{\mu} = \frac{1}{(2.0 \times 10^{-6}\ \text{m}^{-1})} =$$

5.0×10^5 m (= 500 km).

(b) From Equation 2, $\frac{I}{I_0} = e^{-\mu x}$.

When $x = 1.0 \times 10^5$ m, $\mu x = 2.0 \times 10^{-6}\ \text{m}^{-1} \times 1.0 \times 10^5\ \text{m} = 0.20$.

$\frac{I}{I_0} = e^{-0.2} = 0.82$; ie the intensity has fallen to 0.82 of its initial value.

5 From Equation 2 or Equation 4, we have:

$$\log_e\left(\frac{I}{I_0}\right) = -\mu x$$

Putting $I = \frac{I_0}{2}$, $\log_e\left(\frac{1}{2}\right) = -\mu x$, ie $-0.693 = -\mu x$

With $x = 50$ km $= 5.0 \times 10^4$ m

$$\mu = \frac{0.693}{(5.0 \times 10^4\ \text{m})} = 1.386 \times 10^{-5}\ \text{m}^{-1}$$

6 Around 1.7 μm would be best, but other possibilities are near 1.35 μm and 1.22 μm – the other troughs on the graph.

7 (a) Number of electrons $= \frac{Q}{e}$

$$= \frac{0.16 \times 10^{-12}\ \text{C}}{1.6 \times 10^{-19}\ \text{C}}$$

$= 1 \times 10^6$ electrons

(b) From Equation 5,

$$V = \frac{Q}{C} = \frac{0.16 \times 10^{-12}\text{C}}{1 \times 10^{-12}\ \text{F}} = 0.16\ \text{V}$$

8 From Equation 5, $Q = CV = 1\ \text{F} \times 0.45\ \text{V} = 0.45\ \text{C}$

9 From Question 8 $Q = 0.45$ C

Current $I = 50\ \text{mA} = 50 \times 10^{-3}\ \text{C s}^{-1}$

$$\text{Time to charge} = \frac{0.45\ \text{C}}{50 \times 10^{-3}\ \text{C s}^{-1}} = 9\ \text{s}$$

10 The most convenient expression is Equation 8: $W = \frac{1}{2}CV^2$ Substituting values:

(a) $W = \frac{1}{2} \times 10^4 \times 10^{-6}\ \text{F} \times (3\text{V})^2 = 4.5 \times 10^{-2}\ \text{J}$

For parts (b) and (c) you can multiply the answer to (a) by 4 and then by 9, since W is proportional to V^2. This gives:

(b) $W = 4 \times 4.5 \times 10^{-2}\ \text{J} = 0.18\ \text{J}$

(c) $W = 9 \times 4.5 \times 10^{-2}\ \text{J} = 0.405\ \text{J}$

11 From Equation 11: $\lambda = \frac{hc}{\Delta E}$

indium/antimony:

$$\lambda = \frac{6.63 \times 10^{-34}\ \text{J s} \times 3.00 \times 10^8\ \text{m s}^{-1}}{0.18 \times 1.60 \times 10^{-19}\ \text{J}}$$

$= 6.91 \times 10^{-6}$ m

gallium/nitrogen:

$$\lambda = \frac{6.63 \times 10^{-34}\ \text{J s} \times 3.00 \times 10^8\ \text{m s}^{-1}}{3.4 \times 1.60 \times 10^{-19}\ \text{J}}$$

$= 3.66 \times 10^{-7}$ m

12 Using Equation 11: $\Delta E = \frac{hc}{\lambda}$

$\lambda = 400$ nm

$$\Delta E = \frac{6.63 \times 10^{-34}\ \text{J s} \times 3.00 \times 10^8\ \text{m s}^{-1}}{400 \times 10^{-9}\ \text{m}}$$

$= 5.0 \times 10^{-19}$ J

$$= \frac{5.0 \times 10^{-19}}{1.6 \times 10^{-19}}\ \text{eV} = 3.1\ \text{eV}$$

Similarly for $\lambda = 600$ nm,

$\Delta E = 3.3 \times 10^{-19}\ \text{J} = 2.1\ \text{eV}$

13 (a) $P = VI = 2.15\ \text{V} \times 15 \times 10^{-3}\ \text{A} = 3.23 \times 10^{-2}\ \text{W}$

(b) $P = VI = 1.3\ \text{V} \times 100 \times 10^{-3}\ \text{A} = 0.13\ \text{W}$

(c) Taking the answer to (a) to be typical of visible LEDs, 10^6 of them would have a rating of $3 \times 10^{-2} \times 10^6\ \text{W} = 3 \times 10^4\ \text{W} = 30$ kW.

14 $1\ \text{V} = 1\ \text{J C}^{-1}$, and $1\ \text{J} = 1\ \text{N m}$

so $1\ \text{V m}^{-1} = 1\ \text{J C}^{-1}\ \text{m}^{-1} = 1\ \text{N m C}^{-1}\ \text{m}^{-1} = 1\ \text{N C}^{-1}$

Maths reference

Derived units
See Maths note 2.3

15 (a) From Equation 16,

$$E = \frac{1 \times 10^3\ \text{V}}{0.1\ \text{m}}$$
$$= 1 \times 10^4\ \text{V m}^{-1}$$
$$= 1 \times 10^4\ \text{N C}^{-1}$$

(b) From Equation 12,

$$F = qE = 3 \times 10^{-6}\ \text{C} \times 1 \times 10^4\ \text{N C}^{-1}$$
$$= 3 \times 10^{-2}\ \text{N}$$

16 $P = IV = 140 \times 10^{-12}\ \text{A} \times 2.5\ \text{V} = 3.5 \times 10^{-10}\ \text{W}$ (= 0.35 nW)

17 (a) From Equation 16,

$$E = \frac{200\ \text{V}}{4.0 \times 10^{-2}\ \text{m}} = 5.0 \times 10^3\ \text{V m}^{-1}$$

(b) (i) $E = \dfrac{400\ \text{V}}{4.0 \times 10^{-2}\ \text{m}} = 1.0 \times 10^4\ \text{V m}^{-1}$

(ii) $E = \dfrac{200\ \text{V}}{8.0 \times 10^{-2}\ \text{m}} = 2.5 \times 10^3\ \text{V m}^{-1}$

(iii) $E = \dfrac{100\ \text{V}}{2.0 \times 10^{-2}\ \text{m}} = 5.0 \times 10^3\ \text{V m}^{-1}$

The field lines should be drawn as in Figure 2.58 so that their spacing indicates the field strength.

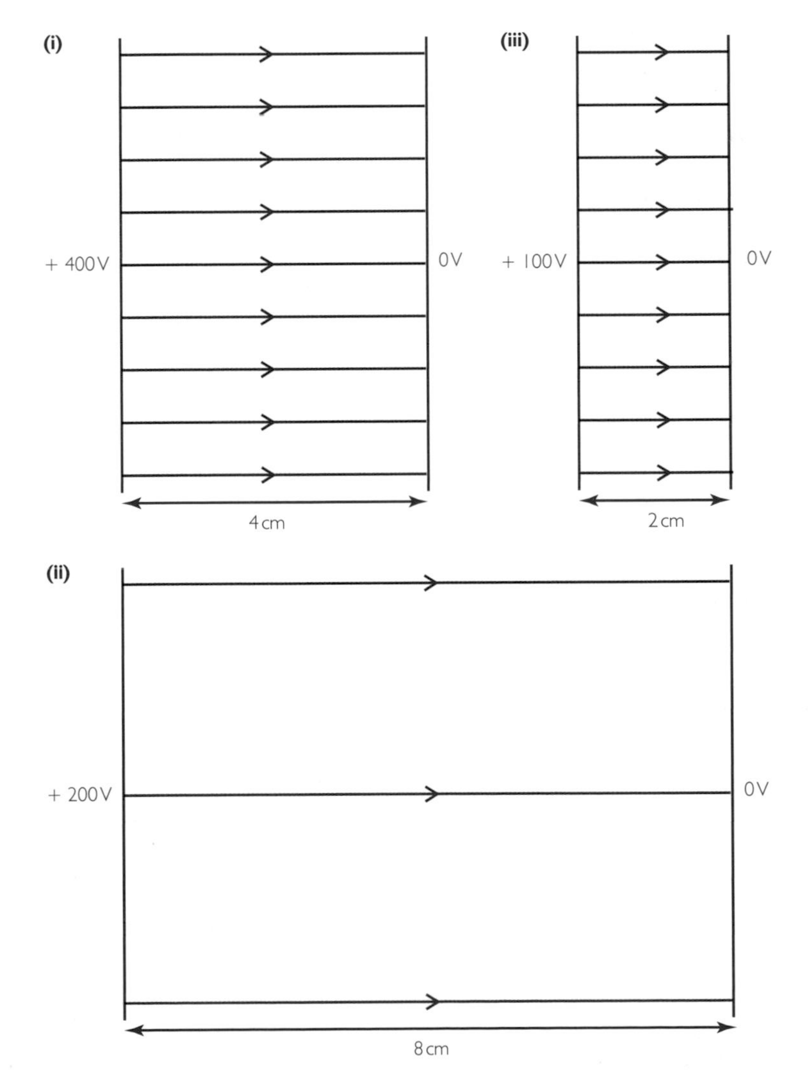

Figure 2.58 Answers to Question 17(b)

18 $E = \frac{V}{d} = \frac{4.9\ \text{V}}{8.0 \times 10^{-3}\ \text{m}} = 610\ \text{V m}^{-1}$

19 (a) Using Equation 13, $\Delta W = q\Delta V$,

energy transferred to electron

$= 1.60 \times 10^{-19}\ \text{C} \times 5000\ \text{V} = 8.0 \times 10^{-16}\ \text{J}$

(b) (i) Assuming that the kinetic energy gained by the electron is equal to the energy transferred by the electric field in the gun, $E_k = 8.0 \times 10^{-16}$ J.

(ii) $E_k = \frac{1}{2}mv^2$, and so

$$v = \sqrt{\left(\frac{2E_k}{m}\right)}$$

$$= \sqrt{\left(\frac{2 \times 8.0 \times 10^{-16}\ \text{J}}{9.11 \times 10^{-31}\ \text{kg}}\right)}$$

$$= 4.19 \times 10^{7}\ \text{m s}^{-1}$$

(c) Electrons would collide with air molecules, transferring energy to them, so the electrons' kinetic energy would be greatly reduced. (They would also be scattered and the beam would not remain focused.)

20 (a) The spot would be brighter as the kinetic energy of each electron would be increased, so each electron would transfer more energy to the phosphor screen.

(b) The spot would be brighter as more electrons would be produced by thermionic emission, more would be accelerated, and so there would be a greater overall rate of energy transfer to the phosphor.

(c) The spot would be less bright as the electrons would be decelerated to some extent by the grid.

21 From Equation 21 with $\theta = 90°$, $\sin\theta = 1$,

$F = 0.10\ \text{T} \times 1.60 \times 10^{-19}\ \text{C} \times 8.0 \times 10^{7}\ \text{m s}^{-1}$
$= 1.3 \times 10^{-12}\ \text{N}$

22 From Fleming's left-hand rule, the field must be horizontal and directed across the tube.

23 They would take a lot of power that might otherwise be used for manoeuvring. Also, the screens become heated by the bombardment of high-energy electrons, so might provide an enhanced target for heat-seeking missiles.

24 From Equation 16,

$$E = \frac{200\ \text{V}}{150 \times 10^{-6}\ \text{m}} = 1.3 \times 10^{6}\ \text{V m}^{-1}$$

25 From Equation 11,

$$E = \frac{hc}{\lambda} = \frac{6.63 \times 10^{-34}\ \text{J s} \times 3.00 \times 10^{8}\ \text{m s}^{-1}}{147 \times 10^{-9}\ \text{m}}$$

$= 1.35 \times 10^{-18}\ \text{J}$

26 See Figure 2.59. The gas will be ionised near the top of the cell where the field strength is high, shown by the field lines being close together.

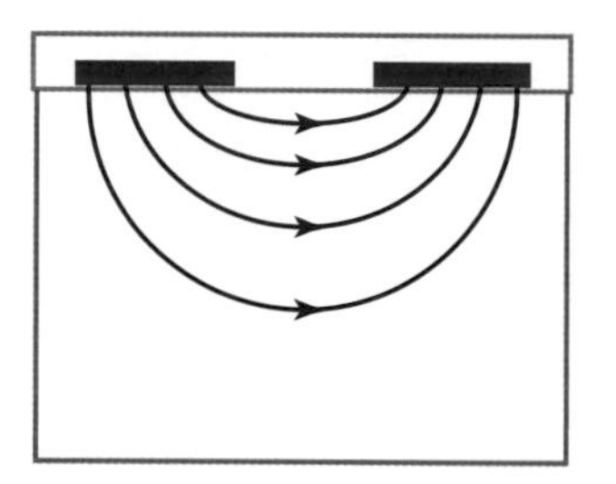

Figure 2.59 Answer to Question 26

27 (a) From Equation 16,

$$E = \frac{1.5\ \text{V}}{10 \times 10^{-6}\ \text{m}}$$

$= 1.5 \times 10^{5}\ \text{V m}^{-1} = 1.5 \times 10^{5}\ \text{N C}^{-1}$

(b) From Equation 12, $F = eE$

$= 1.60 \times 10^{-19}\ \text{C} \times 1.5 \times 10^{5}\ \text{N C}^{-1}$

$= 2.4 \times 10^{-14}\ \text{N}$

CHAPTER 3

Probing the Heart of Matter

Why a chapter called *Probing the Heart of Matter*?

Science has followed a long quest to probe the structure of matter in an attempt to discover what we are all made of. In recent years this story has taken a new twist. Two different strands of physics have come together to provide a picture of how the fundamental structure of matter and the universe itself are linked.

In the 1970s, theoreticians working on the physics of fundamental particles developed theories that were far ahead of the experimental physicists' means to test them out, so they started searching for new ways of seeing if their theories made sense. At the same time, physicists working on the relatively new subject of cosmology (the study of how the Universe came into being) discovered that they needed to understand particle physics in order to chart the very early history of the Universe.

The particle physicists realised that one way to test their theories was to see if they made sensible predictions about the way in which the early universe evolved. This helped the cosmologists to solve some of their most difficult problems regarding the development of the universe after the Big Bang. As a consequence, physicists think that they have a fairly complete theory telling them how the universe evolved from 10^{-35} seconds after the Big Bang right up to the present day. After 10^{-12} s, theory is supported by experimental evidence.

You may have read about the latest developments in this field in newspapers and science magazines, or you may have seen television documentaries on the subject. For example, in February 2000 scientists at CERN (the European particle-physics lab) announced that they had created a type of matter called quark-gluon plasma (Figure 3.1) thought to have existed in the very early universe. Particle physics is a subject that is alive with new possibilities and many of the most able young physicists starting their research careers choose to work in this exciting field. In this chapter you will study some of the ideas that underpin this theory. You will also see where some of the problems lie and the work that is being done by experimenters and theoreticians to try to produce solutions.

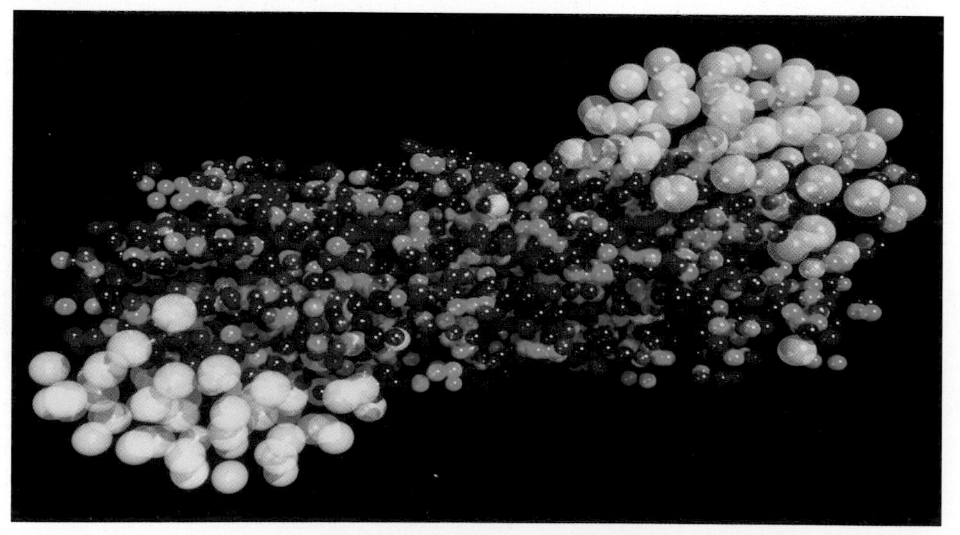

Figure 3.1 (a) Particle tracks produced from a quark-gluon plasma

In doing so, you will be introduced to some aspects of modern particle physics and will also study the physics of circular motion and particle acceleration that underlies many of the experiments.

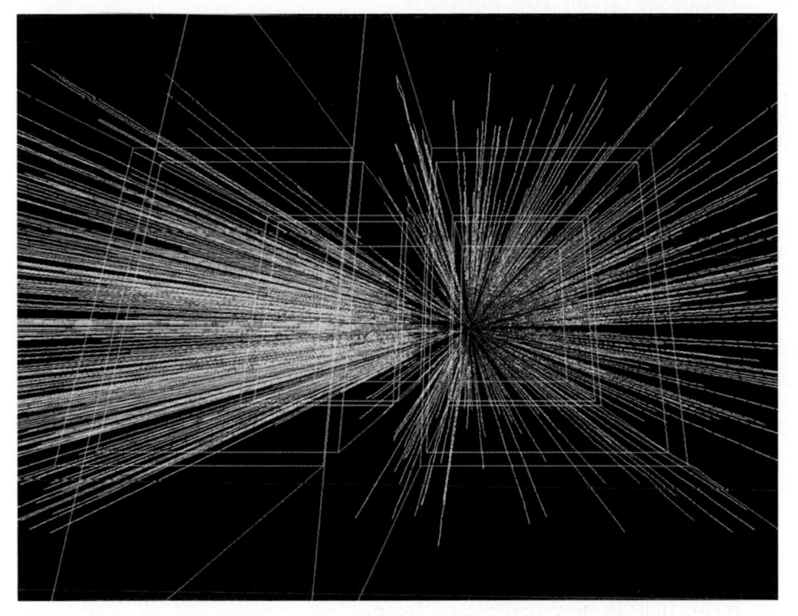

Figure 3.1 (b) Computer-generated model of a quark-gluon plasma

Overview of physics principles and techniques

You will begin by looking at the Big Bang theory of the creation of the universe and how it manages to take us from a short time after the creation event right up to the present day. Along the way, some of the physics of fundamental particles will be introduced together with some key ideas about conservation laws and forces. Next you will look at the use of beams of particles to probe matter, which will involve learning about electrostatic and magnetic forces and about circular motion. In the final part of the chapter you will see some of the most important results that these devices have produced. In the course of the chapter, you will also carry out several activities that involve communication and IT skills: researching information, and discussing and communicating what you have found.

In this chapter you will extend your knowledge of:

- momentum from *Transport on Track*
- kinetic energy and work from *Higher, Faster, Stronger* and *Transport on Track*
- wave–particle duality from *The Sound of Music, Technology in Space* and *Digging Up the Past*
- electric and magnetic fields from *Transport on Track* and *The Medium is the Message*
- vectors from *Higher, Faster, Stronger* and *Transport on Track*
- using graphs from *Higher, Faster, Stronger, Digging Up the Past* and *The Medium is the Message*.

Elsewhere you will do more work on:

- forces and motion in *Build or Bust?*
- motion in a circle in *Reach for the Stars*
- kinetic energy and work in *Reach for the Stars*
- inverse-square law fields in *Reach for the Stars*
- nuclear physcics in *Reach for the Stars*.

1 In the beginning

1.1 Very large and very small

You will have read in the Introduction about the **Big Bang theory** of the origin of the Universe. The idea is that at some time in the past, about 13 700 million years ago, the Universe came into being. All of the matter and energy in the observable Universe, which is now spread out to the furthest distance we can see (about 10^{20} km), was compressed into a very small volume. From that tiny, super-dense state at the instant of creation, the Universe grew. As it expanded, it cooled, and matter clumped together to form galaxies, stars and planets. The evidence for this model of the history of the Universe comes from observations of the light from distant galaxies.

Activity 1 The expanding universe

What ideas do you already have, from GCSE science and from general reading, about the history of the universe? Share your ideas with a partner, and then with the rest of your class. Here are some phrases that may help you to recall what you have studied or read about previously:

- galaxies moving apart
- red shifts
- distant galaxies – looking into the past
- the age of the Universe.

You know that we think of matter as being made up of atoms – the building blocks of the chemical elements (see Figure 3.2). But atoms are not the smallest, most basic particles of matter, and not all matter is made up of atoms. For example, there is a beam of electrons inside old-style television sets (when they are producing a picture); electrons are matter, and they are much smaller than atoms.

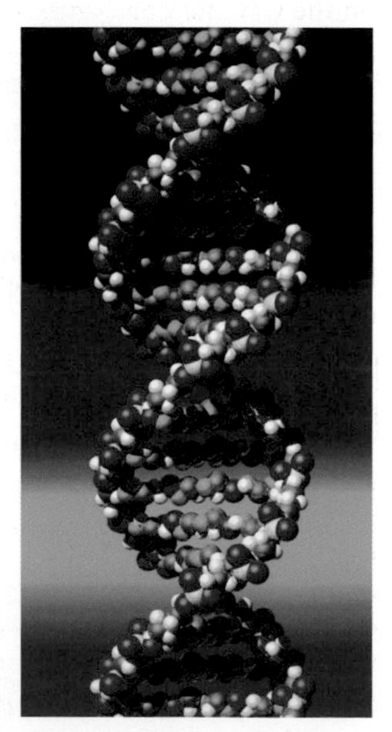

Figure 3.2 Model of a DNA molecule showing the atoms from which it is made

Activity 2 The particles of matter

In the next part of this chapter, you will learn about the current picture which physicists have developed of the sub-atomic particles of matter. What ideas do you already have about particles? Share your ideas with a partner, and then with the rest of the class. Here are some phrases which may help you to recall what you have studied or read about previously:

- atoms, nuclei, protons, neutrons and electrons
- quarks
- particle accelerators – 'atom smashers'
- radioactivity, fission and fusion.

2 Theories of everything

One of the things that scientists like to do is to develop theories. They want their theories to account for as many different observations as possible. For example, Newton developed a theory of gravity. He managed to come up with a theory that explained how things fall on the Earth, and how the Moon orbits the Earth. Until then, people felt that there must be one set of laws for what happens on Earth, and another for what happens in the heavens. So Newton's theory was a dramatic advance, unifying two previously separate realms of ideas.

Today, we know much more about the laws of physics, especially about the particles of which matter is made and the forces that act between them. Physicists would like to come up with a single theory that could explain everything we know about these things, and they call this a 'theory of everything'. It's perhaps a rather arrogant idea. Such a theory won't explain everything such as evolution, human consciousness, poetry or love. But such a theory would still be a remarkable achievement, and in this part of the chapter you will learn something about how physicists have made progress towards this end during the last few decades.

2.1 Ideas in cosmology

Late on 6 December 1979, Alan Guth (Figure 3.3), then a researcher at SLAC (the Stanford Linear Accelerator Centre in California), sat down to start some work. He was accustomed to working late at night, once his young son had gone to bed, as he could concentrate better. He had been working on a paper about the early universe with some colleagues and had decided to check to see if their ideas had any influence on the way the universe expanded at about 10^{-35} seconds after the Big Bang. Guth had been doing such calculations for some time and the work that night was quite routine. The results he discovered were not routine, however. They were to change thinking about cosmology completely.

Figure 3.3 Alan Guth

Not yet realising the full implications of what he had done, Guth cycled into the lab the next morning. In his book *The Inflationary Universe*, Guth records that he set a personal best time of nine minutes, 32 seconds on that ride. Arriving at this office he opened his notebook, and at the top of one of the pages wrote:

Spectacular realization

and then set out to refine the rough work that he had done the night before. Guth records that he cannot find any other instances in his notes where he has used a double box to emphasise a point.

At the time Guth was coming to the end of his period of employment at SLAC and was looking to find another job. He decided to take the risk of not immediately setting down his new ideas in a published paper. Instead he travelled round several research centres in America presenting his ideas in a seminar and seeing if he could impress people enough to offer him a job. The risk paid off. At one conference he was given 10 minutes to explain his theory to the attending physicists. Part way through his presentation Murray Gell-Mann, a Nobel Prize winner and important character in this chapter, jumped up and shouted 'You've solved the most important problem in cosmology!' On the Monday after, Guth took the plunge and rang MIT (the Massachusetts Institute of Technology) asking if they were interested in offering him a job. Twenty-four hours later he was offered an assistant professorship.

What Guth had done was to apply some of the very latest ideas in particle physics to the universe shortly after the Big Bang. Particle physicists had started working on **Grand Unified Theories** (GUTs), which brought the fundamental forces that act between fundamental particles together into one theory. Unfortunately, these theories cannot be directly checked by experiments. The effects that they predict can only be seen if the particles involved have very high energies – much higher than could be produced in experiments. However, particles with high enough energies would have existed early in the history of the universe when all matter and energy were compressed into a tiny space, so GUTs should have something to say about what was going on at the time. What Guth's calculations told him was that, instead of expanding at a relatively gradual rate, as everyone had assumed, the universe must have exploded in size at a gigantic rate for a short period of time. So fast would be this expansion that in a time interval of 10^{-35} seconds, the universe would become 10^{50} times bigger. This is like magnifying a proton to the current size of the visible universe in less time than it takes for a light ray to travel from one side of a proton to another. Guth named this effect **inflation**, and it has become a central part of our thinking about the early history of the universe.

2.2 Building blocks

You probably picture an atom as being like a miniature solar system, with a nucleus made up of protons and neutrons, surrounded by orbiting electrons. This is the picture that scientists developed in the early decades of the 20th century, and for many purposes it is still a very useful model. Electrons, protons and neutrons were at first thought to be **fundamental particles**, ie particles that cannot be divided into smaller constituents. However, it was inevitable that physicists would question whether, just as the atom was found to be divisible, these particles too could be subdivided.

Later in this chapter, you will learn something of how this question was investigated. In the meantime, we will look at some of the results of these investigations.

The **standard model of particle physics** (Figure 3.4), which is currently our best theory of how the universe works, identifies 12 fundamental particles from which all matter is made and four fundamental forces that govern reactions between the particles. The 12 fundamental particles divide into two distinct groups according to their properties. There are:

- six types of **quark**, the particles of which protons and neutrons (and some other particles) are made; the name *quark* was coined by Murray Gell-Mann
- six types of **lepton**, including the electron; the name *lepton* comes from Greek and means 'light'; leptons have very small masses.

Although electrons are thought to be truly fundamental, physicists have shown that protons and neutrons are not – each is made up of three quarks.

The particles are generally grouped into three **generations** – so-called because there is some 'family' resemblance between particles of one generation and another (not because one group precedes another). Table 3.1 lists the names (which are often slightly weird), symbols and some properties of the quarks. All three quarks in a horizontal row have the same electrical charge, shown in units of the proton charge e. There are also three generations of lepton (Table 3.2). In each generation, one of the leptons is a neutrino. Many of the names in Tables 3.1 and 3.2 will be new to you. Probably the only familiar one is the electron. As this chapter progresses you will start to see how the others fit into our description of the universe. Do not get worried if you feel that you are not entirely sure what these particles are. This was the problem that faced physicists as they started to discover them!

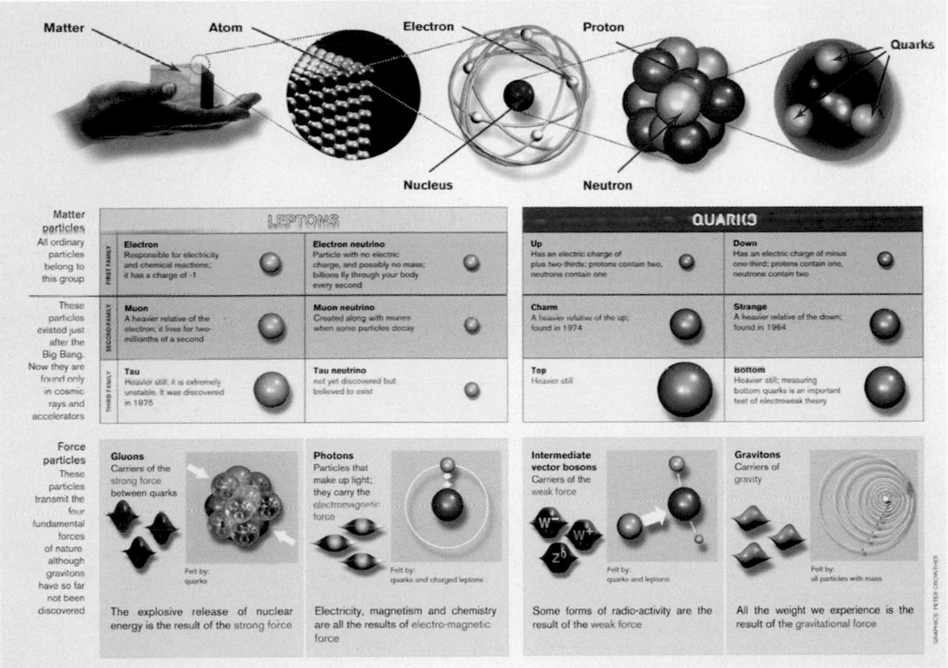

Figure 3.4 Constituents of matter

1st generation	2nd generation	3rd generation	Charge
up, u	charm, c	top, t	$-\frac{2}{3}e$
down, d	strange, s	bottom, b	$+\frac{1}{3}e$

Table 3.1 Quarks

1st generation	2nd generation	3rd generation	Charge
electron, e	muon, μ	tau, τ	$-1\ e$
electron-neutrino, ν_e	muon-neutrino, ν_μ	tau-neutrino, ν_τ	0

Table 3.2 Leptons

Table 3.3 shows where leptons are found in nature. We cannot provide a similar table for quarks, because it is believed that quarks always occur bound together in twos or threes, never separately.

Lepton	Where it is found
Electron	found in atoms important in electrical currents produced in beta radioactivity
Muon	produced in large numbers in the upper atmosphere by 'cosmic rays'
Tau	so far only seen in lab experiments
Electron-neutrino	produced in beta radioactivity produced in large numbers by atomic reactors produced in huge numbers by the nuclear reactions in the Sun
Muon-neutrino	produced by atomic reactors produced in upper atmosphere by cosmic rays produced in the Sun by nuclear reactions
Tau-neutrino	so far only seen in lab experiments

Table 3.3 Leptons in nature

It is worth mentioning something about the masses of the fundamental particles.

- The top quark is the heaviest (most massive) of all fundamental particles. Its mass is almost 200 times the mass of a proton. That makes it much heavier than many molecules, such as a water molecule, H_2O.
- An electron has a mass roughly 1/2000 times that of a proton.
- For many years it was suggested that electron-neutrinos (which appear during beta decay of a radioactive material) have absolutely no mass at all. However recent experiments have shown that this isn't the case and that the neutrinos have a very tiny mass, less than 10^{-9} of the mass of a proton.

Activity 3 Charge and no charge

A proton is made up of three quarks: two up quarks and a down (written uud). A neutron is also made of three quarks, a different combination of up and down quarks.

- Show that the charge of a proton is correctly given by adding the charges of two up quarks and one down quark.
- What combination of up and down quarks make up a neutron?

A proton is made of two up quarks and a down (uud); a neutron is an up and two downs (udd). All other possible collections of three quarks also correspond to particles, although all of them apart from the proton (which has the lowest mass) are unstable. This means that they will decay into other (less massive) particles. That is why they are unfamiliar to you. Particles such as the $\Delta+$ (pronounced 'delta plus' and made from three up quarks, uuu) and the $\Delta-$ ('delta minus', ddd) typically last about 10^{-25} seconds before they decay so they are not seen in the everyday world. We only understand their properties to the extent that we do because we have been able to make and study them in experiments. However, in the early moments after the Big Bang such particles were much more common than they are now, and so our attempts to discover their properties also help us to understand the early universe.

Symmetry and prediction

You will no doubt have noticed the similarities between the patterns of the lepton and quark families. Each has six members in three generations. The particles' masses increase as you move up the generations. The model is said to have **symmetry**. When this model was first proposed by Murray Gell-Mann (Figure 3.5), there was no experimental evidence for the top and bottom quarks. He suggested that they must exist, because of this symmetry. The bottom quark was identified in 1977, a great vindication of his theory. The top quark was eventually found in 1994, in an experiment at Fermilab in the USA.

Figure 3.5 Murray Gell-Mann

Activity 4 Top quark

Research the story of how the bottom and top quarks were predicted and discovered. Identify the key events in the story and produce an annotated timeline showing what happened.

2.3 Fundamental forces and interactions

Physicists have identified four **fundamental forces** that operate in nature. Two of these, gravity and the electromagnetic force, will be familiar to you from your studies in physics so far. By a fundamental force we mean a force that cannot be explained in terms of another force acting. For example, the force of friction is not a fundamental force as it is caused by the electrostatic forces between atoms as one object rubs against another. The four fundamental forces are shown in Table 3.4.

Force	Range	Relative strength	Acts between ...
Gravity	no limit	10^{-34}	all objects
Electromagnetic	no limit	10	charged objects
Strong force	10^{-15} m	10^{3}	quarks
Weak force	10^{-18} m	10^{-10}	fundamental particles

Table 3.4 The four fundamental forces of nature

Range and strength

The **range** of a force is the maximum distance by which two objects can be separated and still feel the force acting. The **relative strength** of the force is a way of comparing how big an influence each force would have on the same pair of objects. The relative strengths quoted correspond roughly to the force in newtons between two protons separated by 10^{-15} m.

The so-called strong and weak forces have not come into your study of physics before as they only act over very short distances – smaller than the diameter of a nucleus. For this reason physicists did not discover them until they started to experiment with nuclei.

As this chapter progresses you will start to get a feel for the properties of the strong and weak forces. For the moment, just think of them as new forces that we do not see in our everyday life as their size is so small at the sort of distances we deal with (though as you can see from Table 3.4, over small enough distances even the weak force is not as weak as gravity). You would not worry about the force of gravity pulling you towards the Moon as you cross the road because it is so small; likewise when we hit a tennis ball the strong and weak forces do not influence its motion.

Study note

The electromagnetic force combines both magnetic and electrostatic effects; if you have studied the chapter *Transport on Track*, you will have seen that these effects are closely interconnected.

Activity 5 Forces – fundamental or not?

Make a list of the forces you have come across in physics so far – Figure 3.6 illustrates several examples. Divide them into two categories – contact forces, which require two objects to be touching each other for the force to exist (such as friction), and force fields, which do not require contact (such as gravity). Try to explain each of the forces you have listed in terms of a fundamental force from Table 3.4. Which category of force do you think is the more fundamental – force fields or contact forces?

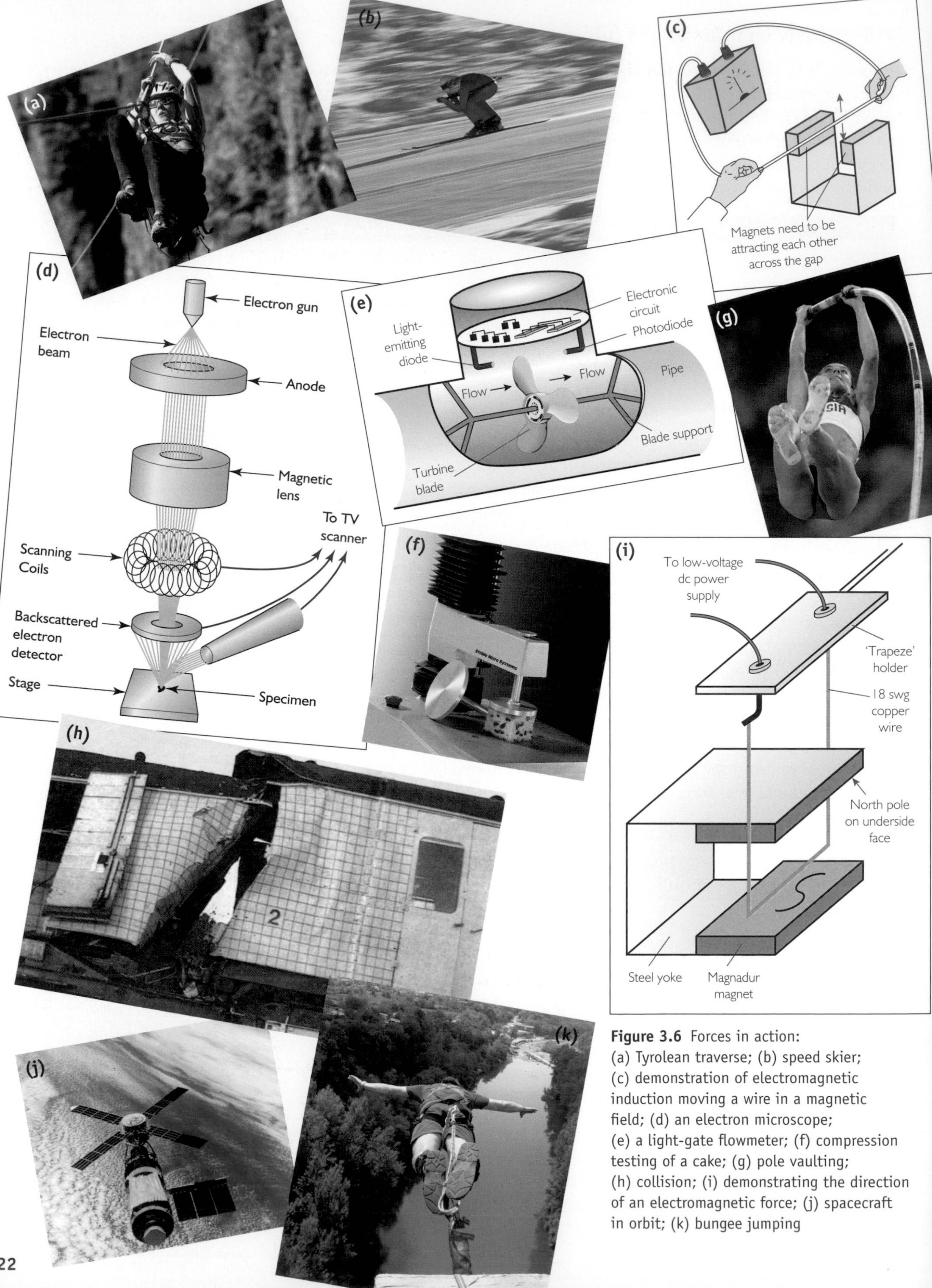

Figure 3.6 Forces in action:
(a) Tyrolean traverse; (b) speed skier; (c) demonstration of electromagnetic induction moving a wire in a magnetic field; (d) an electron microscope; (e) a light-gate flowmeter; (f) compression testing of a cake; (g) pole vaulting; (h) collision; (i) demonstrating the direction of an electromagnetic force; (j) spacecraft in orbit; (k) bungee jumping

Binding particles together

Not all of the fundamental forces act on each of the fundamental particles. For example, the main difference between quarks and leptons is that quarks experience the strong force, but leptons do not. As you can see from Table 3.4, between two quarks the strong force would completely dominate over the weak force. In fact the strong force is so strong that it pulls quarks together into groups from which they can never be separated. These groups are essentially particles in their own right, called **hadrons** (see Figure 3.7). As far as we can tell, quarks are never found on their own. They are always bound together into hadrons.

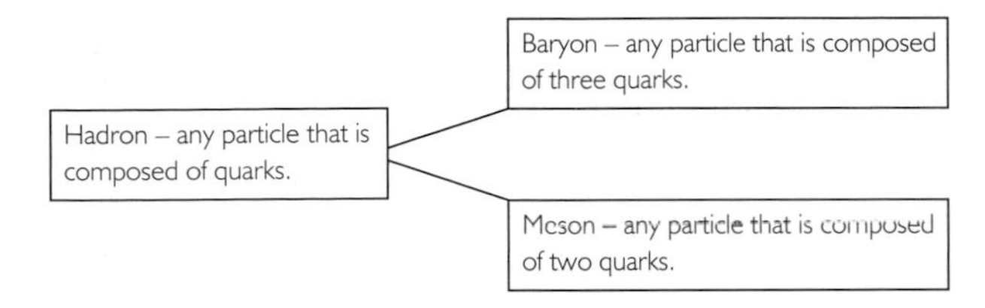

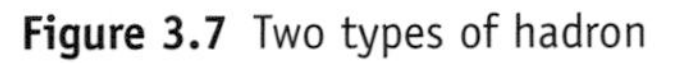

Figure 3.7 Two types of hadron

The strong force allows only two types of quark combination to exist. One type, the **baryon**, is always composed of three quarks. The most familiar examples of such a combination are the particles found in the nucleus – the proton and the neutron. The second type of hadron is the **meson**. Mesons are made of just two quarks bound together (actually a quark and an anti-quark – see Section 2.4).

Question

1 In each of (a) to (e), say which of the fundamental forces is or are involved:

(a) a quark and an anti-quark are held together to make a meson

(b) the Earth is held in its orbit around the Sun

(c) a compass needle points north

(d) protons and neutrons are close together in an atomic nucleus

(e) an electron-neutrino is absorbed by an atomic nucleus.

Changing the nature of particles

The action of the weak force is to cause particles to change from one type into another. For example, if an electron e^- happens to fly near to a muon-neutrino ν_μ then the weak force acting between them can turn the particles into a muon μ^- and an electro-neutrino ν_e. We would write this down as a **reaction equation** in the following way:

$$e^- + \nu_\mu \rightarrow \mu^- + \nu_e$$

The charges of the electron and muon are shown by the superscripts, which give charge in multiples of the proton charge. By looking at the charges of the particles involved, you can see that, in this reaction, charge is conserved:

$$-1e + 0 \rightarrow -1e + 0$$

The total charge on each side of the reaction equation is always the same. Charge is *always* conserved.

Activity 6 Equations and conservation

Write a reaction equation to represent the following event:

- a muon reacts with an electron-neutrino to form an electron and a muon-neutrino.

Show that change is conserved in this reaction.

2.4 Anti-matter

Physicists have known about the existence of **anti-matter** since the **positron** (the anti-matter version of the electron – see below) was discovered in 1932. Since then anti-matter versions of all the fundamental particles have been discovered. Anti-matter has become standard stuff used in experiments.

Study note

The $\overline{\mu}$ and $\overline{\tau}$ are also written μ^+ and τ^+, respectively.

The first thing to understand about anti-matter is that it is not very special stuff. Despite what you may have read in science fiction, it does not have anti-gravity properties or anything weird like that. For every type of particle of matter that exists, there is a type of particle of anti-matter with the same mass but with the opposite electrical charge (if it is electrically charged). Thus as well as there being electrons in the universe, there are also objects called anti-electrons, also known as positrons. These have the same mass as electrons, but they have a positive charge identical in size to that of a proton. For every quark that we mentioned earlier there is an anti-quark; for every lepton, there is an anti-lepton. These are shown in Tables 3.5 and 3.6. Each anti-particle is represented by the same symbol as its corresponding particle, with a bar above it – except for the positron, which has its own symbol, e^+.

Figure 3.8 CERN experiment that produced the first atoms of anti-hydrogen

There is nothing in the laws of physics to suggest that a position in orbit around an anti-proton will not be as stable as a familiar hydrogen atom (an electron in orbit around a proton). The problem is what happens when this hydrogen anti-atom meets a hydrogen atom – the two annihilate one another, leaving only a flash of gamma rays. Physicists believe that matter and anti-matter behave in exactly the same manner in most circumstances. But, as always, physicists like to check such things just in case they happen to be wrong. A few atoms of anti-hydrogen were first produced in an experiment called PS210 at CERN in 1995 (Figure 3.8). More recent experiments have produced tens of thousands of anti-hydrogen atoms. PS210 was designed to produce and detect such anti-atoms as a first step to confirming their properties.

Activity 7 Making anti-atoms

Read about the PS210 experiment and write a summary of 80–100 words, explaining what was done and why it was significant.

1st generation	2nd generation	3rd generation	Charge
$\bar{u}$	$\bar{c}$	$\bar{t}$	$-\frac{2}{3}e$
$\bar{d}$	$\bar{s}$	$\bar{b}$	$+\frac{1}{3}e$

Table 3.5 Anti-quarks

1st generation	2nd generation	3rd generation	Charge
$\bar{e}$ or e^+	$\bar{\mu}$	$\bar{\tau}$	+1 e
electron-anti-neutrino $\nu_{\bar{e}}$	muon-anti-neutrino $\bar{\nu}_\mu$	tau-anti-neutrino $\bar{\nu}_\tau$	0

Table 3.6 Anti-leptons

Mesons

In Section 2.3, we discussed the action of the strong force on quarks and said that it is so strong that it binds them together into particles. One such grouping combines three quarks together (such as in the proton and neutron) to form a baryon. We also mentioned another possible combination; a quark can bind to an anti-quark to make a meson. An example of such a quark–anti-quark pairing is a u quark with a $\bar{d}$ anti-quark, $u\bar{d}$. Such an object would have a charge equal to the sum of the two quark charges:

$$\text{u charge} = +\frac{2}{3}e$$

$$\bar{\text{d}}\text{ charge} = +\frac{1}{3}e$$

$$\text{total charge} = +1e$$

Any pairing is possible. Table 3.7 shows the possibilities if we just take the lightest quarks, the up and the down.

Quarks	Total charge	Name of bound particle	Symbol
$u\bar{u}$	0	pi zero	π^0
$d\bar{d}$	0	pi zero	π^0
$u\bar{d}$	+1e	pi plus	$\pi+$
$\bar{d}u$	−1e	pi minus	$\pi-$

Table 3.7 Combining up and down quarks to make pions

All these combinations correspond to one of a family of three particles called pi-mesons or **pions** (notice that there are two ways to make a pi-zero meson). Pions are the lightest of the meson family of particles and are very commonly produced in particle-physics experiments. They are also produced high in the Earth's atmosphere when cosmic rays (high-energy protons) from the Sun and beyond the solar system hit the nuclei of atoms in the atmosphere.

Figure 3.9 shows a schematic diagram of a typical shower of particles produced by a single cosmic ray particle, together with some of the detectors and observations used to study the effects of cosmic rays. N is used to represent a light nucleus. Notice that the shower contains many muons, neutrinos and positrons as well as pions.

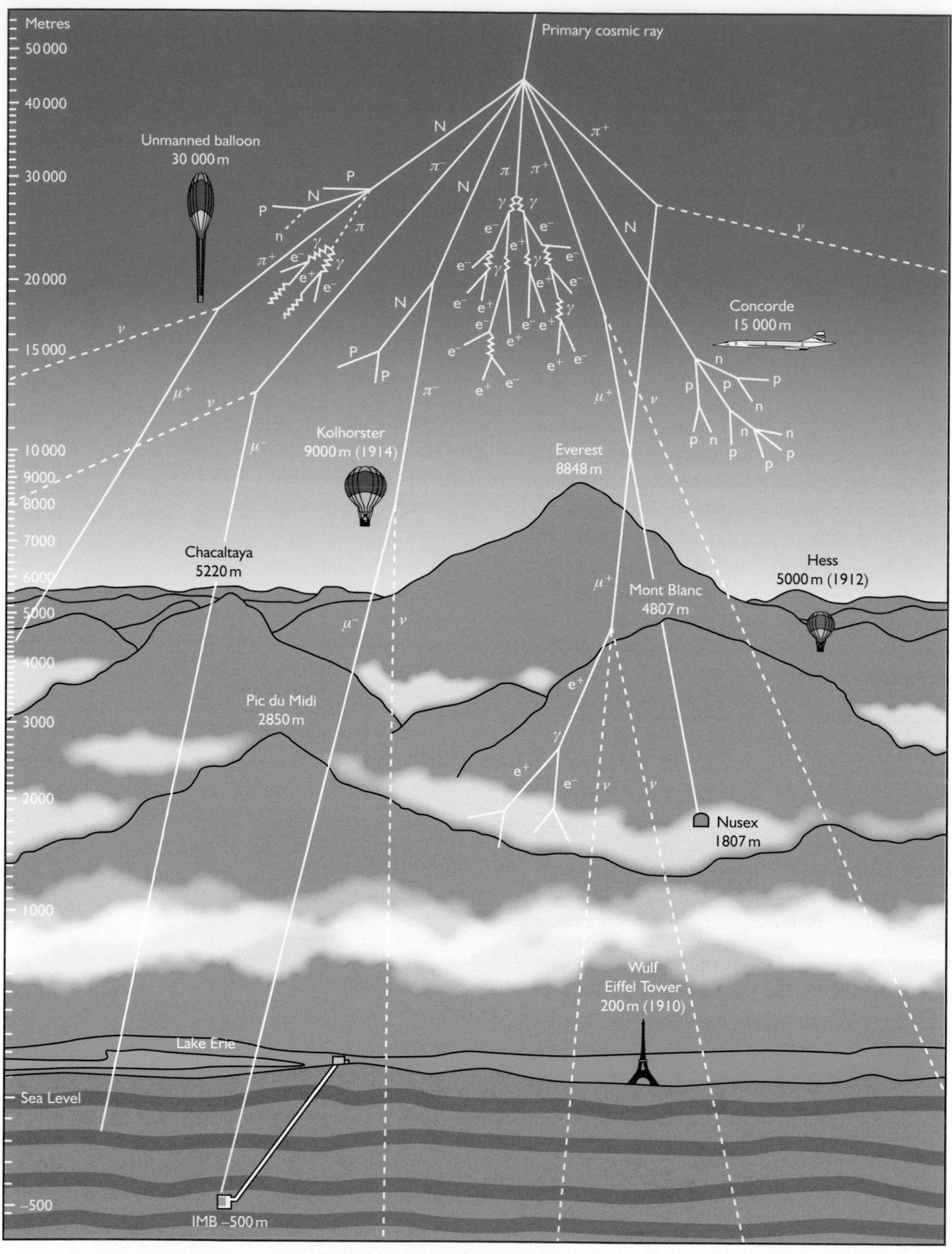

Figure 3.9 Shower of particles produced by cosmic radiation

Activity 8 Making mesons

All mesons are made up of a quark and an anti-quark.

- A phi meson (ϕ) is made of a strange quark and a strange anti-quark. What will be its charge?
- Four types of K meson are possible: $u\bar{s}$, $\bar{u}s$, $d\bar{s}$, $\bar{d}s$. Which of these is/are neutral?

Playing with fire

It is very difficult to experiment with anti-matter. If an anti-matter particle comes into contact with a matter particle of the same type (eg an electron and a positron or a proton and an anti-proton), then they will react with each other and turn into electromagnetic radiation (usually gamma-ray photons). This is called an **annihilation reaction**. For instance, an up quark and its anti-particle may come together to form a meson but this is very unstable and decays very rapidly; a π^0 meson decays in about 10^{-17} s. The amount of energy released in a single reaction of this sort is tiny and poses no safety risk – the problem is keeping the antimatter particles away from the matter particles for long enough to experiment on them. (Remember, any equipment used will be made of matter particles.) Here is the equation that represents an electron and an anti-electron (positron) annihilating:

$$e^- + e^+ \rightarrow \text{electromagnetic radiation}$$

The reverse can also occur:

$$\text{electromagnetic radiation} \rightarrow e^- + e^+$$

Figure 3.9 shows such reactions occurring naturally in the atmosphere as a result of cosmic-ray bombardment. A wiggly line labelled γ (gamma) represents a photon of electromagnetic radiation. Physicists can exploit these reactions to create new particles to study. The idea is to take a beam of matter particles (say electrons) and a beam of anti-matter particles (positrons in this case) and accelerate them to very high energy. The beams are then allowed to collide with one another. This process sounds simple, but steering the beams is actually a highly complex affair.

When the matter and anti-matter come within range of each other, they annihilate, producing a burst of electromagnetic radiation that will rapidly materialise into new particles, perhaps ones that have never been seen before. Here are three equations that show possible reactions:

$$e^- + e^+ \rightarrow \text{electromagnetic radiation} \rightarrow \mu^- + \mu^+$$

$$e^- + e^+ \rightarrow \text{electromagnetic radiation} \rightarrow \tau^- + \tau^+$$

$$e^- + e^+ \rightarrow \text{electromagnetic radiation} \rightarrow u + \bar{u}$$

In the first reaction, two muons are produced. In the second, two tau leptons are produced – this is how the tau was discovered. In the third, two quarks are produced; they cannot exist freely like this, but bind together to form a pion (see Table 3.7).

Questions

2 Write an equation to show what happens when a muon annihilates with its anti-particle.

3 Some mesons are made from a quark bound to its own anti-quark. Explain why such a meson has no electric charge.

Mass from energy

It is surprising to find that particles of matter can disappear, leaving only energy. It is also surprising to find that particles can appear where before there was only energy. This sounds as if the laws of conservation of energy and mass are both being broken, and in a sense that's true. When a particle and anti-particle collide and annihilate one another, their mass disappears and we are left with photons of electromagnetic radiation. If mass, m, disappears and energy, E, appears, then these two are related by Einstein's equation:

$$E = mc^2 \qquad (1)$$

where c is the speed of light in a vacuum (also called 'free space'). Similarly, when a particle–anti-particle pair are created out of energy, we can use the same expression to deduce the possible mass that might be created. To produce particles 'out of nothing', the photons of electromagnetic radiation must have a great deal of energy; the more massive the particles produced, the greater the energy of the photons.

Equation 1 has a more general application. If ever the mass of particles changes by an amount, Δm, there is a corresponding change in the amount of energy, ΔE, and these two are related by:

$$\Delta E = c^2 \Delta m \qquad (1a)$$

Questions

Use the following data to answer Questions 4–6:

mass of positron = mass of electron, $m_e = 9.11 \times 10^{-31}$ kg

mass of proton, $m_p = 1.67 \times 10^{-27}$ kg

speed of light in vacuum $c = 3.00 \times 10^8$ m s^{-1}

4 Explain why electron and positron beams must be accelerated to high energies in order to produce particles such as tau leptons, which are more massive than electrons.

5 Using Equation 1, calculate the energy equivalent to the mass of a proton.

6 An electron–positron pair can be created from a single photon, if it is energetic enough. Any excess energy is shared equally between the two particles, as kinetic energy.

(a) What energy of photon is required to create an electron–positron pair?

(b) A photon of energy 2×10^{-13} J disappears, and an electron–positron pair is created. Calculate the kinetic energies of the electron and the positron.

Anti-matter space fuel?

It has often been suggested that anti-matter might be the answer to providing fuel for future space missions. The idea is that a spacecraft would be accelerated by a rocket in which matter and anti-matter annihilated one another. This would power it so that it could flash across the galaxy, just like the Starship *Enterprise*, at speeds approaching the speed of light (Figure 3.10). Scientists estimate that just 20 kg of anti-matter would do the job. The problem is making the stuff, and containing it. With present technology physicists at CERN can now produce about 10 ng of anti-matter particles a year, so it would take 100 million years to make 1 g. So anti-matter drives are still a long way off.

Figure 3.10 Starship *Enterprise*

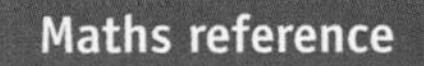

Maths reference

SI Prefixes
See Maths note 2.4

Activity 9 Physics in fiction: *Angels and Demons*

Angels and Demons by Dan Brown is a detective story about a secret society that wants to destroy the Vatican using an antimatter bomb. In the book, the anti-matter has been stolen from CERN.

Chapter 21 of the book contains lots of physics. Read the chapter, identify the physics and write an explanation for a reader who has studied science to AS level. (You might need to refer to work that you did for earlier chapters including *The Medium is the Message*.) Comment on whether the situation described in the book is plausible.

2.5 History of the universe

We started this part of the chapter by discussing the Big Bang and the idea of an expanding universe. You may have felt that we have taken a bit of a diversion by considering the current view of the fundamental particles and forces that govern the universe. Now we need to bring these two topics together.

As the universe expands, it cools downs; energy and matter are spread more thinly, and temperature is related to the concentration of energy. Out in space, it's cold. The COBE satellite (Cosmic Background Explorer, Figure 3.11) measured the temperature out there to be about 2.7 K. (Stars are of course much hotter; so is some of the gas between them. The 2.7 K refers to the temperature of 'empty space' where there is essentially no matter, just radiation.) Matter is very thinly spread – on average, less than one proton per cubic metre.

Figure 3.11 Cosmic Background Explorer satellite

When the universe was much younger than it is today, the energy was concentrated in a smaller volume, and so the temperature was much higher. At high temperatures, the particles of matter have much higher energies, and they rush around, colliding and interacting. (These are the conditions that physicists try to reproduce in their giant accelerators.) These temperatures are too high for atoms and molecules to exist – they knock one another into their constituent parts.

So the history of the universe can be summarised as follows:

- in its early stages: small, hot, dense; matter in the form of quarks, hadrons, leptons
- today: vast, cold, matter spread thinly; matter in the form of atoms and molecules.

Study note

2.7 K = −270.3°C

The Kelvin temperature scale is discussed in the chapter *Reach for the Stars*.

Using a log–log graph

Figure 3.12 shows in more detail how the temperature of the universe has changed with time. Here you can see the different phases that the universe has gone through since the end of the inflationary phase. Notice that *both* scales are logarithmic. This is necessary in order to show the great ranges of these quantities.

Study note

You met logarithmic scales in the AS chapter *Digging Up the Past*.

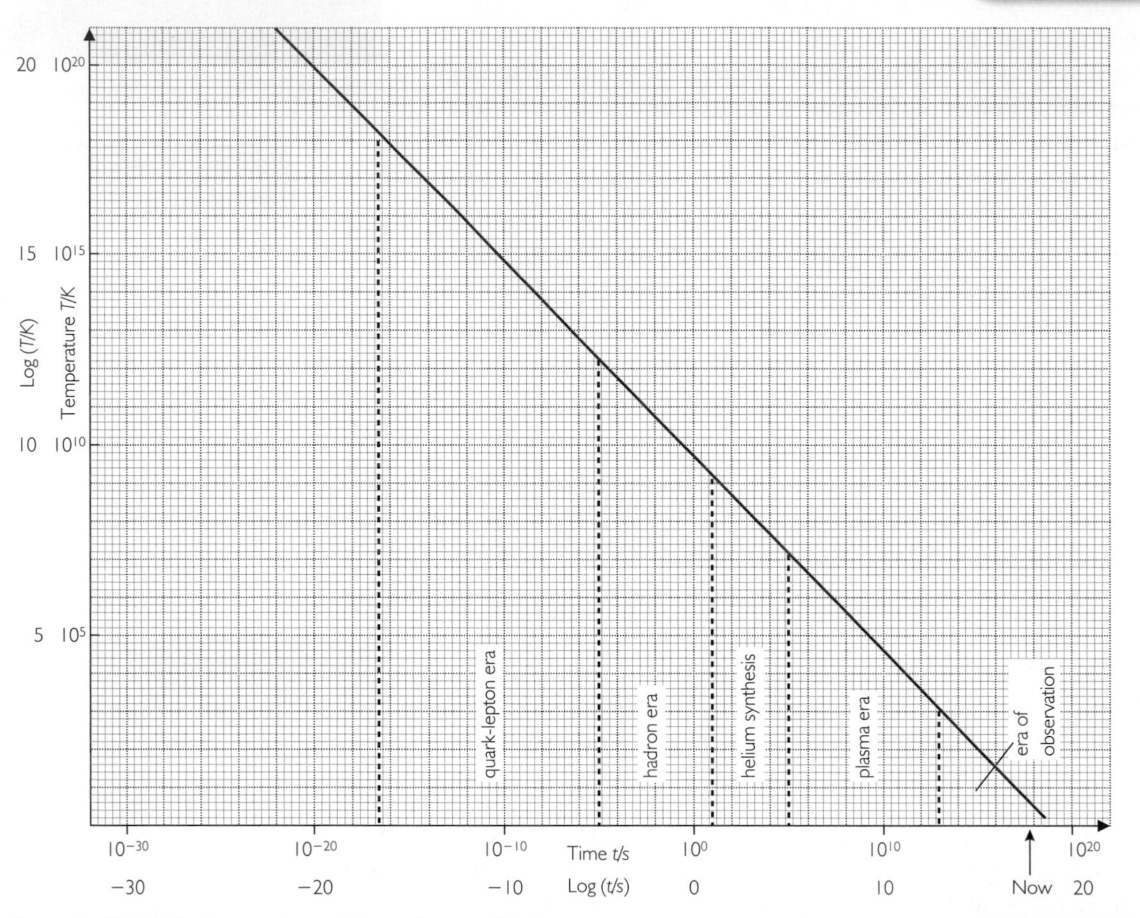

Figure 3.12 Fall in temperature of the universe with time

Study note

Figure 3.12 is an example of a log–log graph. Both scales are logarithmic.

The shape of the graph has been worked out from theoretical calculations that predict a relationship between temperature and time:

$$T = \frac{k}{\sqrt{t}} = kt^{-\frac{1}{2}} \qquad (2)$$

where T is the absolute temperature (ie in kelvin), t is the time since the Big Bang, and k is a constant. Equation 2 is an example of a **power-law relationship** – where one variable (T in this case) is directly proportional to another variable (t) raised to a power or **exponent** (here, the exponent is $-\frac{1}{2}$).

When numbers related by a power law are plotted in a **log–log graph** (ie one where both axes are logarithmic), then the resulting graph is a straight line. We can illustrate this by taking logs of Equation 2:

$$\log(T) = \log(k) + \log(t^{-\frac{1}{2}})$$

$$\log(T) = \log(k) - \tfrac{1}{2}\log(t) \qquad (3)$$

Plotting log (T) on the y-axis against log (t) on the x-axis gives a graph of the type $y = mx + c$, where the gradient, m, is equal to the exponent.

Activity 10 The universal story

Use Figure 3.12 to answer the following questions.

- How many seconds is it since the origin of the universe?
- At what time did the 'hadron era' start? What was the temperature at that time?
- According to the graph, what is the temperature (of 'space') now?

Find the gradient of the graph in Figure 3.12 and check that the graph is described by Equations 2 and 3.

Use Figure 3.12 to find the value of the constant, k, in Equations 2 and 3, and use Equation 2 to deduce its SI units.

Maths reference

Index notation and powers of ten
See Maths note 1.1

Powers that are not whole numbers
See Maths note 1.5

Gradient of a linear graph
See Maths note 5.3

Logs and powers
See Maths note 8.4

Log scales
See Maths note 8.6

Using logarithmic graphs
See Maths note 8.7

Step-by-step history

Now we can look in more detail at the different periods in the history of the universe, as shown in Figure 3.12. Physicists like to divide the history of the universe up into distinct periods, depending on what was happening at the time.

After inflation

We will start just after the end of the period of inflation, 10^{-35} s after the origin of the universe. At this time, the particles of matter and antimatter that were in the universe had been scattered over vast distances. If nothing had happened to fill the universe again then matter would now be very scarce indeed. Certainly life could never have evolved, as it is difficult to see how stars could ever form in such a 'thin' universe. However, at the end of the inflation period the universe was topped up with matter and antimatter by the decay of particles called 'Higgs particles'. These are very important in the theory of the universe and particle physics in general. Unfortunately they have never been seen in our experiments. New experiments are being undertaken now which should be able to detect the Higgs – if they don't, a lot of people are going to have to do some very hard thinking!

In summer of 2000, scientists at CERN announced results that gave some evidence for the detection of Higgs particles in the ALEPH detector at CERN's Large Electron–Position Collider (LEP) (a machine used between 1989 and 2000 to study W and Z bosons). However the accelerator was turned off shortly after in order to build the Large Hadron Collider, which began operation in 2008 and continues the search.

After inflation had stopped, the universe was filled with matter, antimatter and radiation again. You may think that nothing important has therefore happened, but that is not true. The amount of new matter and antimatter that appeared seems to be just enough to slow the universe down to a gentle expansion – its gravity stops it from expanding so fast that stars and galaxies would never get the chance to form. But with much more matter, the universe would have collapsed under its own gravity back to a 'big crunch' long ago. Inflation seems to have been just what was needed to get the right amount of matter into the universe.

Soon after inflation ended, most of the anti-matter in the universe disappeared. This is a very interesting puzzle that has had physicists thinking for decades. Where did the anti-matter go? You may ask how we know that it vanished at all. The answer is simple – we are here (see Figure 3.13)! Everything that we touch is made of matter (even the Moon – we know because people have walked on it). In reactions such as those you saw in Section 2.4, matter and anti-matter are created and destroyed in equal amounts. But to account for the predominance of matter, anti-matter must in fact decay more rapidly than matter, so physicists are having to think how their ideas might need to be modified or even overturned completely.

Figure 3.13 Earth and the living creatures that inhabit it are made from matter

Quark-gluon plasma

Between 10^{-18} and 10^{-12} seconds after the Big Bang no hadrons existed. The universe was composed of a soup of quarks, leptons and various particles involved with the four fundamental forces (see Figure 3.1). As the universe expanded, so the average distance between the quarks got bigger until at the end of this period in history they started to group into the hadrons that we find now.

The hadron era

Unsurprisingly, the next period, 10^{-12} seconds to 3 minutes after the Big Bang, is called the hadron era. As well as leptons, the universe was now full of hadrons – mostly protons and neutrons. Various reactions were going on and the numbers of protons and neutrons were being kept in balance. However, the protons and neutrons were not able to join together to make nuclei as the electromagnetic radiation in the universe was hot enough (had enough energy) to break them apart again straight away. As the end of this era drew near, the radiation had cooled down enough to let hadrons start to combine.

Nucleosynthesis

Three to 13 minutes after the Big Bang, nuclei started to form. Protons and neutrons came together to make deuterium, an isotope of hydrogen. Deuterium nuclei came together to make helium nuclei. As this process took place, all the free neutrons were effectively swept up into nuclei. A detailed calculation shows that at the end of nucleosynthesis the universe had about 78% hydrogen and 22% helium (measured by mass), which is almost exactly what we see now.

> **Study note**
>
> A very similar nuclear process goes on in the Sun now – you will learn more about this in the chapter *Reach for the Stars*.

Plasma era

For the next 3000 years, the universe was basically made up of leptons (especially electrons), electromagnetic radiation, helium nuclei and protons. The electrons could not stick to the nuclei to make atoms as the energy in the radiation was great enough to blast them away again. The universe was effectively an ionised gas – a plasma.

After about 3000 years the temperature had dropped (the radiation was less energetic) to such an extent that electrons that did stick to nuclei could stay there. This is a very important step in the history of the universe. Until then, the electromagnetic force had been very important as all the objects in the universe were charged. After the loose charges had been stuck together and the universe had become neutral, gravity became the most important of the fundamental forces.

Towards the present

After the plasma era, stars and galaxies formed, planets revolved in orbit about some stars (Figure 3.14). One planet (that we know of) took the radical step of evolving life. After some time, one complex form of life started to wonder about the universe it lives in. And that's where you come in…

Figure 3.14 Planets of the solar system

2.6 Summing up Part 2

In this part of the chapter you have learned about two important areas of physics:

- cosmology – the way the universe has evolved since the Big Bang
- particle physics – the standard model of the fundamental particles and the forces which act between them.

You have also seen how these two areas, which might at first seem unrelated, are in fact intimately connected.

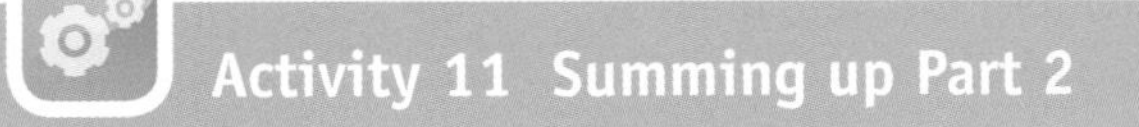

Activity 11 Summing up Part 2

Look back at your notes on Activities 1 and 2 in Part 1 of this chapter. How might you respond to these activities, now that you know a lot more about cosmology and particle physics?

Questions

7 (a) What two particles are represented by e^+ and e^-?

(b) What difference is there between them?

(c) What happens when these two particles meet?

8 Why was Murray Gell-Mann able to predict the existence of the top and bottom quarks, even though there was no experimental evidence for their existence? What could he have predicted about their charges and masses?

9 For each of the particles in the list below, say whether it is a hadron, meson, baryon, quark or lepton (it may fall into more than one category).

proton	pion	electron
charm	neutron	anti-electron
up	electron-neutrino	

3 Towards the standard model

The standard model of particle physics, with its 12 fundamental particles and four fundamental forces, has been very successful. As we have seen in Part 2, it has allowed physicists to link their picture of the tiniest particles to their picture of the history of the whole universe – particle physics has been united with cosmology. This has been one of the great achievements of physics in the second half of the 20th century. But the standard model hasn't always been with us. In this part of the chapter, we will look at what went before.

3.1 The discovery of the atomic nucleus

In the 19th century, most scientists came to accept that matter is made of atoms, although they never expected to see any. The idea of atoms allowed scientists to explain all sorts of things – the molecular kinetic theory, for example, in which the behaviour of gases was explained in terms of the movement of the particles that make up a gas. Much of chemistry depends on our understanding of the ways in which atoms arrange and rearrange themselves. It was thought that atoms were indivisible, the fundamental building blocks of matter. Then came two discoveries that changed all that.

> **Study note**
>
> You will study the molecular kinetic theory in the chapter *Reach for the Stars*

- In 1896, Henri Becquerel discovered radioactivity. Atoms of some elements, including uranium and radium, were capable of giving out smaller particles. Atoms were not indivisible.
- In 1897, Joseph (J. J.) Thomson discovered the electron. These were tiny particles, much lighter than atoms, that could be emitted by many different metals; this suggested that they must be more fundamental than atoms.

Thomson devised a model of the atom based on his discovery of the negatively-charged electron. He suggested that each atom was made up of thousands of electrons, spinning in such a way that they stuck together to form a neutral atom. This was not a very successful model. You have probably come across two other models of the atom, the plum-pudding model and the nuclear model, like a mini solar-system. These are shown in Figure 3.15.

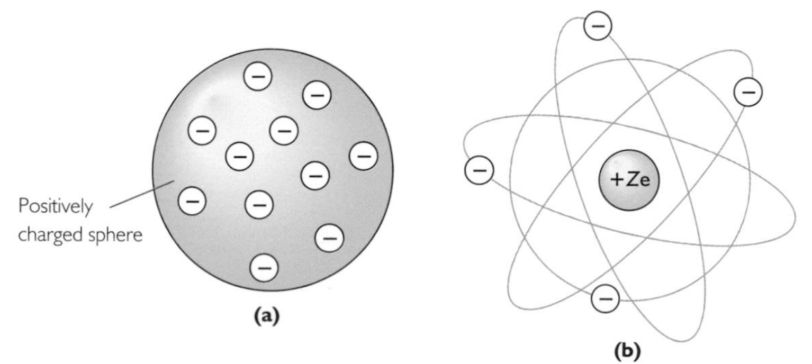

Figure 3.15 (a) Plum pudding and (b) nuclear models of the atom

> **Study note**
>
> Z (the atomic number) represents the number of protons in the nucleus.

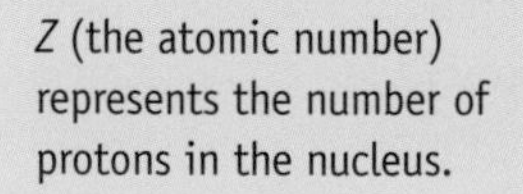

Activity 12 Model atoms

Discuss the two models shown in Figure 3.15, and J. J. Thomson's all electron model. Here are some points to consider.

- How are the positive and negative electric charges arranged in each model?
- What can you say about the attraction and repulsion between electric charges in each model?
- What do we mean by the term 'model' here?

The discovery of radioactivity was important not just because it showed that atoms could fall apart; it also became a useful research tool. The radiation produced by radioactive substances could be used as a probe to find out more about matter on the scale of individual atoms. Someone who made profound use of this tool was Ernest Rutherford (1871–1936).

Figure 3.16 shows the apparatus used by Rutherford's co-workers Geiger and Marsden to investigate the scattering of alpha radiation by a thin foil of gold. It was Rutherford who suggested the experiment and who interpreted the results. The source of radiation, R, is a thin-walled glass tube containing purified radium. The gold foil F is less than a thousandth of a millimetre thick. The tube T is used to pump the air out of the chamber. Radiation passing through the foil hits the scintillator S and produces a flash of light. The flashes are counted by the experimenter looking through the microscope M.

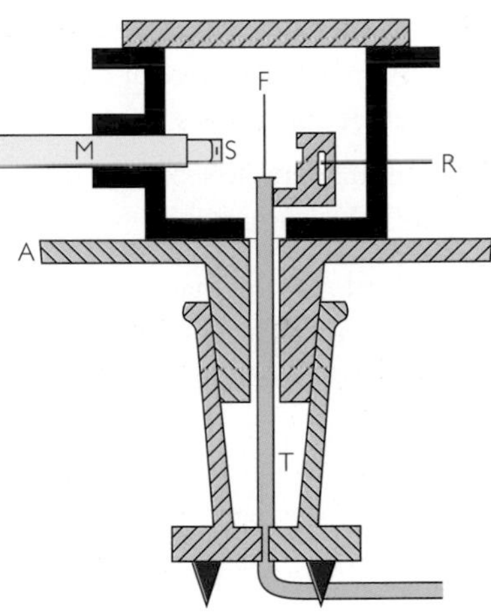

Figure 3.16 Geiger's and Marsden's apparatus (based on a drawing in the paper that reported their results)

When the microscope was aimed directly at the source of radiation, Geiger and Marsden saw frequent flashes of light, showing that most of the alpha particles passed straight through the gold foil. When they moved the microscope round through a small angle, they found that the number of flashes became much less. Relatively few alpha particles were deflected by the foil. This was what they had anticipated. However, when they moved the microscope round through large angles, they were surprised to find that they could still see some flashes even when the microscope was on the same side of the foil as the source. A very few alpha particles were being reflected back from the gold foil towards their original source. It was this 'back-scattering' that Rutherford had to explain.

Study note

An alpha particle is made up of 2 protons and 2 neutrons.

Rutherford explains back-scattering

To understand how alpha particles are scattered by gold foil, we need to know something about electrostatic forces. Alpha particles are positively charged, and so they are repelled by other positive charges and attracted by negative charges. Rutherford pictured an alpha particle as a tiny, positively charged bullet moving at high speed. If it encountered a gold 'plum-pudding atom', it would be deflected slightly as the positive and negative charges of the atom pushed and pulled the alpha particle. But because the positive charge of the pudding is spread out, and the embedded electrons have very small mass, it would not be deflected very much.

To explain the back-scattering of alpha particles, Rutherford came up with the nuclear model of the atom. He suggested that all of the atom's positive charge was concentrated at the centre of the atom in a tiny **nucleus**. As well as encountering a concentration of positive charge, the alpha particles need to meet something with large mass; a fast-moving bullet will not bounce back off a low-mass object in its path; it will simply knock it out of the way. Rutherford therefore proposed that most of the atom's mass, as well as the charge, was in the nucleus. The electrons orbited around this nucleus, providing a negative charge to balance the positive charge of the nucleus. Figure 3.17 shows how Rutherford's model explained Geiger's and Marsden's results.

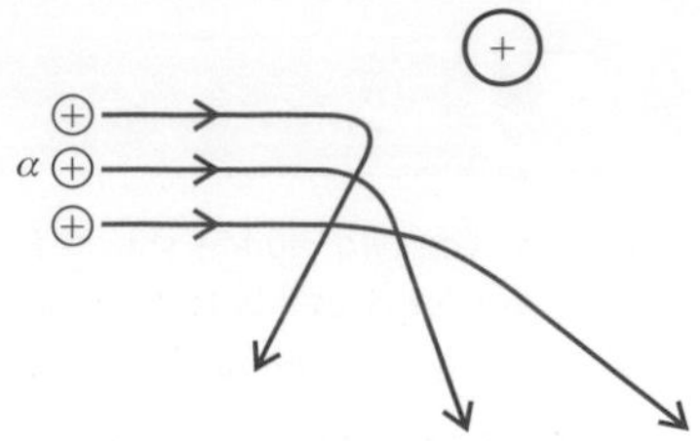

Figure 3.17 Alpha particles (α) are deflected as they pass close to a nucleus

- An alpha particle that passes the nucleus at a large distance is barely affected because the electrons it encounters are very light (their mass is tiny compared to the alpha particle).
- An alpha particle passing close to the nucleus will be deflected because it is repelled by the positive charge of the nucleus. The degree of deflection depends on how closely it approaches the nucleus of a gold atom.
- An alpha particle scoring a direct hit on the nucleus is repelled back in the direction from which it came.

Rutherford realised that the nucleus of the atom must be very tiny compared to the whole atom – this explains why most alpha particles whiz straight through without ever coming close to a nucleus.

Geiger and Marsden measured the numbers of alpha particles scattered through different angles, and Rutherford produced a mathematical model that was able to explain the pattern of their results. His model also predicted the results that would be obtained if the alpha particles had different energy, and if foils of different thickness, and made from different metals, were used. Geiger and Marsden carried out a series of experiments to test Rutherford's model – and every time the results were as predicted. This was enough to convince most physicists that the nuclear model of the atom was correct. Table 3.8 shows some results obtained using gold foils of different thickness.

Thickness, t, of gold foil/10^{-6} m	number, N, of scintillations per minute
0.23	21.9
0.46	38.4
1.07	84.3
1.70	121.5
1.89	145.0

Table 3.8 Some of Geiger's and Marsden's results

Activity 13 A model for Rutherford scattering

Use a model such as the one shown in Figure 3.18 to investigate Rutherford scattering. A marble runs down a ramp, and then past a plastic or metal 'hill'. The closer it gets to the hill, the more it is deflected.

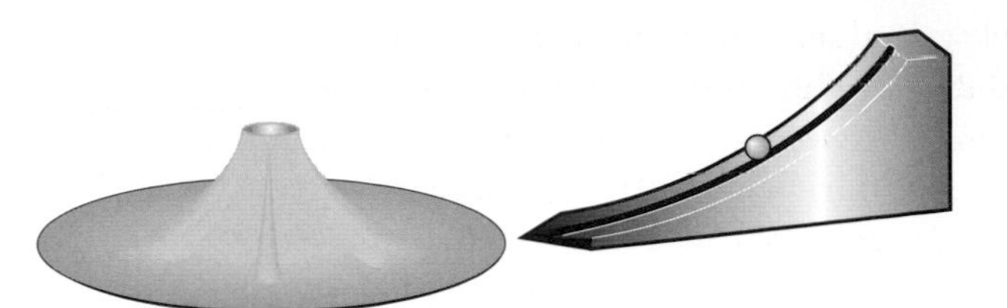

Figure 3.18 Model to represent the scattering of an alpha particle by a nucleus

Activity 14 Analysing the results

By drawing suitable graphs using the data from Table 3.8, find a mathematical relationship between N, the number of scintillations per minute, and t, the thickness of the foil.

Questions

10 Alpha radiation is easily absorbed. An alpha particle has a range of about 5 cm in air, because it collides with innumerable molecules of the air and loses its energy. Alpha particles travel even shorter distances through solid materials, because they are much denser than air. Use these ideas to explain some of the features of the design of Geiger's and Marsden's apparatus.

11 Geiger and Marsden tested Rutherford's model using thicker gold foils. Would this make it more or less likely that an alpha particle would score a direct hit on a gold nucleus? How would you expect their results to change?

12 Geiger and Marsden used thin sheets of mica to slow down the alpha particles so that they would not pass so rapidly through the gold foil. How would you expect their results to change?

13 Silver atoms are the same size as gold atoms, but their nuclei are smaller and have lower charge. Geiger and Marsden tried silver foil instead of gold. What do you think they observed?

Study note

Mica is a silicate mineral that naturally forms very thin sheets.

Charge repelling charge

Rutherford used his model to say something about the size of the nucleus of the gold atom. To see how he did this, think about an alpha particle that approaches the nucleus of a gold atom head-on, as shown in Figure 3.19. As the alpha particle approaches, it begins to feel the repulsion of the gold nucleus. (The two positive charges repel one another.) The repulsive force slows it down. The closer the alpha particle gets, the greater the force, and so the greater its deceleration. Eventually it comes instantaneously to a halt, and then starts to move back towards where it came from. You can compare this motion with that of a ball thrown up in the air. The ball moves fast as it leaves your hand; it gradually slows as it rises in the air; it stops instantaneously, and then it accelerates downwards. In this case, the force is caused by the gravitational attraction between the ball and the Earth; for the alpha particle, the force is the electrostatic repulsion.

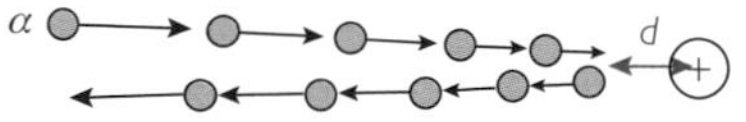

Figure 3.19 A fast-moving alpha particle slows down as it approaches a gold nucleus, until it goes into reverse

If we can find out how close the alpha particle gets to the gold of the nucleus, that could give us an idea of how big the nucleus is. To do this, Rutherford needed to know about the law that describes the electrostatic force between two charged particles. He was able to show that the alpha particles reached within about 10^{-14} m of the centre of the atom. (This is the distance shown as *d* in Figure 3.19.) So, although he could not say how big the nucleus was, he could deduce that its diameter must be no greater than about 10^{-13} m. That's no more than a thousandth of the diameter of an atom – today we know that it's even smaller, about 10^{-15} m across. In the next section we will look at the law that allowed Rutherford to make his estimate.

3.2 Electrical forces

No doubt you have carried out some simple experiments on static electricity. If you rub a balloon on your jumper, it becomes charged and you can do interesting things with it – stick it to the wall, for example. Here you are making use of the attractive force between opposite charges. Suppose the balloon gains a positive charge when you rub it. It will then attract negative charges in the wall, and the two stick together.

If you rub two balloons together so that they both become positively charged, you can observe the repulsive force between like charges (charges with the same sign). Hang the balloons side by side and they repel one another, as shown in Figure 3.20(a).

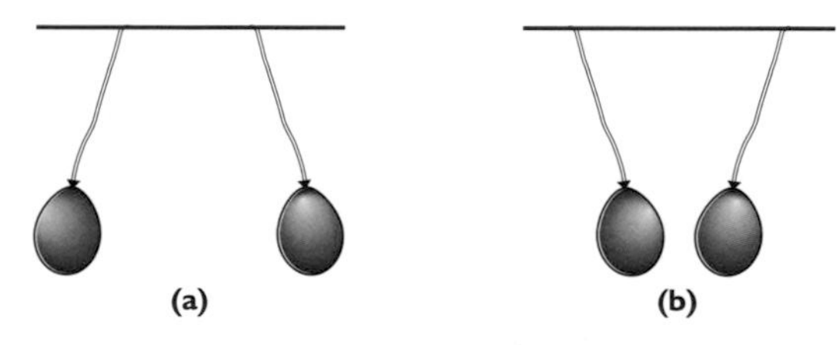

Figure 3.20 (a) Like charges repel one another (b) unlike charges attract one another

The fact that there are two types of electric charge, and the rules of attraction and repulsion, were known in the 18th century. Today we understand that a balloon becomes charged when you rub it because electrons are transferred from one object to another. If electrons are transferred from the balloon to your jumper, the balloon will become positively charged because it has lost electrons, which are negatively charged.

The French physicist Charles Coulomb (1736–1806) went beyond these simple rules of attraction and repulsion. He wanted to know more about the nature of the force, and in particular how the force between two charged objects depended on their separation. He used balls of pith, because pith is a good insulating material which retains an electric charge long enough for measurements to be made. Today we might use expanded polystyrene. Figure 3.21 shows two charged balls suspended so they repel one another. This is a clever arrangement because, provided we know the weight of each ball, we can determine the force between them.

Study note

Pith is a low-density material that is found in plants; eg under the skin of oranges and in the centre of hollow stems such as bamboo.

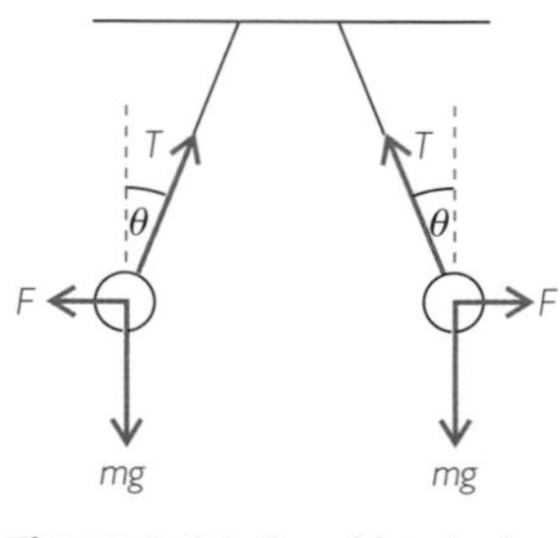

Figure 3.21 Two identical charged balls suspended by insulating threads

Study note

You met Newton's third law and force vector diagrams in the AS chapter *Higher, Faster, Stronger*.

Note that Figure 3.21 is symmetrical. Each ball feels a horizontal electrostatic force of magnitude *F* because it is repelled by the charge of the other ball. The pair of electrostatic forces is an example of Newton's third law of motion:

- they are equal in magnitude and opposite in direction
- they act on different bodies (the two balls)
- they are the same type of force (electrostatic).

Now look at the forces acting on just one of the balls, as shown in the vector diagram in Figure 3.22. There are three forces:

- the ball's weight mg
- the electrostatic force F
- the tension of the string T.

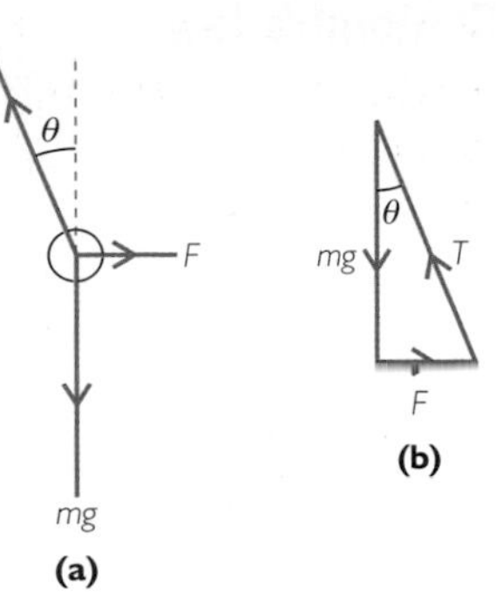

Figure 3.22 The three forces acting on one of the balls shown in Figure 3.21

The easiest of these to measure is the weight mg. We can use this, along with the angle θ to find the electrostatic force F. The ball is stationary, ie it is in equilibrium, so we know that the forces are balanced. The weight mg downwards is balanced by the vertical component of the tension T acting upwards:

$$mg = T\cos\theta \qquad (4)$$

The electrostatic force F to the right is balanced by the horizontal component of the tension T acting to the left:

$$F = T\sin\theta \qquad (5)$$

We can eliminate T from these equations by dividing Equation 5 by Equation 4:

$$\frac{F}{mg} = \frac{T\sin\theta}{T\cos\theta} = \tan\theta$$

Rearranging gives:

$$F = mg\tan\theta \qquad (6)$$

So, if we can measure mg and the angle θ, we can calculate F. Alternatively, you could measure F from a scale drawing of Figure 3.22(b).

Questions

14 Figure 3.21 shows two charged balls repelling one another. Sketch diagrams to show how the situation would change (a) if the balls had more charge; (b) if the balls had opposite charge.

15 Sketch two charged balls hanging close to one another, one having twice as much positive charge as the other. (Hint: remember Newton's third law!)

16 Calculate the electrostatic force on each of the balls shown in Figure 3.23, using data from the diagram.

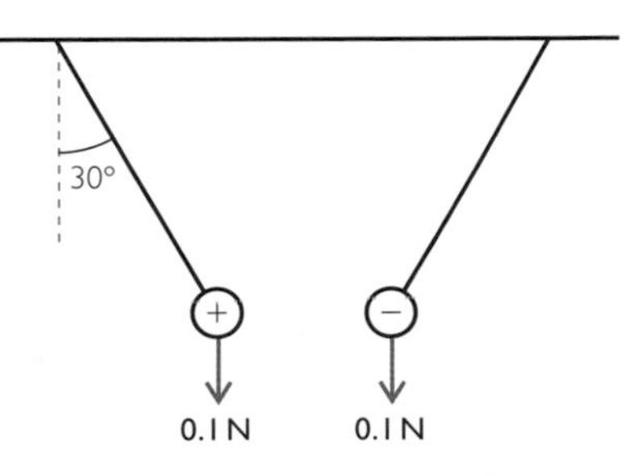

Figure 3.23 Diagram for question 16

Coulomb's law

Coulomb investigated the force between two charged spheres. Two spheres with charges Q_1 and Q_2 are separated by a distance r; note that r is measured from the centre of one sphere to the centre of the other (Figure 3.24). It is not surprising to find that the force depends on the size of each of the two charges, and that the force decreases as the spheres are moved apart. Coulomb found that the force is proportional to each of the charges:

$$F \propto Q_1 \text{ and } F \propto Q_2 \qquad (7)$$

And that the force is inversely proportional to the square of the distance r between the two charges – the force obeys an **inverse square law**:

$$F \propto \frac{1}{r^2} \qquad (8)$$

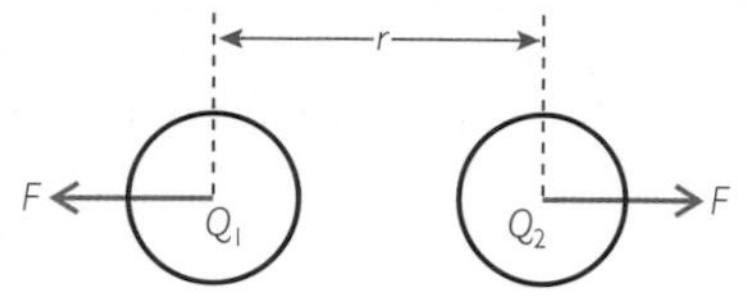

These relationships can be combined:

$$F \propto \frac{Q_1 Q_2}{r^2}$$

$$F = \frac{k\,Q_1 Q_2}{r^2} \qquad (9)$$

where k is a constant of proportionality. Equation 9 represents **Coulomb's law** of electrostatic force. In SI units, with F in newtons, r in metres and Q in coulombs:

$$k = 9.0 \times 10^9 \text{ N m}^2 \text{ C}^{-2}$$

The constant k is sometimes written in a different form as

$$k = \frac{1}{4\pi\varepsilon_0}$$

Where the quantity ε_0 is known as the **permittivity of free space** and has the value 8.854×10^{-12} F m^{-1} (farads per metre).

Coulomb's law is written as:

$$F = \left(\frac{1}{4\pi\varepsilon_0}\right)\frac{Q_1 Q_2}{r^2} \qquad (9a)$$

Coulomb's law tells us about one of the four fundamental forces of nature, part of the standard model. One remarkable feature of this law is that, while Coloumb found out about it by making measurements on charged pith balls in his laboratory, it applies equally to the charged particles of the subatomic world – protons, electrons and the other members of the subatomic menagerie.

Figure 3.24 Defining quantities: the force between two charged spheres

Study note

If you have studied either of the chapters *Transport on Track* or the *The Medium is the Message*, you will have met the farad as a unit of capacitance. 1 F = 1 C V^{-1}

Questions

17 Show that units of k and ε_0 are consistent with one another, ie that the units F m^{-1} and (N m^2 C^{-2})$^{-1}$ are equivalent.

18 Calculate the force that acts between two charges, each of 10 nC, separated by a distance of 5 cm.

19 The average separation of a proton and an electron in a hydrogen atom is about 0.037 nm.

(a) Calculate the force each exerts on the other.

(b) Draw a diagram to show the directions in which these forces act.
($e = 1.60 \times 10^{-19}$ C, $k = 9.0 \times 10^{9}$ N m^2 C^{-2})

Activity 15 Exploring Coulomb's law

Explore the inverse square relationship of the force between two charged objects. This requires care, as well as dry atmospheric conditions.

Force, work and energy

Now that we know an expression for the electrostatic force between two charged particles, we can go back to Rutherford's problem. How close can an alpha particle get to the nucleus of a gold atom?

Figure 3.25 shows, in graphical form, how the force depends on the distance of the alpha particle from the gold nucleus. As an alpha particle ($Q_1 = +2e$) speeds directly towards a gold nucleus ($Q_2 = +79e$), it feels a force which is at first small, but which increases more and more rapidly. When the values are plotted on a logarithmic graph, the gradient is –2, showing that there is an inverse square relationship between F and r. The force slows the alpha particle until it stops, goes into reverse, and accelerates back towards where it came from.

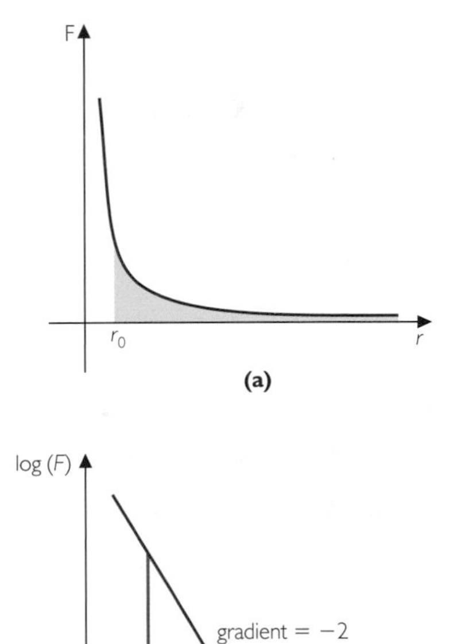

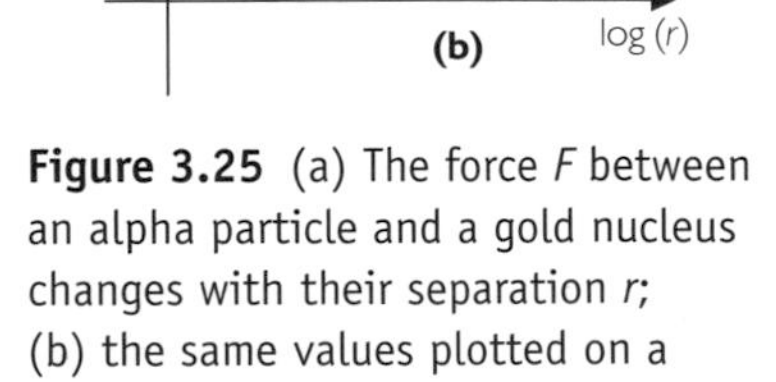

Figure 3.25 (a) The force F between an alpha particle and a gold nucleus changes with their separation r; (b) the same values plotted on a logarithmic graph

As the alpha particle approaches the nucleus, it loses kinetic energy and gains electrostatic potential energy; the repulsive force does work. At the point of closest approach, in a head-on encounter, the kinetic energy is zero; all of the initial kinetic energy has been transformed to potential energy. This is similar to the situations that you met in the AS chapter *Higher, Faster, Stronger*, where a vertically launched projectile gains gravitational potential energy at the expense of its kinetic energy, and where a bungee jumper loses kinetic and gravitational potential energy as the cord gains elastic potential energy.

The work done on the alpha particle as it approaches to a distance r_0 is represented by the shaded area under the graph in Figure 3.25. The alpha particle stops and goes into reverse when the area is equal to its initial kinetic energy. Knowing the initial kinetic energy of the alpha particles, Rutherford was able to calculate how closely they must have approached the nucleus.

Using calculus, it is possible to show that, as an alpha particle approaches from a very large distance to a distance r, the increase in electrostatic potential energy ΔE, is given by:

$$\Delta E = \frac{kQ_1Q_2}{r} \qquad (10)$$

Questions

20 Sketch and explain how the graphs shown in Figure 3.25 would change if the nucleus used was that of a copper atom, containing 29 protons.

21 Alpha particles typically have a kinetic energy of 8.0×10^{-13} J. If such an alpha particle travelled directly towards a gold nucleus ($Q = +79e$), what would be its distance of closest approach?
($e = 1.60 \times 10^{-19}$ C, $k = 9.0 \times 10^{9}$ N m^2 C^{-2})

3.3 Force fields

Coulomb's law is an example of an inverse square law, so-called because the force decreases in inverse proportion to the square of the distance between the two charged spheres. At twice the distance, the force drops to one-quarter its previous value, at three times the distance the force drops to one-ninth its previous value – and so on.

Figure 3.26 shows the electric field lines of the field around a charged sphere. The lines get further apart as they spread out, showing that the force gets weaker with distance. At twice the distance the lines are spread out over four times the area, showing that the force is a quarter of the strength. This is an inverse square relationship – doubling the distance gives a quarter of the force.

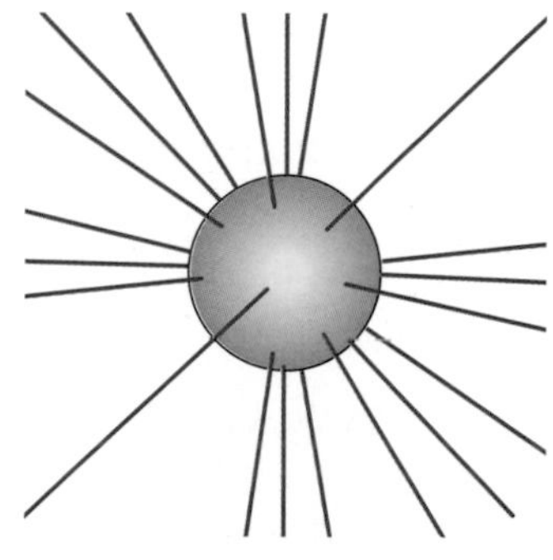

Figure 3.26 Electric field around a charged sphere

An electric field is defined as a region in which a charged object experiences a force, and the electric field strength, E, is defined as the force per unit charge. If a charge q experiences a force of magnitude F, then the magnitude of field strength is:

$$E = \frac{F}{q} \qquad (11)$$

Study note

Electric field strength and field lines were introduced in the chapter *The Medium is the Message*.

The direction of an electric field is defined as that of the force experienced by a positive charge. A diagrammatic representation of a field using electric field lines tells us two things about the electric field near a charged object:

- the strength of the field (closer together means stronger)
- the direction of the field (a small positive charge will feel a force in the direction of the arrow – note that the arrows go from positive to negative charge).

Figure 3.27 shows how field lines can be used to represent the electric field in a variety of situations. Notice that we can still say that a field exists around an isolated charged sphere (Figure 3.27c), even if it is not exerting a force on another charged object. The field lines show us that there would be a force if we placed another charged object nearby.

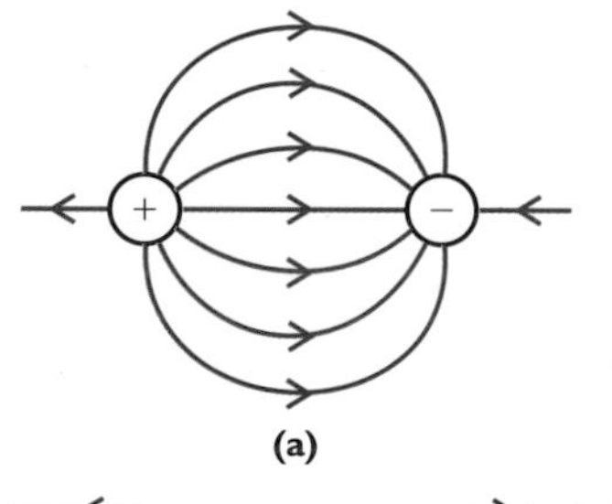

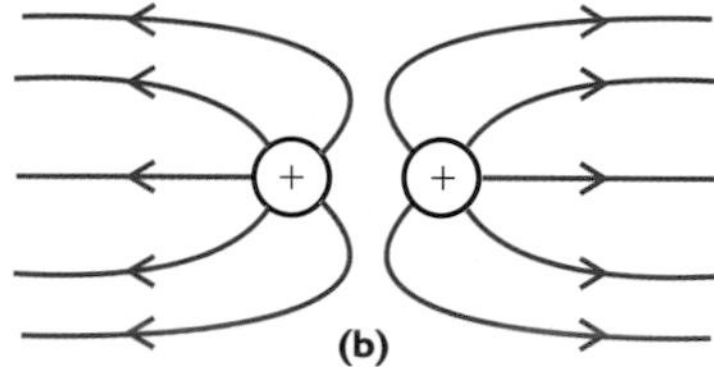

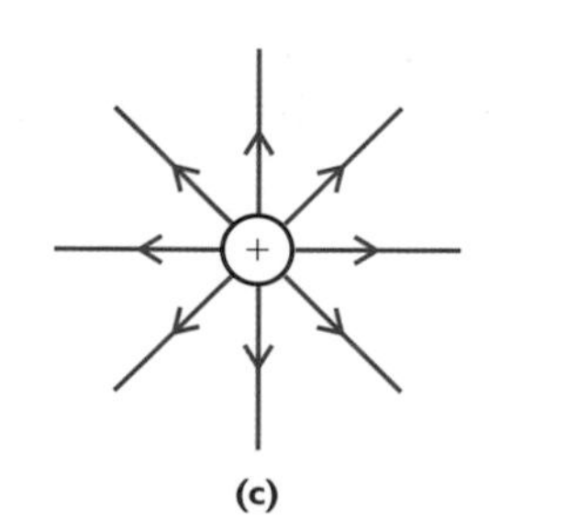

Figure 3.27 Electric fields around (a) two charged spheres with opposite charges, (b) two charged spheres with like charges, and (c) an isolated positively charged sphere

To find out about the field strength at point X, a distance r from the charge Q, we place a small test charge q at X (see Figure 3.28). Outside a charged sphere, we have to measure r from the centre of the sphere. The force F on q is given by Coulomb's law, from Equations 9 and 11:

$$F = \frac{kQq}{r^2}$$

and electric field strength E is then:

$$E = \frac{kQ}{r^2} \qquad (12)$$

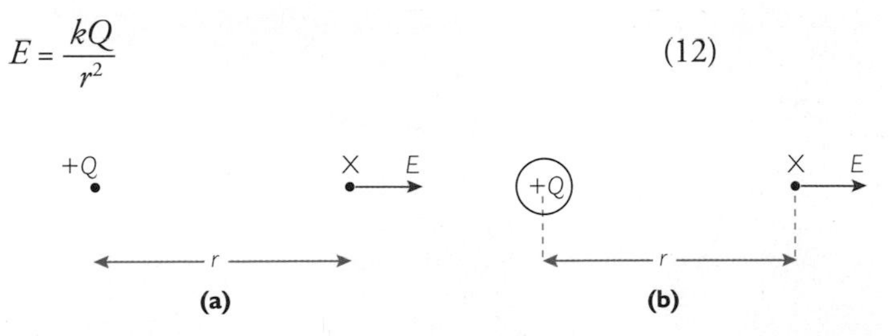

Figure 3.28 Calculating the electric field strength near (a) a point charge Q and (b) a charged sphere

Activity 16 Probing the field

Investigate the size and direction of the electric field around a charged sphere.

Questions

22 Draw a diagram to show the field lines around an isolated negatively charged sphere. Explain why your diagram is different from Figure 3.27(c).

23 Copy the diagram of Figure 3.27(a). Mark and label a point where the electric field is strong, and another point where the field is weak. Draw arrows to show the electrostatic forces acting on the two spheres. Repeat for Figure 3.27(b).

24 Calculate the electric field strength at a point where a charge of 20 mC feels a force of 10 N.

25 An electron, charge $e = 1.60 \times 10^{-19}$ C, passes between two charged plates; the electric field strength between the plates is 2000 N C^{-1}.

What force acts on the electron and what is its acceleration?
(Electron mass $m_e = 9.11 \times 10^{-31}$ kg)

26 The approximate distance of closest approach of an alpha particle and a gold nucleus (charge = $+79e$) is 1.0×10^{-13} m. What is the electric field strength at this distance from the nucleus? ($e = 1.60 \times 10^{-19}$ C, $k = 9.0 \times 10^9$ N m^2 C^{-2}.)

3.4 Collisions

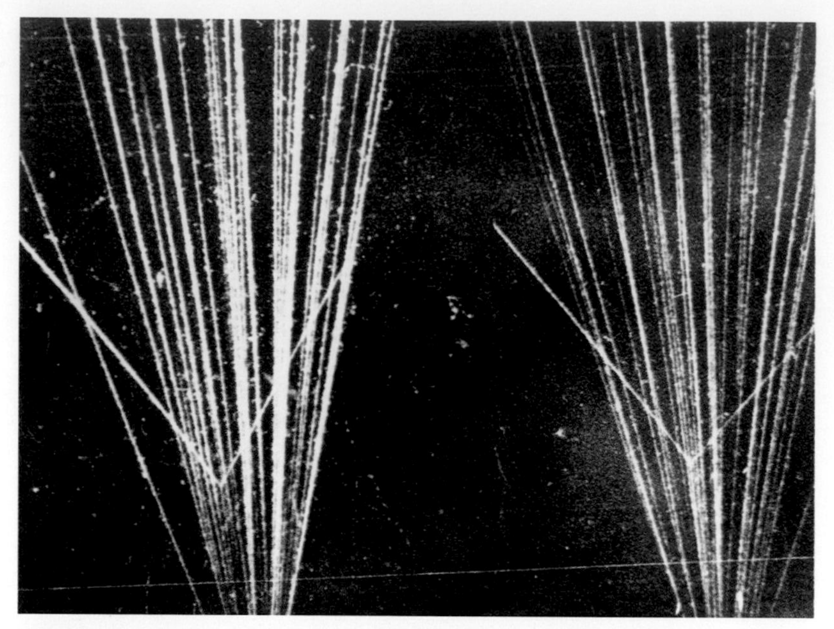

Figure 3.29 Collision between an alpha particle and a helium nucleus

Rutherford used collisions to probe the structure of atoms. Since then, particle physicists have used collisions as one of their main ways to deduce the nature of subatomic particles.

A gold nucleus is much more massive than an alpha particle – about 50 times as massive. So in Rutherford's experiments, a small particle was colliding with a much bigger one that remained more or less at rest. If there is less of a difference in the masses, the 'target' particle is set in motion; this is what happens in most particle-physics experiments. The diagram in Figure 3.29 shows what happens when an alpha particle collides with a much smaller nucleus – that of a helium atom.

Figure 3.29 was obtained using a cloud chamber, which is a device that can show up the tracks of charged subatomic particles. In this case, the cloud chamber was filled with helium gas, and a source of alpha radiation was then inserted. The tracks of alpha particles appear as straight lines extending out from the source. One alpha particle has hit a helium nucleus, and the two particles have gone off at right angles to each other.

The nucleus of a helium atom consists of two protons and two neutrons, so it is identical to an alpha particle. You can investigate a collision between two identical particles using pucks containing dry ice, so that they glide on a layer of carbon dioxide gas, or two identical 'hover-footballs'. When a moving puck (or football) collides with an identical stationary one they move off at 90° to one another, as shown in Figure 3.30, just like the alpha particle tracks in Figure 3.29.

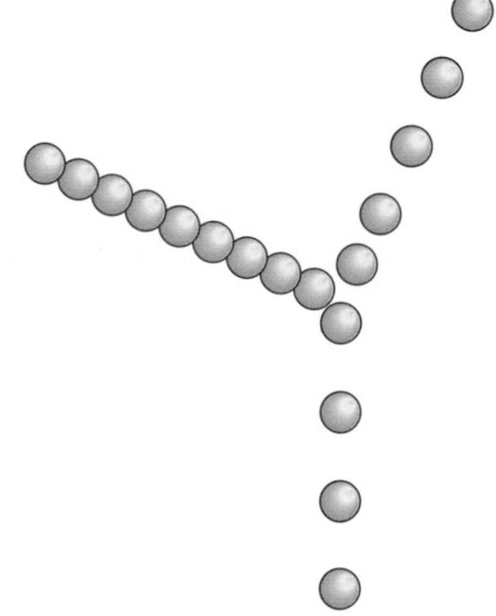

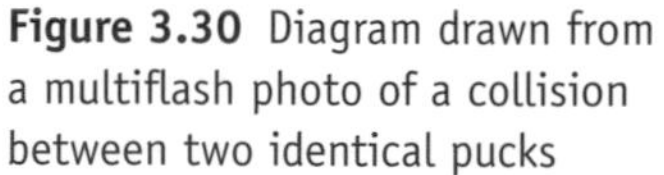

Figure 3.30 Diagram drawn from a multiflash photo of a collision between two identical pucks

Momentum and energy in collisions

By analysing the tracks left in particle detectors, particle physicists can make deductions about the mass and energy of subatomic particles. In order to do this, they use some fundamental physical laws – the conservation of momentum and of energy.

In the chapter *Transport on Track*, you explored one-dimensional collisions and met the vector quantity, *mass* × *velocity* known as **momentum** (which is usually given the symbol p, and has SI units kg m s^{-1}). You saw that, in a collision (or an explosion), momentum is always conserved – that is, the total momentum of the interacting objects is unchanged. We can now generalise this to all collisions. As momentum is a vector, we can draw a vector diagram to represent the momentum of the objects before and after they collide. Figure 3.31 shows such a diagram drawn from a multiflash photo such as that shown in Figure 3.30; the speed was deduced from the spacing of the images in the multiflash photographs. You can see that the vector sum of momenta afterwards is equal to the momentum of the single moving puck before the collision. Momentum is conserved in *all* collisions.

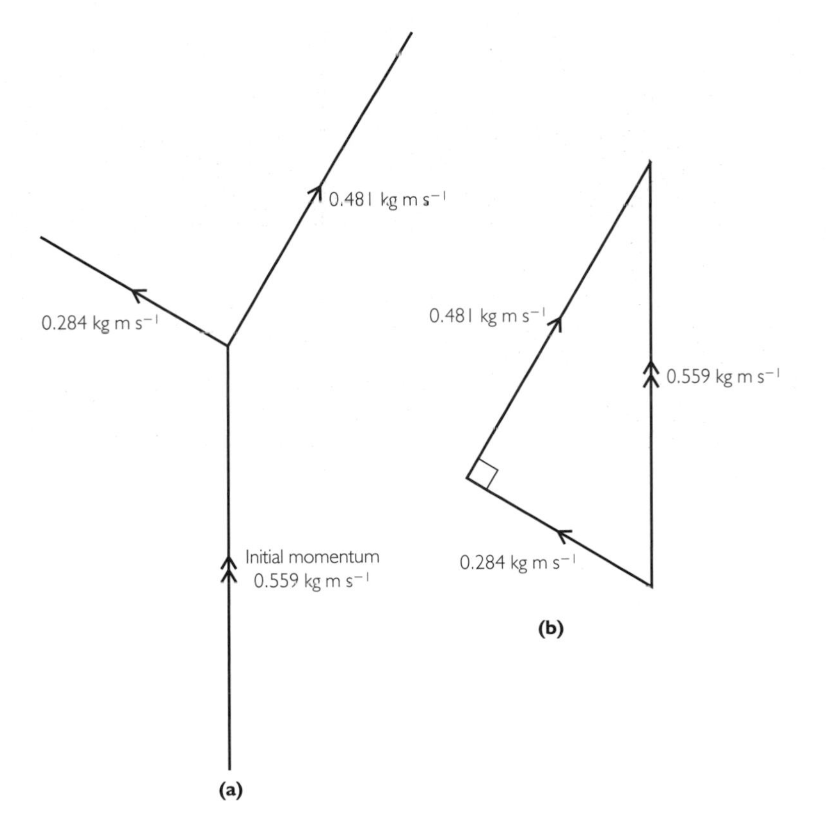

Figure 3.31 Momentum vectors of the pucks shown in Figure 3.30

As you saw in the chapter *Transport on Track*, an **elastic collision** is one in which kinetic energy is conserved. By analysing the momentum and energy in a collision, particle physicists can make deductions about the nature of the particles involved and the interactions between them. You will explore some aspects of this in Activities 17 and 18. For these activities, it is useful to be able to relate kinetic energy directly to momentum without needing first to find the speeds of the moving objects.

An object of mass m moving at speed v has momentum of magnitude p given by:

$$p = mv \qquad (13)$$

and the same object's kinetic energy, E_k, is given by:

$$E_k = \frac{1}{2}mv^2 \qquad (14)$$

To express E_k in terms of p, square Equation 13:

$$p^2 = m^2v^2$$

then divide both sides by $2m$:

$$\frac{p^2}{2m} = \frac{m^2v^2}{2m} = \frac{mv^2}{2}$$

and so we have:

$$E_k = \frac{p^2}{2m} \qquad (15)$$

In fact these equations only apply to objects moving at speeds much less than that of light. They are appropriate for analysing collisions such as those in Activities 17 and 18, but in particle-physics experiments, the particles travel very close to the speed of light and so the equations of **special relativity** are needed.

It turns out that an *elastic* collision between a moving and a stationary object of equal mass always results in the two moving at right-angles to each other. Analysis of Figure 3.31 illustrates why this is the case. Since momentum is conserved, we can write the vector sum:

$$\boldsymbol{p} = \boldsymbol{p}_1 + \boldsymbol{p}_2 \qquad (16)$$

Since both particles have the same mass, m, conservation of kinetic energy gives:

$$\frac{p^2}{2m} = \frac{p_1{}^2}{2m} + \frac{p_2{}^2}{2m} \qquad (17)$$

Multiplying Equation 17 by $2m$ leads to a relationship between the magnitudes of the vectors (in other words, between the lengths of the sides of the triangle):

$$p^2 = p_1{}^2 + p_2{}^2 \qquad (18)$$

Equation 18 is Pythagoras's relationship between the sides of a *right-angled* triangle — so the angle θ must be 90°, ie in this particular case the particles must move at right-angles to one another after they collide.

In Activities 17 and 18, you will explore collisions which give rise to different angles between the tracks.

Activity 17 Nuclear dodgems

Use frictionless pucks or hover-footballs to model collisions between subatomic particles. Explore what happens when a moving puck collides with a stationary puck whose mass is the same as, greater than, or less than its own. If possible, record the collisions in multiflash photographs or on video.

Activity 18 Accounting for momentum and energy

By making measurements on multiflash photographs such as that in Figure 3.30 or your own from Activity 17, draw vector diagrams to represent momentum before and after the collision and also investigate how the kinetic energy changes.

Questions

27 Estimate your own mass, and the fastest speed at which you can run. Use Equations 13 and 14 to calculate your momentum and kinetic energy. Confirm that your answers are related by Equation 15.

28 Calculate the kinetic energy of an alpha particle (mass 6.7×10^{-27} kg) whose momentum is 5.0×10^{-20} kg m s^{-1}.

29 Many particle collisions result in the creation of new particles; for example when an electron and positron collide and interact to produce a muon/anti-muon pair. Muons have greater mass than electrons.

(a) Would momentum be conserved in such a collision?

(b) Would such a collision be elastic? If not, would the total kinetic energy increase or decrease?

30 Particles with no electric charge (such as neutrons and neutrinos) leave no tracks in particle detectors. How might particle physicists deduce the presence of such particles from a record of tracks left by charged particles before and after a collision?

3.5 Particle diffraction

Rutherford's alpha-particle scattering experiment, carried out in 1908, showed how the underlying structure of matter could be investigated by firing a beam of charged particles at the matter and looking at the pattern that the particles formed after they had interacted with the matter. This is the basic approach used in much of particle physics; an approach that was developed and extended throughout the 20th century. In section 3.4, we analysed the results of such experiments in terms of collisions. In this section we see how ideas about diffraction and wave–particle duality can be brought to bear on such experiments.

In the AS chapter *Digging Up the Past*, you saw how diffraction of X-rays enables archaeologists to probe the structure of materials, and you saw that electron beams, too, behave like waves and can be diffracted. Electron diffraction is used to probe matter on an even smaller scale than X-rays – electron beams are used to measure the size of nuclei and to reveal the presence of quarks within protons. After a brief review of diffraction in Activity 19, this section looks in more detail at electron diffraction and you will see how it comes about that electron beams can be used to probe matter on such a minute scale.

Activity 19 Diffraction of light

Observe what happens when light is diffracted by:

- silk, chiffon or a piece of net curtain
- a grating of parallel lines
- a regular grid of crossed lines
- a random array of fine dust particles.

Explore the effects of changing the separation of the lines and the wavelength of the light.

Figure 3.32 shows apparatus that you can use in the laboratory to demonstrate electron diffraction by graphite, and Figure 3.33 shows the pattern produced on the screen with such apparatus when polycrystalline graphite is placed in the electron beam. With a single crystal the pattern is an array of dots, like the pattern you get if you shine a laser beam through crossed diffraction gratings. Just as laser light is diffracted by the grating, so the beam of electrons is diffracted by the regular array of atoms that make up the graphite crystal. The photograph in Figure 3.33 shows what happens when a beam of electrons shines through a piece of polycrystalline graphite. (It's difficult to obtain a single crystal.) This material is made of many tiny crystals, all at different orientations. Because of this, the spots of the diffraction pattern are smeared out into rings.

Study note

You also met electron beams ('cathode rays') in the chapter *The Medium is the Message*.

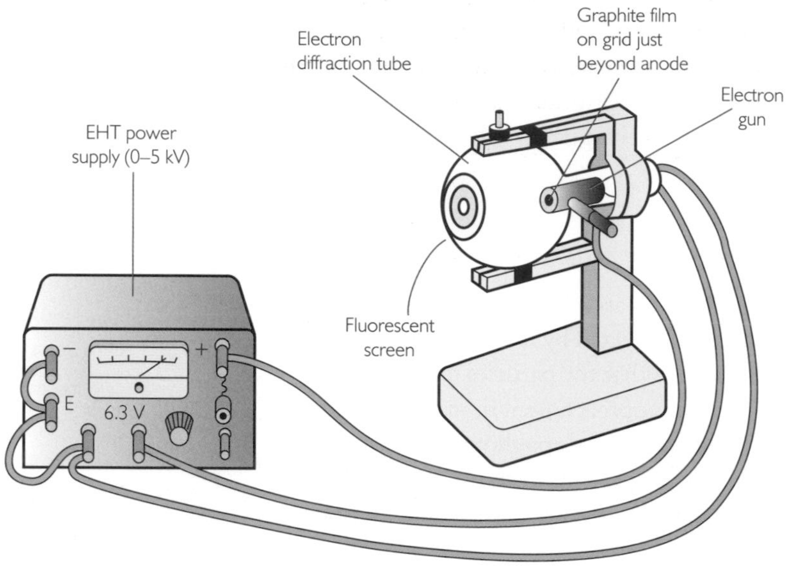

Electron diffraction tube
Graphite film on grid just beyond anode
Electron gun
EHT power supply (0–5 kV)
Fluorescent screen
E
6.3 V

Figure 3.32 Using an electron diffraction tube to demonstrate electron diffraction by graphite

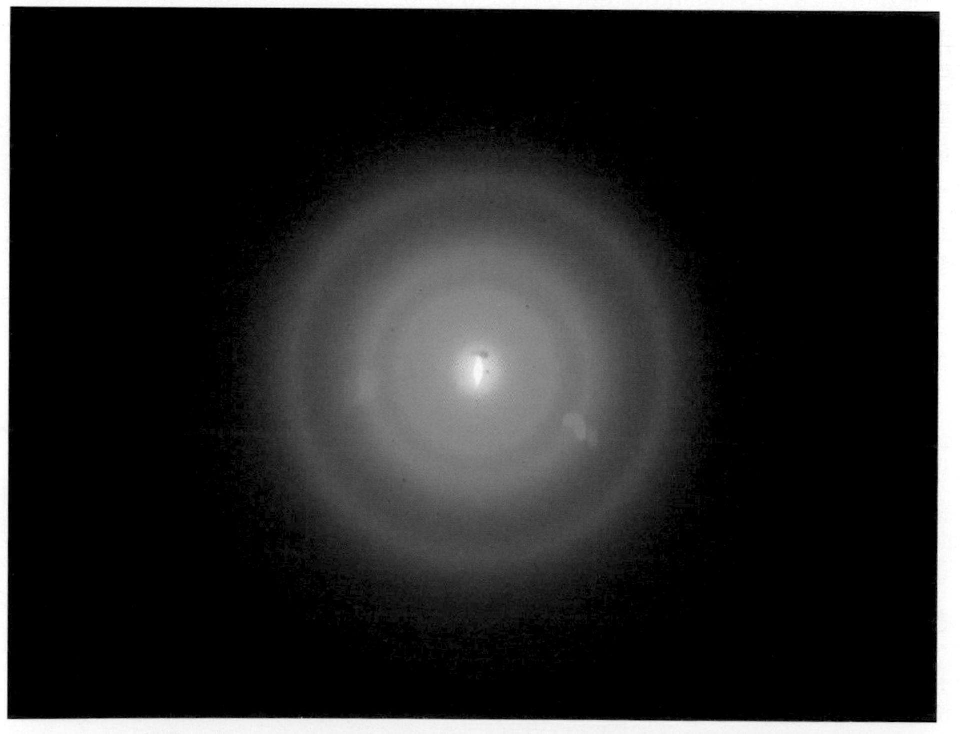

Figure 3.33 Diffraction pattern produced when an electron beam passes through polycrystalline graphite

Wave–particle duality

Study note

You met diffraction in the AS chapter *The Sound of Music*

Diffraction is something that happens to waves. When waves pass through a narrow gap, they spread out. The bright spots you see in the laser-light diffraction pattern are formed when many waves arrive at the same point in step with one another.

The fact that electrons can be diffracted tells us something striking about electrons: sometimes they seem to behave as particles, sometimes as waves. When they are moving freely as a beam, and when they strike the screen at the end of the tube, they behave as particles. When they pass between the atoms of a crystal of graphite, they behave like waves. This double behaviour is not like anything we experience on the macroscopic scale of things, but it is a phenomenon we have to come to terms with when considering the behaviour of matter and energy on the microscopic scale.

The rule is that sometimes we have to use wave ideas to explain our observations, and sometimes particle ideas. We can't say that under certain circumstances electrons are waves; they are neither waves nor particles, and all we can do is use our ideas of waves or particles to explain their behaviour as best we can. What we need is a way of translating back and forth between the two ways of describing electrons. In the AS chapter *Digging Up the Past*, you met this 'translation' – it is the de Broglie equation:

$$\lambda = \frac{h}{p} \qquad (19)$$

where p is the electron's momentum and λ is its wavelength (more correctly, its **de Broglie wavelength**, after Louis de Broglie (1892–1987) who found this relationship). The quantity h is the Planck constant:

$$h = 6.63 \times 10^{-34}\ \text{J s}$$

Equation 19 thus allows us to translate from a particle property (momentum) to a wave property (wavelength). Figure 3.34 shows a log–log graph relating λ to p. Notice that the gradient of the graph is negative (because the two quantities vary in inverse proportion). The gradient is –1 (because $\lambda \propto p^{-1}$).

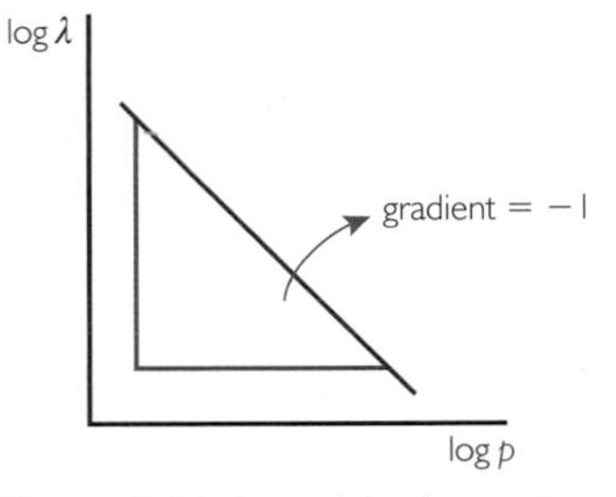

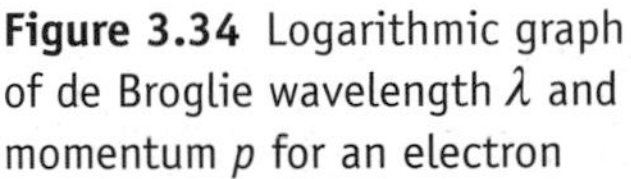
Figure 3.34 Logarithmic graph of de Broglie wavelength λ and momentum p for an electron

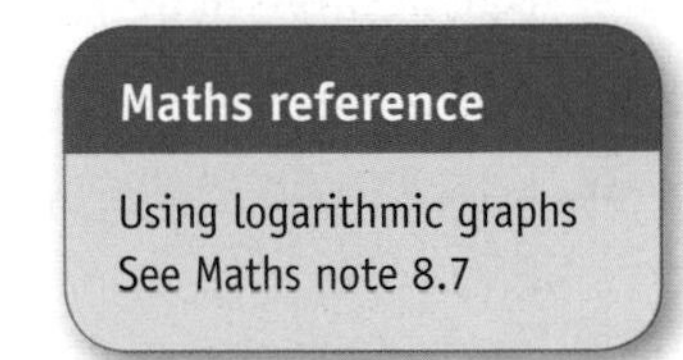
Maths reference

Using logarithmic graphs
See Maths note 8.7

In Activity 19 of this chapter and in the AS chapter *Digging Up the Past* you saw that, in order to analyse the structure of a material using a diffraction technique, the wavelength of the radiation should be comparable with the separation of the diffracting objects. So probing matter on a subatomic scale requires a shorter wavelength than that used to investigate structure on the scale of whole atoms or molecules. In Activity 20, you will explore how the de Broglie wavelength of electrons can be controlled. You will need to draw on ideas about charge and voltage that you met in earlier chapters (eg the AS chapter *Technology in Space*).

The electron beam is produced in an electron gun (Figure 3.35). Electrons are first released from a heated cathode by thermionic emission, and then accelerated by a high voltage applied between cathode and anode. They pass through a hole in the anode and then travel at constant speed as they pass through the diffracting object (a graphite crystal, for example) and then strike the fluorescent screen.

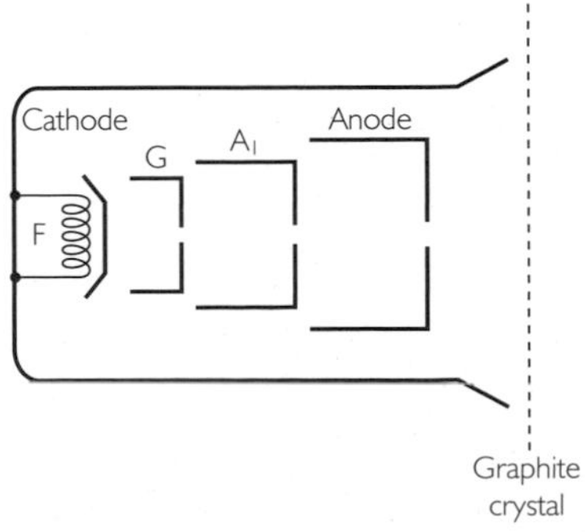

Figure 3.35 Electron gun

So long as there is no other energy transfer taking place, the kinetic energy, E_k, acquired by an electron as it is accelerated in the gun is given by:

$$E_k = eV \qquad (20)$$

where e is the electron's charge and V the potential difference between cathode and anode. Provided the electron's speed does not approach the speed of light, we can use the non-relativistic Equations 15, 19 and 20 to derive an expression for the de Broglie wavelength. From Equations 15 and 20:

$$p = \sqrt{(2mE_k)} = \sqrt{(2meV)}$$

and so, using Equation 19:

$$\lambda = \frac{h}{p} = \frac{h}{\sqrt{(2meV)}} \qquad (21)$$

Activity 20 Diffraction of electrons

Use apparatus such as that shown in Figure 3.32 to explore electron diffraction and the relationship between the de Broglie wavelength and accelerating voltage.

Questions

31 Calculate the kinetic energy, the speed and the momentum of an electron which is accelerated through a pd of 500 V.
(electron mass $m_e = 9.11 \times 10^{-31}$ kg; electron charge $e = 1.60 \times 10^{-19}$ C)

32 What potential difference is required to accelerate an electron to a speed of 1.0×10^6 m s^{-1}?

33 As electrons are accelerated from the cathode to the anode of an electron beam tube, does their de Broglie wavelength increase or decrease?

34 If the cathode–anode voltage is increased in an electron beam tube, does the de Broglie wavelength of the electrons in the beam increase or decrease?

35 Calculate the de Broglie wavelength of an electron moving at 1.0×10^6 m s^{-1}.
(electron mass $m_e = 9.11 \times 10^{-31}$ kg; Planck constant $h = 6.63 \times 10^{-34}$Js)

36 Calculate the momentum of an electron whose de Broglie wavelength is 3.0×10^{-10} m.

37 Estimate your own mass and your speed when walking. Find the order of magnitude of your own de Broglie wavelength, and hence explain why you do not notice the wave-like aspect of your nature — for example, you are not noticeably diffracted when walking through a metre-wide doorway.

Faster electrons

When electrons are diffracted as they pass through a graphite crystal, they are interacting with the electrons of the carbon atoms that make up the graphite. Electrons can be used to look at the nuclei of atoms, too, but if they are to be diffracted by something as small as an atomic nucleus, they must have a very short wavelength, similar to the dimensions of the nucleus itself.

Since atomic nuclei are of the order of 10^{-15} m in diameter, we need electrons whose wavelength is of this order too. To achieve this, the electrons must be accelerated through many millions of volts.

At this point, we need to be aware of the limitations of some of the equations used earlier. In particular, consider what happens if we try to calculate the speed of an electron that has been accelerated through 100 MV (= 1.00×10^8 V):

$$E_k = eV = 1.60 \times 10^{-19}\ \text{C} \times 1.00 \times 10^8\ \text{V}$$
$$= 1.60 \times 10^{-11}\ \text{J}$$

(No problem here.)

Assuming $E_k = \frac{1}{2}mv^2$:

$$v = \sqrt{\left(\frac{2E_k}{m}\right)}$$

$$= \sqrt{\left(\frac{2 \times 1.60 \times 10^{-11}\ \text{J}}{9.11 \times 10^{-31}\ \text{kg}}\right)}$$

$$= 5.9 \times 10^9\ \text{m s}^{-1}$$

This is greater than the speed of light ($c = 3.00 \times 10^8$ m s^{-1}), impossible (according to the theory of relativity). For speeds approaching the speed of light, we need to use a different expression for the particle's energy, derived from relativity. Particles that are moving at speeds that are greater than, say, $0.1c$ are described as relativistic, and some of the relationships between quantities such as velocity, kinetic energy and momentum have to be revised. In particular, we can no longer calculate a particle's kinetic energy using $E_k = \frac{1}{2}mv^2$.

Diffraction patterns

Figure 3.33 shows the diffraction pattern of electrons that have been diffracted by polycrystalline graphite. You saw a similar pattern in Activity 19 when light is diffracted by tiny dust particles. There is a bright central spot, surrounded by rings of decreasing brightness. The graph in Figure 3.36 shows how the brightness varies across the centre of this diffraction pattern. The central hump of the graph corresponds to the bright central spot; then there is a dip where the pattern is dark, then a lower peak for the first bright ring, and so on.

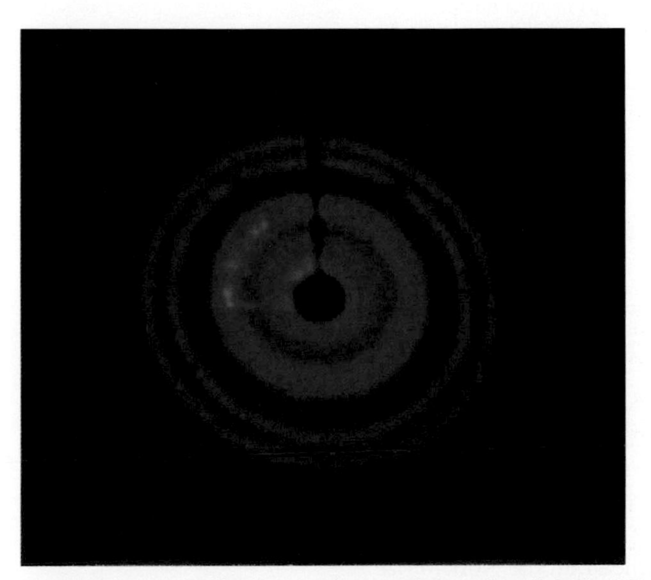

(b)

Figure 3.36 (a) Diffraction pattern produced when light is diffracted by a random array of fine particles; (b) graph showing how the brightness varies along the diameter of the pattern

The diffraction pattern can change in two ways: the electrons' de Broglie wavelength can be changed, as can the size of the particles that are causing the diffraction. We will concentrate on the size of the particles. With smaller and smaller particles, the diffraction rings get wider and wider. So measuring the diameter of the rings can tell us about the size of the tiny particles that are doing the diffracting.

Figure 3.37 shows what happens when high-energy electrons are diffracted by samples of carbon and oxygen. The de Broglie wavelength of the electrons is similar to the diameter of the nuclei, so the electrons are strongly diffracted. In this case, the results don't appear as a set of rings on a screen. Instead, the detector is moved round to different positions to measure the intensity of the diffracted electrons. The result is a graph of intensity against angle. You can see the similarity in shape between these graphs and the graph for the diffraction of light (Figure 3.36). In Figure 3.37(a) there is a central peak, then a minimum at an angle of about 50°, and then a smaller peak. From this, the diameter of the carbon nuclei can be determined.

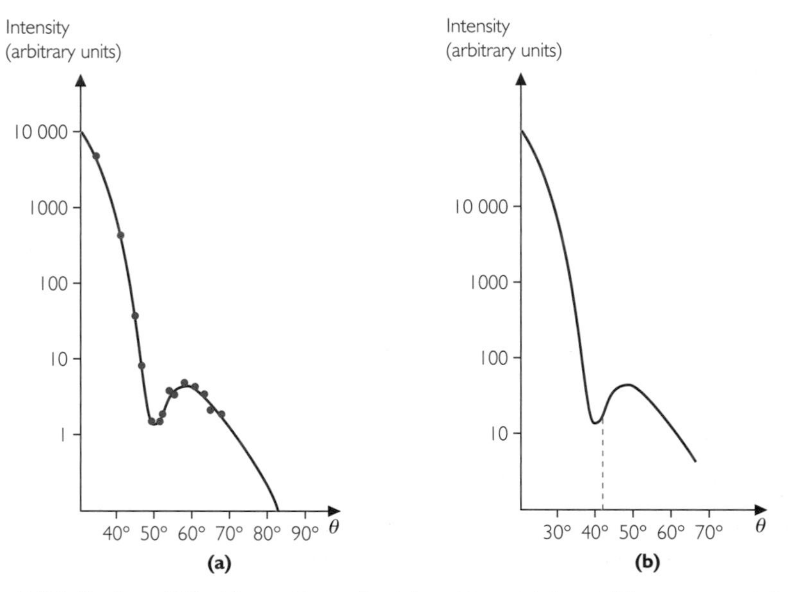

Figure 3.37 Electron diffraction patterns for (a) carbon nuclei and (b) oxygen nuclei.

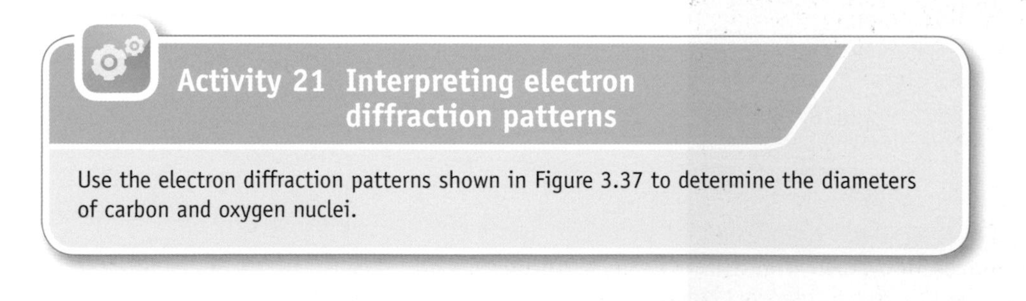

Activity 21 Interpreting electron diffraction patterns

Use the electron diffraction patterns shown in Figure 3.37 to determine the diameters of carbon and oxygen nuclei.

'Seeing' protons

The smallest and lightest of all atoms is the hydrogen atom. Its nucleus is a single proton. In the 1950s, experiments were done using electron diffraction with liquid hydrogen as the target. With electrons accelerated through nearly 1000 MV (10^9 V), Robert Hofstadter of Stanford University in California was able to estimate the diameter of the proton as 5.6×10^{-15} m. This result gained him the Nobel Prize for Physics in 1961.

As far as Hofstadter could tell, a proton was simply a single particle, not much different from a rather fuzzy billiard ball. The electrons he used interacted with the protons via the Coulomb force – their charges interacted. Hofstadter could find no evidence that the proton was anything other than a sphere of charge.

Today, as we have seen in Part 2 of this chapter, our picture of protons is rather different. High-energy electron accelerators have been used to probe right inside protons, and this has shown that the charge inside a proton is not uniformly distributed; each proton is made of three quarks tightly bound together. This three-fold structure was revealed by 'deep inelastic scattering' experiments in the late 1980s in which beams of high-energy electrons were fired at protons, rather as Rutherford had fired alpha particles at gold foil many years previously.

Activity 22 Deep inelastic scattering

Write a paragraph to explain why very high-energy electrons were needed to reveal the quarks which make up a proton. Include the following terms:

energy wavelength diffraction.

3.6 Summing up Part 3

In this part of the chapter, you have seen that the story of our understanding of the particles that make up matter has developed rapidly during the 20th century. This has been achieved by probing matter with increasingly energetic particles – electrons, alpha particles and others. The electrons used in deep inelastic scattering experiments had energies equivalent to being accelerated through tens or hundreds of billions of volts. In Part 4, we will go on to see how such high energies can be achieved.

You have looked at the development of our understanding of the structure of matter, starting from Rutherford's alpha particle scattering experiment – an understanding of electrostatic forces, developed in the 19th century, helped make this possible. You have seen how the conservation of momentum (a fundamental law of physics) is brought into the analysis of collisions, and how a knowledge of the wave-like nature of electrons (and other particles) explains their usefulness as a tool for probing the heart of matter.

Activity 23 Summing up Part 3

During the course of the 20th century, experiments using a variety of subatomic 'probes' were carried out to investigate the structure of atoms, and to find out about other aspects of the fundamental nature of matter. Draw a timeline to show this, giving dates and brief descriptions of the experiments, along with any other details you wish to include. Include as many as possible of the terms printed in bold – and make sure you know their meanings. Refer back to Part 2 for information about some other experiments that you might show on your timeline.

4 Particle beams and accelerators

In this part of the chapter, you will find out how charged particles are accelerated to high energies, so that they can be used in experiments to recreate the conditions in the early stages of the history of the universe.

4.1 Big experiments

CERN is the European centre of nuclear research. It sits between the Alps and the Jura mountains, near Geneva in Switzerland. It's a great place to work if you're a particle physicist and you enjoy skiing. The photograph in Figure 3.38 shows an aerial view of CERN, with the positions of the underground tunnels marked. The main particle accelerator is circular in shape. It is so large – 8.6 km across – that it extends under the border into France.

Figure 3.38 The CERN lab near Geneva

One reason for building this vast particle-physics laboratory was political. During the Second World War, scientists had been fighting each other, competing to develop the nuclear bomb. After the war, it was hoped that a European centre could be built so that scientists from different countries could work together for non-military research. The Conseil Européen pour le Recherche Nucléaire (its original name) was established in 1952. It was the first international organisation that Germany joined after the war. There are now (2008) 20 European member states. CERN employs around 2500 people. The laboratory's scientific and technical staff design and build the particle accelerators, ensure their smooth operation, and also help prepare, run, analyse and interpret the data from complex scientific experiments. Some 8000 visiting scientists – half of the world's particle physicists – come to CERN for their research. They represent 580 universities and 85 nationalities.

Particle accelerators such as those at CERN are so expensive that they can only work if many countries contribute. The member states' contribution to the CERN budget in 2008 was £540 million, of which the UK contributed 17.35%. It sounds a lot, but works out at about 75p per person in Europe. Other non-member countries such as the USA, China and Japan also contribute in exchange for their scientists taking part in the experiments. The CERN council has to work hard to justify the money spent and tries to keep costs down. An example of this is the closing down of the Large Electron Position collider (LEP) in order use the tunnel for the Large Hadron Collider (LHC).

Activity 24 Funding physics

Discuss the following in small groups. You may find that you have opinions, but need to look up information to back up your arguments. A visit to the *Physics and Ethics Education Project* website might help (see **www.shaplinks.co.uk**).

- Is it ever right for politics to interfere with scientific research?
- Are there any examples from the 20th or 21st centuries where politics has played a part in the development of physics research?
- Do the developments of physics ever affect political decisions? Try to use some contemporary examples.

People and ideas

Physicists like to develop theories that are simple. They are currently trying to come up with one theory that can explain the four fundamental forces (electromagnetic, weak, strong and gravitational) in terms of a single theory – rather as, in the 19th century, electrostatics and magnetism were unified through the theory of electromagnetism. In the 1960s, the electromagnetic and weak forces were unified in this way. This theory then predicted the existence of so-called W- and Z-boson particles. These particles have a lot of mass. This led teams of scientists to try to accelerate particles to high energies, collide them and try to create the Z and W particles. In 1983, Italian physicist Carlo Rubbia headed a team of 100 physicists at CERN that detected these particles. This helped to confirm the electroweak unified theory. Carlo Rubbia and his colleague Simon van de Meer got the Nobel Prize for this work.

Scientists are now using the LHC to help them probe the mysterious dark matter and dark energy of the universe, to search for a particle called the Higgs boson (which has been predicted by theory and is believed to be responsible for the masses of other particles) and explore possible extra dimensions of space–time.

If you ever visit CERN, you will see people from many different countries. Because particle-physics research involves using expensive machines, single universities and research institutes rarely do research in isolation. Organisations collaborate not just within one country, but internationally. If you look at scientific papers written today, they list all the scientists who have contributed to the project, and in particle physics this may mean scores of names (Figure 3.39).

Someone once joked that broken English, not English, is the international language at CERN. In the cafeteria (Figure 3.40) you can hear physicists arguing about the meaning of the results they have just analysed. Much of this work involves mathematical modelling, computer programming and careful, detailed discussion. All of the skills that they learn and practise are transferable to other disciplines.

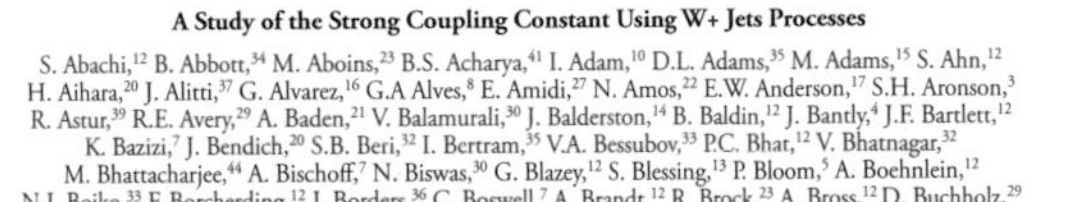

A Study of the Strong Coupling Constant Using W+ Jets Processes

S. Abachi,[12] B. Abbott,[34] M. Aboins,[23] B.S. Acharya,[41] I. Adam,[10] D.L. Adams,[35] M. Adams,[15] S. Ahn,[12] H. Aihara,[20] J. Alitti,[37] G. Alvarez,[16] G.A Alves,[8] E. Amidi,[27] N. Amos,[22] E.W. Anderson,[17] S.H. Aronson,[3] R. Astur,[39] R.E. Avery,[29] A. Baden,[21] V. Balamurali,[30] J. Balderston,[14] B. Baldin,[12] J. Bantly,[4] J.F. Bartlett,[12] K. Bazizi,[7] J. Bendich,[20] S.B. Beri,[32] I. Bertram,[35] V.A. Bessubov,[33] P.C. Bhat,[12] V. Bhatnagar,[32] M. Bhattacharjee,[44] A. Bischoff,[7] N. Biswas,[30] G. Blazey,[12] S. Blessing,[13] P. Bloom,[5] A. Boehnlein,[12] N.I. Bojko,[33] F. Borcherding,[12] J. Borders,[36] C. Boswell,[7] A. Brandt,[12] R. Brock,[23] A. Bross,[12] D. Buchholz,[29] V.S. Burtovoi,[33] J.M Butler,[12] D. Casey,[36] H. Castilla-Valdez,[9] D. Chakraborty,[39] S.M. Chang,[27] S.V. Chekulaev,[33] L.-P. Chen,[20] W. Chen,[39] L. Chevalier,[37] S. Chopra,[32] B.C. Choudhary,[7] J.H. Christenson,[12] M. Chung,[15] D. Claes,[39] A.R. Clark,[20] W.G. Cobau,[24] J. Cochran,[7] W.E. Cooper,[12] C. Cretsinger,[36] D. Cullen-Vidal,[4] M.A.C. Cummings,[14] D. Cutts,[4] O.I. Dahl,[20] K. De,[42] M. Demarteau,[12] R. Demina,[27] K. Denisenko,[12] N. Denisenko,[12] D. Denisov,[12] S. P. Denisov,[33] W. Dharmaratna,[13] H.T. Diehl,[12] M. Diesburg,[12] G. Di Loreto,[23] R. Dixon,[12] P. Draper,[42] J. Drinkard,[6] Y. Ducros,[37] S.R. Dugad,[41] S. Durston-Johnson,[36] D. Edmunds,[23] J. Ellison,[7] V.D. Elvira,[12] R. Engelmann,[39] S. Eno,[21] G. Eppley,[35] P. Ermolov,[24] O.V. Eroshin,[33] V.N. Evdokimov,[33] S. Fahey,[23] T. Fahland,[4] M. Fatyga,[3] M.K. Fatyga,[36] J. Featherly,[3] S. Feher,[39] D. Fein,[32] T. Ferbel,[36] G. Finnochiaro,[39] H.E. Fisk,[12] Yu. Fisyak,[24] E. Flattum,[23] G.E. Forden,[2] M. Fortner,[28] K.C. Frame,[23] P. Franzini,[10] S. Fuess,[12] A.N. Galjaev,[33] E. Gallas,[42] C.S. Gao,[42] S. Gao,[12],T.L. Geld,[23] R.J. Genik II,[23] K. Genser,[12] C.E. Gerber,[12] B. Gibbard,[3] V. Glebov,[36] S. Glenn,[5] B. Gobbi,[29] M. Goforth,[13] A. Goldschmidt,[20] B. Gomez,[1] P.I. Goncharov,[33] H. Gordon,[3] L.T. Goss,[43] N. Graf,[3] P.D. Grannis,[39] D.R. Green,[12] J. Green,[28] H. Greenlee,[12] G. Griffin,[6] N. Grossman,[12] P. Grudberg,[20] S. Grunedahl,[36] W. Gu,[12] G. Guglielmo,[31] J A. Guida,[39] W. Guryn,[3] S.N. Gurzhiev,[33] P. Guierrez,[31] Y.E. Gutnikov,[33] N.J. Hadley,[21] H. Haggerty,[12] S. Hagopian,[13] V. Hagopian,[13] K.S. Hahn,[36] R.E. Hall,[6] S. Hansen,[12] R. Hatcher,[23] J.M. Hauptman,[17] D. Hedin,[28] A.P. Heinson,[7] U. Heintz,[12] R. Hernandez-Montoya,[9] T. Heuring,[13] R. Hirosky,[13] J.D. Hobbs,[12] B.Hoeneisen,[1] J.S. Hoftun,[4] F. Hsieh,[22] Ting Hu,[39] Tong Hu,[16] T.Huehn,[7] S. Igarashi,[12] A.S. Ito,[12] E. James,[2] J. Jaques,[30] S.A. Jerger,[23] J.Z-Y Jiang,[39] T. Joffe-Minor,[29] H. Johari,[27] K. Johns,[2] M. Johnson,[12] H. Johnstad,[40] A. Jonckheere,[12] M. Jones,[44] H. Jostlein,[12] S. Y. Jun,[29] C.K. Jung,[39] S. Kahn,[3] G. Kalbfleisch,[34] J. S. Kang,[18] R. Kehoe,[39] M.L Kelly,[30] A. Kernan,[7] L. Kerth,[20] C.L. Kim,[18] S.K. Kim,[38] A. Klatchko,[13] B. Klima,[12] B.I. Klochkov,[33] C Klopfenstein,[39] V.I. Klyukhin,[33] V.I. Kochetkov,[33] J.M. Kohli,[32] D. Koltick,[34] A. V. Kostritskiy,[33] J. Kotcher,[3] J. Kourlas,[24] G. Landsberg,[12] R.E. Lanour,[4] J-F. Lebrat,[37] A. Leflat,[24] H. Li,[39] J. Li,[42] Y.K. Li,[29] Q.Z. Li-Demarteau,[12] J.G.R. Lima,[8] D. Lincoln,[22] S.L. Linn,[13] J. Linnemann,[23] R. Lipton,[12],Y.L. Liu,[29] F. Lobkowicz,[36] S.C. Loken,[29] S. Lokos,[39] L. Lucking,[12] A.L. Lyon,[21] A.K.A Maciel,[8] R.J. Madaras,[20] R. Madden,[13] I.V. Mandrichenko [33] Ph. Mangeout,[37] S. Mani,[5] B. Mansouie,[37] H.S Mao,[12] S Margulies,[15] R. Markeloff,[28] L. Markosky,[2] J. McKinley,[23] T. McMahon,[34] H.L. Melanson,[12] J.R.T. de Mello Neto,[8] K.W. Merritt,[12] H. Miettinen,[35] A. Milder,[2] A. Mincer,[26] J.M. de Miranda,[8] C.S. Mishra,[12] M. Mohammadi-Baarmand,[39] N. Mokhov,[12] N.K. Mondal,[41] H.E. Montgomery,[12] P. Mooney,[1] M. Mudan,[26] C. Murphy,[16] C.T. Murphy,[12] F. Nang,[4] M. Narain,[12] V.S. Narasimhan,[41] A. Narayanan,[2] H.A. Neal,[22] J.P. Negret,[1] E. Neis,[22] P. Nemethy,[26] D. Nesic,[4] D. Norman,[43] L. Oesch,[22] V. Oguri,[8] E. Oltman,[29] N. Oshima,[12] D. Owen,[23] P. Padley,[35] M. Pang,[17] A. Para,[12] C.H. Park,[12] Y.M. Park,[19] R. Partridge,[4] N. Parua,[41] M. Paterno,[36] J. Perkins,[12] A. Peryshkin,[12] M. Peters,[44] H. Piekarz,[13] Y. Pischalnikov,[34] A. Pluquet,[37] V.M. Podstavkov,[33] B.G. Pope,[23] H.B. Prosper,[13] S. Protopescu,[3] D. Puseljic,[20] J. Qian,[22] P.Z. Quintaz,[12] R. Raja,[12] S. Rajagopalan,[39] O. Ramirez,[15] M.V.S Rao,[41] P.A. Rapidis,[12] L. Rasmussen,39 A.L. Read,[12] S. Reucroft,[27] M. Rijssenbeek,[39] T. Rockwell,[23] N.A. Roe,[20] P. Rubinov,[39] R. Ruchti,[30] S. Rusin,[24] J. Rutherfoord,[2] A. Santoro,[8] L. Sawyer,[42] R.D. Schamberger,[39] H. Schellman,[29] J. Sculli,[26] E. shabalina,[24] C. Shaffer,[13] H.C. Shankar,[41] R.K. Shivpuri,[11] M.Shupe,[2] J.B. Singh,[32] V. Sirotenko,[28] W. Smart,[12] A. Smith,[2] R.P. Smith,[12] R. Snihur,[29] G.R. Snow,[25] S. Snyder,[39] J. Solomon,[45] P.M. Sood,[32] M. Sosebee,[42] M. Souze,[8] A.L. Spadafora,[20] R.W. Stephens,[42] M.L. Stevenson,[20] D. Stewart,[22] D.A. Stoianova,[33] D. Stoker,[6] K. Streets,[26] M. Strovink,[20] A. Taketani,[42] P. Tamburello,[21] J. Tarazi,[6] M. Tartaglia,[12] T.L. Taylor,[29] J. Teiger,[37] J. Thompson,[21] T.G. Trippe,[20] P.M Tuts,[10] N. Varelas,[23] E.W. Varnes,[20] P.R.G Virador,[20] D. Vititoe,[2] A.A. Volkov,[33] A.P. Vorobiev,[33] H.D. Wahl,[13] G. Wang,[13] J. Wang,[12] L.Z. Wang,[12] J. Warchol,[30] M. Wayne,[30] H. Weerts,[23] F. Wen,[13] W.A. Wenzel,[20] A. White,[42] J.T. White,[43] J.A. Wightman,[17] J. Wilcox,[27] S. Willis,[28] S.J. Wiimpenny,[7] J.V.D. Wirjawn,[43] J. Womersley,[12] E. Won,[36] D.R. Wood,[12] H. Xu,[4] R. Yamada,[12] P. Yamin,[3] C. Yanagisawa,[39] J. Yang,[26] T. Yasuca,[27] C. Yoshikawa,[14] S. Yousef,[43] J. Yu,[36] Y. Yu,[38] Y. Zhang,[12] Y.H. Zhou,[12] Q. Zhu,[26] Y.S. Zhu,[12] Z.H. Zhu,[36] D. Zieminska,[16] A. Ziminski,[16] and A. Zylberstejn[37]

Figure 3.39 Particle-physics research involves collaboration between physicists of many different nationalities

Most young physicists do not remain in physics after their PhD at CERN. Some of them go on to work in the City as merchant bankers, stockbrokers, managers etc. In these jobs they are using many of the same skills that they needed in their scientific research.

Figure 3.40 Discussions in the CERN cafeteria

Over the years, it became increasingly important to be able to send data from CERN back to the scientists' home universities overseas very quickly to help speed up analysis. The World Wide Web was established by scientists at CERN to help them communicate across different continents as well as countries. As the Web was CERN's response to a new wave of scientific collaboration at the end of the 1980s, so the Grid is the answer to the need for increased data analysis required by the world particle-physics community. With the LHC, the CERN experiments have petabytes (10^{15} bytes) of information to analyse, and as a result they have developed a so-called Grid, which allows access to the processing power of many computers distributed around the world. The Enabling Grids for E-sciencE (EGEE) infrastructure has been developed for Europe and there is another Grid system for the US.

Helenka Przysiezniak completed a master's degree in Canada as a theoretician in astrophysics and decided to do her PhD in particle physics at CERN. In the *Times Higher Education Supplement*, 24 April 1998, she described the impromptu debates that happen in corridors, in the cafeteria (Figure 3.40) and in offices at CERN:

> *It sounds like you're having a big argument, but actually you're just discussing things. You want to prove that something is right if you believe in it. That's just how it works when you're discussing the 'truth'. You can't survive as a physicist unless you have a bit of ego.*

Przysiezniak went on to suggest that physicists tend to be just as passionate about their other interests – many are accomplished musicians and concerts are frequently held at CERN. The mountains and lakes around CERN attract physicists who are keen skiers, mountaineers and sailors.

Activity 25 Particle physics today

Use the internet to find some information about current and planned particle-physics experiments.

With a small group of students, share your opinions on the way current particle-physics research is organised and funded. Is this different from other areas of scientific research?

4.2 Achieving high energy

In the 1960s as particle physics developed there grew a demand for higher and higher energy accelerators. Partly this demand was fuelled by the desire of the engineer and practical physicist to see what could be built and discovered at higher energies. At the same time, theoreticians wanted experiments that could test the predictions of the standard model of fundamental particles and forces. The theory predicted the existence of quarks; it also suggested that there were much more massive particles. To discover these particles required bigger accelerators capable of delivering more energy.

Particle physics is often known as high-energy physics (HEP). As you have seen in Parts 2 and 3, there are three main reasons why high energies are used by particle physicists.

- If you want a positively charged particle to get closer to a nucleus of an atom, a lot of energy must be supplied.
- The more energy that can be given to particles, the shorter their wavelength and the smaller the detail that can be investigated using them as a probe.
- By colliding particles together, the energy is re-distributed, producing new particles. The higher the collision energy, the larger the mass of particles that can be produced.

As theory predicted different particles, this drove the advancement of the technology – more energetic machines were needed. Sometimes unexpected results were found and this led to refinements in the theory, which sometimes needed even higher energies in order to be tested.

You have already encountered particle accelerators that operate on essentially the same principle as those used in particle-physics research. In the electron gun that forms part of any cathode-ray tube, electrons are accelerated in an electric field produced by applying a potential difference between the cathode and anode. In such a tube used in an old-style TV set, or in demonstrating electron beams in the laboratory, electrons are accelerated through a few thousand volts at most. In particle-physics experiments, much higher energies are required. This has several consequences. One is that the required energies cannot be achieved all in one go; accelerators have to be designed that 'kick' the particles many times over, increasing their energy each time. Another consequence is that the equations of classical physics break down at speeds close to that of light, and to describe and predict the behaviour of such energetic particles, the equations of **special relativity** must be used. The second of these is related to some of the units that particle physicists use, and that you are likely to come across in your reading about particle physics.

Relativistic effects

In Section 3.5, you saw that, according to classical physics, an electron accelerated through 100 MV would travel faster than light. This is actually impossible, and the incorrect prediction arose because we used inappropriate equations. In particle-physics accelerators, electrons (and other charged particles, such as protons and positrons) are accelerated through potential differences of several gigavolts (1 GV = 10^9 V) or even teravolts (1 TV = 10^{12} V).

Accelerators give particles energy. When particles are accelerated, they go faster, but they can never reach the speed of light. Einstein was able to show that any particle that has mass when at rest can never be accelerated to the speed of light. Even if we keep giving the particle energy when it is close to the speed of light, its speed will keep increasing but it will never reach that elusive 3×10^8 m s^{-1}. The reason for this is that the particle's mass keeps increasing, the more energy it is given. (You might guess this from the famous equation $E = mc^2$. The energy you give the particle reappears as mass.) And so the particle becomes more and more massive, and more and more difficult to accelerate. That's why, unless you can find a loophole in Einstein's theory of relativity, faster-than-light travel will remain in the realm of science fiction.

In 1964, US physicist William Bertozzi conducted an experiment to find the speeds of electrons accelerated by electric fields. Figure 3.41 shows Bertozzi's apparatus. The electrons were effectively accelerated through potential differences up to 15 MV. The 'time-of-flight' of the electrons between the two detectors was measured using an oscilloscope, enabling their speeds to be found. To check that their energies were as expected, the electrons were allowed to strike a target, which heated up. Bertozzi measured their speeds and found that the electrons did not exceed the speed of light and the relationship between their speed and energy was as predicted by Einstein.

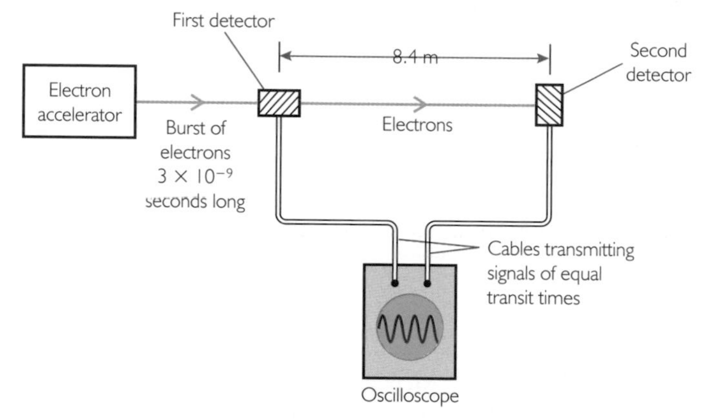

Figure 3.41 Bertozzi's apparatus

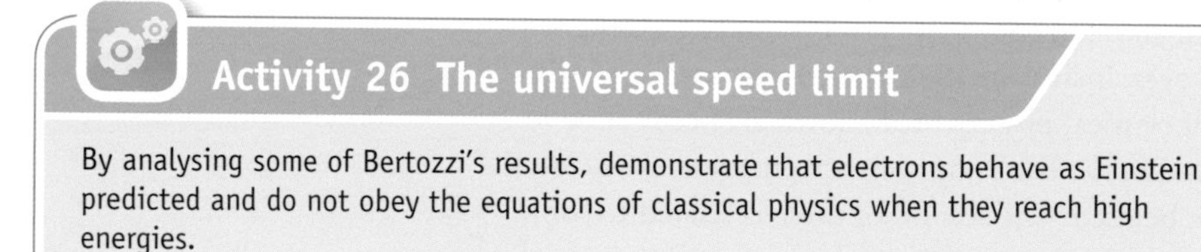

Activity 26 The universal speed limit

By analysing some of Bertozzi's results, demonstrate that electrons behave as Einstein predicted and do not obey the equations of classical physics when they reach high energies.

Units

When a particle has been accelerated to a very high energy, it is more useful to talk about its energy than its speed – its speed is in any case very nearly the speed of light, regardless of any additional energy it acquires, and if you want to know what a particle can do (eg create new particles in a collision) then its energy is the most important thing to know. For this reason, particle physicists nearly always use a system of units that relates directly to particle energies.

When talking about the energy of an electron (or other subatomic particles), it's rather awkward to work in joules, with all those negative powers of ten. Instead, we often state its energy in units of electronvolts (eV).

Study note

You met the eV in the AS chapter *Technology in Space*.

The kinetic energy of an electron accelerated through 1 V is 1 electronvolt (1 eV). The magnitude of electron charge, $e = 1.60 \times 10^{-19}$ C, so:

$$1 \text{ eV} = 1.60 \times 10^{-19} \text{ J}$$

The most energetic accelerators can produce particles with energies in the giga-electron-volt (GeV) or even tera-electron-volt (TeV) range. Remember that the electron-volt is a unit of energy, not of voltage. It can be used for any particle, not just electrons – and they do not have to be electrically accelerated.

Particle physicists commonly express mass in units that, at first, look a little odd, but do in fact make their life simpler. In Part 2 you met Einstein's equation:

$$E = mc^2 \qquad \text{(Equation 1)}$$

We can use this to find the energy associated with a particle when it is at rest – the energy that would be liberated if its entire **rest mass** were to dematerialise or, equivalently, the energy needed to create the particle 'from nothing'. This energy, E_0, is called the particle's **rest-mass energy**, (or, sometimes, just its rest energy). An example shows how this leads to new (non-SI) units for mass.

An electron at rest has mass $m_0 = 9.11 \times 10^{-31}$ kg. Its energy equivalent is:

$$E_0 = 9.11 \times 10^{-31}\ \text{kg} \times (3.00 \times 10^8\ \text{m s}^{-1})^2$$
$$= 8.199 \times 10^{-14}\ \text{J} = 5.12 \times 10^5\ \text{eV} = 0.512\ \text{MeV}$$

We can express the mass in terms of its energy equivalent, again using Equation 1:

$$m_0 = \frac{E_0}{c^2} = 0.512\ \text{MeV}/c^2$$

Rather than putting in numbers and doing any calculations, particle physicists leave this expression just as it is! The mass of the electron is now expressed in the rather odd-looking units of MeV/c^2.

Now imagine the electron has been accelerated so that its kinetic energy is 1 GeV. Its mass is still given by Equation 1, where E is its total energy (rest energy and kinetic energy combined – though now its kinetic energy vastly exceeds its rest energy):

$$m = \frac{E}{c^2} = 1\text{GeV}/c^2$$

Without needing to convert back to kg, you can see that the mass equivalent to the electron's total energy is now about two thousand times its rest mass (as 1 GeV = 10^3 MeV).

Now suppose this 1 GeV electron were to collide head-on with a positron that had also been accelerated to 1 GeV. Given that a muon has rest mass 0.106 GeV/c^2, would the collision be able to produce a muon-anti-muon pair? We can see straight away that it would be possible to produce *many* such pairs, as the energy of the electron and positron greatly exceeds the energy associated with a muon and an anti-muon at rest.

So by expressing particle masses in units of MeV/c^2 or GeV/c^2, particle physicists can tell whether there is enough energy in a given collision to produce a given particle, without having to do any arithmetic at all.

Another non-SI unit that particle physicists find convenient to use is the **atomic mass unit** (u). This is defined as one-twelfth of the mass of a carbon-12 atom. The hydrogen atom was the first choice for the standard of atomic mass, but as there are several isotopes of hydrogen that are difficult to separate out, carbon 12 was chosen. The atomic mass unit is useful because experimentally it is much easier, and more precise, to compare the masses of atoms and molecules (relative masses) than to measure their absolute masses.

Study note

The atomic mass unit is widely used in nuclear physics, which is one of the topics that you will study in the chapter *Reach for the Stars*. In that chapter you will learn more about isotopes.

The mass of a ^{12}C atom is 1.99×10^{-26} kg, so:

$$1\text{ u} = \frac{1.99 \times 10^{-26}}{12} = 1.66 \times 10^{-27}\text{ kg}$$

Questions

38 A proton has rest mass $m = 1.67 \times 10^{-27}$ kg.

(a) What is its rest energy (i) in J (ii) in GeV?

(b) What is its rest mass expressed in units of GeV/c^2?

39 A pi-zero meson has a rest mass of 0.14 GeV/c^2. How many such particles could, in principle, be produced in a collision between an electron and a positron each with energy 1 GeV?

40 Suppose an electron is accelerated to an energy of 20 GeV. Express its mass in units of GeV/c^2.

41 Show that 1 u is equivalent to 931 MeV.

42 Express the rest mass of an electron in atomic mass units.

Designing accelerators

In 1928, a Norwegian, Rolf Wideroe, invented the **linear accelerator** (LINAC). A LINAC accelerates charged particles to very high energies without the need for high voltages. Instead of being accelerated once, charged particles travel along a series of tubes separated by gaps, and are given voltage kicks at each gap.

In a LINAC there is a series of tubular electrodes (see Figure 3.42). These are connected to an alternating voltage supply, so that the voltage of each electrode switches back and forth between positive and negative. The aim is that the particle should emerge from the end of one tube and find that the next tube attracts it.

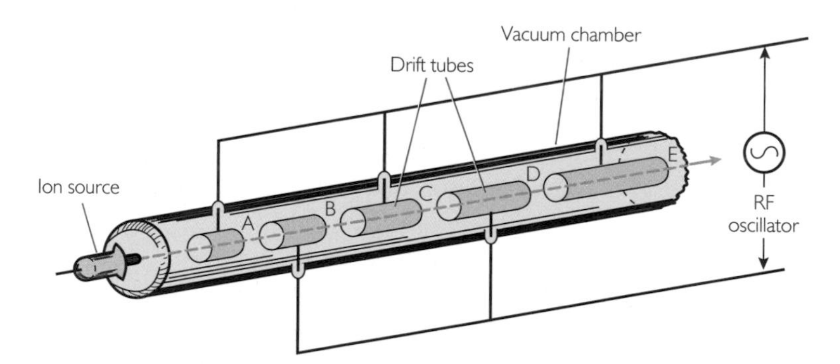

Figure 3.42 Construction of a linear accelerator

A proton leaves the ion source when electrode A is negative. It accelerates towards A, and then travels at a steady speed inside the tube. While it is inside the electrode, the voltage switches so that A is positive and B is negative. Now, when the proton enters the gap between A and B, it accelerates towards B. When it is inside B, again the voltage is switched on so that on exiting, B is positive and C is negative.

Each time, the same magnitude of voltage V is used and so the energy of the particle is built up in steps. After n electrode gaps, the proton's kinetic energy is equal to $n \times e \times V$.

An alternating voltage is used, so it can easily be stepped up using transformers. The frequency of the oscillating voltage is 'radio frequency' (ie a few MHz) and is kept constant. What does this mean for the design of the electrode drift tube tubes? They need to get longer as the proton gains speed so that the time to move between gaps is constant.

However, to achieve very high energies, very long accelerators are needed. The longest linear accelerator is at Stanford, California (Figure 3.43). The Stanford Linear Accelerator (SLAC) is 3 km long and can produce 20 GeV–50 GeV electrons. This has been used to smash electrons into protons and neutrons to help probe their structure. It was using this accelerator that physicists managed to show that protons and neutrons have a substructure of three quarks (see Section 3.5). High-energy physicists are now collaborating on a new project, the International Linear Collider (ILC) to be built in the USA. It will consist of two linear accelerators facing each other and producing collisions between beams of electrons and positrons at energies around 500 GeV.

Figure 3.43 Stanford Linear Accelerator

LINACs have the advantage that high voltages are not required. The main limitation of a LINAC, however, is that each accelerating section is only used once, and so to achieve higher energies the machines must be made longer and so are more expensive.

Although LINACs are still the most common type of accelerator, another type of accelerator was conceived that could achieve higher energies – a circular accelerator, in which charged particles went round and round. The same accelerating sections could be used again and again, giving the particles more energy each time they came around. In this way, not only could the energy of a particle be increased, but two particles (oppositely charged) could be accelerated in opposite directions around the circle and then made to collide – increasing the total energy of interactions further.

The first circular accelerator was the **cyclotron**, invented by the American Ernest Lawrence in 1929. He used the idea (developed by Wideroe for the linear accelerators) of accelerating charged particles in a gap between electrodes, then allowing them to drift to the next gap. But rather than let the particles drift to the next gap along a straight line, requiring more electrodes, he used a magnetic field to bring the particles round in a circle to be accelerated through the same gap every half-turn. Figure 3.44 shows a simplified illustration of the cyclotron.

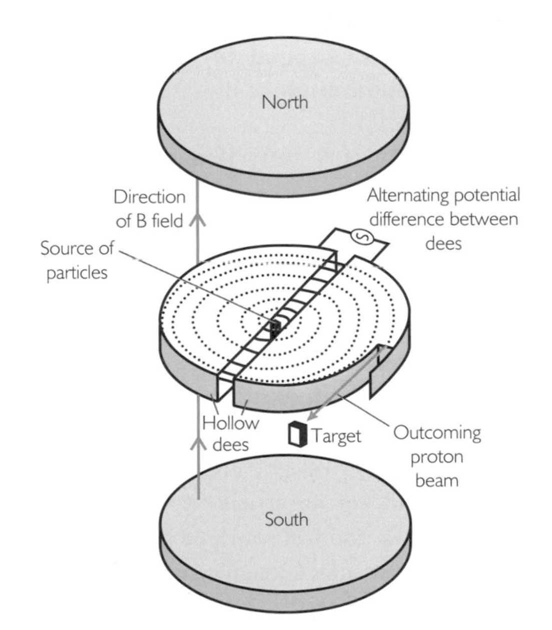

Figure 3.44 Principle of the first cyclotron

The electrodes were two hollow semi-circular pieces of metal – called 'dees' because of their shape – inside which the particles orbited. In 1931 Lawrence achieved a proton energy of 80 keV with a cyclotron 11 cm in diameter. The following year he achieved 1 MeV with a 26 cm diameter instrument. To reach energies much beyond 25 MeV, a new type of cyclotron had to be developed. As they reach higher kinetic energies, the particles' mass increases, which means that they take longer to complete each orbit. In a synchro-cyclotron the frequency of the accelerating voltage is reduced as the kinetic energy of the particle increases – the accelerating 'kicks' are synchronised with the particles' orbital motion. To reach still higher energies, the design has to be modified again, and in modern large accelerators such as those at CERN, the particles are steered around in a path of constant radius as their energy is increased.

4.3 Circular motion

To see how particles can be steered in a circular path, we need to consider circular motion in general and also to see how charged particles behave in magnetic fields. These two aspects of particle accelerators are the subjects of this section.

Study note

You met Newton's first law of motion in the AS chapter *Higher, Faster, Stronger*.

Finding the force

How can an object be made to travel in a circular path at constant speed? In the absence of any net force, a moving object continues along a straight line at constant velocity, as described by **Newton's first law of motion**. To produce circular motion, then, a net force must be acting. An object in circular motion is constantly changing its direction – in other words, it is accelerating in the sense that its velocity is changing, but the change is one of direction not of speed.

A few examples will help to identify the forces that produce motion in a circle. Imagine that you are the hammer thrower shown in Figure 3.45. The hammer is massive. You have to whirl it around so that it builds up speed before you release it. If your grip isn't strong enough, the hammer will fly off before you mean to let go. To keep the hammer moving round, you need to keep pulling on it. You can feel the hammer pulling on you outwards as you pull it inwards. So an inwards force is needed to keep the hammer moving along its circular path.

Figure 3.45 Throwing the hammer

As a second example, consider the motion of the Moon or an artificial satellite around the Earth (Figure 3.46). (The motion is not quite circular, but it is very nearly so.) What forces are acting on a satellite as it orbits? There is no air resistance slowing the satellite down, as space is essentially a vacuum. The only force is the gravitational attraction between the Earth and the satellite. If this is the only force acting on the satellite, then it must be the force causing the satellite to accelerate as it orbits. The force is directed towards the Earth – towards the centre of the circle. In fact, for any object to move in a circle, a force towards the centre of the circle is required.

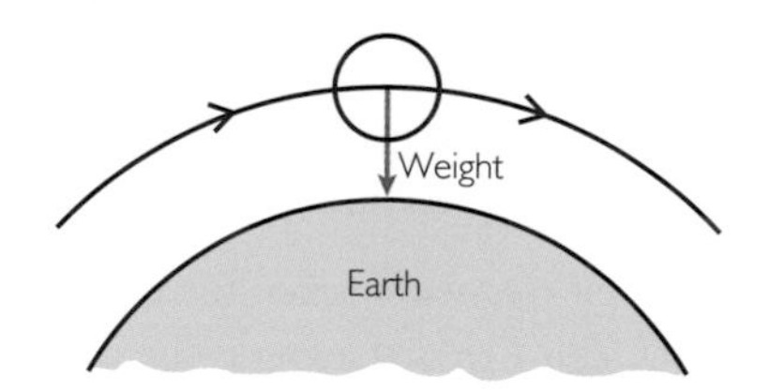

Figure 3.46 Satellite orbiting Earth

Centripetal, not centrifugal

Any force that acts towards the centre of a circle is described as a **centripetal force**. The word 'centripetal' means 'directed towards the centre' or 'centre seeking'.

Because of our experience of circular motion, we can believe that such motion involves a force acting radially outwards from the centre of a circle – a centrifugal ('centre-fleeing') force. When you are in a car that is turning a bend quickly, you feel thrown outwards, away from the centre of the circle. But there is no force actually pushing you outwards. The illusion arises because you are experiencing your motion relative to the car. Seen relative to a car cornering, you may seem to be moving outwards, but relative to the ground, you are actually trying to continue in a straight line – just as Newton's first law states. You will continue to move in a straight line until a resultant force acts on you that provides enough centripetal force to move you in the same circle as the car. This is normally provided by a seat-belt and friction from the seat, or by the side of the car if you're not belted up.

Table 3.9 lists some further examples of circular motion, with the name of the force (or forces) that act as the centripetal force in each case. In Activities 27 and 28, you explore two more examples of circular motion.

Example of circular motion	Inward force
Earth orbiting the Sun	Gravitational attraction between the Sun and Earth
Car rounding a bend on a flat road	Frictional force of the road on the tyres
Plane turning in a horizontal circle	Lift force on the plane's tilted wings
Car rounding a bend on a banked track (Figure 3.47)	Normal reaction of the road (and friction)

Table 3.9 Some examples of circular motion

Figure 3.47 Car rounding a bend on a banked track

Activity 27 Whirling a bung

Take a rubber bung and tie it to one end of a piece of string about 1.5 m long. Find a suitable safe space and whirl the bung at a controlled speed in a horizontal circle around your head, initially with a radius of about 1 m. You can feel a force on your hand through the string as the bung whirls around. The string is pulling outwards on you because you are pulling inwards on the string.

List the forces that are acting on the bung as it moves and describe the direction it acts on the bung. Which of these forces is affecting the horizontal motion of the bung?

Investigate qualitatively how different factors affect the size of the force on your hand from the string. You could start by changing the radius of the circle, or by using bungs of different mass. How do they affect the size of the force? What else affects the size of the force?

In the chapter *The Medium is the Message*, you saw that a charged particle moving in a magnetic field experiences a force at right angles to its direction of motion. This is just the sort of force that is required to produce motion in a circle. In Activity 28, you can see how a magnetic field can be used to steer an electron beam into a circular path as required in a particle accelerator.

Activity 28 Electrons in orbit

Use a fine-beam tube (Figure 3.48) to observe electrons moving in circular paths under the influence of a magnetic field.

Figure 3.48 Fine-beam tube

We will return to magnetic fields and particle accelerators later. In order to understand the situation properly, we must first take a more precise look at circular motion.

Describing circular motion

If we want to analyse motion in a circle, it is useful to be able to describe the position of a body around a circle, and how fast it is moving. Figure 3.49 shows how we do this.

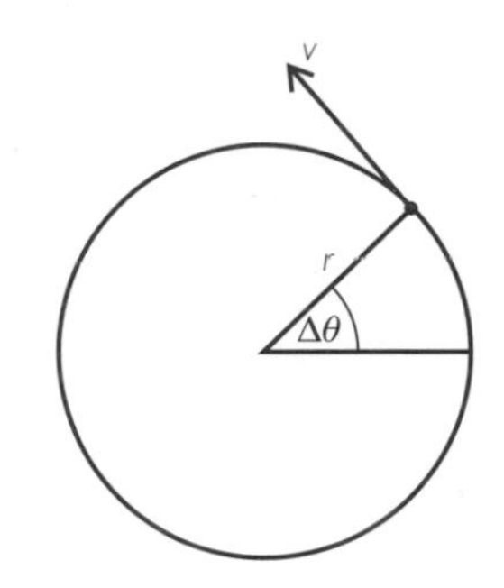

Figure 3.49 Defining quantities in circular motion

The important quantities are as follows:

- The **radius** of the circle, r, which has SI units of metres, m.
- The object's **angular displacement**, $\Delta\theta$. This is the angle through which the object has moved, relative to some fixed position. Although degrees are most commonly used to measure angles, they are not the SI unit. The unit used for angle (or angular displacement) is the radian, rad.
- The object's speed, v, with SI units m s^{-1}. Notice that the direction of motion is along the tangent to the circle.
- The **angular velocity**, ω (omega). Measured in rad s^{-1}. The relationship between angular velocity and angular displacement is similar to that between linear velocity and linear displacement:

$$\omega = \frac{\Delta\theta}{\Delta t} \qquad (22)$$

Study note

You met radians in the AS chapter *The Sound of Music* and used them to describe the phase of waves.

We will consider angular velocity in a little more detail. The rate at which many devices turn in a circle or spin is measured by counting how many revolutions they complete in a certain time. The most common everyday unit is 'revolutions per minute' (rpm). Most car engines rotate at between 2000 rpm and 3000 rpm in normal operation. A compact disc rotates at between 200 rpm and 500 rpm (depending on which part of the disc is being read.

In order to have consistent units, we do not measure angular velocity in rpm. Instead, ω is measured in radians per second. For example, if a wheel is rotating at a constant rate of 1 revolution per second, it turns through 2π rad (360°) every second. Its angular velocity, $\omega = 2\pi$ rad s^{-1}.

Maths reference

Degrees and radians
See Maths note 6.1

Worked example

Q A compact disc is rotating at 300 rpm. How many revolutions does it complete every second? What is its angular velocity?

A The disc completes 300/60 = 5 revs per second = 5 s^{-1}

In each revolution, the compact disc turns through 2π rad.

So its angular velocity, ω is:

$\omega = 5\ \text{s}^{-1} \times 2\pi\ \text{rad} = 10\pi\ \text{rad s}^{-1} = 31.4\ \text{rad s}^{-1}$

Another important quantity associated with circular motion is the **period** (symbol T) which is the time for one complete revolution. The angular velocity and the period are related:

$$\omega = \frac{2\pi}{T} \text{ or } T = \frac{2\pi}{\omega} \qquad (23)$$

For example, if a body in circular motion has a period $T = 0.5$ s, it completes two revolutions every second, so the angle it turns through every second is 4π rad. Thus, $\omega = 4\pi$ rad s^{-1}.

Speed and angular velocity

Imagine two athletes jogging around a circular running track, side by side. The athlete in the outside lane has to run slightly faster than the athlete in the inside lane in order to complete a lap in the same time. Both athletes have the same angular velocity, ω, but the one whose track has a greater radius, r, has to run with a greater speed, v. How are these three quantities (v, ω and r) related?

In one complete circuit, the athlete runs a distance $2\pi r$. The time taken to run around 2π radians with angular velocity ω is $\frac{2\pi}{\omega}$. So the athlete's speed is given by:

$$v = \frac{\Delta s}{\Delta t} = 2\pi r \times \frac{\omega}{2\pi} = \omega r$$

Hence:

$$v = \omega r \qquad (24)$$

Worked example

Q Turbine-generators in a power station rotate at 3000 rpm. The longest blades in a turbine are about 1.5 m long. Calculate the linear speed of the rotor tip.

A First, calculate the angular velocity ω:

$$\omega = \frac{3000}{60} \text{ rev s}^{-1} = 50 \text{ rev s}^{-1} = 50 \times 2\pi \text{ rad s}^{-1} = 314 \text{ rad s}^{-1}$$

Now calculate the speed of the tip:

$$v = \omega r = 314 \text{ rad s}^{-1} \times 1.5 \text{ m} = 471 \text{ m s}^{-1}$$

(The rotor tip is moving faster than the speed of sound in air.)

Questions

43 A car engine is rotating at 3000 rpm.

(a) How many revolutions does it complete in 1 second?

(b) What is its angular velocity?

(c) What is the period of its motion?

44 A CD has an angular velocity of 30 rad s^{-1}.

(a) How many revolutions does it complete in 1 second?

(b) What is the period of its motion?

45 A positron is travelling around the circular accelerator at CERN. Its speed is close to the speed of light ($c = 3.00 \times 10^8$ m s^{-1}). The accelerator's diameter is 8.6 km. Calculate the positron's angular velocity.

46 A spinning roundabout of radius 2 m has a period $T = 1.25$ s.

(a) What is the angular velocity of the roundabout?

(b) (i) What is the speed of a boy sitting on the edge of the roundabout?

(ii) If the boy moves halfway to the centre of the roundabout, what is his speed there?

(iii) If he then moves to the centre of the roundabout, what is his speed there?

47 A garden strimmer cuts grass with a fast rotating piece of cord. The cord is 15 cm long and the strimmer spins at 100 revs per second.

(a) What is the angular velocity of the cord?

(b) What is the speed of the tip of the cord as it spins?

48 The head of a golf club has a speed of approximately 30 m s^{-1} when it strikes a ball. Estimate the angular velocity of a golf club at this point of a golf swing.

Centripetal acceleration

We will now return to the idea that an object moving in a circle at constant speed is accelerating. Imagine a car travelling at a steady speed around a bend in the road. Although its speed is constant, its direction is changing all the time, so its velocity is not constant. Figure 3.50 shows that the car's velocity is changing all the time as it moves around the bend. If its velocity changes, it must be accelerating, even though its speed is constant. Our car is moving at a constant speed, but it is not moving in a straight line. It is constantly changing direction.

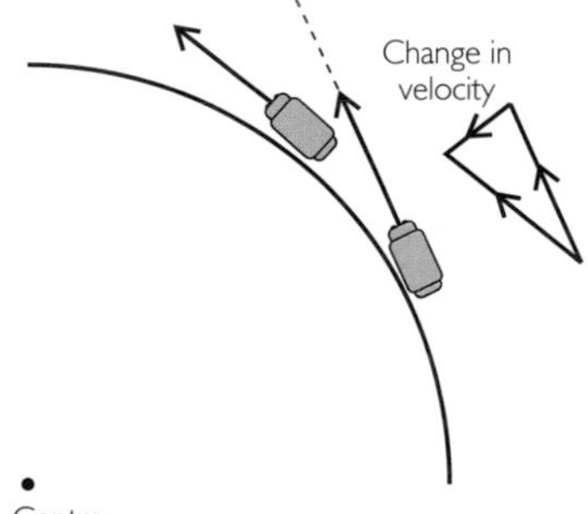

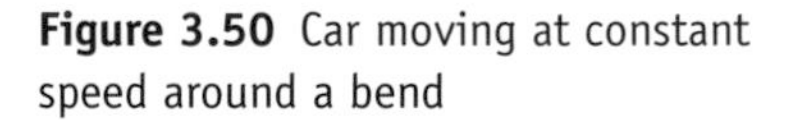
Figure 3.50 Car moving at constant speed around a bend

In Figure 3.50, you can see that the change in velocity of the car, as it moves a short distance around the bend, is directed towards the centre of the circle. It is a **centripetal acceleration**, produced by the centripetal force that we identified with earlier. We can derive an equation for this acceleration, with the help of Figure 3.51.

The object shown in Figure 3.51(a) moves at a steady speed v from A to B in time Δt. It moves through angle $\Delta\theta$, so its angular velocity ω is:

$$\omega = \frac{\Delta\theta}{\Delta t} \qquad \text{(Equation 22)}$$

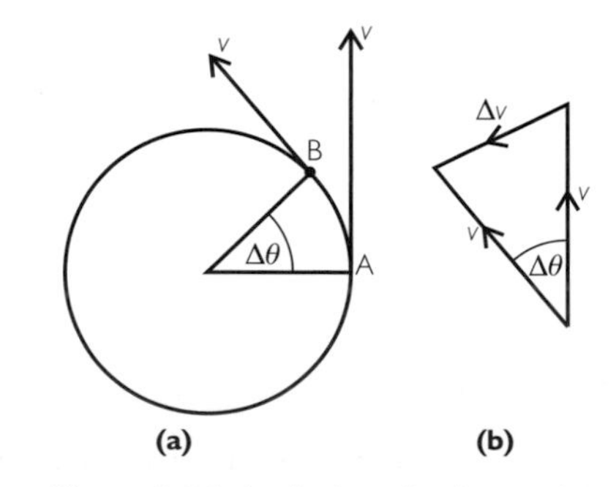

Figure 3.51 Analysing circular motion

In this time, its velocity changes direction, as shown. Its velocity vector also moves through angle $\Delta\theta$.

Figure 3.51(b) shows how we work out the object's change in velocity. We draw a vector triangle; the short third side is the change in velocity Δv. From this triangle *provided* $\Delta\theta$ *is small* (and using the definition of an angle in radians) we can write:

$$\Delta v = \Delta\theta \times v$$

Dividing both sides by Δt gives:

$$\frac{\Delta v}{\Delta t} = \left(\frac{\Delta\theta}{\Delta t}\right) \times v$$

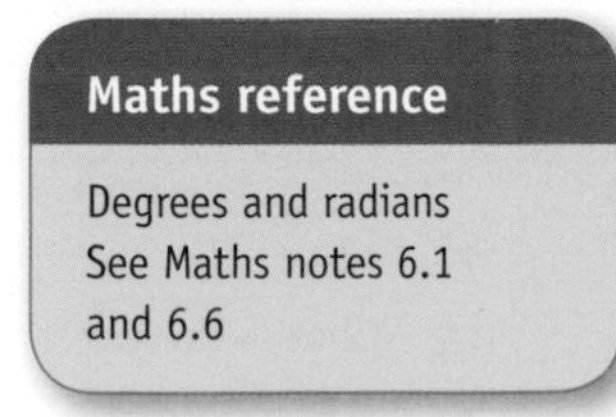
Maths reference

Degrees and radians
See Maths notes 6.1 and 6.6

But:

$$\frac{\Delta v}{\Delta t} = a, \text{ and } \frac{\Delta \theta}{\Delta t} = \omega$$

so we have:

$$a = \omega v$$

Now, if we substitute $\omega = \frac{v}{r}$, we get a useful expression for the centripetal acceleration:

$$a = \frac{v^2}{r} \qquad (25)$$

and if we substitute $v = \omega r$ we get another useful expression:

$$a = r\omega^2 \qquad (26)$$

Question

49 Go back to Questions 45–48 and calculate the centripetal acceleration in each case.

Centripetal force again

We can now work out what force is needed to make an object move in a circle, because we have two expressions for centripetal acceleration. Using the familiar relationship $F = ma$, we can simply multiply Equations 25 and 26 by the object's mass m to find the centripetal force:

$$F = \frac{mv^2}{r} \qquad (27)$$

and:

$$F = mr\omega^2 \qquad (28)$$

Questions

50 You are cycling along, and decide to turn off to the left.

(a) Which requires a greater sideways force, turning sharply left or following a more gentle path?

(b) If you were travelling faster, would you need a bigger or smaller sideways force?

51 What force is required to keep a 5 kg object moving at 10 m s^{-1} in a circle of radius 2 m?

52 What force is required to keep a 1 kg object moving with angular velocity 10 rad s^{-1} in a circle of radius 10 m?

53 A man's competition hammer has a mass of 7.26 kg. The length of the hammer chain is 1.2 m and the length of the man's arms is 0.9 m. Before throwing, the man whirls the hammer round his head in a horizontal circle a few times. The average period for each revolution is 0.75 s.

(a) What is the average angular velocity of the hammer?

(b) What is the average tension required in the hammer chain if it is horizontal?

54 A girl of mass 35 kg is swinging on a tyre of mass 10 kg attached to a tree branch by a rope of length 6 m.

(a) What is the weight of the tyre and the girl? (use $g = 9.81$ N kg^{-1}.)

(b) If her speed at the bottom of the swing is 4 m s^{-1}, what centripetal force is required at this point by the tyre and the girl?

(c) Use your answers to (a) and (b) to calculate the tension in the rope at this point. Remember that the resultant force on the child and tyre must equal the value of the centripetal force and act in the right direction.

55 A car rounding a bend requires a centripetal force of 2 kN. This is provided by friction between the tyres and the road.

(a) If the same car at the same speed rounds another bend of twice the radius, what centripetal force is required?

(b) If the same car rounds the original bend at twice the original speed, what centripetal force is required?

4.4 Steering and tracking charged particles

Round the bend

Now we can return to thinking about making charged particles move in circular orbits; you will recall that this is how we can avoid the problem of making very long linear accelerators. In Activity 28 you saw how a magnetic field can steer a beam of electrons into a circular path. In the chapter *The Medium is the Message*, you saw that a particle of charge q moving at speed v in a magnetic field of flux destiny B will experience a force of magnitude:

$$F = Bqv \sin\theta \qquad (29)$$

where θ is the angle between the directions of the particle's motion and the magnetic field.

Fleming's left-hand rule (Figure 3.52) tells us how the directions of the magnetic field, the current and the force are related; the force always acts at right angles to the other two. But take care! Recall that for moving electrons, whose charge is negative, the conventional direction of current is in the opposite direction to the motion of the charges.

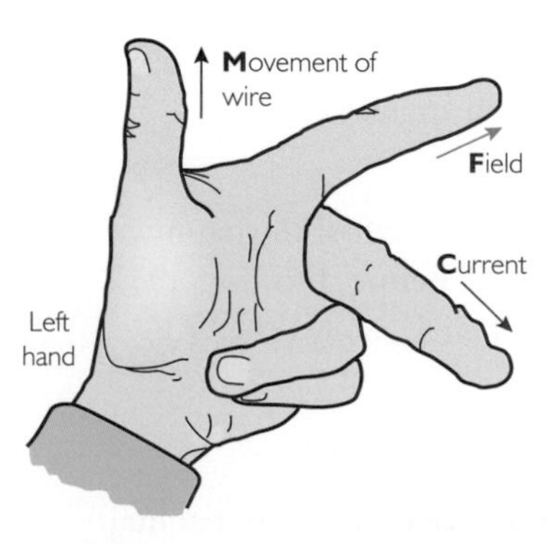

Figure 3.52 Fleming's left-hand rule

Figure 3.53 shows what happens when a positive charge $+q$ enters a region where there is a magnetic field; the crosses indicate that the flux is directed into the page. The direction of the current is left-to-right. From the left-hand rule, the force on the charge is directed upwards, at right angles to the motion of the charge. This force will alter the direction of motion of the charge (but will not affect its speed). Even when the direction of motion is changed, the force will still be at right angles to it. Because the force is always perpendicular to the motion, it will produce a circular path. The effect of the magnetic field on the charge is providing a centripetal force that enables the charge to move in a circle.

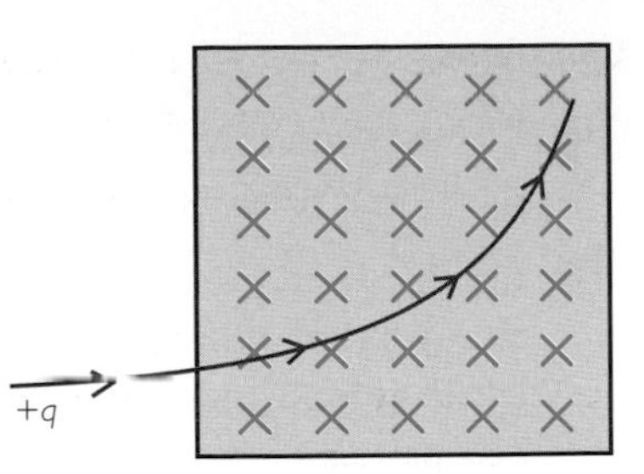

Figure 3.53 Positively charged particle moving in a magnetic field

In a circular particle accelerator or in a fine-beam tube, the magnetic force provides the centripetal force needed to keep the electrons moving in a circle. The field is applied at right-angles to the direction of motion, ($\theta = 90°$, $\sin\theta = 1$) so from $ma = F$ we have:

$$\frac{mv^2}{r} = Bqv$$

Now we can rearrange this to deduce an expression for the radius of the particle's orbit:

$$r = \frac{mv}{Bq} \qquad (30)$$

Substituting p for the momentum mv of the particle, we can write this expression as:

$$r = \frac{p}{Bq} \qquad (30a)$$

So, the greater the momentum of a particle, the greater the radius of its path in a given field; put another way, the stronger the field needed to keep it in a given orbit.

Study note

Equation 30(a) has the advantage that it applies equally to relativistic and non-relativistic particles.

Activity 29 The ISIS accelerator

ISIS, at the Rutherford Appleton Laboratory near Oxford is the most intense source of pulsed neutrons in the world.

Via **www.shaplinks.co.uk** go to the ISIS website's accelerator page.

Download the video animation that explains how the ISIS facility works. In particular, notice the use of both a linear accelerator to accelerate H^- ions and a synchrotron to accelerate the protons before they collide with a target and produce neutrons.

Look at the 'about ISIS' page to find out the advantages of using neutrons to probe the structure of materials.

Study note

H^- ions are hydrogen atoms with an extra electron.

Activity 30 The fine-beam tube

Use a fine-beam tube to explore how the accelerating voltage and the magnetic field affect the radius of the electrons' path.

Questions

56 Redraw Figure 3.53, showing how the situation would change if the charge shown was negative.

57 Beta particles, produced in radioactive decay, are fast-moving electrons. If a narrow beam of beta radiation from a source is passed into a magnetic field, it curves around and spreads out. What does this tell you about the energies of the beta particles?

58 An electron is travelling at 1% of the speed of light ($c = 3.00 \times 10^8$ m s^{-1}). If a magnetic field, $B = 0.2$ T is applied at right angles to its motion, what radius of path will result? ($m_e = 9.11 \times 10^{-31}$ kg; $h = 6.63 \times 10^{-34}$ Js)

59 Electrons in a fine-beam tube are accelerated through a pd of 2.5 kV. A magnetic field of 1 T is applied at right angles to the electron beam.

(a) What is the speed of the electrons as they leave the electron gun? (Ignore relativistic effects.)

(b) What force is exerted on each electron because of the magnetic field? Describe the direction of the force.

(c) What acceleration will the electrons have as a result of this force? Does their speed increase?

(d) What will be the radius of the path that the electrons follow?

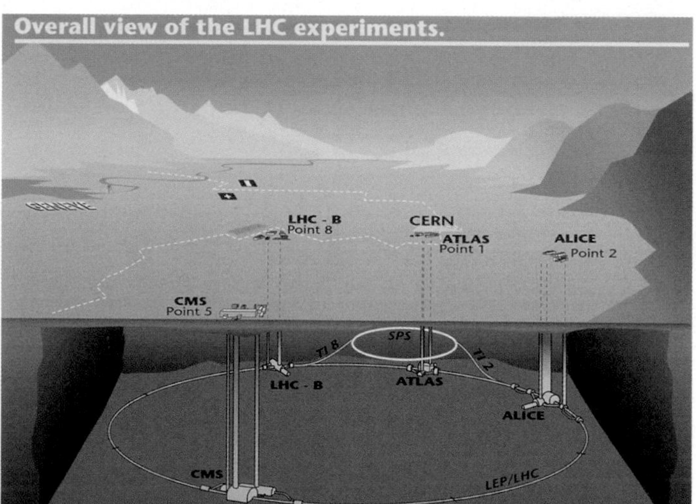

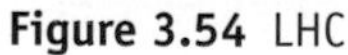
Figure 3.54 LHC

Making tracks

To finish this section, you will see how what you have learned about circular motion and about electric and magnetic fields is applied to another aspect of particle physics – that of studying the results of collisions between particles.

Figure 3.54 shows the layout of the LHC, which was switched on in 2008. There are four main experiments (ALICE, ATLAS, CMS and LHCb) located between 50 m and 150 m underground in huge caverns. An older accelerator called SPS (super proton synchrotron) is also shown; its function is to inject energetic particle beams into the LHC where they undergo their final acceleration before colliding.

In many particle-physics experiments, such as those carried out at the LHC, beams of particles are steered so that they collide at particular locations. In the LHC, each of the four collision points is surrounded by a vast detector to record the resulting debris. Figure 3.55 shows the LHC detector known as ATLAS. Apart from LHCb, all the detectors have a similar layered structure.

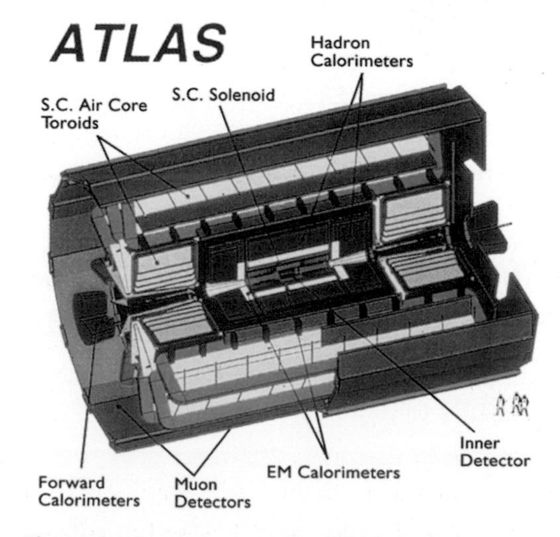

Figure 3.55 Structure of the ATLAS detector

The different layers are designed each to be sensitive to different types of particle, and each works in a slightly different way. But in each layer, passage of a charged particle is detected by the ionisation it produces in the low-density gas that fills the detector. To detect ionisation, most modern detectors use so called multiwire chambers or drift chambers. A multiwire chamber consists of an array of fine wires, with potential differences applied between them. When ions are created in the detector, the electric fields between the wires sweep the ions onto the wires, and the resulting small pulse of current is recorded electronically (rather as a Geiger tube records the ionisation produced by the passage of an alpha or beta particle). The record of current pulses provides a record of a particle's passage through the chamber. A drift chamber works in a very similar way, except that it measures the time taken for ions to drift towards the nearest wire, enabling a more accurate record to be built up.

Figure 3.56 shows Peter Glassel, the technical coordinator for the ALICE time projection chamber, sitting inside the detector. Thousands of wires are connected to read out electronic data produced as particles are created in collisions at the centre of the detector.

Figure 3.56 Inside the ALICE detector

A detector such as ALICE or ATLAS incorporates some layers whose function is to measure the energy deposited – these are the calorimeters shown in Figure 3.55. Such measurements provide information about the energy of the particles produced in a collision, but that on its own is not enough for particle physicists to unravel what went on. Figure 3.57 shows Louis Rose-Dulcina, a technician from the ATLAS collaboration, working on the ATLAS tile calorimeter. Special manufacturing techniques were developed to mass produce the thousands of elements in this detector. Tile detectors are made in a sandwich-like structure; scintillator tiles (which emit tiny flashes of light along the tracks of a particle) are placed between metal sheets.

Figure 3.57 Making detectors for ALICE

As well as using electric fields to detect ionisation, detectors such as ALICE and ATLAS use magnetic fields to obtain information about particles. A strong magnetic fields is applied along the axis of the detector. In order to make the field as strong as possible, superconducting magnets are used. The colliding particles travel in the same direction as the field, so their paths are not affected, but charged particles produced in a collision are likely to be travelling across the field and so they are forced into curved paths. The sign of their charge can be deduced from the direction of curvature of their tracks, and further measurements enable their momentum to be found.

The magnetic field produced by the ATLAS toroids (Figure 3.58) wraps around the beam path. It provides the field in the outer muon detectors. There is an axial field (parallel to the beam path) provided by a solenoid in the centre of ATLAS. This combination of magnetic fields improves resolution, as it is not possible to build a solenoid large enough to cover the whole area of the detector.

Figure 3.58 View of ATLAS during assembly looking along the direction of the beam, showing huge toroidal magnets, muon chamber and the inner detector

In Activities 31 and 32 you will need to apply ideas from earlier in this chapter. There are several key points to keep in mind:

- only *charged* particles cause ionisation
- hence only charged particles make tracks in detectors
- the direction of curvature of a particle's track in a magnetic field indicates the sign of its charge
- the radius of curvature of a track provides information on the particle's charge and momentum
- momentum is always conserved in any collision, interaction or decay
- the presence of uncharged particles can be deduced by applying momentum conservation
- in collisions and interactions, there is interconversion between mass and energy
- a particle's kinetic energy can be deduced from the amount of ionisation it produces along its track.

Activity 31 Particle tracks

Many particle-physics websites include examples of particle tracks. To find out how such records are analysed, go to the Manchester University Particle Physics Group's website (via **www.shaplinks.co.uk**) where there are lots of useful links.

You might also like to try the simulations in the Lancaster Particle Physics Package (also via **www.shaplinks.co.uk**). The animations aim to give you a feel for the real physics and experimental techniques involved in particle-physics research.

Activity 32 Track and field events

Write a short account of how electric and magnetic fields are used in particle detectors. Include an explanation of how momentum could be measured from a particle track (refer back to Section 4.3).

4.5 Summing up Part 4

In this part of the chapter, you have seen how an understanding of the way in which charged particles behave in electric and magnetic fields can be used to accelerate particles to high energies. High-energy particles are needed to recreate the conditions of the early universe, on a small scale.

Use Activity 33 to help you look back over your work for this part of the chapter and see how the key ideas are being applied to the latest particle accelerators.

Activity 33 Accelerator energy

STOP PRESS ...On 6 April 2008, CERN opened its doors to the general public, offering a unique chance to visit its newest and largest particle accelerator, the Large Hadron Collider (LHC), before it went into operation later that year. At full power, two beams of protons will each travel at a maximum energy of 7 TeV (tera-electronvolt), corresponding to head-to-head collisions of 14 TeV (1 TeV = 10^{12} eV).

Find out the latest from CERN and other accelerator labs by checking their websites. (See **www.shaplinks.co.uk**)

Find out how the energy available from different particle accelerators has increased as the technology has developed. In doing this you may read about many types of accelerator: LINACs, cyclotrons, synchro-cyclotrons etc. For each accelerator, state the type of particles it accelerates, how it accelerates them, how it directs the particle beam, and the energies it can achieve.

Quote all energies in electronvolts.

5 On Target

5.1 Big ideas

This chapter has looked at some of the most profound ideas in physics: the nature of the universe, the fundamental particles of nature and the forces which act between them. Along the way, you have learnt about circular motion, and the way charged particles behave in electric and magnetic fields. Use the following activity to remind yourself how these ideas fit together.

Activity 34 Unification

Copy the words below on to separate, small pieces of paper. With a partner, discuss the ideas that link groups of words. (Move the pieces of paper around to show how they can be grouped together.)

Now on a single sheet of paper, draw a concept map to show your thoughts. Write the words on the paper, and add arrows showing connections between them. Annotate the arrows to explain the connections.

fundamental particles	quarks	electrons
leptons	gravitation	electromagnetism
nuclear forces	magnetic fields	electric fields
electric charge	moving charges	circular motion
angular velocity	centripetal force	the Big Bang
particle accelerators	high-energy particles	the universe
dark matter	Higgs boson	neutrinos

5.2 Questions on the whole chapter

You will need the following information when you answer these questions:

permittivity of free space, $\varepsilon_0 = 8.85 \times 10^{-12}$ F m^{-1}

electron charge, $e = 1.60 \times 10^{-19}$ C

electron mass, $m_e = 9.11 \times 10^{-31}$ kg

proton mass $m_p = 1.67 \times 10^{-27}$ kg

speed of light in vacuum, $c = 3.00 \times 10^8$ m s^{-1}

Questions

60 We can picture an anti-atom of 'anti-hydrogen' as a positron (an anti-electron) in a circular orbit around an anti-proton. The average separation of the two particles in the anti-atom is 0.037 nm.

(a) What force holds the positron in its orbit? Sketch a diagram of the anti-atom. Draw an arrow to show the force acting on the positron.

(b) Explain why a force is needed to keep the positron in its orbit.

(c) Calculate the force exerted on the positron by the anti-proton, and the force exerted on the anti-proton by the positron.

(d) How many times per second does the positron orbit the anti-proton?

61 In an experiment to investigate the annihilation of particles, beams of high-energy electrons and positrons are produced, and then caused to collide with one another. In one collision, a 500 keV electron collides head-on with a 500 keV positron. The two particles annihilate one another, and two photons of electromagnetic energy are produced.

(a) What is the combined momentum of the two particles, just before they collide?

(b) What is the energy of a 500 keV electron, in joules?

(c) What is the combined energy of the two photons, in joules?

62 This question is about the acceleration of protons in the Proton Synchrotron (PS) at CERN. See Figure 3.59. Table 3.10 lists some data about the PS and the Super Proton Synchrotron (SPS), also at CERN.

	PS	SPS
diameter of ring	200 m	2.2 km
circumference of ring	528 m	
number of accelerating points	14	
average accelerating voltage	4 kV	
proton energy at injection	50 MeV	
final energy of proton	28 GeV	
final momentum of proton	1.7×10^{-17} kg m s^{-1}	
final speed of proton	almost 3×10^8 m s^{-1}	almost 3×10^8 m s^{-1}

Table 3.10 Data for Question 62

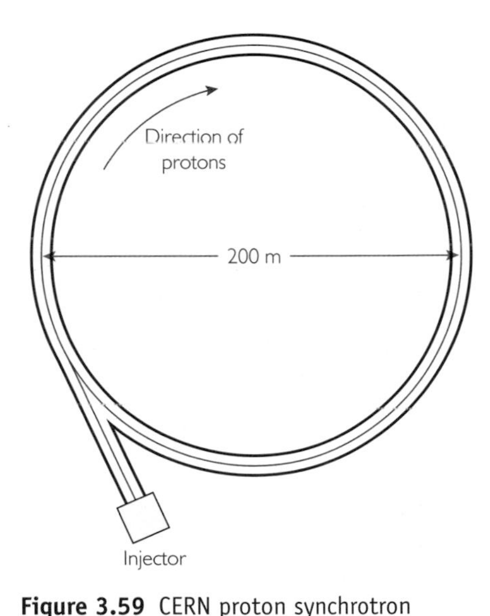

Figure 3.59 CERN proton synchrotron

Particle accelerators are used to increase the energy of charged particles such as protons and electrons. The accelerated particles are made to collide with other particles on a 'target' in order to investigate the structure of matter. Subatomic particles created in these collisions can be studied.

One type of particle accelerator is the synchrotron. In this machine a magnetic field causes charged particles to move in a circular path. The particles are accelerated to higher energies by an electric field. As their momentum increases, the magnetic flux density is increased to keep them travelling in a path of constant radius.

(a) Ignoring relativistic effects, show that:

(i) the speed of a proton at injection into the PS is about 1×10^8 m s^{-1}

(ii) a proton takes about 6 μs to travel round the ring at this speed

(iii) the momentum of a proton at injection is about 1.6×10^{-19} kg m s^{-1}.

(b) The accelerator ring is a pipe maintained at very low pressure. It is 'filled' with protons by injecting a proton current of 100 mA for 6 μs.

(i) Calculate the number of protons injected.

(ii) Explain why the ring must be maintained at very low pressure.

(c) Before the protons are accelerated, an electric field is used to group the protons in the ring into a number of bunches. The bunches are then accelerated as they pass through the acceleration points which are spaced equally around the ring. An acceleration point is essentially a pair of electrodes between which an alternating voltage is applied.

(i) Explain why an *alternating* voltage is needed.

The proton bunches pass through an acceleration point when the potential difference between its electrodes is about 4 kV.

(ii) By how much does the energy of one proton increase in one revolution? (Give your answer in eV.)

(iii) Estimate the number of times a proton must travel round the ring in order to reach its final energy.

(iv) Explain briefly why *linear* accelerators are not used to accelerate protons to this final energy.

(d) (i) Show that the magnetic flux density, B, required to maintain a proton in a circular path of radius r is proportional to its momentum, p.

(ii) Estimate the magnetic flux density needed to maintain 50 MeV protons within the PS.

(iii) Explain why the frequency of the alternating voltage must change as the protons are accelerated to higher energies.

(e) Estimate:

(i) by what factor the magnetic field must be increased during the acceleration

(ii) by what factor the mass of the proton increases during acceleration.

Protons from the PS can be injected into the SPS for further acceleration.

(iii) Explain why the magnetic field in the SPS is increased as protons are accelerated, but the frequency of the accelerating voltage is kept almost constant.

5.3 Achievements

Now you have studied this chapter you should be able to achieve the outcomes listed in Table 3.11.

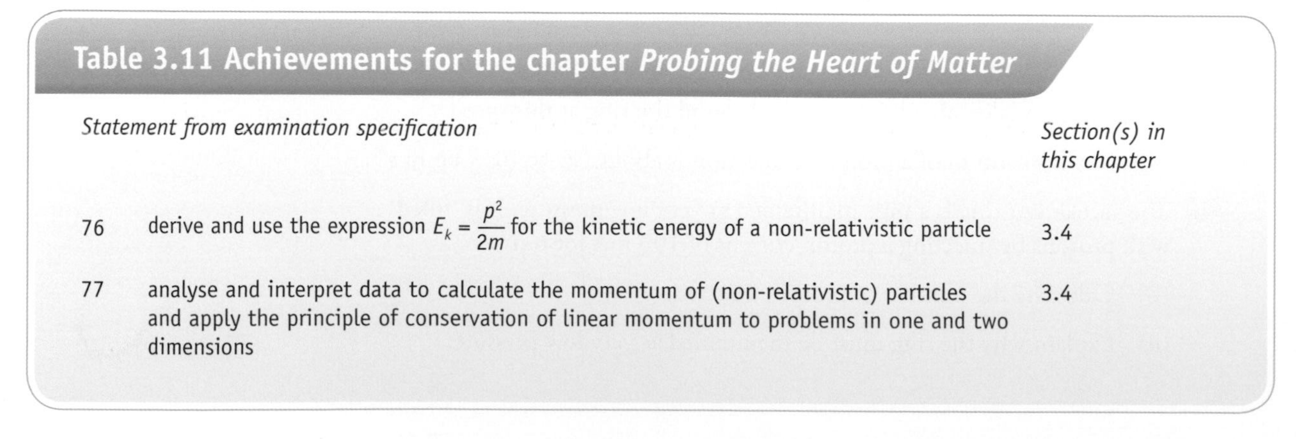

Table 3.11 Achievements for the chapter *Probing the Heart of Matter*

	Statement from examination specification	*Section(s) in this chapter*
76	derive and use the expression $E_k = \frac{p^2}{2m}$ for the kinetic energy of a non-relativistic particle	3.4
77	analyse and interpret data to calculate the momentum of (non-relativistic) particles and apply the principle of conservation of linear momentum to problems in one and two dimensions	3.4

Table 3.11 Achievements continued

	Statement from examination specification	*Section(s) in this chapter*
79	express *angular displacement* in radians and in degrees, and convert between those units	4.3
80	explain the concept of *angular velocity*, and recognise and use the relationships $v = \omega r$ and $T = \frac{2\pi}{\omega}$	4.3
81	explain that a resultant force (*centripetal force*) is required to produce and maintain circular motion	4.3
82	use the expression for centripetal force $F = ma = \frac{mv^2}{r}$ and hence derive and use the expressions for *centripetal acceleration* $a = \frac{v^2}{r}$ and $a = r\omega^2$	4.3
85	use the expression $F = \frac{kQ_1Q_2}{r^2}$, where $k = \frac{1}{4\pi\varepsilon_0}$ and derive and use the expression $E = \frac{kQ}{r^2}$ for the *electric field* due to a point charge	3.2, 3.3
96	use the terms *nucleon number* (mass number) and *proton number* (atomic number)	1.1, 2.3
97	describe how large-angle *alpha particle scattering* gives evidence for a nuclear atom	3.1
99	explain the role of *electric* and *magnetic fields* in particle accelerators (LINAC and cyclotron) and detectors (general principles of ionisation and deflection only)	3.5, 4.2, 4.4
100	recognise and use the expression $r = \frac{p}{BQ}$ for a charged particle in a magnetic field	4.4
101	recall and use the fact that charge, energy and momentum are always conserved in interactions between particles and hence interpret records of particle tracks	2.3, 3.4, 4.4
102	explain why high energies are required to break particles into their constituents and to see fine structure	4.2
103	recognise and use the expression $\Delta E = c^2\Delta m$ in situations involving the *creation* and *annihilation* of matter and *anti-matter* particles	2.4
104	use the non-SI units MeV and GeV (energy) and MeV/c^2, GeV/c^2 (mass) and *atomic mass* unit *u*, and convert between these and SI units	4.2
105	be aware of *relativistic effects* and that these need to be taken into account at speeds near that of light (use of relativistic equations not required)	3.5, 4.2
106	recall that in the *standard quark-lepton model* each particle has a corresponding *anti-particle*, that *baryons* (eg neutrons and protons) are made from three *quarks*, and *mesons* (eg pions) from a quark and an anti-quark, and that the *symmetry* of the model predicted the top and bottom quark.	2.2, 2.4
107	write and interpret equations using standard nuclear notation and standard particle symbols (eg π^+, e^-)	2.3
108	use *de Broglie's wave equation* $\lambda = \frac{h}{p}$	3.5

Answers to questions

1 (a) strong; (b) gravity; (c) electromagnetic; (d) strong and electromagnetic; (e) weak.

2 $\mu^+ + \mu^- \rightarrow$ electromagnetic radiation

3 A quark and its anti-quark have equal but opposite charges, so they cancel out.

4 Creating the mass of the tau leptons requires much more than the energy equivalent of an electron–positron pair at rest; additional energy must be supplied in the form of kinetic energy of the colliding particles.

5 $E = mc^2 = 1.67 \times 10^{-27}\ \text{kg} \times (3.00 \times 10^8\ \text{m s}^{-1})^2$
$= 1.50 \times 10^{-10}\ \text{J}$

6 (a) $E = mc^2$
$= (2 \times 9.11 \times 10^{-31}\ \text{kg}) \times (3.00 \times 10^8\ \text{m s}^{-1})^2$
$= 1.64 \times 10^{-13}\ \text{J}$

(b) Energy 'left over' $= 2.00 \times 10^{-13}\ \text{J} - 1.64 \times 10^{-13}\ \text{J}$
$= 0.36 \times 10^{-13}\ \text{J}$

This is shared equally between the two particles, so the kinetic energy of each is 0.18×10^{-13} J (1.8×10^{-14} J).

7 (a) Electron and positron (anti-electron).

(b) They have opposite charges.

(c) They annihilate, producing electromagnetic radiation.

8 He guessed they showed the same pattern as the leptons, ie six quarks in three generations of two each — they had the same symmetry. He could predict their charges, and that their masses were much greater than those of the other quarks.

9 hadrons: proton, neutron, pion
meson: pion
baryons: proton, neutron
quarks: charm, up
leptons: electron, anti-electron, electron-neutrino

10 To avoid the absorption of alpha particles: air is pumped out through tube T; source R and detector S are close to foil F; foil F is thin.

11 Direct hits are more likely, so more alpha particles will be back-scattered.

12 Slower alpha particles wouldn't whiz straight past so easily, so more would be scattered through bigger angles.

13 There is less chance of a direct hit, and the force is weaker, so fewer alpha particles will be back-scattered.

14 See Figure 3.60.

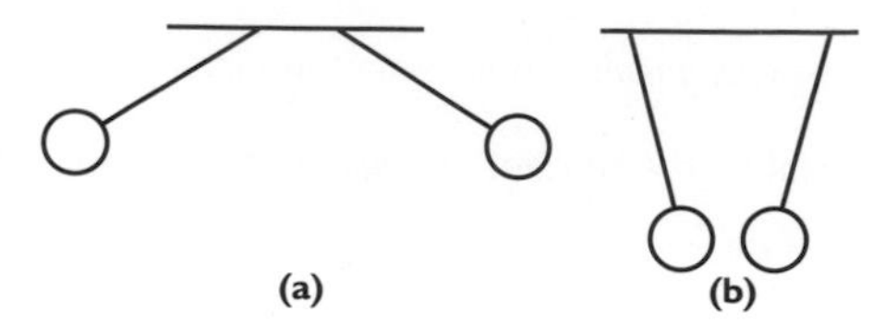

Figure 3.60 Answer to Question 14

15 See Figure 3.61

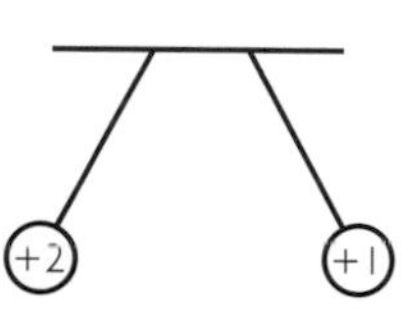

Figure 3.61 Answer to Question 15

16 Figure 3.62 shows the forces acting on one ball.

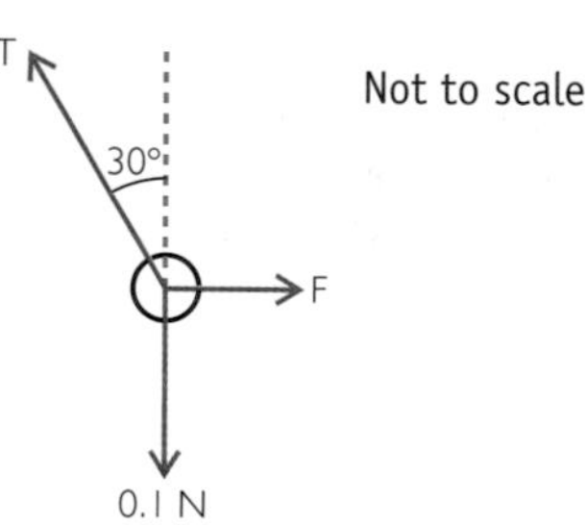

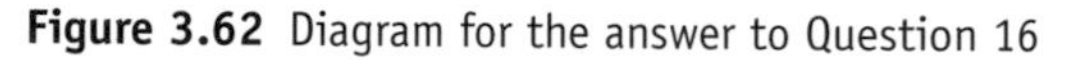
Figure 3.62 Diagram for the answer to Question 16

Resolving horizontally: $F = T\sin30°$
Resolving vertically: $W = T\cos30°$

Dividing one equation by the other and cancelling T:

$$\frac{F}{W} = \tan30°$$

so $F = 0.10\ \text{N} \times \tan30° = 0.058\ \text{N}$

17 $1\ \text{F} = 1\ \text{C V}^{-1}$, and $1\ \text{V} = 1\ \text{J C}^{-1}$, so $1\ \text{F} = 1\ \text{C}^2\ \text{J}^{-1}$.
$1\ \text{J} = 1\ \text{N m}$, so $1\ \text{F} = 1\ \text{C}^2\ \text{N}^{-1}\ \text{m}^{-1}$

Hence $1\ \text{F m}^{-1} = 1\ \text{C}^2\ \text{N}^{-1}\ \text{m}^{-2} = (1\ \text{N m}^2\ \text{C}^{-2})^{-1}$ as required.

18 $F = \left(\frac{1}{4\pi\varepsilon_0}\right)\frac{Q_1 Q_2}{r^2}$

$= \frac{9.0 \times 10^9\ \text{N m}^2\ \text{C}^{-2} \times (10^{-8}\ \text{C})^2}{(0.05\ \text{m})^2}$

$= 3.6 \times 10^{-4}\ \text{N}$

19 (a) $F = \left(\frac{1}{4\pi\varepsilon_0}\right)\frac{Q_1 Q_2}{r^2}$

$$= \frac{9.0 \times 10^9\ \text{N m}^2\ \text{C}^{-2} \times (1.60 \times 10^{-19}\ \text{C})^2}{(0.037 \times 10^{-9}\ \text{m})^2}$$

$= 1.68 \times 10^{-7}$ N

(b) See Figure 3.63

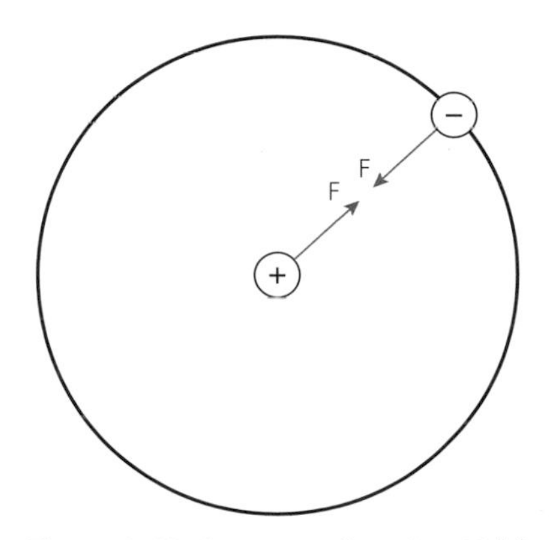

Figure 3.63 Answer to Question 19(b)

20 The nuclear charge is smaller, so the force is weaker at a given distance (about one third of that with gold). As shown in **Figure 3.64**, the graph plotted on a linear scale is scaled down by a factor of about 3, and the log graph is *shifted* downwards.

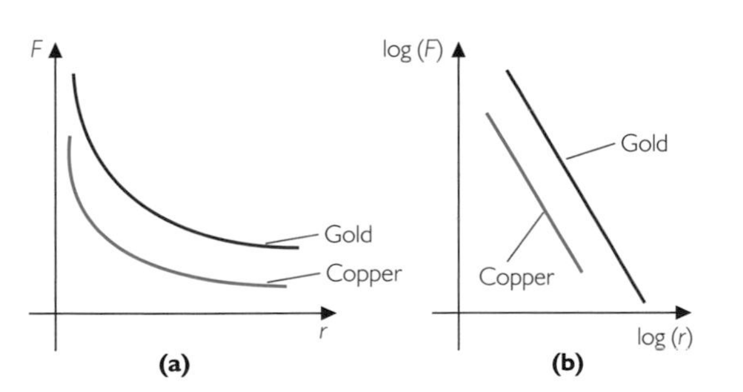

Figure 3.64 Answer to Question 20

21 Initial $E_k = \Delta E = \frac{kQ_1Q_2}{r}$ (Equation 10),

so

$$r = \frac{kQ_1Q_2}{E_k}$$

$$= \frac{9.0 \times 10^9\ \text{N m}^2\ \text{C}^{-2} \times 79 \times 1.60 \times 10^{-19}\ \text{C} \times 2 \times 1.60 \times 10^{-19}\ \text{C}}{8.0 \times 10^{-13}\ \text{J}}$$

$= 4.6 \times 10^{-14}$ m

22 Lines of force go *towards* a negatively-charged sphere. See Figure 3.65.

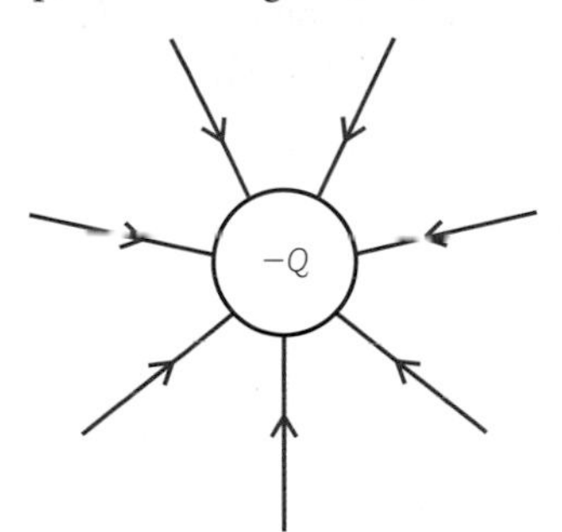

Figure 3.65 Answer to Question 22

23 See Figure 3.66.

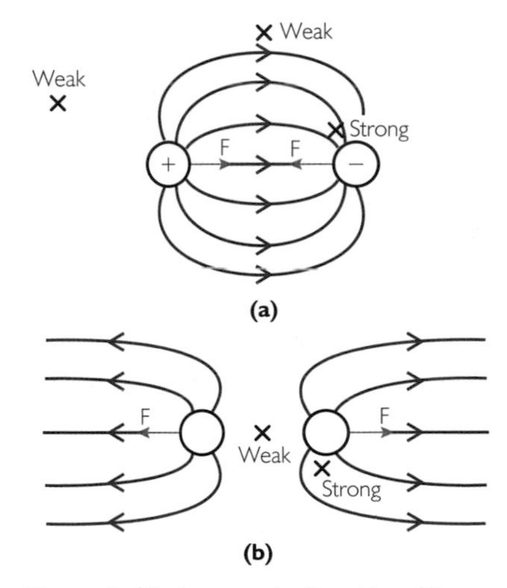

Figure 3.66 Answers to Question 23

24 $E = \frac{F}{q} = \frac{10\ \text{N}}{20 \times 10^{-3}\ \text{C}} = 500\ \text{N C}^{-1}$

25 $F = Eq$

$= 2000\ \text{N C}^{-1} \times 1.60 \times 10^{-19}$ C

$= 3.2 \times 10^{-16}$ N

Acceleration $a = \frac{F}{m} = \frac{3.2 \times 10^{-16}\ \text{N}}{9.11 \times 10^{-31}\ \text{kg}}$

$= 3.5 \times 10^{14}\ \text{m s}^{-2}$

(Note that this is more than a million million times the acceleration due to gravity.)

26 $E = \frac{kQ}{r^2}$

$$= \frac{9.0 \times 10^9\ \text{N m}^2\ \text{C}^{-2} \times 79 \times 1.60 \times 10^{-19}\ \text{C}}{(1 \times 10^{-13}\ \text{m})^2}$$

$= 1.1 \times 10^{19}\ \text{N C}^{-1}$

27 For someone of mass 60 kg capable of running at 10 m s^{-1} (close to the world record for the 100 m sprint!):

$p = mv = 60 \text{ kg} \times 10 \text{ m s}^{-1} = 600 \text{ kg m s}^{-1}$

$E_k = \frac{1}{2} mv^2 = \frac{1}{2} \times 60 \text{ kg} \times (10 \text{ m s}^{-1})^2 = 3 \text{ kJ}$

$\frac{p^2}{2m} = \frac{(600 \text{ kg m s}^{-1})^2}{2 \times 60 \text{ kg}} = 3 \text{ kJ}$

28 $E_k = \frac{p^2}{2m}$

$= \frac{(5.0 \times 10^{-20} \text{ kg m s}^{-1})^2}{2 \times 6.7 \times 10^{-27} \text{ kg}}$

$= 1.9 \times 10^{-13} \text{ J}$

29 (a) Momentum would be conserved – it is *always* conserved in any interaction.

(b) It is *not* elastic. Some of the initial kinetic energy would provide the additional mass of the muons (see Section 2.4 of this chapter) so there would be an overall decrease in the kinetic energy.

30 They would measure the momentum before and after the collision (you will see in Part 4 how this can be done). Any discrepancy can be accounted for in terms of the momentum of 'unseen' particles.

31 $E_k = eV = 1.60 \times 10^{-19} \text{ C} \times 500 \text{ V} = 8.00 \times 10^{-17} \text{ J}$

$p = \sqrt{(2meV)} = 1.21 \times 10^{-23} \text{ kg m s}^{-1}$

$v = \frac{p}{m} = \frac{1.21 \times 10^{-23} \text{ kg m s}^{-1}}{9.11 \times 10^{-31} \text{ kg}}$

$= 1.33 \times 10^7 \text{ m s}^{-1}$

32 $E_k = \frac{1}{2}mv^2 = \frac{1}{2} \times 9.11 \times 10^{-31} \text{ kg} \times (10^6 \text{ m s}^{-1})^2$

$= 4.55 \times 10^{-19} \text{ J}$

$V = \frac{E_k}{e} = \frac{4.55 \times 10^{-19} \text{ J}}{1.60 \times 10^{-19} \text{ C}} = 2.84 \text{ V}$

33 Greater kinetic energy means greater momentum and hence shorter wavelength.

34 Greater voltage means greater kinetic energy and momentum, so shorter wavelength.

35 De Broglie wavelength $\lambda = \frac{h}{p}$

$= \frac{6.63 \times 10^{-34} \text{ J s}}{9.11 \times 10^{-31} \text{ kg} \times 10^6 \text{ m s}^{-1}}$

$= 7.3 \times 10^{-10} \text{ m}$

(This wavelength is comparable to the spacing of atoms in a solid.)

36 $p = \frac{h}{\lambda} = \frac{6.63 \times 10^{-34} \text{ J s}}{3.0 \times 10^{-10} \text{ m}}$

$= 2.21 \times 10^{-24} \text{ kg m s}^{-1}$

37 For a person of mass 60 kg walking at 1 m s^{-1}:

$p = 60 \text{ kg m s}^{-1}$

$\lambda = \frac{h}{p} = \frac{6.63 \times 10^{-34} \text{ J s}}{60 \text{ kg m s}^{-1}} \approx 10^{-35} \text{ m}.$

This wavelength is *extremely* small. You would only notice diffraction effects when passing through an aperture of comparable size – which is about 35 orders of magnitude smaller than your own physical size. A doorway is some 35 orders of magnitude too large, so any diffraction effects are unobservable.

38 (a) $E_0 = m_0c^{-2}$

$= 1.67 \times 10^{-27} \text{ kg} \times (3.00 \times 10^8 \text{ m s}^{-1})^2$

$= 1.50 \times 10^{-10} \text{ J}$

$= 9.39 \times 10^8 \text{ eV} = 0.939 \text{ GeV}$

(b) $m = 0.939 \text{ GeV}/c^2$

39 Total energy available = 2 GeV. Energy needed to create a single pi-zero at rest is 0.14 GeV.

$\frac{2}{0.14} = 14.28.$

So at most 14 pi-zeros could be created. (This assumes the pi-zeros are created at rest, and that no other particles are created. In practice, neither of those assumptions is likely to be correct.)

40 $m = 20 \text{ GeV}/c^2$

41 $1 \text{ u} = 1.66 \times 10^{-27} \text{ kg}$

$E = mc^2 = 1.66 \times 10^{-27} \text{ kg} \times (3.00 \times 10^8 \text{ m s}^{-1})^2$

$= 1.49 \times 10^{-10} \text{ J}$

$= \frac{1.49 \times 10^{-10} \text{ J}}{1.60 \times 10^{-19} \text{ J eV}^{-1}}$

$= 9.31 \times 10^8 \text{ eV} = 931 \text{ MeV}$

42 $m_e = 9.11 \times 10^{-31} \text{ kg}$

$= \frac{9.11 \times 10^{-31} \text{ kg}}{1.66 \times 10^{-27} \text{ kg u}^{-1}}$

$= 5.49 \times 10^{-4} \text{ u}$

43 (a) $\frac{3000 \text{ rpm}}{60 \text{ s}} = 50 \text{ s}^{-1}$

(b) Angular velocity, $\omega = 2\pi \times 50 \text{ s}^{-1} = 314 \text{ rad s}^{-1}$

(c) Period, $T = \frac{1}{50 \text{ s}^{-1}} = 0.02 \text{ s}$

44 (a) Number of revolutions per second = $\frac{30}{2\pi}$ = 4.8

(b) Period $T = \frac{1}{4.8\ \text{s}^{-1}} = 0.21$ s

45 Radius of accelerator r = 4300 m

Angular velocity $\omega = \frac{v}{r} = \frac{3.00 \times 10^8\ \text{m s}^{-1}}{4300\ \text{m}}$

$= 6.98 \times 10^4$ rad s^{-1}

46 (a) Angular velocity $\omega = \frac{2\pi}{T} = \frac{2\pi}{1.25\ \text{s}} = 5.0$ rad s^{-1}

(b) (i) Speed $v = \omega r = 5.0$ rad s^{-1} × 2 m = 10 m s^{-1}

(ii) Speed is halved, ie 5 m s^{-1}

(iii) Speed = 0 m s^{-1}

47 (a) Angular velocity $\omega = 100 \times 2\pi$ rad s^{-1}

= 628 rad s^{-1}

(b) Speed $v = \omega r$

= 628 rad s^{-1} × 0.15 m

= 94.2 m s^{-1}

48 Radius $r \approx$ distance from shoulder to ground

$\approx$ 1.5 m, say.

Angular velocity = $\omega = \frac{v}{r} = \frac{30\ \text{m s}^{-1}}{1.5\ \text{m}} = 20$ rad s^{-1}

49 From Q45: $a = \frac{v^2}{r} = \frac{(3.00 \times 10^8\ \text{m s}^{-1})^2}{4300\ \text{m}}$

$= 2.1 \times 10^{13}$ m s^{-2}

From Q46: **(i)** $a = \frac{v^2}{r} = \frac{(10\ \text{m s}^{-1})^2}{2\ \text{m}} = 50$ m s^{-2}

(ii) $a = \frac{v^2}{r} = \frac{(5\ \text{m s}^{-1})}{1\ \text{m}} = 25$ m s^{-2}

From Q47: $a = \frac{v^2}{r} = \frac{(94.2\ \text{m s}^{-1})^2}{0.15\ \text{m}} = 5.9 \times 10^4$ m s^{-2}

From Q48: $a = \frac{v^2}{r} = \frac{(30\ \text{m s}^{-1})^2}{1.5\ \text{m}} = 600$ m s^{-2}

50 (a) Turning sharply.

(b) A bigger force is needed.

51 $F = \frac{mv^2}{r} = \frac{5\ \text{kg} \times (10\ \text{m s}^{-1})^2}{2\ \text{m}} = 250$ N

52 $F = mr\omega^2 = 1$ kg × (10 rad s^{-1})^2 × 10 m = 1000 N

53 (a) $\omega = \frac{2\pi}{T} = \frac{2\pi}{0.75\ \text{s}} = 8.4$ rad s^{-1}

(b) $F = mr\omega^2$ = 7.26 kg × (8.4 rad s^{-1})^2 × 2.1 m

$= 1.1 \times 10^3$ N

54 (a) $W = mg$ = 45 kg × 9.81 N kg^{-1} = 441 N

(b) $F = \frac{mv^2}{r} = \frac{45\ \text{kg} \times (4\ \text{m s}^{-1})^2}{6\ \text{m}} = 120$ N

(c) Tension in rope = 441 N + 120 N = 561 N

(The rope must support the weight of the child + tyre, *and* supply the centripetal force.)

55 (a) $F \propto \frac{1}{r}$, so twice the radius requires half the force; ie 1 kN.

(b) $F \propto v^2$, so twice the speed requires four times the force; ie 8 kN.

56 See Figure 3.67.

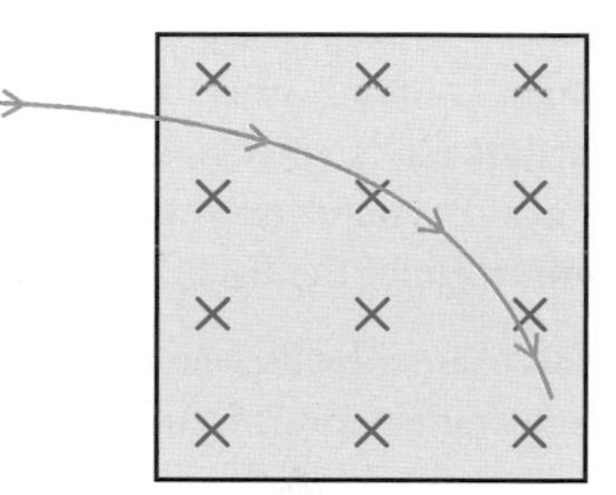

Figure 3.67 Answer to Question 56

57 The electrons do not all have the same momentum (or, therefore, the same kinetic energy). The most energetic electrons will curve least.

58 $r = \frac{mv}{Bq}$

$= \frac{9.11 \times 10^{-31}\ \text{kg} \times 0.01 \times 3.00 \times 10^8\ \text{m s}^{-1}}{(0.2\ \text{T} \times 1.60 \times 10^{-19}\ \text{C})}$

$= 8.5 \times 10^{-5}$ m

59 (a) $v = \sqrt{\left(\frac{2eV}{m}\right)}$

$= \sqrt{\left(\frac{2 \times 1.60 \times 10^{-19}\ \text{C} \times 2500\ \text{V}}{9.11 \times 10^{-31}\ \text{kg}}\right)}$

$= 2.96 \times 10^7$ m s^{-1}

(b) $F = Bqv$

= 1 T × 1.60 × 10^{-19} C × 2.96 × 10^7 m s^{-1}

$= 4.7 \times 10^{-12}$ N

(c) $a = \frac{F}{m} = \frac{4.7 \times 10^{-12}\ \text{N}}{9.11 \times 10^{-31}\ \text{kg}}$

$= 5.2 \times 10^{18}$ m s^{-2}

The electrons change direction but not speed.

(d) $r = \frac{mv}{Bq} = 1.70 \times 10^{-4}$ m

or use $a = \frac{v^2}{r}$

$r = \frac{v^2}{a} = 1.70 \times 10^{-4}$ m.

CHAPTER 4

Build or Bust?

Why a chapter called *Build or Bust?*

In March 2008, Fulstow brewery, Lincolnshire, produced a special beer, named 'Epicentre'. The brew had an alcohol content of exactly 5.2% by volume, and was brewed to commemorate a most unusual event (for the UK). In this country we rarely feel earthquakes (although they can be measured), but the one that occurred at 00:56 on 27 February, near Market Rasen was the strongest British quake for many years.

It is difficult to imagine the feeling of being in a severe earthquake – the terror as whole buildings collapse around you without warning, like a pack of cards, and the ground beneath your feet just opens up. Figure 4.1(a) shows the devastation that an earthquake can bring. An earthquake consists of shock waves travelling through the Earth. Physicists can study these waves, and discover a great deal about the structure and origin of the Earth and how it is evolving.

This chapter focuses on two aspects of building design: earthquake resistance and temperature control. An understanding of waves enables us to design buildings and create materials that will withstand most earthquakes – and to locate valuable deposits of oil. Another aspect of modern building design involves heating and temperature control (Figure 4.1(b)). An understanding of material properties is important for all of this. Knowledge of how energy travels is vital in designing energy-efficient buildings, so that they both make the best use of resources and are pleasant to live and work in.

Figure 4.1 (a) A major earthquake caused widespread devastation in China in 2008

Figure 4.1 (b) Many modern buildings are designed with efficient temperature control in mind

Overview of physics principles and techniques

In this chapter you will continue your study of waves, energy and materials. Much of the work will build on and extend ideas you have met in previous chapters. You will also study the type of oscillation known as simple harmonic motion, consider why objects vibrate most at certain frequencies (resonance) and what that means for building design. Simple harmonic motion underlies most studies of waves and vibrations, and is the most important topic in this chapter. In addition you will extend your understanding of how materials absorb and transmit heat, and see how that understanding is useful in building design.

In this chapter you will extend your knowledge of the following from the AS chapters:

- travelling waves from *The Sound of Music* and *Spare Part Surgery*
- energy and its conservation from *Higher, Faster, Stronger*
- refraction and reflection from *The Sound of Music*
- bulk properties of solids from *Good Enough to Eat* and *Spare Part Surgery*.

You will do more work on:

- waves in *Reach for the Stars*
- energy and temperature in *Reach for the Stars*.

1 Earthquakes

1.1 Shaking the Earth

Earthquakes – the wrath of the gods?

Earthquakes are killers. In 2008, a major earthquake devastated much of China's Sichuan province; official estimates put the death toll at nearly 70 000. Modern cities in earthquake zones represent a particular hazard; in 1995 6000 people died in the earthquake in Kobe in Japan. In cities that are at high risk of earthquakes, such as Kobe or San Francisco, the major concerns are to predict accurately when an earthquake will occur so that people can leave, and to build so that the buildings, bridges and roads will not collapse. A lot of effort has been put into predicting earthquakes, but with little success. Building earthquake-resistant cities makes prediction less important and has been more successful. In Kobe, one building was able to withstand the earthquake while all those around were completely destroyed. If we specify that we want a building to be 'earthquake proof', how do engineers meet that specification?

Figure 4.2 Minoan bull statue

In Bronze Age Crete, the Minoans revered the bull (Figure 4.2) as the incarnation of Poseidon, in his role as god of earthquakes. There is even evidence of human sacrifice, with the remains of a youth found on an altar, his body having been drained of blood to offer to the gods. This failed to avert the earthquake: the bodies of the priest, the sacrificial knife and the vessel for containing the victim's blood were all found where they were crushed by the falling masonry as the earthquake shook the Earth.

We now know that the Earth is not solid and it is constantly moving and changing. Figure 4.3 shows the Earth's structure. The central solid core is made of iron and nickel and is held at immensely high temperature and pressure. The liquid core of molten iron and nickel surrounding this is continuously moving in convection currents and there are also convection currents in the mantle, which is soft and liquid in places. The Earth's crust is relatively thin (only a few kilometres in places) and is made of a number of continental and oceanic plates that float on the mantle. It is the movement of these plates that causes earthquakes.

As the plates are forced slowly towards, or past, one another, the forces increase and the strain gradually builds up until the plates suddenly move: an earthquake. Seismic waves (shock waves) spread out and travel all round the world. As rocks settle after the earthquake there are smaller aftershocks. Seismologists study how earthquake waves travel in order to establish more about the Earth and even use the information to locate oil and gas deposits.

Figure 4.4 shows the edges (margins) of the plates. At so-called constructive margins, the plates are spreading or diverging and new oceanic crusts form mid-ocean ridges with volcanoes. There are two types of destructive margin: subduction zones, where the oceanic crust moves towards the light continental crust, sinks and is destroyed; and collision zones, where two continental crusts collide and are forced up into mountains. At conservative margins, two plates move sideways past each other, and land is neither formed nor destroyed.

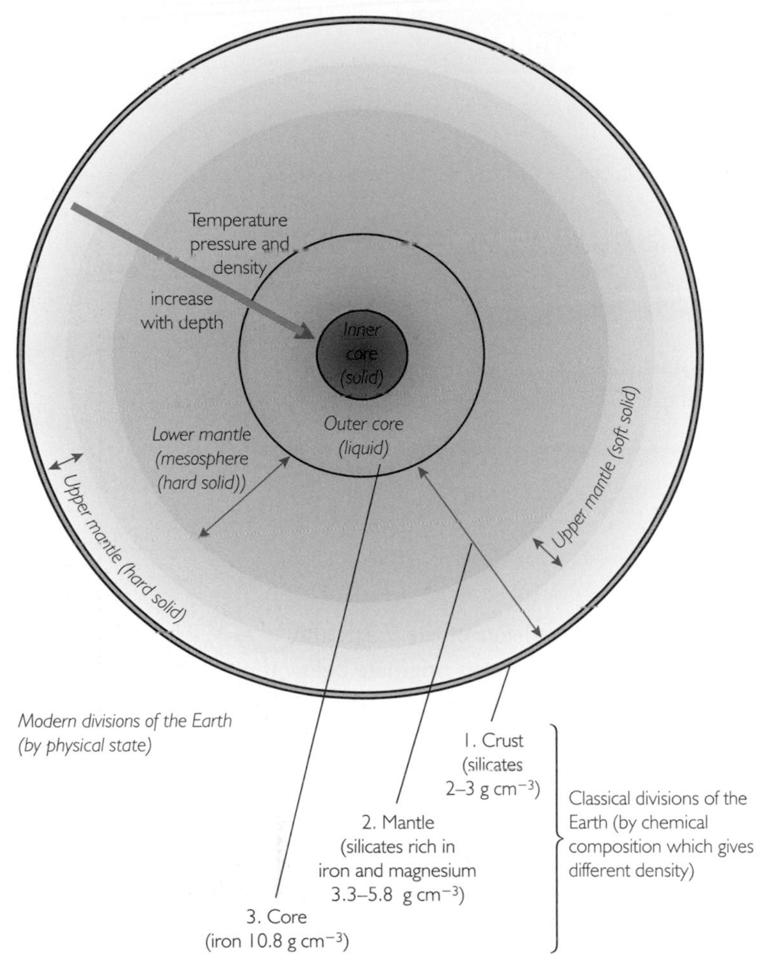

Figure 4.3 Structure of the Earth

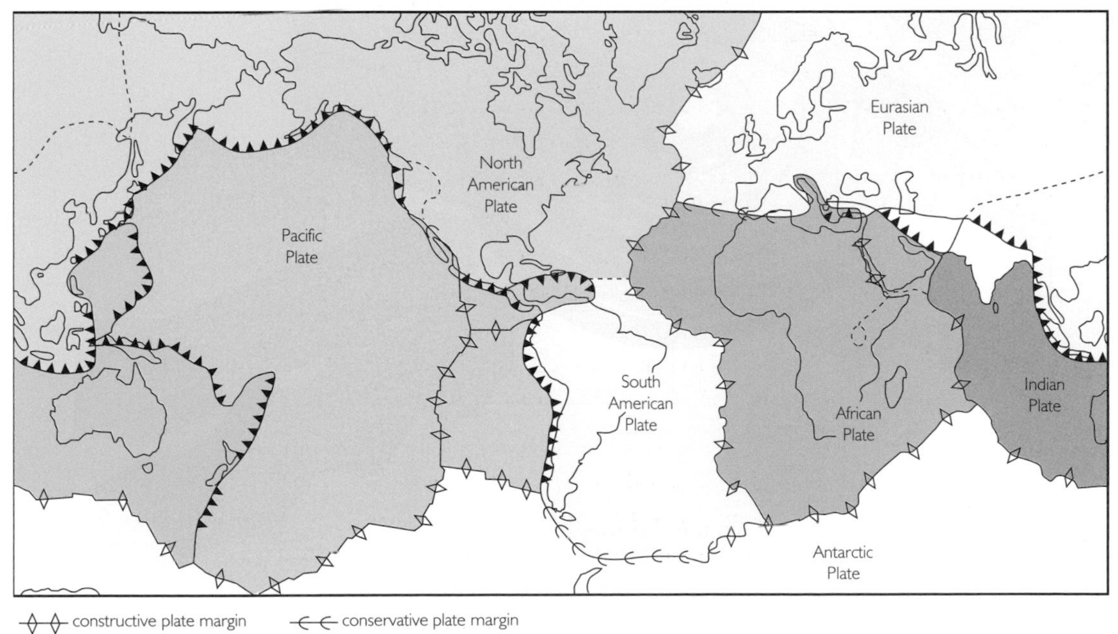

Figure 4.4 Plate margins of the Earth

Activity 1 Earthquakes on the internet

Use the internet to find out about one or more of the following topics. Some useful links can be found at **www.shaplinks.co.uk**.

- Where was the most recent earthquake recorded in the UK? In the world? What can you say about the sizes of these earthquakes?
- What was the largest earthquake recorded in the UK? Roughly how many earthquakes occur in the UK each year?
- How are instruments designed to detect and measure earthquakes? What principles of physics are involved?
- How do engineers and architects choose materials, and use principles of physics, when designing earthquake-resistant buildings?

1.2 Seismology

Did the Earth move for you?

The study of earthquakes – seismology – increases our knowledge of the Earth and enables engineers to design earthquake-resistant cities. Earthquakes originate below the surface of the Earth at a point called the focus, or hypocentre; the point on the Earth's surface directly above the earthquake is known as the epicentre. The seismic waves spread out from the focus in all directions (see Figure 4.5).

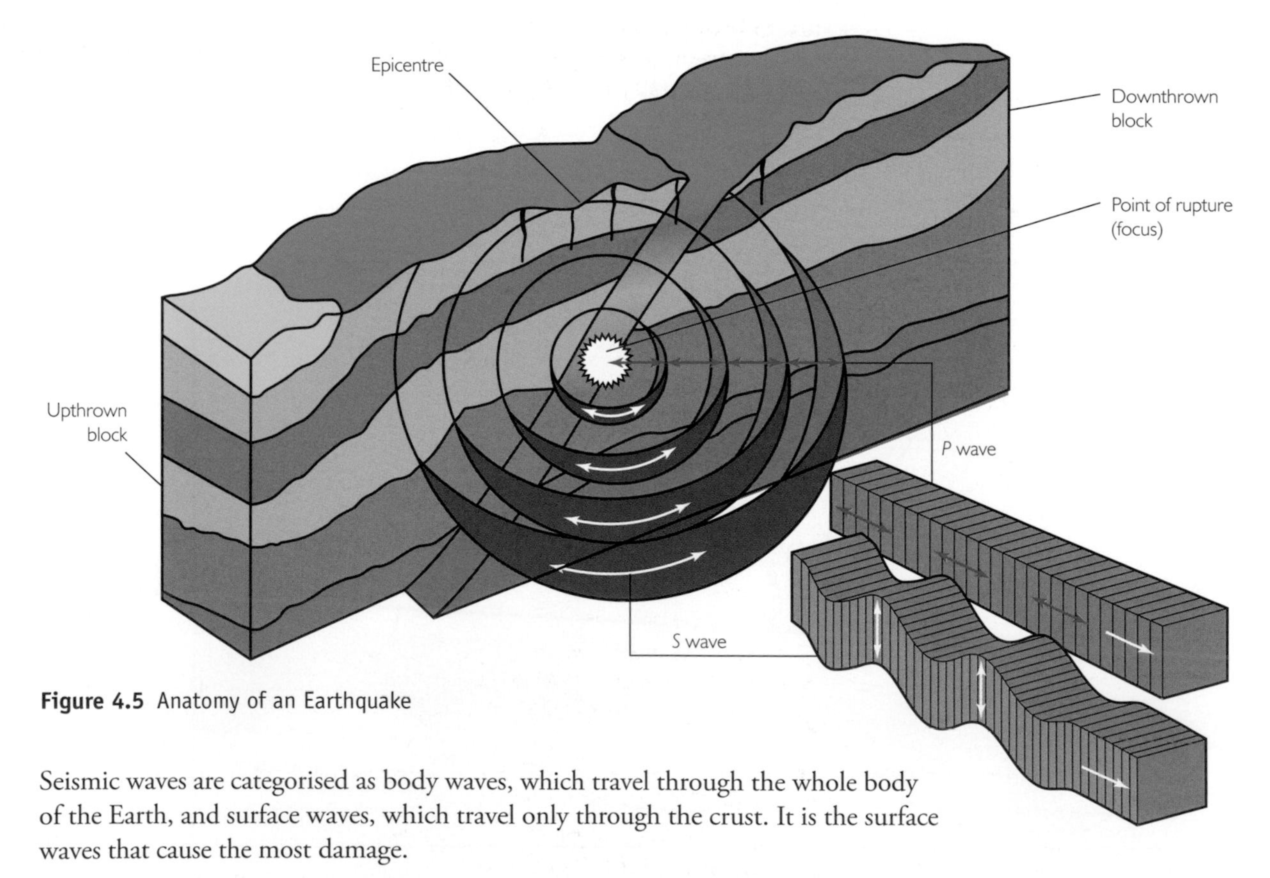

Figure 4.5 Anatomy of an Earthquake

Seismic waves are categorised as body waves, which travel through the whole body of the Earth, and surface waves, which travel only through the crust. It is the surface waves that cause the most damage.

There are two types of body waves (Figure 4.6), called primary waves (P waves) and secondary waves (S waves). They typically have frequencies of a few hertz.

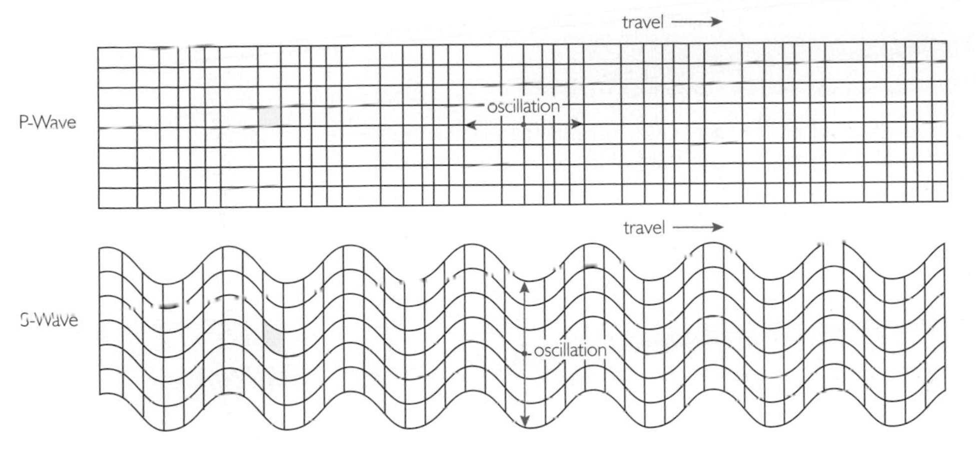

Figure 4.6 Body waves

P waves are **longitudinal waves** and can travel through the whole Earth, including the molten core. They travel through the rocks as a series of compressions and rarefactions. They are called primary waves because they travel faster than the S waves and arrive first at seismic monitoring stations.

S waves (sometimes also known as shear waves or shake waves) are **transverse waves** and cannot travel through the liquid parts of the Earth. The particles oscillate in a direction that is perpendicular to the direction of wave travel, so there is a shearing movement in the rock. The delay between the arrival of the P and S waves can be used to calculate the distance of the station from the earthquake.

> **Study note**
>
> You have met longitudinal and transverse waves before, notably in the AS chapter *The Sound of Music*. Animations of these waves may be viewed via **www.shaplinks.co.uk.**

There are also two types of surface waves, named Rayleigh waves and Love waves after the scientists who studied them (see Figure 4.7). Surface waves occur only in the Earth's crust, when one boundary of the material in which the wave is travelling is a free boundary. They have longer wavelengths (hundreds of kilometres) and lower frequencies than body waves (typically 0.1 Hz or less), and the displacement is largest at the surface, decreasing exponentially with the depth.

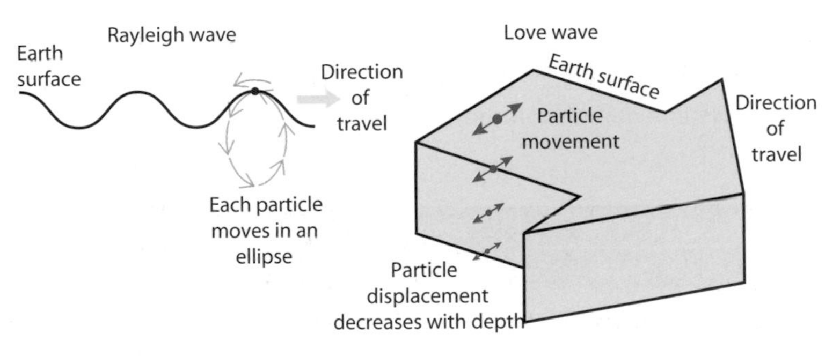

Figure 4.7 Surface waves

Love waves are transverse (or shear) waves where there is a strong horizontal displacement at right angles to the direction of travel. These waves are like water waves that have been turned on their side, through 90°.

Rayleigh waves are transverse waves, similar to water waves, where there is a strong vertical movement at right angles to the direction of travel. The particles that are being displaced have what is known as retrograde elliptical motion. This means that they are moving in the opposite direction to the wave at the crest. The path of each particle moved by the wave is an ellipse. Rayleigh waves travel more slowly than S or P waves.

Activity 2 Revision of waves

On a set of plain postcards, generate a set of Key Facts revision cards that summarise what you have previously learned about waves. Include definitions of key terms such as longitudinal, wavelength, phase and so on.

Activity 3 Revision of material properties

Earthquakes result from the exposure of rocks to compressive stresses in three dimensions. Generate a concept map that summarises everything you need to know about Young modulus, including the definitions of terms such as stress and strain. Find out what is meant by **bulk modulus** and **shear modulus** and include these terms in your concept map.

Study note

You learned about the Young modulus in the AS chapter *Spare Part Surgery*.

Questions

1 If the speed of waves in the Earth's crust is about 3 km s^{-1}, estimate how long it would take for seismic waves to travel through the crust to reach the opposite side of the world.

(Radius of Earth = 6.378×10^6 m)

2 A quartz wire 10 cm long and 6 μm in diameter is used to suspend a small metal sphere of mass 2 g.

(a) Calculate the extension of the wire.

(b) What is the maximum mass that can be suspended from the wire?

(Use g = 9.81 N kg^{-1}. Data for quartz: Young modulus E = 73.1 GN m^{-2}; tensile strength 1000 MN m^{-2})

3 Table 4.1 shows the speeds of seismic waves in materials commonly found in and on the Earth.

Material	Elastic modulus/10^9 N m^{-2}		Density/ 10^3 kg m^{-3}	Seismic wave speed/10^3 m s^{-1}	
	Bulk	Shear		P waves	S waves
air at 273 K	1.0×10^{-4}	0.0	1.0×10^{-3}	0.3	0.0
water at 298 K	2.2	0.0	1.0	1.4	0.0
ice	3.0	4.9	0.92	3.2	2.3
shale	8.8	17	2.4	3.6	2.6
sandstone	24	17	2.5	4.3	2.6
salt	24	18	2.2	4.7	2.9
limestone	38	22	2.7	5.0	2.9
quartz	33	39	2.7	5.7	3.8
granite	88	22	2.6	6.7	2.9
peridotite (compressed)	1.4×10^2	58	3.3	8.1	4.2

Table 4.1 Waves speeds, densities and elastic moduli for various materials

(a) From the data in Table 4.1, suggest, qualitatively, how elastic modulus will affect wave speed. (Does increasing the modulus increase or decrease the wave speed for a given density?)

(b) Suggest how density affects wave speed.

Journey to the centre of the Earth

In recent years the study of earthquakes has provided us with a great deal of information about the Earth. It is now apparent that the Earth is made of many layers that have different characteristics (see Figure 4.3), so that seismic waves have different speeds when travelling through these layers. Figure 4.8 is based on the Preliminary Earth Reference Model which was developed in the 1980s using seismic data. It shows how density and wave speed vary with depth.

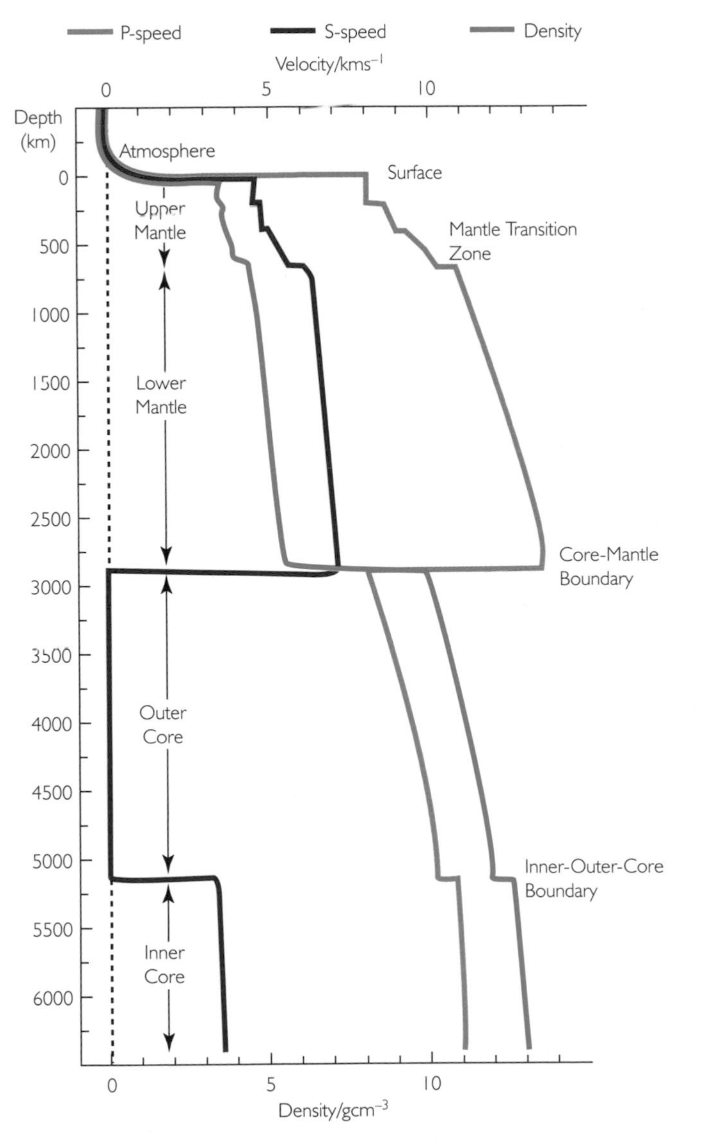

Figure 4.8 Variation in density and earthquake wave speed with depth

The surface of the Earth is made of rigid plates that consist of three layers: the crust, the Moho and the upper mantle. The crust has P-wave speeds of about 2 to 5 km s^{-1} for sedimentary rocks and about 6 km s^{-1} for igneous rocks. The Moho (short for

Mohorovicic discontinuity) separates the crust from the upper mantle, and is about 10 km down under the oceans and 20–70 km down under continents; here P-wave speeds increase to about 8 km s^{-1}.

Below the plates, the lower part of the mantle is a soft plastic solid. Some of it may be partially melted. Because of this the P and S waves have lower speeds, about 7 to 8 km s^{-1}, gradually increasing to about 10 km s^{-1} with depth.

Pressure increases with depth, so the lower mantle is more rigid. Then there is a sudden increase in P and S wave speeds marking the boundary with the core.

Wave speeds suddenly decrease at the boundary and then slowly increase further with depth. The outer core transmits no S waves, which results in a shadow zone (see Figure 4.9(a)). This tells us that the outer core is a fluid. There is also a P-wave shadow zone (Figure 4.9(b)) due to the sudden change in speed. The inner core is assumed to be solid since some weak P waves arrive on the other side of the Earth sooner than expected.

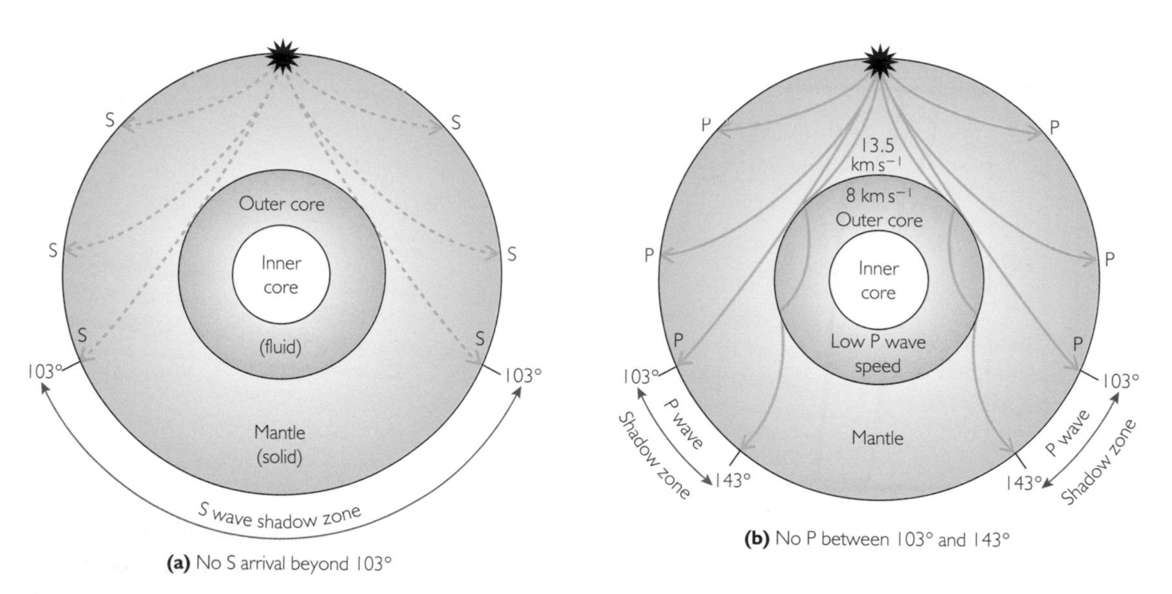

Figure 4.9 Shadow zones (a) for S waves and (b) for P waves

The layered structure of the Earth has been deduced from measurements of seismic waves together with knowledge of the waves' speeds in different types of rock. But what properties of a material will determine the speed of waves within it? Question 3 showed that density cannot be the only factor and Activity 3 gave a hint concerning elastic moduli. As Activity 4 demonstrates, the density and elastic modulus together determine the wave speed.

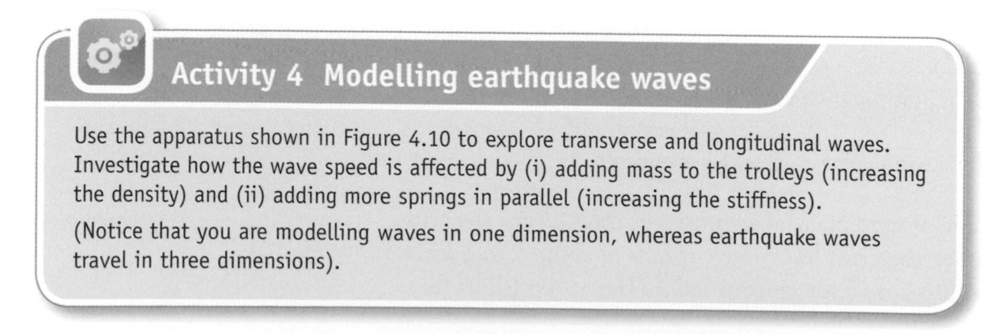

Activity 4 Modelling earthquake waves

Use the apparatus shown in Figure 4.10 to explore transverse and longitudinal waves. Investigate how the wave speed is affected by (i) adding mass to the trolleys (increasing the density) and (ii) adding more springs in parallel (increasing the stiffness).

(Notice that you are modelling waves in one dimension, whereas earthquake waves travel in three dimensions).

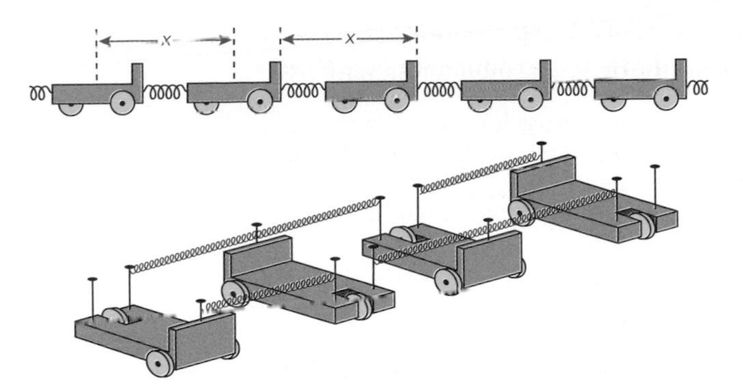

Figure 4.10 Modelling earthquake waves

As you will see in Section 2.3 of this chapter, the speed, v, of longitudinal waves is given by:

$$v = \sqrt{\left(\frac{E}{\rho}\right)} \qquad (1)$$

where E is the elastic modulus (Young modulus) and ρ the density.

> **Study note**
>
> For shear waves E is replaced by the shear modulus.

At a boundary between different materials, the wave speed changes – the waves are refracted. If the waves do not meet the boundary head-on, their direction also changes. The speeds and directions are related by Snell's law:

$$\frac{\sin i}{\sin r} = \frac{v_1}{v_2} = {}_1\mu_2 \qquad (2)$$

where ${}_1\mu_2$ is the refractive index between the two materials, and i and r are the angles between the normal to the boundary and the incident and refracted wave directions, respectively (Figure 4.11).

> **Study note**
>
> You have met examples of wave refraction before, in the AS chapter *The Sound of Music* and in the A2 chapter *The Medium is the Message*. You may wish to look back at this earlier work as you tackle Questions 4–7.

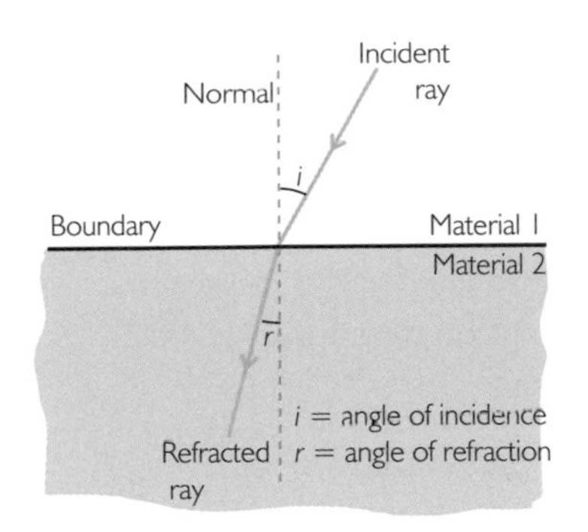

Figure 4.11 Refraction of waves at a boundary

Questions

4 At the Moho there is a sudden increase in the velocity of seismic waves. Draw a ray diagram to show what happens to a wave when it crosses a boundary and its velocity increases. What is this phenomenon called? Also draw a wavefront diagram to show how the change in wavelength results in the change in direction of travel.

5 Figure 4.12 shows the paths of several waves that have been formed by a controlled explosion. The waves are detected by several monitoring stations.

(a) Describe what has happened to each wave to cause it to travel in the path shown.

(b) Explain how information from the monitoring stations could be used to find out about the structure of the Earth below the stations.

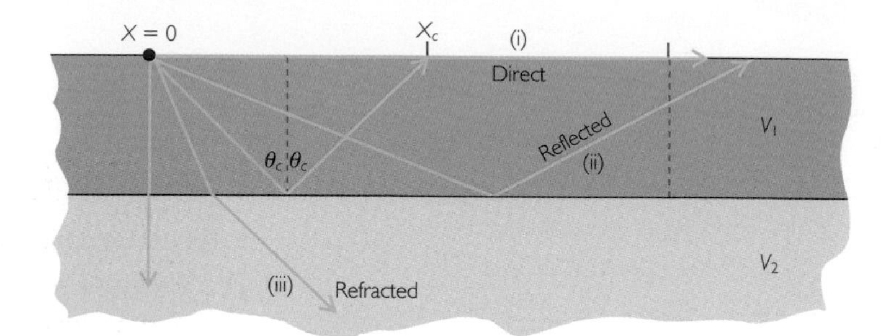

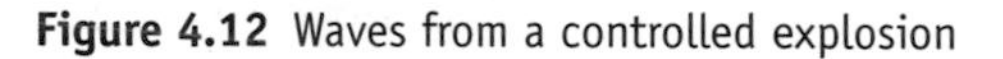

Figure 4.12 Waves from a controlled explosion

6 A P wave crosses the boundary from the lower mantle into the outer core at an angle of incidence, $i = 30°$. Using information from Figure 4.8, calculate a value for the refractive index between the two layers and hence calculate the angle of refraction.

7 (a) The waves' paths shown in Figure 4.9 are curved. What does this imply for the way that wave speed changes with depth in a given material?

(b) Deep within the Earth, the pressure is very high so materials are highly compressed. What effect will this have on the density of a given material?

(c) In view of your answers to (a) and (b), what can you deduce about the way elastic modulus varies with depth (and hence with pressure)?

Seismology in action

Mike Evans is an oil and gas exploration geophysicist who graduated with a degree in geological geophysics from the University of Reading and went on to manage an E&P (Exploration and Prospection) consultancy, Focus Exploration.

It can be very exciting as it's a high-risk game with companies committing millions of pounds to sink a well, often based primarily on your interpretation of seismic data.

Geophysics plays an important role in locating oil and gas deposits and deciding the best place to sink a well. The most important geophysical tool is the seismic method, which is a three-stage process requiring plenty of communication between each stage. The first stage is the seismic acquisition where acoustic waves are sent into the Earth. As the waves travel through the different rock types they are variably reflected, refracted and transmitted at each rock boundary. Hydrophones and geophones are used to pick up the reflected waves, building up a mass of data covering the whole area. Data processing is the next stage, where analysts use computers to produce a representation of the sub-surface – seismic profiles (2D seismic) or a seismic volume (3D seismic).

Mike was talking about his work at the third stage, as a seismic interpreter. He had to use his interpretive experience and knowledge of geology to develop an understanding of the present-day geometries of rock in the sub-surface, and their structural history. He was looking for evidence of mature hydrocarbon source rocks and structural or stratigraphic features where oil or gas may be trapped.

It's satisfying when all the bits of the jigsaw come together, and we're fortunate that our search for oil takes us all over the world from the remotest jungles of the Far East to downtown Los Angeles.

Seismometers

Earthquakes have long been the subject of human study. Figure 4.13 shows an early Chinese seismometer. Inside the large jar is a pendulum. An earthquake starts the pendulum swinging and rods inside move and opens a dragon's mouth so that the ball falls into the mouth of one of the frogs, thus indicating the direction of the earthquake.

Figure 4.13 Early Chinese seismometer

Close to the epicentre, ground movements can be quite large. But at large distances the movements are very small. Modern seismometers are instruments that are able to detect very small movements from earthquakes thousands of kilometres away.

In the 1960s, seismometers like the one in Figure 4.14 began to be used. This one records vertical movements. It has a heavy mass inside, which is mounted on springs. The instrument is bolted firmly to the ground. When the ground moves the instrument moves with it, but the mass will 'try' to stay in its original position due to inertia. A coil inside the mass will register changes in the magnetic field and an electrical signal shows the movement of the mass and the instrument. This electrical output is digitised and datalogged by computer. Networks of seismic stations are linked in real time via satellites to enable global monitoring of earthquakes. The output can be displayed visually in a record known as a seismogram. Seismic monitoring stations usually have three seismometers to measure oscillations in three directions at right angles to each other.

Figure 4.14 Modern seismometer

Activity 5 Seismic waves

Explore animations and recordings of earthquake waves that can be found on the internet. If you have the opportunity, download the Seismic Waves Software (Figure 4.15) via **www.shaplinks.co.uk**

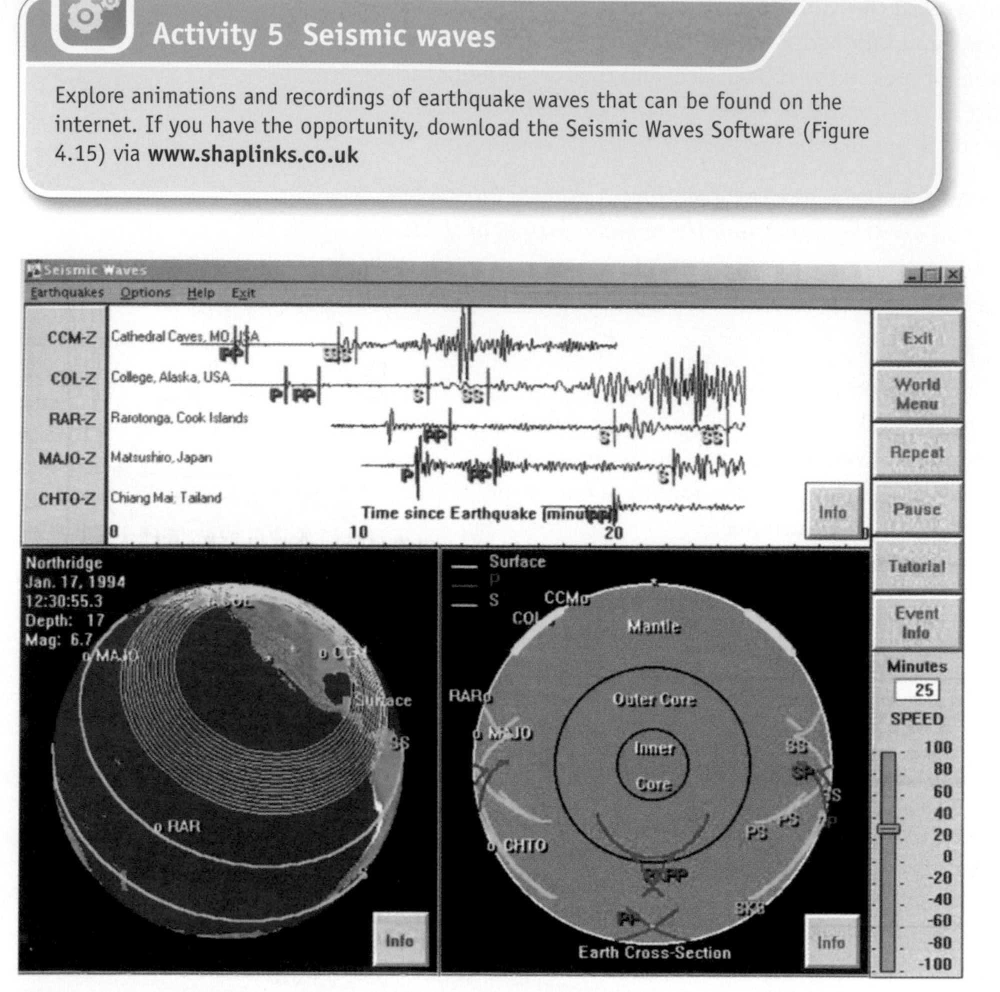

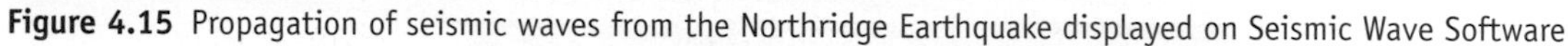

Figure 4.15 Propagation of seismic waves from the Northridge Earthquake displayed on Seismic Wave Software

Seismic alert

Like hurricanes in Hampshire, you might think that earthquakes in the UK hardly ever happen.

Market Rasen is a quiet market town in rural Lincolnshire (Figure 4.16). On 27 February 2008 at 00:56 GMT, the British Geological Survey (BGS) recorded an earthquake with a magnitude of 5.2 on the Richter scale with an epicentre just 4 km north of the town. A magnitude 5 earthquake is equivalent to an explosion of 1000 tonnes of TNT. This was a deep earthquake; measurements suggest that the hypocentre was 17.8 km below the Earth's surface.

SHAP physics teachers John Fox and Norman Palmer from De Aston School, Market Rasen described their experiences.

> *I was woken up from a sound sleep just before 1 a.m., thinking an articulated lorry or train was about to come through the bedroom wall! The floor was moving up and down, and I was wobbling about drunkenly as I walked across the room, watching the walls shake and things move! There was a really loud roaring sound, as if there was a railway tunnel the other side of the wall, with a heavy train going through it. My other thought as the ceiling was shaking was that there were*

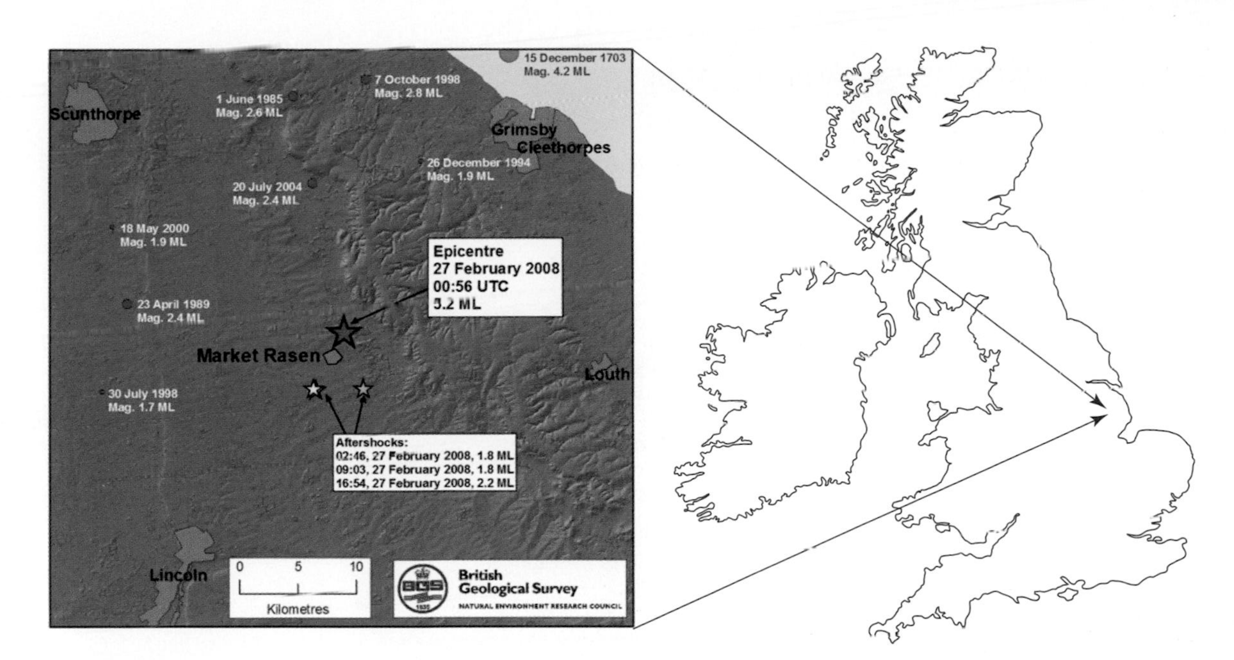

Figure 4.16 Market Rasen earthquake

elephants on the roof. It really was that severe and frightening. I went outside the front door, as I honestly thought that a heavy goods train had fallen off the railway embankment which passes through the centre of town. Our neighbours were all outside too, wondering what was going on.... we were relieved when we switched on the radio and heard that there was a countrywide earthquake. [JF]

My wife and I were woken by a loud noise, sounding like a possible derailment of the tanker train which runs a few hundred metres from the house. As I climbed out of bed I realised that it was shaking, then I found that the floor was like jelly. We had experienced a smaller earthquake in Durham in the early '70s and also some tremors in San Francisco on holiday. My wife ... the brains and the memory ... suggested that it probably was an earthquake. Imagine our surprise when Market Rasen suddenly had SKY TV helicopters and loads of reporters in the Market Place, shouting to interview the vicar and others about chimney pots and the stone cross which fell down from the church vestry roof, damaging the roof. The front street was closed while chimneys were made safe [Figure 4.17]. [NP]

In fact, 20–30 earthquakes are felt each year by people in the UK. Most of these are very small and have little effect. However, some British earthquakes have caused damage, although nothing like the devastation caused by large earthquakes in other parts of the world.

Figure 4.17 Earthquake damage in Market Rasen

Activity 6 Seismic alert

Study the seismograms from the Market Rasen earthquake (Figure 4.18) and answer these questions.

- Which station did the seismic waves reach first?
- Estimate the distance from the epicentre of the earthquake to each of the monitoring stations (Figure 4.19).
- Which waves always arrive first, P or S waves?
- The seismogram changes shape at the moment of the P and S arrivals due to the difference in the waves' frequencies. On a copy of Figure 4.18, add labels to the charts to indicate where you think the P and S waves arrive.
- How does the difference in arrival time of P waves and S waves change as the distance between the epicentre and the seismic monitoring station increases?
- These seismograms are all from stations relatively close to the focus of the earthquake. Why is it harder to locate earthquakes from SP times over longer distances on a flat map?

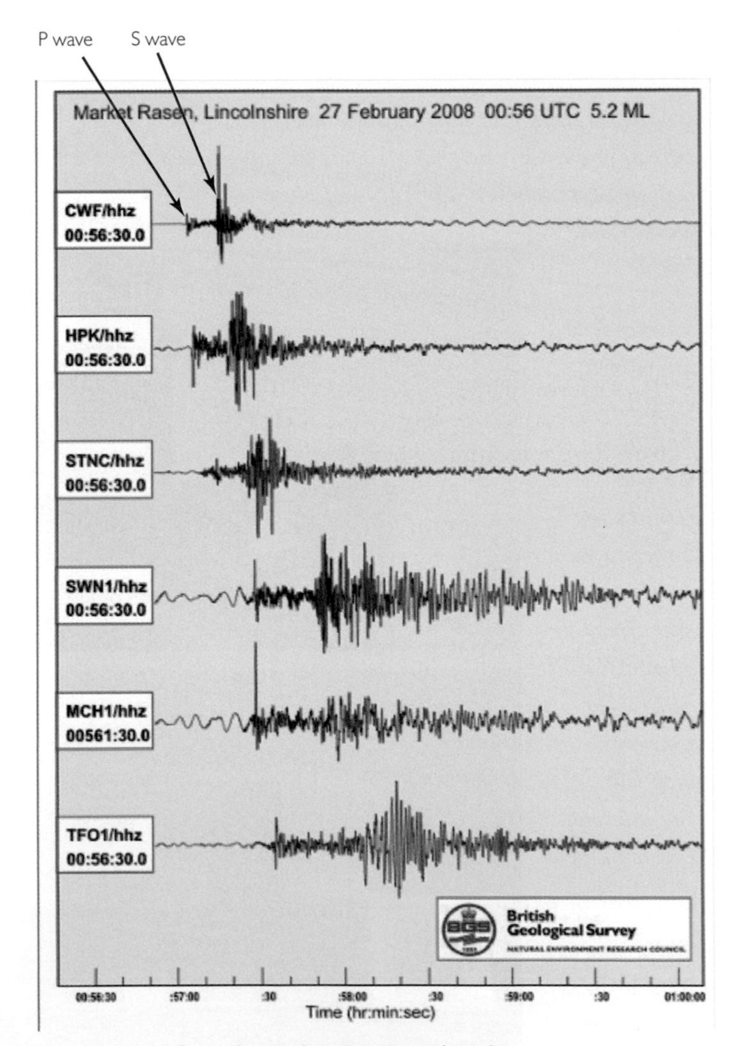

Figure 4.18 Seismograms from the Market Rasen earthquake

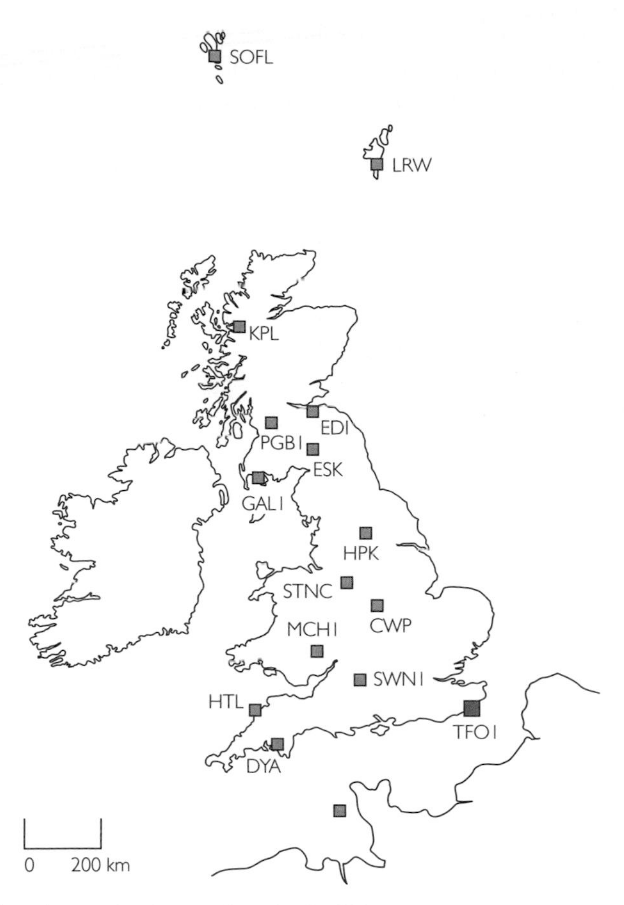

Figure 4.19 Earthquake monitoring stations in the UK

Earthquake scales

There are two scales commonly used to characterise the strength of an earthquake: the Richter scale and the Mercalli scale. The Richter scale describes the earthquake itself, whereas the Mercalli scale describes its observable effects at locations where it is detected.

Richter scale

The Richter scale is summarised in Table 4.2.

Magnitude on Richter scale	Effects
0.0 to 1.9	recorded by instruments
2.0 to 2.9	felt by very sensitive people; suspended objects swing slightly
3.0 to 3.9	felt by some people; vibration like a heavy vehicle
4.0 to 4.9	felt by most people; hanging objects swing; dishes and windows rattle and may break
5.0 to 5.9	people frightened; chimneys topple, furniture moves
6.0 to 6.9	buildings may suffer substantial damage
7.0 to 7.9	few buildings remain standing; large landslides and fissures
8.0 to 8.9	complete devastation; ground waves

Table 4.2 The Richter scale for earthquakes

Ground movements in an earthquake can be as large as several metres, but movements of only a few micrometres can be detected with modern instruments. With such a wide range of values, a linear scale would be inconvenient. The Richter scale, devised in 1935, is a logarithmic scale that is based on the amplitude of the P waves. On this scale, earthquakes are given a magnitude, defined such that a magnitude 3 earthquake has movements ten times that of a magnitude 2 earthquake, and so on. The greater the distance from the epicentre the less the ground will move, so the scale must take into account the distance from the epicentre.

Maths reference

Logarithmic scales
See Maths notes 8.6 and 8.7

Charles Richter (1900–1985) devised the scale that bears his name in the 1930s. He defined an earthquake of zero magnitude to be one that produced a zero-to-peak deflection on his seismometer of 0.001 mm at a distance of 100 km from the epicentre. A magnitude 1 earthquake would cause a disturbance with 10 times this amplitude, and a magnitude 6 earthquake would cause a disturbance a million times as big. For an earthquake of magnitude M, the deflection D at 100 km is given by:

Study note

You might like to look back at your work on logarithmic scales in *The Medium is the Message*.

$$D = 0.001 \text{ mm} \times 10^{M} \qquad (3)$$

As different seismometers produce different deflections for a given disturbance of the ground, each instrument has to be calibrated so that its output corresponds to the Richter scale.

Seismometers are very sensitive instruments, and the deflection of the instrument is generally much greater than the actual ground movement. For this reason, it is very difficult to measure a major earthquake close to its epicentre, and observations from very distant monitoring stations must be used.

Once the distance has been allowed for, each earthquake can be described in terms of a single magnitude. The largest recorded earthquake, with a magnitude of 9.5, occurred in Chile in 1960.

Study note

The software mentioned in Activity 5 includes information about the magnitudes of earthquakes measured on the Richter scale.

Question

8 What would be the deflection on Richter's seismometer 100 km from the magnitude 5.2 Market Rasen earthquake? Comment on your answer.

The Mercalli scale

The Mercalli scale uses the effects at the surface, as seen by people there at the time, to define the 'intensity'. Like the Beaufort scale for wind conditions, it is empirical – in other words, it is defined initially in terms of observable effects. Intensity will vary according to distance from the epicentre and the type of ground you are standing on. The Mercalli scale ranges from 1, which is detected by instruments, but not felt by people except in special circumstances, to 12, where there is almost total destruction, objects are thrown in the air, and waves are seen on the Earth's surface.

Activity 7 Earthquake Data

Use a resource such as the US Geological Survey's Earthquake Center, available via **www.shaplinks.co.uk**, to research a number of earthquakes and plot them on the Richter scale (log scale). Plot them according to the magnitude of the Earth movement resulting from each one (mm scale). You will find that the linear scale has to be so big that it will not fit around the room – a logarithmic scale is more convenient.

Maths reference

Using log scales
See Maths note 8.6

1.3 Summing up Part 1

This part of the chapter has been mainly revision of earlier work. You have reviewed some key ideas about waves, and have seen how the layered structure of the Earth affects the passage of seismic waves. You have also seen how logarithmic scales are used to describe earthquakes. Use the following activities to check your knowledge and understanding of waves in materials and of log scales.

Activity 8 Exam howlers

In an exam, the question 'Why do P and S waves follow curved paths?' produced the following answers:

(i) ... because the Earth spins and causes the waves to spin to the surface.

(ii) ...because the waves are affected by the gravity of the core.

(iii) ...because the magnetic core repels the seismic waves because they are of opposite charge.

(iv) ...because the waves cannot travel through a vacuum.

For each one discuss and explain why the answer is not correct. Then give the correct explanation in answer to the question.

Activity 9 Virtual earthquake

Visit the Virtual Courseware for Earth and Environmental Science website via **www.shaplinks.co.uk**.

Set off a virtual earthquake. Locate your earthquake and find its magnitude on the Richter scale.

2 Shaken not stirred

Earthquakes are not the only cause of major building collapse. In 1940 the famous Tacoma Narrows Bridge in the USA (Figure 4.20) finally shook itself to bits. For many years previously the fact that it would sway and twist violently when the wind blew up the river drew people from miles around to drive up and down this natural 'fairground' ride. You can see the collapse of the bridge for yourself on YouTube – search for 'Tacoma Narrows Bridge Collapse'.

Figure 4.20 Tacoma Narrows Bridge

In fact all buildings will sway if the wind is blowing at the right speed, but taller ones are more impressive – the Empire State Building sways with a frequency of $\frac{1}{8}$Hz when the wind blows while the old Severn Bridge vibrates at $\frac{1}{7}$Hz. Exactly how these wind-induced oscillations get started is quite complicated and depends on the complex pattern of air flow round a shaped beam.

In this part of the chapter, you will see how oscillations affect buildings and other structures – the oscillations may be produced by earthquakes, but there are other possible sources too. You will also see how one common type of oscillation can be analysed mathematically.

2.1 Forced oscillations

Resonance rings a bell

Every structure has at least one **natural frequency** where it will sway, swing or vibrate most easily and goes into violent oscillation with a large amplitude when driven at this frequency. This phenomenon is called **resonance**, and engineers have to consider this very seriously when designing everything from skyscrapers to cars. A seated human body resonates at around 5 Hz. Car designers must ensure that the car body doesn't vibrate at this frequency; otherwise passengers would be shaken up quite badly by the end of the journey – regardless of the driving abilities of the driver. Examples of resonance include a wineglass 'humming' and then shattering when vibrated by a pure note (although this is very difficult to do with modern glasses, as they are made to be much stronger); a flag flapping audibly; and parts of vehicles that suddenly start vibrating as the revs increase and then stop as the revs climb higher.

Resonance occurs when a structure is driven or forced to vibrate at its natural frequency and consequently absorbs maximum energy from the driving source. It is possible to drive structures to vibrate at other frequencies, but their response will be much less impressive (they vibrate with much smaller amplitude as much less energy is transferred). The University of Salford's A-level *Sounds Amazing* website (available via **www.shaplinks.co.uk**) is an excellent resource on the topic of resonance and includes a video showing the shattering of a wineglass by a pure note. You can use this website to explore the characteristics of simple harmonic motion and the causes and effects of resonance in different objects. Activities 10–12 study resonance in more detail.

Activity 10 Wind-induced oscillations

Explore the oscillations of a simple structure such as that shown in Figure 4.21. The structure is a horizontal beam of semicircular cross-section suspended between four springs. (A rectangular strip would do, provided it is fairly thin compared with its width.) The air from a fan or blower is directed on to it and as the air speed is varied the beam leaps into violent vertical oscillations at a particular air speed. (Are there any other ways – 'modes' – in which you can see the structure vibrating, apart from those in a vertical plane?)

As an extension, devise some means of measuring air speed, and measure the amplitude of vibrations produced at different speeds. Explore the effect of using different beams.

Figure 4.21 Exploring wind-induced oscillations

Activity 11 Barton's pendulums

Barton's pendulums (see Figure 4.22) are a set of pendulums of different lengths all attached to the same horizontal piece of string. One (the driver) is heavier than the rest. Predict and then observe what happens when you set the driver pendulum in motion.

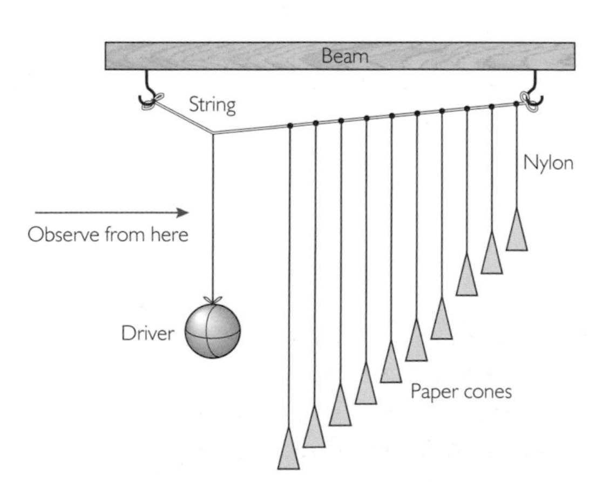

Figure 4.22 Barton's pendulums

Activity 12 Coupled pendulums

Set up two identical pendulums suspended from the same horizontal piece of string as in Figure 4.23. Predict what you will see happen if you set one swinging and wait. As an extension, investigate whether the time of transfer of energy from one pendulum to the other and back depends on the separation of the pendulums.

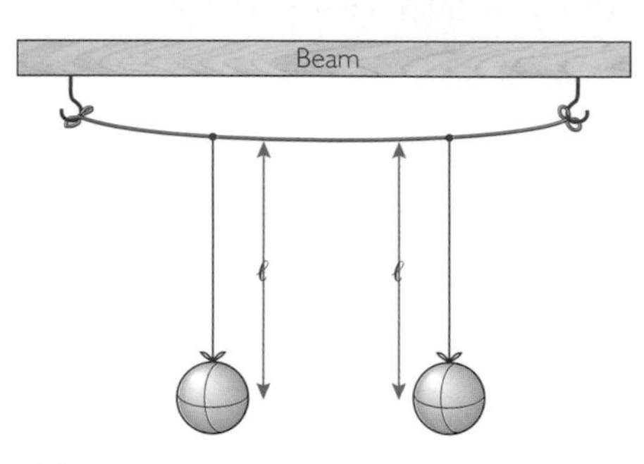

Figure 4.23 Coupled pendulums

Activities 10–12 show just how easy it is to set something oscillating at its natural frequency compared with other frequencies. In Activity 11 you will have seen that the pendulums first begin to oscillate with what appear to be their natural frequencies, but these **transient** oscillations die away and the pendulums settle down to oscillating with the same frequency as the driver, though with different amplitudes and phases. Activities 11 and 12 also show how the phases of the driver and the driven oscillators are related. Perhaps surprisingly, there is a phase difference of a quarter of a cycle – the two oscillators are said to be **in quadrature**.

> **Study note**
>
> You learned about phase difference in the AS chapter *The Sound of Music*.

Marching soldiers will be told to 'break step' when going over a bridge just in case the rhythm of their marching matches the natural frequency of the bridge and sets it vibrating (see Figure 4.24), and anyone can set a rope bridge swinging by timing their bounce just right.

Figure 4.24 Break step when walking over a suspension bridge!

London's Millennium Bridge

Firmly positioned in the 'A list' of London attractions, the Millennium Bridge (Figure 4.25) is a 330 m steel bridge linking the City of London at St Paul's Cathedral with the Tate Modern Gallery at Bankside. Built to an award-winning design, it was the first new pedestrian river crossing over the Thames in central London for more than a century, opening in time for the first year of the new millennium.

Figure 4.25 Millennium Bridge, London

Such was the interest in the new bridge that when it opened to the public on 10 June 2000, an estimated 80 000–100 000 people crossed it. Although the Millennium Bridge, like all bridges, was designed to cope with a degree of movement, it soon became clear that things were going seriously awry as the deck swayed about with motions of amplitude up to 70 mm between piers. Elderly walkers clung on to handrails on the side of the bridge. People reported feeling seasick.

The swaying bridge was beginning to look like a very expensive fairground ride. The bridge was instantly renamed 'The Wobbly Bridge', and after two days of random swaying, swinging and oscillating wildly, it was closed down by embarrassed engineers.

Activity 13 The Millennium Bridge

Research the story of the design and engineering of the Millennium Bridge.

Use the engineer's website (Arup) (available via **www.shaplinks.co.uk**) to research the cause of these oscillations and how engineers overcame the problems that developed.

Prepare a five-minute presentation or design a poster to communicate your findings.

Many major engineering disasters have been attributed to resonance – bridges, buildings and turbine blades in marine and aero engines. What is interesting, though, is that, with the notable exception of the Millennium Bridge, it is difficult to find recent examples of such failures – are we getting better at design? It does seem that experiment, together with theoretical and computer analysis, is increasing engineers' understanding of oscillations and resonance and that is feeding into better design practice. However, complacency could be disastrous; the problems with the Millennium Bridge occurred despite thorough testing and computer modelling. Engineers have to establish the natural frequencies of any structure they propose to build to ensure that it won't be set into violent oscillation, especially if it is in an earthquake zone. Studying how structures oscillate is an important part of this.

During earthquakes in cities, some buildings are destroyed, while others in the same district seem barely affected. One of the reasons for this effect is resonance.

Activity 14 Earthquakes and resonance in buildings

Make and study some simple models that illustrate the effect of resonance in buildings.

Seismometers

In Activity 15 you will explore resonance further and see how it comes into play when designing a seismometer.

A seismometer is an example of a forced oscillator, where a system capable of undergoing free oscillations is driven from the outside by some external force which itself oscillates at a frequency called the **driver frequency**, f_d. For a seismometer, the earthquake itself constitutes the driver. There is further slight complication in that an earthquake is unlikely to have a single frequency; there will be a spectrum of frequencies. However, in the simple model that you are going to use, it will be assumed that an earthquake can be modelled by a single frequency.

A simple seismometer consists of a mass hanging vertically on a spring (Figure 4.26). The top of the spring is driven by a vibrator at fixed amplitude from a variable-frequency generator. The generator–vibrator combination simulates a variable-frequency earthquake. The design has the two essential ingredients for an oscillation to occur – mass and stiffness. How likely is it in practice that the mass will stay still and not oscillate? One way of re-stating that the mass should be large in a 'good' seismometer is to say that the natural frequency should be low – but what does 'low' mean? What can we compare it with?

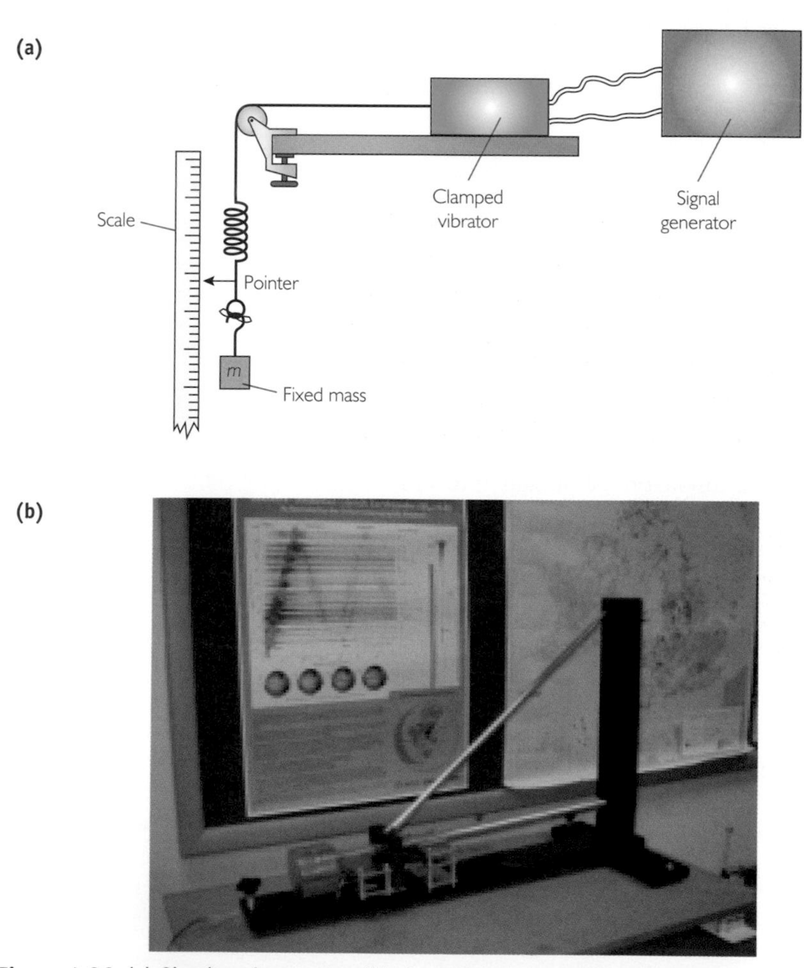

Figure 4.26 (a) Simple seismometer (b) the SEP (science enhancement programme) vibration detector

Activity 15 Make and market a seismometer

Make and test a simple seismometer as shown in Figure 4.26(a).

When you have made and tested your instrument, imagine that you represent a company that markets seismometers. Prepare a sales presentation to convince a client that your instrument has features that will make it work well.

A system which oscillates in response to external vibrations is the beginning of a seismometer, but to record those vibrations electronically requires the use of electromagnetic induction.

Study note

Look back at your work on electromagnetic induction in the chapter *Transport on Track*.

Activity 16 The Earth *did* move

Study the output of a simple vibration detector (Figure 4.26(b)) and use it to extend your understanding of real seismometers.

It's not all bad news...

Vibrations can cause damage, but an interesting development is that vibrations can reduce stress in steel and give stronger welds, leading to safer structures. Specimens of steel strip (formed by rolling steel slabs) are vibrated in a cantilever mode (ie held at one end) over a period of a few minutes. Portable X-ray diffraction equipment can measure the residual stress that the rolling process introduces into the steel. Both for strips and for simple welded structures it was found that the residual stresses were significantly reduced after vibration treatment; importantly, this reduction seems to be permanent.

Vibrations and resonance in machinery

Rotating machinery, particularly if it is slightly out of balance, can set up vibrations in neighbouring components or parts of itself. The driving frequency in some of these examples may be the frequency of rotation of the engine, or possibly of the wheels. It is clearly important to know what rates of rotation are involved if resonance is to be avoided.

Vibration effects on people

In order to protect passengers from excessive noise and vibration, it is important that large out-of-balance forces in a car engine are not transmitted to the vehicle body. Scientists study the effects of vibration on the human body and can then predict levels of passenger comfort or the degree of interference with activities or levels of concentration. They can even predict the likelihood of motion sickness.

Vibration and vomiting

The motion-sickness dose value (MSDV) predicts the probability of vomiting based on the frequency and magnitude of vertical vibrations and the time of exposure:

$$\mathrm{MSDV} = a_{\mathrm{rms}}\sqrt{t}$$

Here a_{rms} is the frequency-weighted acceleration reflecting the greatest sensitivity to vibrations in the range of 0.125–0.25 Hz and t is the time of exposure. The percentage of unadapted adults expected to vomit is $\frac{1}{3}$MSDV. These relationships have been derived from experiments in which up to 70% of subjects vomited during exposures lasting between 20 minutes and 6 hours.

Vibration-induced white finger

A well known occupational hazard is vibration-induced white finger (VWF) in which the sufferer experiences intermittent whitening of the ends of the fingers due to reduced blood flow as a result of the vibrations experienced in the use of machinery. In severe cases amputation is the only treatment.

Activity 17 Further examples of resonance

Discuss, or make notes on, the following questions.

- Machinery is normally operated at speeds well above any possible resonances. What problems might be presented, though, when the machinery is started up or shut down? (This can be a particular problem for ships with a very long transmission shaft from engine to propeller.) What instructions might you give to an operator?
- What does a garage always do – and charge for – when a tyre is changed and why?

2.2 A useful type of oscillation

To ensure that the effects of resonance are controlled, engineers must understand the fundamentals of vibration. Vibrations, and any kind of motion which repeats itself, such as a simple pendulum or a mass bouncing on a spring, are described as a **harmonic motion**. There is a particular form of harmonic motion, called **simple harmonic motion** (SHM) which has the following very important properties:

- the period (or frequency) is independent of the **amplitude** (maximum displacement) of the motion
- the force is always directed towards the central point of the oscillation: if displacement is positive the force is negative, and vice versa. Mathematically this can be expressed as:

  $$F = -kx \qquad (4)$$

  where F and x are the magnitudes of the force and displacement, respectively, and k is a constant.

The second property means that the force – and hence the acceleration – must be zero as the system passes through its central position. This is therefore the **equilibrium position**; the one to which the system gradually returns as the oscillations die away. Because of this change in direction of force either side of equilibrium, it is usually called the **restoring force** and is provided by the '**stiffness**', k, of the system. It is the **inertia** of the system which makes it continue on past its equilibrium position

Activity 18 Oscillation circus

Observe a variety of oscillating systems and decide whether each exhibits simple harmonic motion. Try to identify what provides the inertia and what provides the stiffness of the system. Some systems require detailed measurements, while others simply require observation.

As Activity 19 shows, there is a connection between simple harmonic motion and motion in a circle. This provides us with a useful way to analyse SHM mathematically.

Activity 19 Circular motion and SHM

Observe the shadow of a mass undergoing SHM on the end of a spring alongside the shadow of a ball rotating in a vertical circle, as in Figure 4.27.

Hold your arms out in front of you with index fingers pointing forwards and level with each other. Set your left hand moving anticlockwise at a steady speed and move your right hand vertically so that your index fingers stay at the same horizontal level. While your left hand executes a circle the right is executing SHM.

You have just demonstrated that SHM and circular motion are closely related. Make sure you are familiar with the terms angular velocity, period, centripetal force and acceleration and the relationships between them. Also make sure you are familiar with angular displacements expressed in radians.

A quick internet search reveals many websites that go over this in more detail and provide animations to back up your work. For example, see the University of New South Wales website (available via **www.shaplinks.co.uk**).

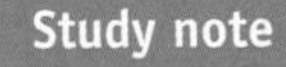

Study note

Look back at your work on circular motion in *Probing the Heart of Matter*

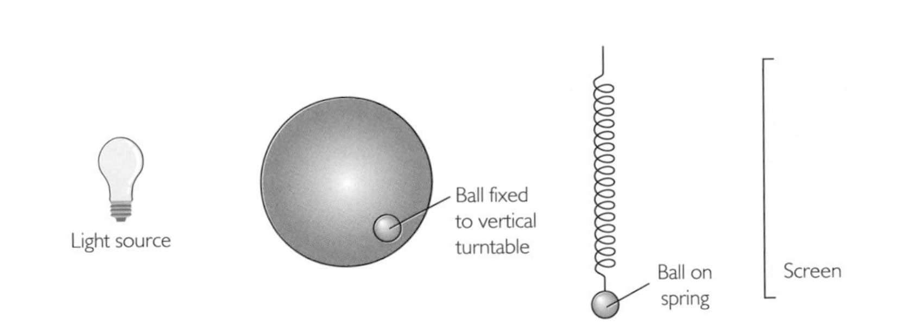

Figure 4.27 Shadows of SHM and circular motion

Maths reference

Angular measurements
See Maths notes 6.1

Analysing SHM

Figure 4.28 shows an object, mass m, moving with angular velocity ω (measured in radians per second) in a circle of radius A. As was demonstrated in Activity 19, its 'shadow' cast on the x-axis performs simple harmonic motion; to describe its motion we need expressions for the x component of displacement, velocity and so on.

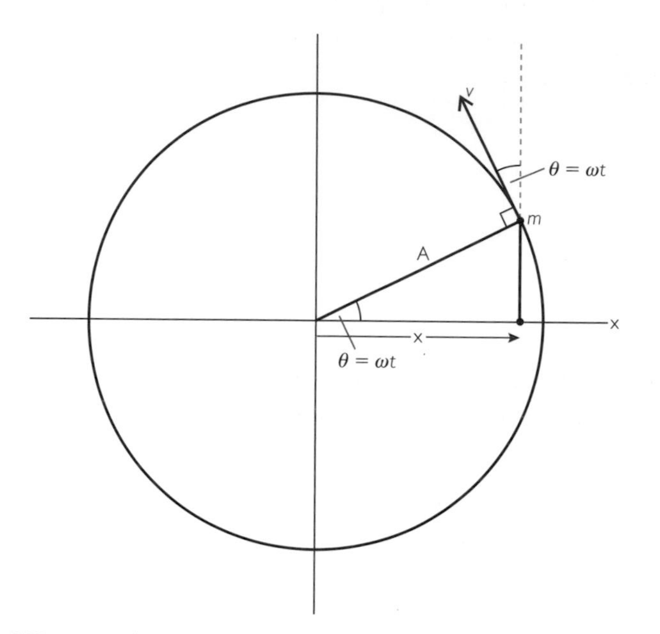

Figure 4.28 SHM as a projection of circular motion

Maths reference

Sin, cosine and tangent of an angle

See Maths note 6.2

The amplitude of the oscillating motion is A. For an angular displacement of ωt the horizontal displacement is:

$$x = A\cos\omega t \qquad (5)$$

The maximum displacement is when $\cos\omega t = \pm 1$; ie $x_{max} = \pm A$.

In circular motion velocity is given by:

$$v = A\omega \qquad (6)$$

at an angle ωt to the vertical. Here we need the horizontal component:

$$v_x = -A\omega \sin\omega t \qquad (7)$$

The negative sign shows that the mass is moving in the negative direction; ie back towards its equilibrium position. The speed (magnitude of velocity) is greatest when $\sin\omega t = \pm 1$; ie:

$$v_{max} = A\omega \qquad (8)$$

For the circular motion, the force acting on m always has magnitude:

$$F = \frac{mv^2}{A} = m\omega^2 A \qquad (9)$$

and is directed towards the centre (it is a *centripetal* force). The x component is thus:

$$F_x = -m\omega^2 A\cos\omega t \qquad (10)$$

Comparing Equations 5 and 10 we have

$$F_x = -m\omega^2 x \qquad (11)$$

Since m and ω are both constants, Equation 11 is equivalent to Equation 4 and describes a key property of SHM – the restoring force is proportional to displacement. Equation 11 is often expressed in terms of the acceleration:

$$a_x = \frac{F_x}{m}$$

and so:

$$a_x = -\omega^2 A\cos\omega t \qquad (12)$$

ie:

$$a_x = -\omega^2 x \qquad (13)$$

Figure 4.29 shows graphs of the variation of displacement, velocity and acceleration with time for a simple harmonic oscillator. We have dropped the subscript x because we are now dealing with motion in just one dimension.

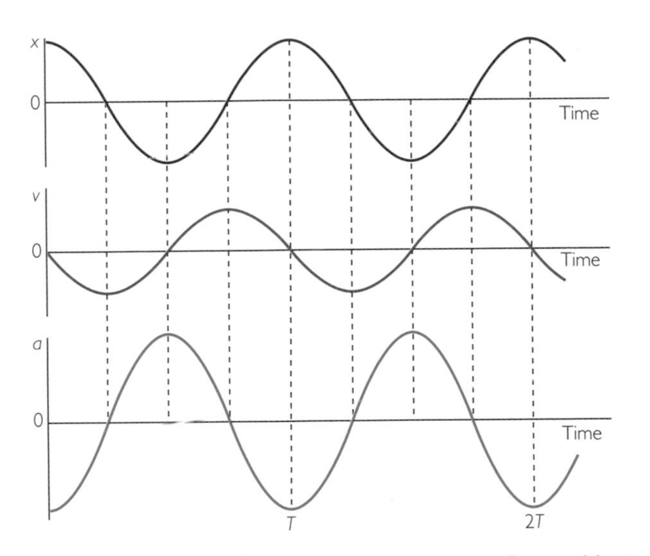

Figure 4.29 Graphs of (a) x versus t (b) v versus t (c) a versus t for an object executing SHM

The circular motion also enables us to deduce the period, T, of the SHM. The time to make one complete oscillation (there and back) is the same as the period of the circular motion; ie:

$$T = \frac{2\pi}{\omega} \qquad (14)$$

Notice that the period, T, is independent of the amplitude, A – the other key property of SHM. We can also write an expression for frequency:

$$f = \frac{1}{T} = \frac{\omega}{2\pi} \qquad (15)$$

Study note

You met Equation 14 in connection with circular motion in *Probing the Heart of Matter*.

Activity 20 Springy oscillations

Use Equations 4–15, and the worked example below, to predict the frequency of a mass oscillating on various combinations of springs (Figure 4.30(a)). Then compare your predictions with experimental measurements of frequency.

Produce a time-trace for the apparatus shown in Figure 4.30(b) (or similar) and compare the results with the theoretical Equations 5, 7 and 14.

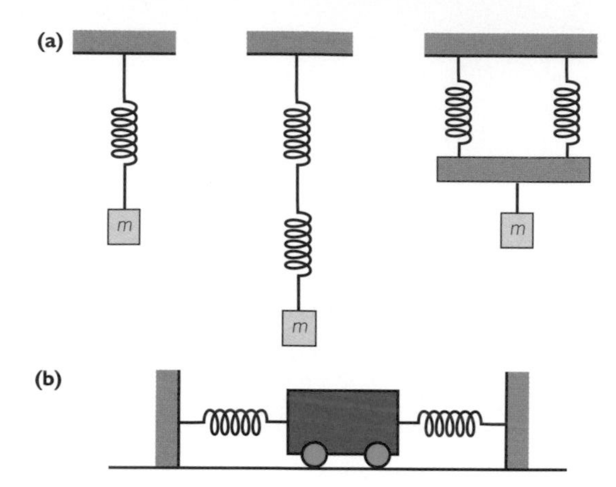

Figure 4.30 Apparatus for Activity 20

A useful recipe

The above theoretical treatment of SHM provides a very useful 'recipe' for analysing the motion of an oscillator. Simply carry out the following steps.

1 Is the restoring force proportional to displacement? If so, then the motion is simple harmonic. (If is it not, the motion is *not* SHM and you cannot proceed with this recipe.)

2 Since restoring force is proportional to displacement, then acceleration must also be proportional to displacement. Find the constant that relates displacement to acceleration. Sometimes it is easy to find this constant directly; sometimes it is easier to find the relationship between displacement and restoring force and then divide by mass.

3 The constant you have found in Step 2 is equal to ω^2 in Equation 13. Find its square root. You now know the value of ω for your oscillator.

4 Using your value of ω, you can now work out the period of the oscillation using Equation 14. If you know the amplitude, you can also find the displacement, velocity and acceleration at any given time using Equations 5, 7 and 13.

Worked example

Q A mass, $m = 0.1$ kg oscillates on the end of a spring that obeys Hooke's law and has a stiffness, $k = 50 \text{ N m}^{-1}$. Is the motion simple harmonic? If so, what is its period?

A Since the spring obeys Hooke's law, $F = -kx$, and so the motion must be SHM:

$$a = \frac{-kx}{m} = -\left(\frac{k}{m}\right)x$$

so:

$$\omega^2 = \frac{k}{m}$$

and:

$$\omega = \sqrt{\left(\frac{k}{m}\right)} = \sqrt{\left(\frac{50 \text{ N m}^{-1}}{0.1 \text{ kg}}\right)} = 22.36 \text{ s}^{-1}$$

$$T = \frac{2\pi}{\omega} = 0.28 \text{ s (to 2 sig. fig.).}$$

Q (Harder!) Suppose the mass in the previous question is displaced by 3.0 cm from its equilibrium position then released. (a) What is its maximum speed? (b) What are its displacement and velocity after 0.05 s?

A We need to use Equations 5 and 7, and so we need to know values of A and ωt. Since 3.0 cm is the maximum displacement, the amplitude, $A = 3.0$ cm.

(a) By inspection of Equation 7, speed is maximum when $\sin\omega t$ is greatest; ie when $\sin\omega t = 1$, hence:

$$v_{max} = A\omega = 3.0 \text{ cm} \times 22.36 \text{ s}^{-1} = 67 \text{ cm s}^{-1} \text{ (2 sig. fig.)}$$

(b) After 0.05 s, $\omega t = 0.05 \text{ s} \times 22.36 \text{ s}^{-1} = 1.118$.

Since ω can be interpreted as an angular velocity in radians per second, ωt can be interpreted as an angle expressed in radians:

$$\cos\omega t = \cos 1.118 = 0.1834$$

From Equation 5:

$$x = A\cos\omega t = 3.0 \text{ cm} \times 0.1834 = 0.55 \text{ cm (to 2 sig. fig.)}$$

$$\sin\omega t = \sin 1.118 = 0.8992$$

From Equation 7:

$$v = -A\omega\sin\omega t = -3.0 \text{ cm} \times 22.36 \text{ s}^{-1} \times 0.8992$$
$$= -60 \text{ cm s}^{-1} \text{ (to 2 sig. fig.)}$$

Study note

Notice the units in the worked example. Since $1 \text{ N} = 1 \text{ kg m s}^{-2}$, the SI units of $\sqrt{\left(\frac{k}{m}\right)}$ reduce to s^{-1}, giving appropriate SI units for T.

Study note

Remember to switch your calculator into 'radian' mode.

Study note

The positive sign tells us that the mass is still on the positive side of the equilibrium position, as is expected after a little less than one-quarter of a complete there-and-back oscillation.

Study note

Notice that v is negative; the mass has not yet reached the equilibrium position, but is travelling towards it. It has almost reached its maximum speed, which it will reach after exactly one-quarter of a cycle as it passes though the equilibrium position.

Questions

9 An object supported by a 'Hooke's law' spring oscillates with SHM and has $A = 0.30$ m and $\omega = 5.0 \text{ s}^{-1}$.

(a) What are (i) its maximum speed and (ii) the frequency of the oscillations?

(b) The object has mass 0.40 kg. What is the stiffness of the spring?

(c) After 0.20 seconds, what are (i) the displacement and (ii) the acceleration?

10 (a) An ultrasonic oscillator has a frequency of 50 kHz and amplitude of 2×10^{-6} m. What are the maximum values of (i) speed and (ii) acceleration?

(b) Expressing the acceleration as a multiple of g (= 9.81 m s^{-2}), do you think the value has any implications for the mechanical strength of the oscillating part?

(c) Gallstones in the body can often be broken up by irradiating them with ultrasonic vibrations. Why is increasing the frequency more effective than increasing the amplitude in this operation?

11 Figure 4.31 shows a simple pendulum of length l displaced through a small angle θ.

(a) Show that, (i) provided θ is small, acceleration is proportional to displacement, with $\omega^2 = \frac{g}{l}$ (ii) and hence show that the pendulum has a period $T = 2\pi\sqrt{\left(\frac{l}{g}\right)}$.

(Hint: use the small angle approximation for angles in radians.)

(b) What is the length of a 1 Hz pendulum? (Use g = 9.81 m s^{-2}.)

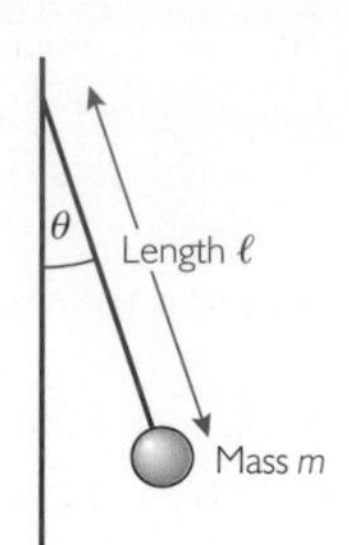

Figure 4.31 Simple pendulum

Maths reference

The small angle approximations
See Maths note 6.6

Activity 21 A less-simple pendulum

Use a computer simulation to explore the factors affecting the period of a simple pendulum. Then apply your understanding as you investigate a less-simple pendulum.

Energy in SHM

The previous analysis can be extended to consider the energy of a simple harmonic oscillator. The general principles can be illustrated by considering a mass oscillating horizontally on the end of a light Hooke's-law spring, resting on a frictionless surface (Figure 4.32) – the only energies that change are the kinetic energy of the mass and the potential energy due to the stretched spring.

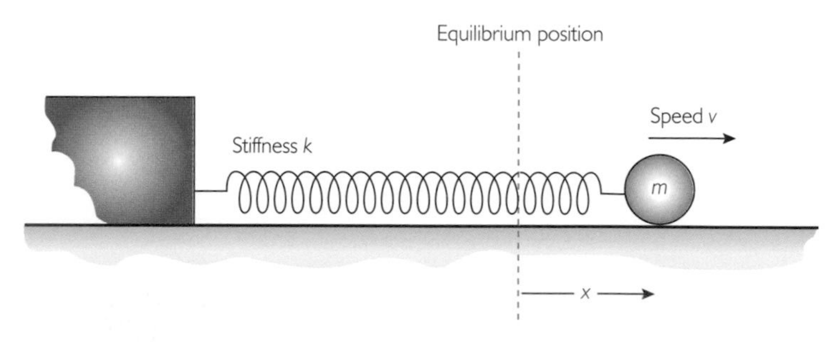

Figure 4.32 'Simple' simple harmonic oscillator

At maximum displacement from equilibrium, the mass is momentarily at rest and the stretched (or compressed) spring has maximum potential energy. As the mass passes through its equilibrium position, it has its greatest speed and hence its greatest kinetic energy, but the potential energy is momentarily zero. At all other positions, the system has both kinetic and potential energy.

The kinetic energy, E_k, of the mass, m, is related to its speed, v, by the familiar equation:

$$E_k = \frac{1}{2}mv^2 \qquad (16)$$

By squaring Equation 7 to obtain an expression for v^2 for the oscillator, we can get an expression showing kinetic energy of the oscillating mass varies with time:

$$E_k = \frac{1}{2}mA^2\omega^2\sin^2\omega t \qquad (17)$$

The (elastic) potential energy, E_p, stored in the stretched spring with stiffness k is given by:

$$E_p = \frac{1}{2}kx^2 \qquad (18)$$

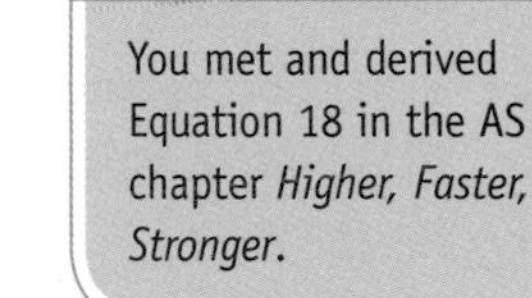

Study note

You met and derived Equation 18 in the AS chapter *Higher, Faster, Stronger*.

By squaring Equation 5 to get an expression for x^2, we can see how the potential energy of the spring changes with time:

$$E_p = \frac{1}{2}kA^2\cos^2\omega t \qquad (19)$$

Figure 4.33 shows graphs of kinetic and potential energy plotted against time.

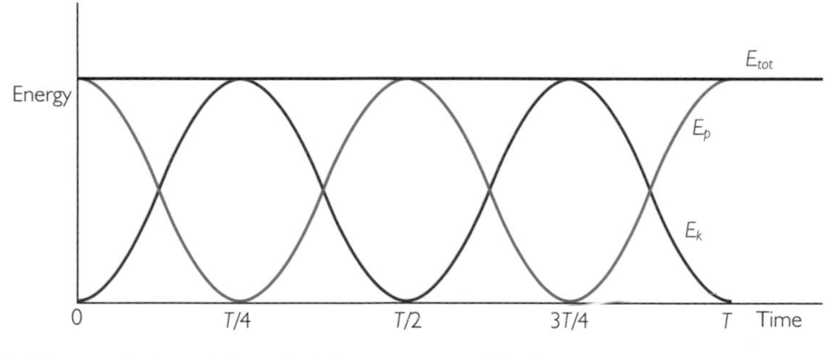

Figure 4.33 Time variation of E_k and E_p for a mass oscillating on a spring

The relationship between Equations 17 and 19 becomes more apparent if we note that $\omega^2 = \frac{k}{m}$, and so Equation 17 becomes:

$$E_k = \frac{1}{2}kA^2\sin^2\omega t \qquad (20)$$

The total energy, E_{tot}, of the system is found by adding Equations 19 and 20:

$$E_{tot} = E_p + E_k$$

$$E_{tot} = \frac{1}{2}k\,A^2(\cos^2\omega t + \sin^2\omega t)$$

$$E_{tot} = \frac{1}{2}kA^2 \qquad (21)$$

because $\cos^2\omega t + \sin^2\omega t = 1$, irrespective of the value of ωt. In other words, the **total energy of a simple harmonic oscillator** is constant. This is true for *all* SHM systems, not just for the simple one we have chosen to study.

Maths reference

Sine, cosine and tangent of an angle
See Maths note 6.2

The energies can also be expressed in terms of m and ω rather than k and A, in which case Equation 21 becomes:

$$E_{tot} = \frac{1}{2}m\omega^2A^2 \qquad (22)$$

Questions

12 Calculate the total energies of (a) the oscillator considered in the previous worked example and (b) the oscillator in Question 9. In each case, show that Equations 21 and 22 give the same answer.

13 (a) A spring of stiffness 1.0×10^2 N m^{-1} is stretched by 0.15 m. What is its elastic potential energy?

(b) It is used (at this stretch) to catapult a small missile of mass 2.0 g. At what speed does the missile leave the catapult? What assumption did you have to make?

(c) Using appropriate symbols k, x and m (mass of missile), turn the reasoning of (b) into a general theoretical argument to produce an algebraic expression for the launch velocity v. Show that the launch velocity is proportional to the initial stretch.

(d) Without completely re-working the problem, find the launch velocity of (i) the same missile for an initial extension of 0.05 m and (ii) a missile twice as massive launched using the original extension of 0.15 m.

2.3 Earthquake waves revisited

Knowing something about SHM, we can now return to the question of waves in a solid material. As you saw in Activity 4, a longitudinal wave in a solid can be modelled with a row of trolleys as in Figure 4.34. Each trolley (or atom in a solid) is held to its neighbours by springy connections, so you can consider each one as an oscillator like that in Activity 20. We can use the equations of SHM to deduce the speed of a compression wave along the row of oscillators.

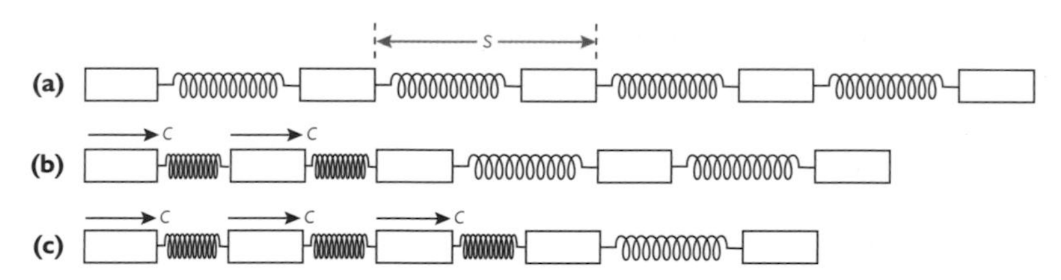

Figure 4.34 Compression pulse travelling along a row of oscillators

Imagine pushing the oscillator at the end of the row. The spring joining it to its neighbour becomes compressed, exerting a force on the next oscillator and setting it in motion so that it passes on the force to its neighbour – and so a compression pulse travels along the row with a time lag between one oscillator and the next. If s is the distance from one oscillator to the next and τ the time interval, then the wave travels at speed v where:

$$v = \frac{s}{\tau} \tag{23}$$

If you assume that each successive oscillator is only affected once its neighbour has reached its maximum displacement, then the time taken for the pulse to be 'handed on' will be one quarter of a cycle. In fact, the time interval is rather shorter – only $\frac{1}{2\pi}$ of a cycle:

$$\tau = \frac{T}{2\pi} \tag{24}$$

Study note

See Activity 20 and the worked example in Section 2.2.

Since we know that, for an oscillator mass m held by springs of stiffness k:

$$T = 2\pi\sqrt{\left(\frac{m}{k}\right)} \tag{25}$$

we now have an expression for the speed at which the pulse travels along the row:

$$\tau = \sqrt{\left(\frac{m}{k}\right)}$$

so:

$$v = s\sqrt{\left(\frac{k}{m}\right)} \tag{26}$$

Activity 22 The speed of a pulse along a row of trolleys

Set up a row of trolleys like that in Figure 4.34. By moving the whole row (without compressing the springs) arrange for the end trolley to crash into a solid obstacle (eg a wall). Observe that, when the end trolley hits the wall, a compression pulse travels along to the free end, where it is reflected as a rarefaction pulse that travels back towards the wall. When this reflected pulse reaches the wall, the trolleys bounce back, losing contact with the wall. So the end trolley remains in contact with the wall for the time that it takes for a pulse to travel along the row and back again.

Use this observation to measure the there-and-back travel time and hence to deduce the speed of the pulse. Compare your result with the speed calculated using Equation 26.

Waves in a solid

The arrangement in Figure 4.34 can be thought of as modelling a row of atoms, and we can use Equation 26 to deduce the speed of longitudinal waves in a solid. Figure 4.35 shows a three-dimensional model of a solid in which atoms are held by spring bonds that, for small extensions and compressions, obey Hooke's law.

For each bond we can write:

$$\Delta f = k\Delta x \qquad (27)$$

where Δf is the force and Δx is the resultant extension. k is the stiffness of each bond. We can relate this interatomic k to the Young modulus of the metal.

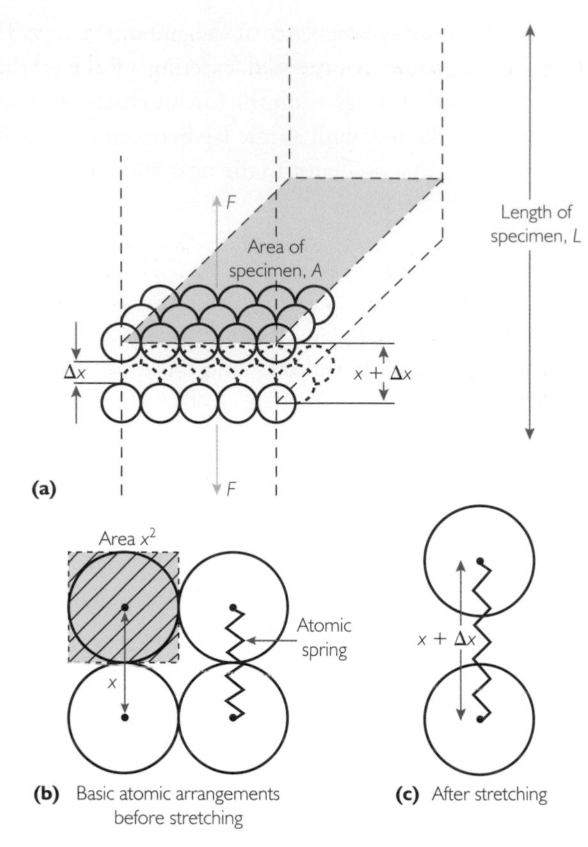

Figure 4.35 Three-dimensional model of a solid

Each atom in Figure 4.35 occupies an area x^2. If the cross-sectional area of the sample is A then the number of atoms in the layer is $\frac{A}{x^2}$ – and this is the number of springs being stretched. Hence the total force F is given by:

$$F = \frac{Ak\Delta x}{x^2} \qquad (28)$$

which leads to an expression for stress:

$$\sigma = \frac{k\Delta x}{x^2} \qquad (29)$$

We can also write down an expression for strain:

$$\varepsilon = \frac{\Delta x}{x} \qquad (30)$$

Combining Equations 29 and 30 leads to an expression for the Young modulus:

$$E = \frac{k}{x} \qquad (31)$$

Putting $s = x$ and $k = Ex$ in Equation 26 we get a new expression for the wave speed:

$$v = x\sqrt{\left(\frac{Ex}{m}\right)} = \sqrt{\left(\frac{Ex^3}{m}\right)} \qquad (32)$$

But as x^3 is the volume occupied by each atom, then the density ρ is:

$$\rho = \frac{m}{x^3}$$

hence we can write the equation that you met in Part 1 of this chapter:

$$v = \sqrt{\left(\frac{E}{\rho}\right)} \qquad \text{(Equation 1)}$$

Study note

Look back at your work on the AS chapter *Spare Part Surgery* for a reminder about stress and strain.

Although we assumed a square arrangement of atoms, Equation 1 applies to longitudinal waves in any solid, regardless of the way its atoms are arranged.

Activity 23 Wave speed in a solid

Measure the speed of a longitudinal wave in a steel or aluminium rod. The principle is the same as that used in Activity 22 – the rod behaves like the row of trolleys and so the contact time between rod and hammer is equal to the there-and-back travel time for a longitudinal pulse.

Compare your result with the speed calculated using Equation 1 and the data given below.

Aluminium: $E = 7.0 \times 10^{10}$ N m^{-2}, $\rho = 2.7 \times 10^3$ kg m^{-3}

Steel: $E = 2.0 \times 10^{11}$ N m^{-2}, $\rho = 7.8 \times 10^3$ kg m^{-3}

Table 4.1 (in Part 1 of this chapter) lists data for density and the speed of P waves in various materials. Seismologists use such data, derived from calculations and laboratory measurements, to help them to interpret observations of seismic waves and hence to make deductions about the internal structure of the Earth. As you will see in Part 3 of this chapter, a knowledge of wave speeds in materials also helps engineers to design buildings that are safe and comfortable for the occupants, and machines that are safe to operate.

Listen to the trees

A research collaboration of Forest Research (part of the Forestry Commission), Fibre-Gen (a New Zealand company), Napier University and the University of Glasgow is striving to improve sustainability and increase efficiency in the UK forest industry.

A large part of the forestry industry is dedicated to the production of wood for the construction of new homes. However, some of the trees cut down are processed before it is realised they are not strong enough for use in construction.

The research team is testing technologies that allow wood stiffness to be determined by measuring the speed at which sound travels through logs and trees by a very similar technique to that seen in Activity 23. This process allows timber quality to be determined before harvesting, processing and transport. You can see a simulation of their work via **www.shaplinks.co.uk**.

Questions

14 (a) Figure 4.36 overleaf shows some results obtained in Activity 23. On the screen trace shown in Figure 4.36 the contact time between a metal bar and the hammer that hit it was estimated at 512 μs. The bar was 0.70 m long. What was the speed of longitudinal waves in the bar?

(b) The trace after 512 μs appears to be exponential. What does that suggest about the circuitry within the computer interface?

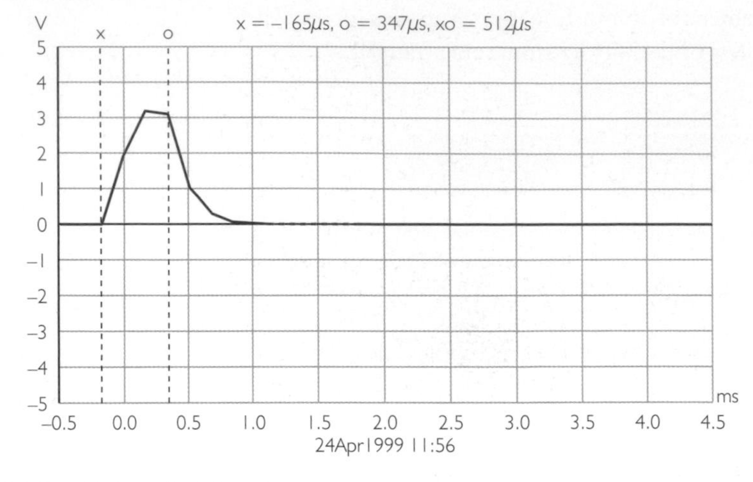

Figure 4.36 Results from Activity 23

15 Table 4.3 lists the speed of longitudinal waves in various common materials along with their densities.

(a) Without doing any calculations, what can you say about the Young modulus of oak compared with that of pine, and the Young moduli of metals compared with those of wood?

(b) Calculate the Young moduli of (i) pine and (ii) copper.

Material	Density, ρ/kg m^{-3}	Wave speed, v/m s^{-1}
pine	500	3313
oak	700	3837
aluminium	2698	5100
steel	7800	5060
copper	8933	3650
lead	11343	1230

Table 4.3 Densities and longitudinal wave speeds in various materials

2.4 Build it better – Summing up Part 2

The physics that you have studied in this part of the chapter is the physics that enables an engineer to design a building for an earthquake zone. You have seen how an oscillator can be driven to oscillate by an external force, and that if the frequency of the driving force matches the oscillator's natural frequency, then resonance occurs and oscillations build up to large amplitude. You then went on to study and analyse simple harmonic motion – a type of motion that occurs in many natural and man-made systems, and which underlies many more complex oscillatory motions.

Figure 4.37 shows a model structure being tested on an earthquake table in an engineering laboratory. As the 'Earth' shakes, the building vibrates. It is the engineer's task to ensure that the amplitude of the vibrations remains fairly small, and also to choose materials that will withstand deformation.
In Activity 24, you will apply your knowledge of resonance and simple harmonic motion to the matter of limiting the vibrations. (In Part 3 you will return to the question of deformation.)

Figure 4.37 Testing a model structure on an earthquake table

Activity 24 Earthquakes in the laboratory

Make some simple model building structures (Figure 4.38). By assuming that each model vibrates like a mass on a spring, calculate its natural frequency of vibration. Test each model by vibrating it on an earthquake table.

The behaviour of buildings in earthquakes was explored initially in Activity 14. The models used in this activity will exhibit more complex modes of vibration.

As you work through this activity, take the opportunity to look back through Part 2 of the chapter and make sure you understand the meaning of the key terms printed in bold.

Figure 4.38 Typical models for use in Activity 24

Activity 25 Resonance of a vibrating blade

Explore the resonance of a thin metal blade vibrated by a vibration generator (Figure 4.39). As is often the case with vibrating machinery, you can only observe the blur of the oscillation. Then construct and test a vibrating blade rotation meter (tachometer).

Safety note

When dealing with an off-balance motor ensure that all parts are firmly fixed together, and that you are not making observations in the plane of rotation.

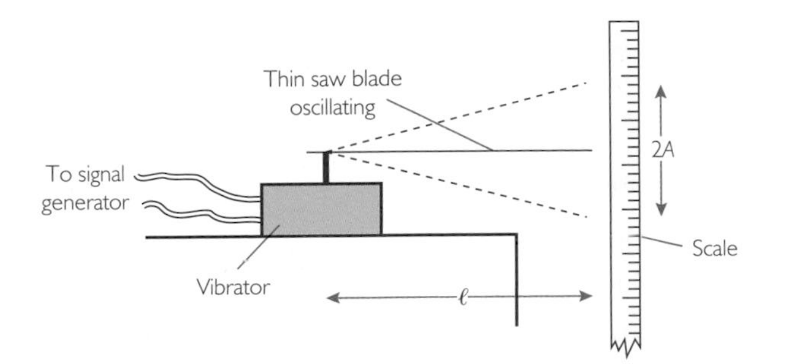

Figure 4.39 Vibrating a metal blade

Further investigations

If violently shaken, a jelly can be made to split apart. Use an earthquake table to explore the response of jellies to various vibrations.

Questions

16 The natural frequency of a building can be dramatically reduced (to remove it from typical earthquake frequencies) by mounting the whole building on rubber bearings. The building then really does oscillate as a single mass without bending.

(a) A 40 tonne model had a natural frequency of horizontal oscillations of 0.60 Hz when mounted on rubber. What is the horizontal stiffness of the bearings? (1 tonne = 1000 kg)

(b) The same model has a vertical natural frequency of 10 Hz.

(i) Using the fact that frequency squared is proportional to stiffness, use your answer to (a) to find the vertical stiffness of the bearings.

(ii) Through what distance will the weight of the model compress the bearings? (Use $g = 9.81 \text{ N kg}^{-1}$.)

(c) On the model *without* rubber bearings the maximum acceleration at third floor level was measured to be 12 m s^{-2} at a frequency of 3.0 Hz.

(i) What is the displacement amplitude?

(ii) The building can be thought of as bending on a fixed base (like your model in Activity 24). If the model is one-third linear scale, what would be the amplitude of oscillation of the real building? Would you find this alarming?

17 The model in Question 16 is built to a one-third linear scale, using the materials that would be used in the actual building.

(a) What would be the mass of the actual building?

(b) What would have to happen to the stiffness of the real building rubber bearings compared to that of the model ones if the natural frequency of horizontal oscillation were to stay the same?

(c) Why are vertical earthquake oscillations likely to cause less damage than horizontal ones?

18 Figure 4.40 shows a pendulum made up of a spring of natural length $l = 0.40$ m attached to a string (which doesn't significantly stretch) of length $L = 0.20$ m. A mass is hung on the spring, which stretches it by 0.20 m. (In your answer use $g = 9.81 \text{ m s}^{-2}$.)

(a) What is the frequency of the 'springy' oscillation?

(b) What is the frequency of the pendulum oscillations – the sideways swinging of the whole arrangement? (Hint: refer to Question 11.)

(c) What is the ratio of these two frequencies? How could you have written down the ratio straight away?

(d) When this arrangement is set oscillating as a pendulum it starts to 'go wild'. Suggest how your answer to (c) might help to explain this. If you have a moment it is worth setting this up and playing with it – it has some interesting properties. The actual values are not important - just adjust L for any mass and spring until you get the ratio of lengths you need: you will know when you're there.

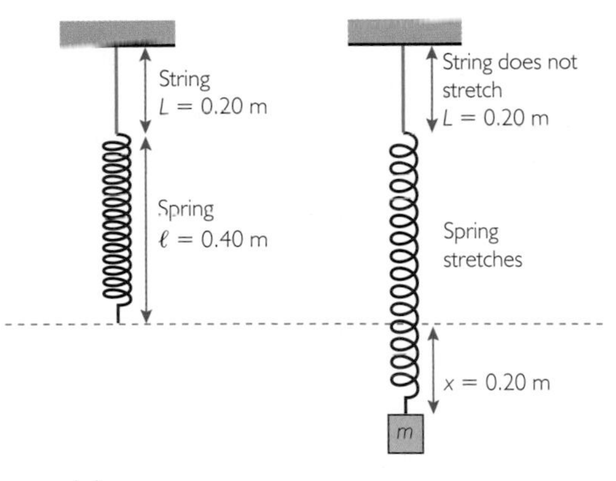

Figure 4.40 Complex pendulum

3 Design for living

In this part of the chapter, you will use and extend what you have learned about vibrations and resonance, and see how buildings (and machines) can be designed to withstand vibrations. Whether the vibrations are the results of earthquakes or something else, the principles are the same.

3.1 Damping the motion

'The wheels on the bus go round and round…' but they also go up and down every time they hit a bump in the road. Passengers expect a smooth ride so shock absorbers perform an essential task – they absorb the energy of the vibration and stop the car body from bouncing up and down long after the bump has been passed.

This is an example of **damping**. The energy of the oscillation is dissipated as heat in oil as pistons are made to move up and down in it. Some modern cars now include 'active suspension systems' whereby the 'bump' is sensed and a restoring force actively applied to stop the wheel pushing the car body up. Whatever the system, all oscillations are damped – a pendulum swinging in the air stops after a while because its energy is dissipated into the air (warming it up slightly). Make the same pendulum swing in syrup and … that's damping to an extreme. There are many cases where damping is the best way of reducing the effects of oscillations – for example the 13th harmonic has to be damped out of notes on a piano otherwise it sounds awful.

Activity 26 Damped oscillations

Use a motion sensor to produce a displacement–time graph of a damped oscillator. Examine the results to see whether the amplitude decays exponentially with time.

Figure 4.41 shows typical results from Activity 26, showing how the amplitude of a damped oscillation dies away exponentially with time.

Study note

It is a common mistake to describe all growths and decays as exponential, but they're not.

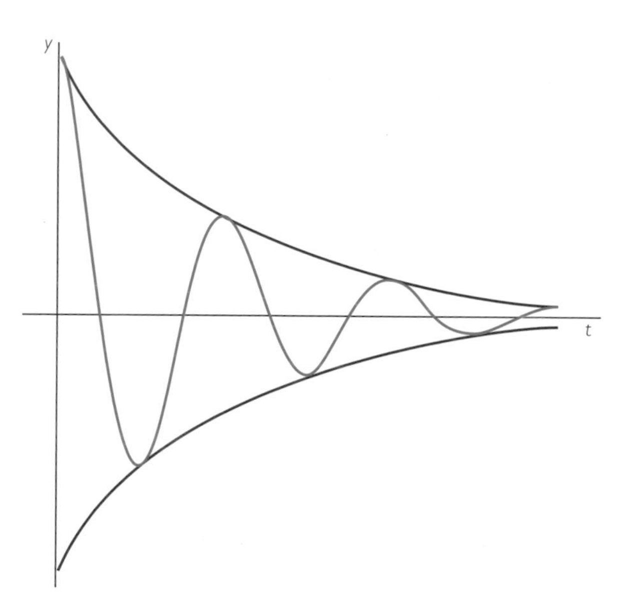

Figure 4.41 Amplitude of damped oscillations decreases exponentially with time

Question

19 A tall tower-like model structure (such as a radio mast) of mass 5.5×10^3 kg is being tested for its vibration characteristics to see if there is any risk of the actual tower being affected by earthquakes. It is initially given a static test by being pulled sideways by a steady force applied near the top. It is found that for the very small displacement possible, the force obeys Hooke's law, a force of 1.0×10^5 N producing a deflection of 50 mm.

(a) At this deflection how much elastic potential energy is stored in the structure?

(b) The load is suddenly removed and the structure 'twangs' like a gigantic tuning fork. As an engineer involved in the test, you decide to make two drastic assumptions:

- damping may be ignored
- the kinetic energy of the tower as it passes through the central position can be represented by the motion of a point mass (equal to the tower mass) at the centre of mass of the tower moving with a speed v.

Using these assumptions, find the speed v.

(c) The dynamic test (the oscillation) now continues and the amplitude of the point of attachment is measured over five complete cycles. The results are listed in Table 4.4.

What evidence is there that the damping is exponential; ie the amplitude decays exponentially with the number of cycles? State as many tests as you can for an exponential variation and apply one of these accurately.

Cycle number, n	Amplitude, A/10–3 m
0	50
1	36
2	26
3	19
4	13
5	10

Table 4.4 Data for Question 19

(d) On a copy of Table 4.4, add a third column headed 'stored energy at end of cycle' and fill in the values. The first entry should be your answer to (a).

(e) Compare the fractional decay in energy per cycle with the fractional decay in amplitude per cycle. Why do you think the energy decay rate is considerably greater?

(f) (i) What is the actual loss in energy in joules over the first cycle?

(ii) Assuming (again not quite correctly) that this damping loss takes place uniformly over the cycle, how much will have been lost over the first quarter cycle?

(iii) Use this value to re-calculate an answer to part (b) which takes account of damping.

(g) Assuming that the velocity found in (b) is not too different from the velocity of the point where the displacement was measured, what are:

(i) the average velocity over the first quarter cycle (remember it starts from rest)

(ii) the time taken for this quarter cycle

(iii) the period, T

(iv) the frequency, f? Does the frequency seem a sensible value? Does it relate to anything you know about earthquakes?

Damping and resonance

Not only can damping make an oscillating system come to rest more quickly, it can also prevent it resonating quite so violently when driven at its natural frequency. A damped system will absorb energy and vibrate when driven (forced vibration), but, as you saw in Activities 10 and 12, the resonant vibrations will have a smaller 'spread' over a wider frequency range; see Figure 4.42.

Damping is therefore one way in which a building can be prevented from shaking itself to bits – even if the earthquake frequency happens to match the natural frequency of the building, it will not vibrate with such large amplitudes.

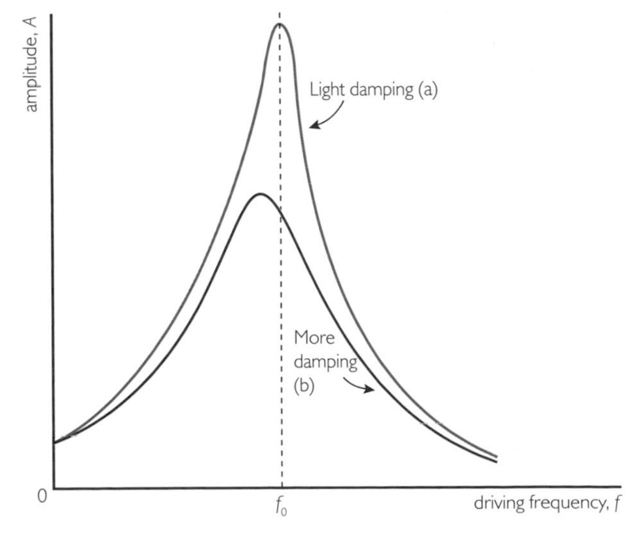

Figure 4.42 Two frequency response curves (a) light damping (b) more damping

Components similar to the shock absorbers in cars, such as an oil-filled piston, can be incorporated into building design. The pistons move with the vibrations of the building, but are slowed by the oil. They absorb the energy of the seismic waves by converting it to heat. They have to be positioned in the building so that they link points that will move relative to each other during an earthquake (Figure 4.43(a)).

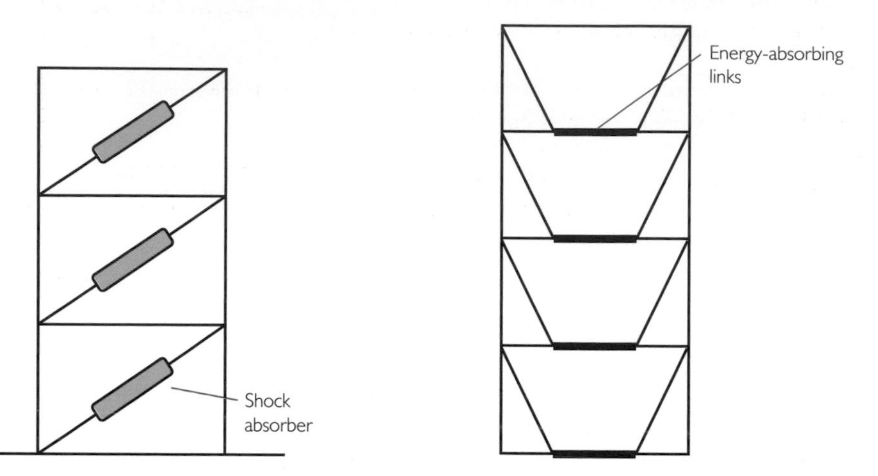

Figure 4.43 (a) Shock absorbers in a building, (b) a knee-bracing system in a building

Other types of energy absorber may deform to absorb the energy, and will be replaced after an earthquake. For example, in the knee-bracing system shown in Figure 4.43(b) there is a large moment set up in links that deform, the energy is absorbed, and since the links are not structural the building is not affected. The links are replaced but the building is otherwise undamaged. Similarly, plastic hinges in the floors of buildings can allow movement without damage.

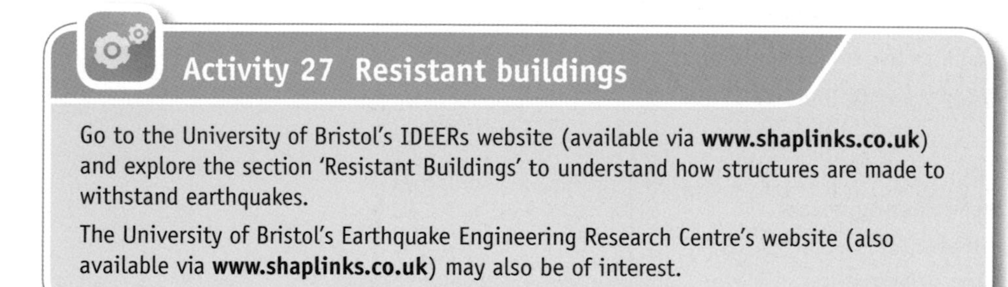

Activity 27 Resistant buildings

Go to the University of Bristol's IDEERs website (available via **www.shaplinks.co.uk**) and explore the section 'Resistant Buildings' to understand how structures are made to withstand earthquakes.

The University of Bristol's Earthquake Engineering Research Centre's website (also available via **www.shaplinks.co.uk**) may also be of interest.

3.2 Other ways to reduce earthquake damage

As well as avoiding resonance and incorporating some damping, there are other methods used by engineers to protect buildings and their occupants from vibrations. While the following information relates mainly to buildings and earthquakes, the same principles are used to prevent damage caused by vibrating machinery in factories and in vehicles.

Structural methods

Isolation

In this method the building is separated from the Earth in some way that prevents the vibrations being transmitted. In other words the structure is 'decoupled' from the ground. A very good example of such a structure is a ship. As water cannot transmit the horizontal components of the earthquake waves the ship will be unaffected.

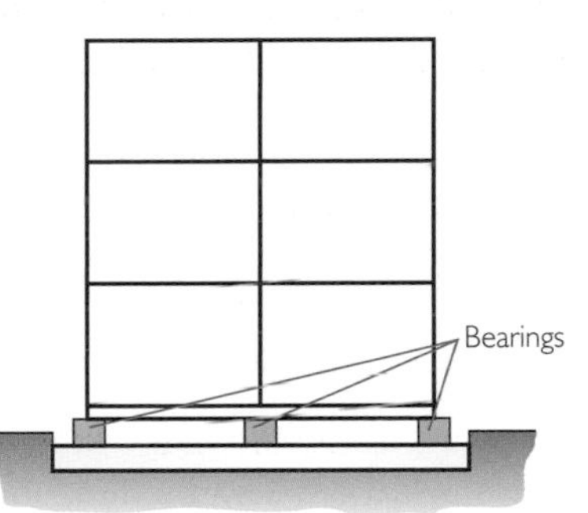

Figure 4.44 Building mounted on bearings

Other possibilities include putting the building on rollers, or using a layer of sand, which would allow the building to slide. One type of construction uses sliders. If the structure can slide on the foundations, because of a low-friction interface, the building will move and so will not suffer a shear force. But there can be problems with high winds causing movement. Also in an earthquake the building may move suddenly so that new vibrations caused by the sudden movement are set up in the structure. Connection of services such as gas, water and electricity would also be a problem. The movement could be in any direction, so the foundations on which the building moves would have to be large and extend in all directions. This could be reduced if the building were anchored with a weak link that would break when the force was large enough that the building needed to move or suffer damage.

Another type of isolation involves using bearings made mainly of rubber placed between the foundations and the building (Figure 4.44). For example, the Museum of New Zealand (Figure 4.45) is protected from large earthquakes by rubber bearings.

Figure 4.45 Museum of New Zealand

Modern designs use steel and rubber laminated bearings that are strong enough to take the weight of the building (rubber alone would bulge outwards). The stiffness of the bearings is less than that of the ground and the building so the fundamental frequency of the whole structure is lower than that of the building on the ground, ensuring that an earthquake cannot set up vibrations at a resonance frequency.

Active control

In this method, the idea is to have sensors measuring the movement and then using a feedback mechanism to move the building back to its equilibrium position. The principle is good, but it is very difficult to realise in practice. The only buildings that have been tested are ones where the forces are due to the wind rather than to ground movement. Problems include what happens if the sensors fail? How do you ensure that they are reliable when needed? An earthquake may not happen for years after the building is completed.

Choice of materials

Every precaution must be taken to protect people from building collapse and the development of better building materials is one more weapon in the armoury.

Reinforced concrete design

Cracking is the main cause of failure. If important structural elements crack, the building may collapse. Reinforced concrete is concrete (which is strong in compression) containing steel bars of rods (which are strong in tension). If a concrete column has bars as shown in Figure 4.46 it will be strong if it is not moved, but in an earthquake the concrete will crack outside the steel links because there is nothing to hold it together. As it falls away the steel reinforcement is no longer held in position, so it can move and the column has lost its strength – often it will collapse. The links must be anchored to the centre of the column by concrete which is held in place by the links themselves (see Figure 4.47).

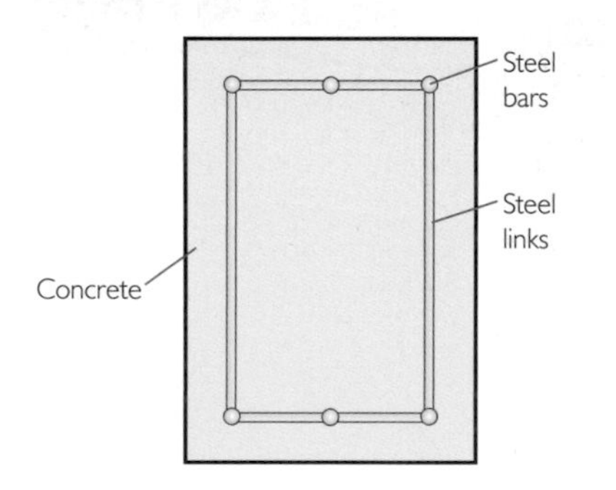

Figure 4.46 Reinforced concrete for normal use

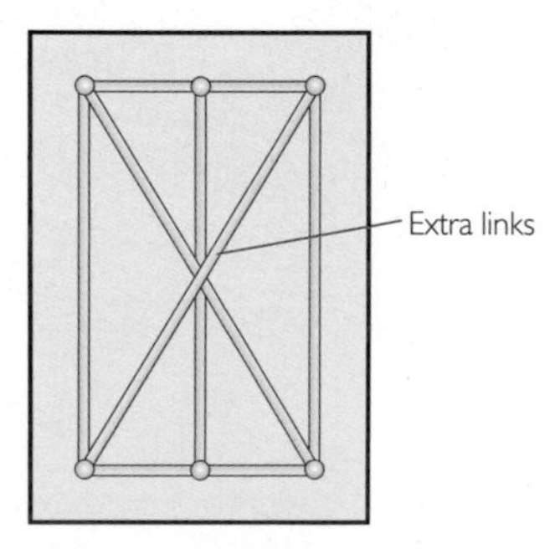

Figure 4.47 Reinforced concrete for use in earthquake zones

Steel frames

Steel is a ductile material and will absorb energy while it is being deformed. In order to produce an earthquake-resistant building, the design must use this property. The design must ensure that there are no sections that are not ductile and that even in a major earthquake a brittle failure will not occur. Floors must always fail before columns so that a whole building will not collapse. There should be extra beams included so that the building will stand if some fail. Figure 4.48 shows a braced steel frame design. There are some redundant parts of the structure. Buckling should occur first in the beams, and not in the diagonals or columns. This design includes energy-absorbing dampers. The Transamerica building in San Francisco (Figure 4.49) is protected by cross bracing at its base.

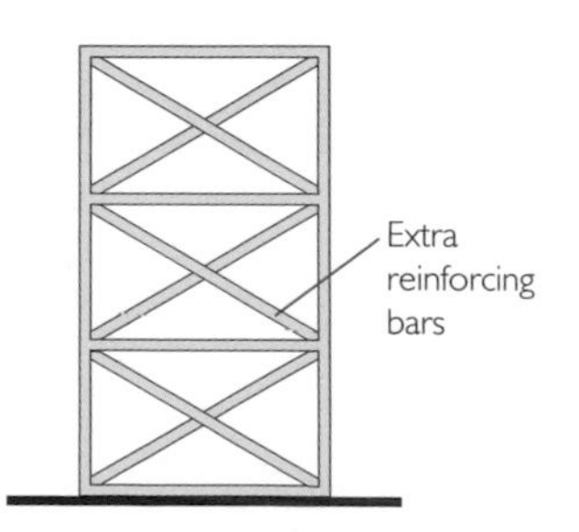

Figure 4.48 Braced steel frame

Polymers protecting people: a story of big vibrations

It's not only in earthquake zones that people suffer problems from the transmission of vibrations. Ever since the 1850s it has been recognised that the special properties of natural rubber enabled it to reduce the transmission of vibration – and one construction that is prone to vibrations is the bridge. Vibration problems occur both because of expansion and contraction of the bridge with changing temperature and because of passing road or rail traffic. It is not surprising, therefore, that as early as 1889 rubber was used to protect the railway viaduct in Melbourne, Australia, with a pad of rubber placed above each plinth to support the bridge deck and absorb impact and noise. In 1957 the Pelham Bridge in Lincoln became the first bridge in the UK to be built using a new type of rubber bearing. The bearing is made of alternating layers of steel and rubber that are securely bonded by chemical means during vulcanisation. This exhibits a strength and stiffness in compression and a softness in shear that easily absorbs vibration and movement in the bridge structure. Visitors to the parliament building in Wellington, New Zealand, are often shown the rubber bearings on which the building is mounted to protect it from earthquake damage.

Figure 4.49 Transamerica building

Often buildings in inner cities need to be protected from the low-frequency vibration caused by adjoining railways. In 1966 a new block of apartments, Albany Court, was built over St. James Park Underground Station, London, and successfully used rubber bearings (the largest measuring only 60 × 50 × 28 cm) to protect it from the ground-borne vibrations from the underground trains. There are now several hundred buildings using similar technology in London alone.

Activity 28 Materials

In the AS chapters *Good Enough to Eat* and *Spare Part Surgery*, you explored ductile and brittle materials. In a small group, discuss the following questions.

How would a structure behave in an earthquake if it were made entirely from materials that were (i) elastic, (ii) brittle and (iii) ductile?

What property is needed for earthquake-resistant building materials? How do you think reinforced concrete could be made so that it had this property?

Further investigations

Explore the use of vibration isolation within a hi-fi system.

3.3 Summing up Part 3

This short part of the chapter has been mainly revision. You have revisited the notion of damping and seen how damping affects resonance. You have seen various ways in which engineers design buildings to withstand earthquakes, and have related building design to your previous knowledge of materials.

Activity 29 Summing up Part 3

Look back through your work on this part of the chapter and make sure you have noted and understood the key points about damping.

4 Keeping buildings warm or cool

According to the Building Research Establishment (BRE), the most frequent complaint that people make about where they work involves the temperature – either too hot or too cold. To design buildings that are pleasant to live and work in, we must know how thermal energy is transferred, and how much heat is required to change the temperature of something. We will look at each of these in turn.

4.1 Energy and temperature change

When an object is heated, the resulting rise in temperature, $\Delta\theta$, depends on the amount of energy transferred to it, ΔE or ΔQ, its mass, m, and the material(s) from which it is made.

$$\Delta E = mc\Delta\theta \qquad (33)$$

The constant c is called the **specific heat capacity** of the material. With energy change in joules (J), mass in kg and temperature change in degrees Celsius or kelvin, K, the units of c are J kg^{-1} °C^{-1} or J kg^{-1} K^{-1}. For many materials, specific heat capacity varies over different temperature ranges so a mean value has to be used.

Study note

The kelvin and the degree Celsius are the same size, so 1 J kg^{-1}°C^{-1} = 1 J kg^{-1} K^{-1}.

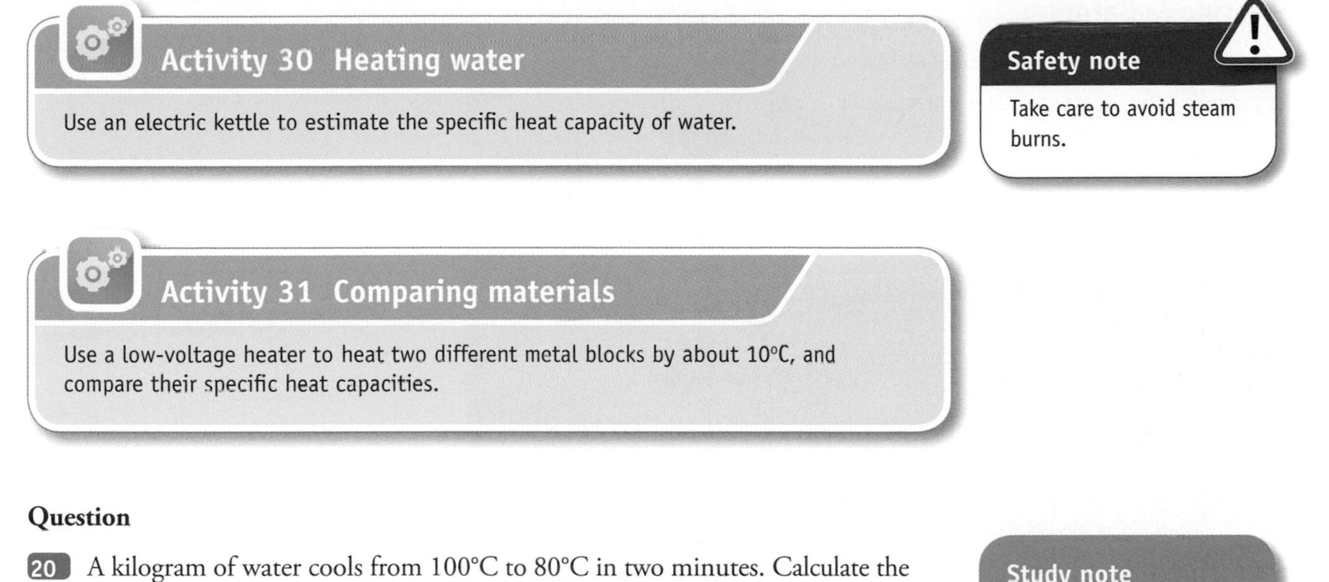

Activity 30 Heating water

Use an electric kettle to estimate the specific heat capacity of water.

Safety note

Take care to avoid steam burns.

Activity 31 Comparing materials

Use a low-voltage heater to heat two different metal blocks by about 10°C, and compare their specific heat capacities.

Question

20 A kilogram of water cools from 100°C to 80°C in two minutes. Calculate the average power transfer from the water. (In this temperature range, water has $c = 4.2 \times 10^3$ J kg^{-1} °C^{-1}.)

The following example based on a domestic 'radiator' shows how to deal with flowing fluids.

Study note

A domestic radiator in fact transfers most energy by convection, so the name is misleading.

Worked example

Q A domestic 'radiator' has water pumped through it at a rate of 6 kg per minute. Before entering, the water temperature is 60°C; it leaves at a temperature of 50°C. Calculate:

(i) the energy transferred each minute by this heater

(ii) its power output.

A Consider a time of 1 minute. In this time, 6 kg of water cools from 60°C to 50°C, so $\Delta\theta = 10$ °C:

$$\Delta Q = 6 \text{ kg} \times 4.2 \times 10^3 \text{ J kg}^{-1} \text{ °C}^{-1} \times 10 \text{°C}$$

$$= 252\,000 \text{ J } (= 2.52 \times 10^5 \text{ J})$$

$$\text{Power transferred} = \frac{\text{energy lost}}{\text{time taken}}$$

$$= \frac{252\,000 \text{ J}}{60 \text{ s}} = 4200 \text{ W}$$

Activity 32 A model radiator

Use a model radiator system (Figure 4.50) to find the energy lost from a hot object. Alternatively, use a continuous flow calorimeter to model the radiator.

Testing radiators

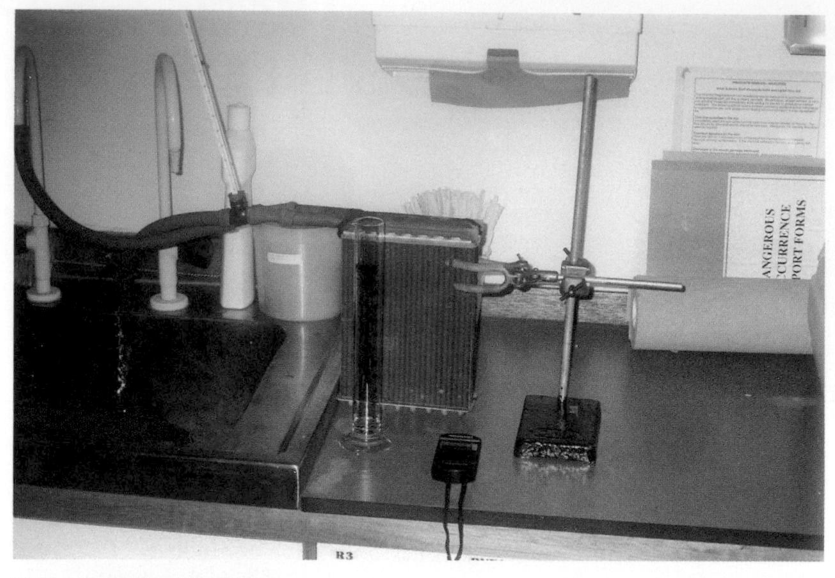

Figure 4.50 Model radiator

The performance of a radiator is related to its temperature, its surface area, the nature of its surface and the temperature of the surroundings. Figure 4.51 shows data obtained by researchers working at ESTEC in the Netherlands (the European Space Research and Technology Centre, run by the European Space Agency – as in Earth-based buildings, temperature control in spacecraft is important). In these tests, the same type of surface was used throughout, and the input power to the radiator was kept constant. The radiator was allowed to reach a 'steady state'; ie a steady temperature at which the power radiated balanced the input power. Figure 4.51 shows how this steady-state temperature depends on the surface area and on the temperature of the surroundings (called the sink temperature). Note that the radiator area axis has a power or logarithmic scale rising in factors of 10.

Maths reference

Using log scales
See Maths note 8.6

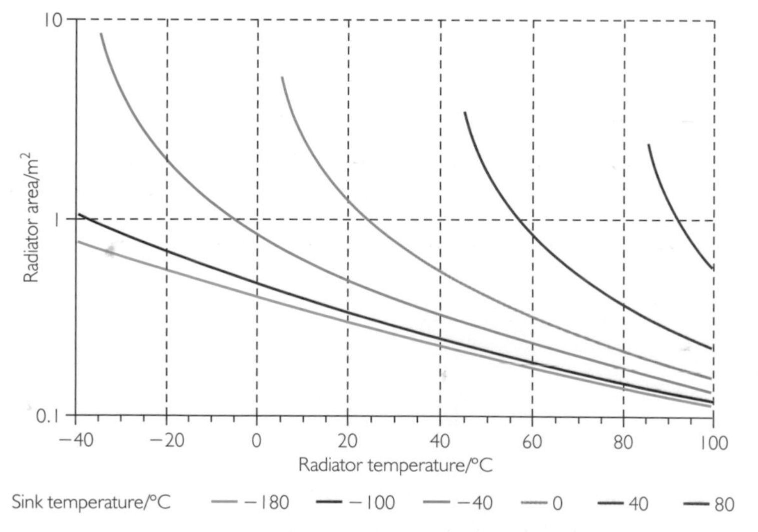

Figure 4.51 Radiator performance data

Questions

21 Figure 4.51 gives some radiator performance data. Use this information to answer the following questions:

(a) Write a short paragraph (about two sentences) saying how the temperature of the radiator is affected by (i) its area and (ii) the temperature of the environment or sink.

(b) If the environment or sink temperature was 0°C, what radiator area would be needed to maintain its temperature at 24°C?

22 In the continuous casting process of steel production, the molten steel has to be cooled before it is straightened, cut and shaped. The steel is poured into a water-cooled mould and further cooling takes place in the curved cooling chamber (Figure 4.52). This question illustrates the large amounts of energy involved in the process.

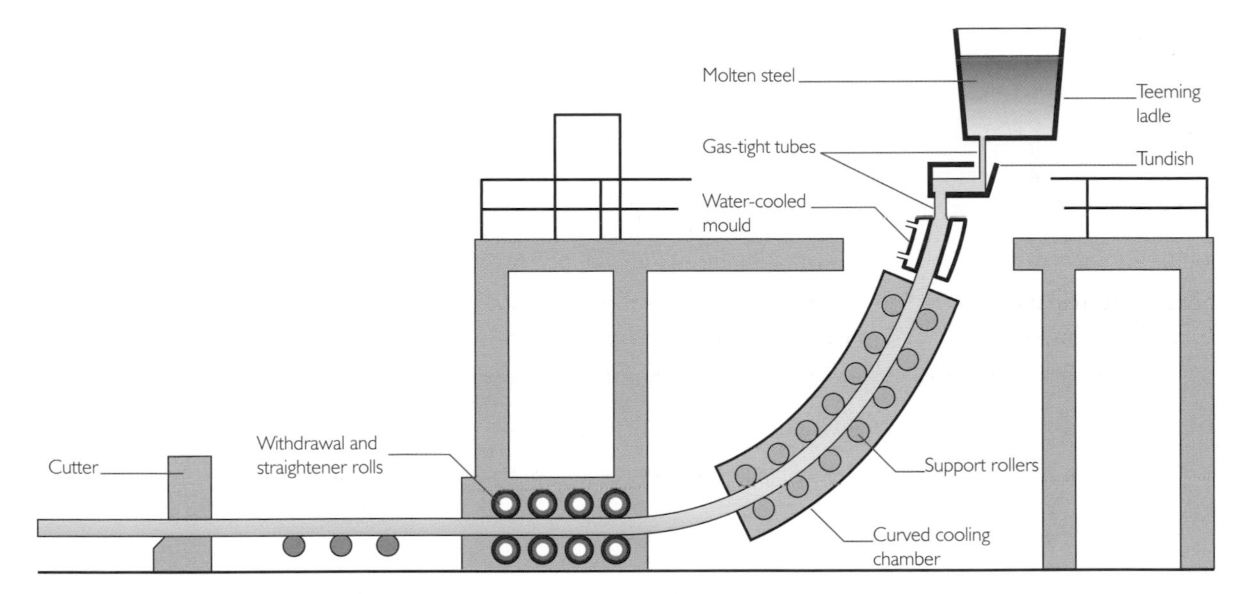

Figure 4.52 Cooling molten steel

Water entering the mould passes through each of two wide and two narrow faces. Flow to each of the wide faces is 2400 kg (2400 litres) per minute and to each of the narrow faces 240 kg per minute. The water enters at a temperature of 40°C and the outlet temperatures are 47°C on each wide face and 48°C on each narrow face. The average specific heat capacity of water over this temperature range is 4179 J kg^{-1} °C^{-1}.

(a) How much energy is transferred to the water in each minute?

(b) What is the total power transfer to the water?

4.2 Thermal energy transfer processes

The thermal transfer of energy, giving rise to heating and cooling, takes place in three basic ways: conduction, convection and radiation. Each of these has its place in a building.

Conduction

Conduction involves energy 'spreading' through a material by virtue of atomic and molecular vibrations, and by the motion of free electrons. Generally, materials that are good electrical conductors are also good thermal conductors as they have a large number of free electrons.

Thermal conduction can be described in a similar way to electrical conduction, but conventionally, rather than using resistance, the thermal **conductance** of building materials is used. The properties of building materials are compared using U-values. The U-value of a building material is defined as the rate of energy transfer through a metre square of material for a 1 K (or 1°C) temperature difference. U-value thus depends on the thickness of a sample as well as on the material from which it is made. U-value has SI units of W m^{-2} K^{-1} or W m^{-2} $°C^{-1}$. If there is a temperature difference $\Delta\theta$ across a sample of area A, then the rate of energy transfer (ie the power), P, is given by:

$$P = UA\Delta\theta \qquad (34)$$

Question

23 For a single-thickness brick wall $U = 2.2$ W m^{-2} $°C^{-1}$. The temperature indoors is 24°C and outdoors is 14°C. What is the rate of thermal energy transfer through such a wall 3 m wide and 2 m high?

You might have studied methods of insulating houses for GCSE. Activity 33 about double glazing will help revise these ideas.

Activity 33 Double glazing

Consider how effective double glazing can be for both heat and sound insulation.

Convection

Convection involves the heating of a fluid (ie a liquid or gas) which expands when heated. Expansion reduces its density below that of the surrounding cooler fluid, and so it experiences an upthrust and rises. A convection current flows as cooler, more dense, fluid sinks to take its place. So-called 'radiators' transfer most energy by convection – they heat the air around them and set up convection currents.

A so-called gravity-fed heating and hot-water system relies on convection to move heated water from a boiler to a water tank and radiators (Figure 4.53). In most buildings however, this movement is assisted by a pump.

Activity 34 Cooling by convection

Use the internet to research how modern building design uses convection in air, and sometimes water, to maintain buildings at a pleasant temperature, while minimising energy input.

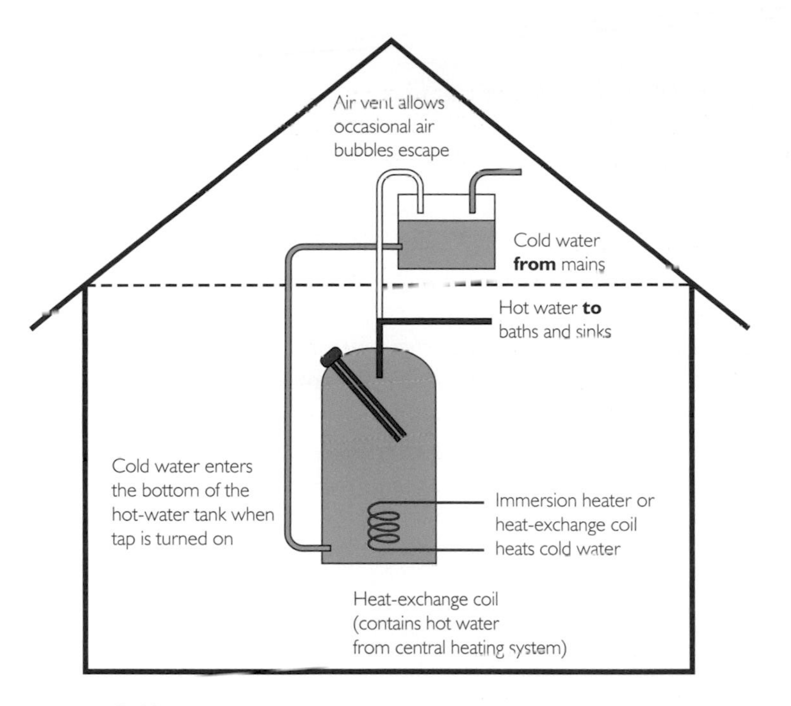

Figure 4.53 Gravity-fed hot-water system

Radiation

All objects, even cool ones, lose energy by emitting electromagnetic **radiation**. The hotter the object, the shorter the wavelength of the radiation it emits, and the greater its radiated power. For objects with temperatures typical of buildings and their surroundings, the radiation is mainly in the infrared part of the electromagnetic spectrum.

Heat losses from buildings can be monitored using infrared-sensitive cameras. Images such as Figure 4.54 show which parts of the building's exterior are warmest and hence indicate where energy is being transferred from the interior (often by conduction and convection).

Buildings can be heated by absorbing solar radiation. Although this 'solar gain' might be welcome in a cold climate, many workplace windows are fitted with infrared reflective glass to prevent overheating on sunny days.

Figure 4.54 Infrared image showing radiation from a building

Study note

The chapter *Reach for the Stars* contains more about electromagnetic radiation.

Activity 35 Radiation

Design and carry out an investigation to compare the effectiveness of different building materials at absorbing and transmitting infrared radiation.

Temperature control

In buildings temperature is often controlled by a combination of *passive* control (buildings cool at night, the Sun warms them during the day), and *active* control (where there is some energy input to cause a required change). Sometimes temperatures are monitored electronically, and controlled by automatic systems, or the system may be manual – like opening the window or closing the blinds in your classroom.

Activity 36 Energy-efficient building design

Use the internet to find out about methods of temperature control in buildings. Write a short report to summarise and explain what you find.

4.3 Summing up Part 4

In this part of the chapter you have learned how temperature control is achieved in modern buildings. You have reviewed the mechanisms of heat transfer and learned how to calculate the energy transfer involved when an object is warmed or cooled.

Activity 37 is designed to help you review your progress, and Questions 24 to 26 are designed to help you put into practice what you have been learning.

Activity 37 Heating and cooling in the laboratory

First, check through Part 4 and ensure you know the meaning of each of the terms printed in bold.

Consider the heating and cooling systems in your school/college laboratory. With the aid of two labelled sketches (or photographs), explain how conduction, convection and radiation are involved in (a) cold and (b) warm weather. Make some recommendations for improvements to the heating and cooling of the room.

Questions

24 A storage heater has a 3.4 kW electric heating element that heats a brick block enclosed in a casing. The specific heat capacity of the brick is 950 J kg °C^{-1}. The U-value of the casing when the flaps are opened is 11.2 W m^{-2} °C^{-1} and its surface area is 0.23 m^2. The mass of the brick block is 152 kg.

At night the brick block is heated electrically, then flaps are opened during the day allowing it to release the stored heat through the casing.

(a) Calculate the average temperature rise of the brick block if the heating element is switched on for three hours. Assume the block emits no heat during this time.

(b) Explain why the value calculated in (a) is an average.

(c) The ventilation flaps are then opened and the heater heats the room. Assuming that the brick block has reached a uniform temperature, calculate the initial power output of the heater.

25 The passage below refers to Figure 4.55 and has been adapted from *Window Solar Collector: Venetian blinds with a new slant* by Rochelle Chadakoff (*Popular Science Magazine*, November 1978). When you have read it and studied the diagram, answer the questions that follow.

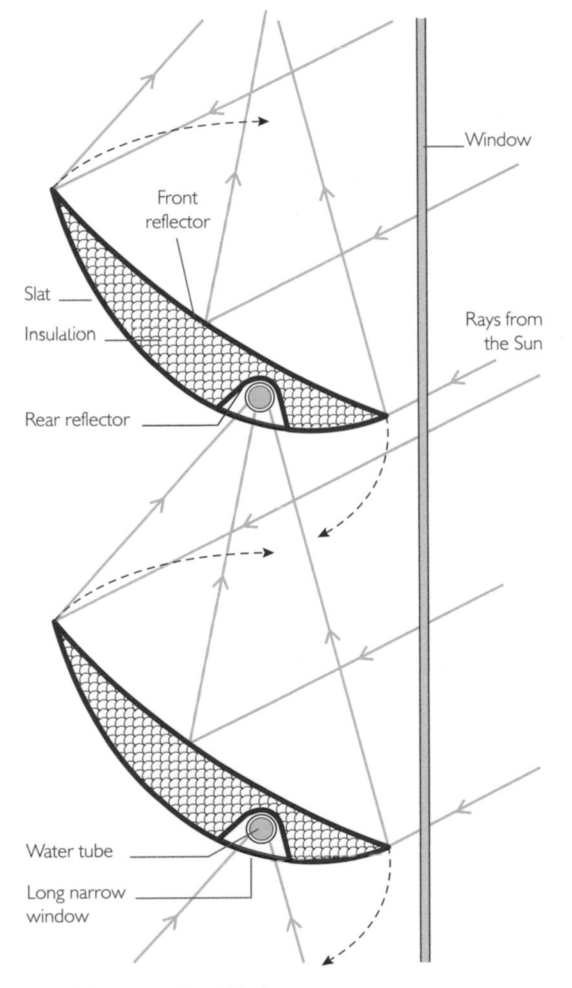

Figure 4.55 Sun-catching venetian blinds

Sun-catching venetian blinds could help to store the energy transferred from solar radiation and help heat buildings. Each slat has a front reflector made of aluminised plastic. Nested in a bed of insulation below the front reflector is a glass tube containing water. This nests in a second reflector and is protected by a long, narrow window of special glass.

Each front curved reflector bounces the Sun's rays up to the slat above. Here the rays are focused through the narrow glass window and their energy transferred to the water. These water tubes are linked by pipes to the building's hot water system. The slats are re-angled occasionally, but do not need to track the Sun precisely.

During the day the Sun-warmed water passes along the pipes and provides some heating for the room as well as adding hot water to the building's heating system. At night, with the blinds shut, the heated water is circulated back through the water tubes to heat the room. The original prototype transferred 70% of the available solar energy.

(a) Why are both the large front reflector and the small rear reflector needed?

(b) Why do the blind slats need insulation behind the reflectors? Suggest a suitable insulating material and explain what makes it a good insulator.

(c) If the hot-water tank is positioned above the level of the blind, explain how the heated water (i) reaches the tank during the day and (ii) circulates through the blind at night.

(d) By what thermal energy transfer process(es) is the room heated by the warm water passing through the blind?

(e) Taking the various energy transfer processes into account, what properties would be ideal for the material making the water tubes? It is suggested that copper tubes might be better than the glass ones. What advantages and disadvantages might the copper tubes have?

(f) In a test of the prototype, the solar flux measured square-on to the blind was 15 MJ m^{-2} per day, the blind measured 2 m × 2 m, and the water circulated at a rate of 6000 kg per day.

Calculate (i) the energy transferred to the water and (ii) the resulting temperature rise of the water. Assume that the blind is 70% efficient at transferring energy to the water and that energy transfers from the heated water to the surroundings can be ignored. The average specific heat capacity of water over this temperature range is 4200 J kg^{-1} $°C^{-1}$.

26 This question is about a technique used in the oil and gas industries to measure rates of fluid flow. It involves measuring the temperature upstream and downstream of a small heater placed in the path of the fluid as shown in Figure 4.56. The apparatus is first calibrated by measuring the temperature difference produced by a known heater power for a known gas flow rate. Measuring an unknown flow rate involves readjusting the flow rate so that it produces the same temperature change. The following example illustrates how this enables the flow rate to be deduced.

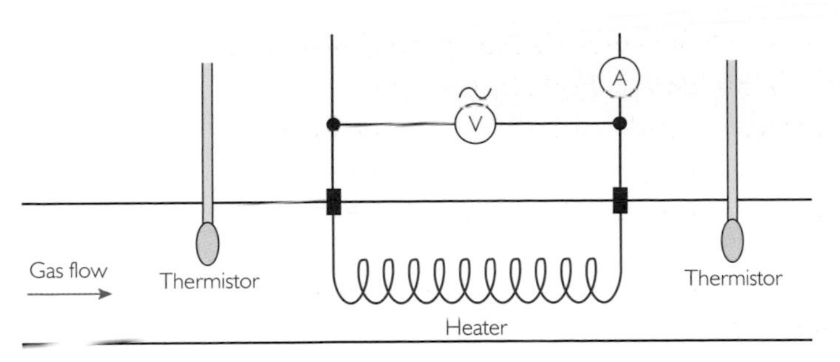

Figure 4.56 Technique for measuring fluid flow rate

(a) If the thermistors are the NTC type, how will their resistance depend on the temperature of the gas?

(b) (i) The voltmeter and ammeter connected to the heater read 240 V and 15 A, respectively. Calculate the power transferred by the heater.

(ii) At a flow rate of 0.10 kg s^{-1}, gas passing the heater rose in temperature by 13°C. Taking the specific heat capacity of the gas to be 2200 J kg^{-1} $°C^{-1}$ over this temperature range, calculate the power transferred to the gas.

(iii) Your answer to (i) should not be the same as your answer to (ii). Explain why this is the case, and calculate the power that is *not* being transferred to the gas. Which conservation law are you using in this calculation?

(c) On another occasion, with the gas at the same initial temperature flowing at a different rate, the power transferred by the heater had to be reduced to 2500 W in order to keep the temperature rise from the same starting temperature to 13°C and thus to keep the power transfer to the surroundings the same as in (b).

(i) Using your answer to (b)(iii), calculate the power transferred to the gas.

(ii) Calculate the flow rate of the gas on this occasion.

5 Rebuilding

5.1 Building up knowledge

Having finished this chapter you are now quite close to the end of your course. In this chapter, you have revisited some earlier work on waves and material properties, and have also used ideas about forces, energy transfer and motion.

Activity 38 Building on firm foundations

Take time to have a thorough look at all the work you have done in this chapter, and at all the related work in other parts of the course.

Go back to Activities 2 and 3 in Part 1 of this chapter and extend your summaries so that they include the new things you have learned.

Activity 39 Why SHM?

Imagine you are designing a course of study for engineers. Write the prospectus entry which will persuade them that studying SHM will be of the value to their future careers. Outline briefly what they will cover in each section and explain:

- why it is important to understand the fundamentals of SHM
- why it is important to know a building's natural frequency
- whether a large structure is likely to have a high or low natural frequency
- what could excite a building at low frequencies
- why it is useful to know about damping.

Activity 40 Bouncy castle

Many principles of physics can be illustrated through everyday situations and events. In a 'bouncy castle', a large pillow is kept inflated by means of an air pump while small children bounce around on it – it illustrates many aspects of the physics of oscillations and materials that you have covered in this chapter, as well as principles from earlier in the course (such as forces and energy).

Imagine you have been asked to write an illustrated article for a newspaper or magazine about the physics of a bouncy castle. Draft a set of headings and notes, with diagrams, equations and order-of-magnitude calculations, showing some of the things that you could include in the article. Include clear statements of principles of physics and show how they are relevant to a bouncy castle.

Activity 41 Build a leisure centre

A city in a semi-tropical earthquake-prone area requires a leisure centre that includes a cinema, bowling alley, swimming pool, gym and a supermarket.

How would you design the centre to be earthquake resistant up to magnitude 6 on the Richter scale?

How would you design the building so as to maintain the interior at a comfortable temperature?

How would you try to balance any conflicting demands that might arise from considering these two requirements?

5.2 Questions on the whole chapter

27 An earthquake occurs at a point X on the surface of the Earth, shown in Figure 4.57. Draw the ray paths to show how the P waves and the S waves would travel within the Earth if (a) the Earth were completely liquid, (b) the Earth were completely solid, and (c) the Earth were solid with a liquid core starting halfway to the centre.

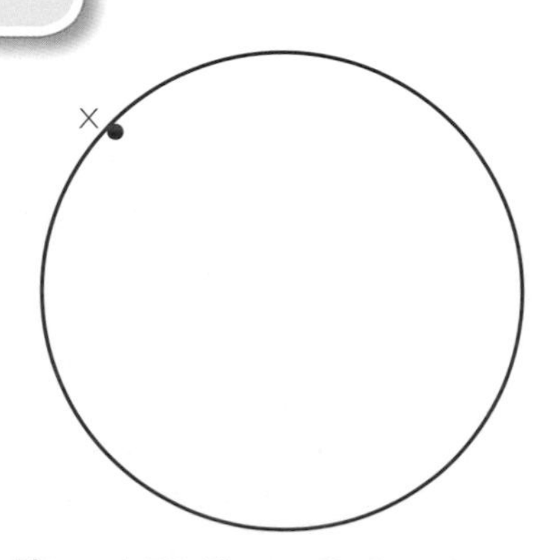

Figure 4.57 Diagram for Question 27

28 Domestic washing machines often incorporate washing, rinsing, spinning and drying. This question is about the spinning.

(a) The inner drum of the machine into which the clothes are placed has quite large holes in it. Explain carefully how, when the clothes are being spin-dried, the water gets from the clothes and out through the holes.

(b) One of the spin speeds in one model of washing machine was listed as 1000 rpm (rpm stands for revolutions per minute). (i) What is this spin speed in radians per second? (ii) If the radius of the spinning drum is 12.5 cm, what would be the highest centripetal force that could be exerted on a wet sweatshirt of mass $m = 0.5$ kg?

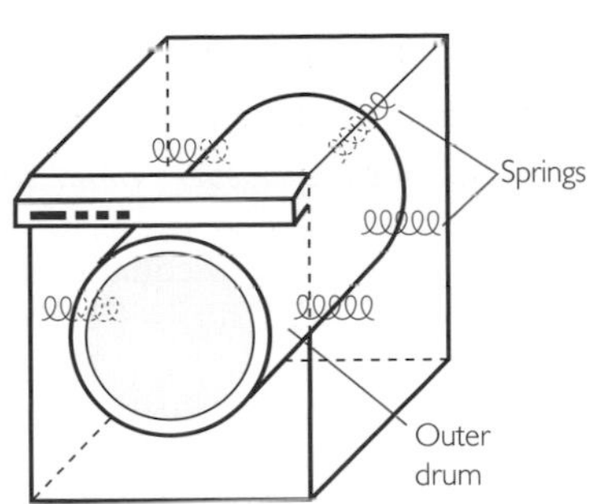

Figure 4.58 Schematic diagram of a washing machine

(c) If clothes are unevenly distributed in the machine, it vibrates slightly as it rotates. The outer drum within which the spinning drum rotates is attached to the rest of the framework of the washing machine by springs (Figure 4.58). What is the purpose of these springs?

(d) For each spring, the spring constant $k \approx 200$ Nm^{-1}. In use, the loading on each spring is effectively 5 kg. Explain, with the aid of a calculation, what is likely to happen when an unevenly-loaded machine begins to spin the clothes.

29 A very accurate method for monitoring the density of a liquid in many industrial processes involves filling a hollow U-tube with the liquid and then measuring the period of natural oscillations by making it resonate to an external variable frequency oscillator. (The arrangement is rather like a hollow tuning fork.) The period of oscillation, T, is given by:

$$T = 2\pi\sqrt{\frac{(m + \rho V)}{c}}$$

where m is the mass of the empty tube of volume V, c is the elastic stiffness of the tube and ρ is the density of the liquid filling the tube.

For a particular tube, $m = 15 \times 10^{-3}$ kg and $V = 2.2 \times 10^{-6}$ m^3.

(a) The evacuated tube has a natural frequency of 800 Hz. What is the value of c, the elastic stiffness of the tube?

(b) The expression for T can be re-written as:

$$T^2 = A\rho + B$$

What are (i) the algebraic expressions for A and B, and (ii) their numerical values and SI units?

(c) A manufacturer of an instrument using this technique quotes numerical values only of A and B for calibration. Assuming frequencies are measured in Hz, why do you not need to ask the manufacturer what units are being used for B, but you do for A?

(d) What is the frequency of oscillation when the tube is filled with a liquid of density 1.2×10^3 kg m^{-3}?

(e) A calibration graph is drawn of T^2 (y-axis) against density, ρ. (i) Express the gradient and intercept in terms of A and B. (ii) Sketch the graph and explain the significance of the point where it crosses the y-axis.

(f) Explain how you could use the calibration graph to find ρ from a measurement of the resonant frequency of the tube.

30 The passage below is taken from the marketing material supplied by a manufacturer of electrically heated showers.

> *Most electric showers draw cold water direct from the main supply and heat it as it is used – day or night. Not only are they particularly useful for those who do not have a stored water supply, but they are versatile because every home can have one.*

(a) Write a word equation to describe the energy transfer that takes place in an electric shower.

(b) Rewrite the equation using the appropriate formulae.

(c) The technical data supplied by one manufacturer states that their most powerful shower system is fitted with a 10.8 kW heating element and can deliver up to 16 litres of water per minute.

Show that the showering temperature is about 25°C if the temperature of the mains water is 15°C and the shower is used at its maximum settings.

Specific heat capacity of water = 4200 J kg^{-1} K^{-1}

Mass of 1 litre of water = 1.0 kg

(d) The marketing material includes the statement:

> *Please remember that during the colder months, flow rates may need to be reduced to allow for the cooler temperature of incoming cold water.*

Calculate the flow rate required for an output of 25°C when the incoming water temperature is 5°C.

31 Next to the 3000-year-old Drombeg Circle in Ireland is a stone-lined pit known as a Fulacht Fiadh. It is believed that this was used as a cooking place for meat caught by hunters. The pit was filled with water. Large stones were heated in a fire and then placed in the water to bring it to the boil and cook the meat.

In experiments to test this idea it was found that the water in the pit started to boil after 22 heated stones had been added. The total mass of the added stones was 198 kg and the mass of water was 513 kg.

(a) Show that this gives a minimum temperature for the fire of about 900°C.

(Specific heat capacity of water = 4200 J kg^{-1}°C^{-1}
Average specific heat capacity of stone = 1100 J kg^{-1}°C^{-1}
Initial temperature of water = 18°C
Temperature of boiling water = 100°C)

(b) Explain why the temperature of the fire would be higher than the calculated value.

5.3 Achievements

Now you have studied this chapter you should be able to achieve the outcomes listed in Table 4.5.

Table 4.5 Achievements for the chapter *Build or Bust?*

	Statement from examination specification	*Section(s) in this chapter*
109	investigate, recognise and use the expression $\Delta E = mc\Delta\theta$	4.1 [also see STA]
119	recall that the condition for simple harmonic motion is $F = -kx$, and hence identify situations in which simple harmonic motion will occur.	2.2
120	recognise and use the expressions $a = -\omega^2 x$, $a = -A\omega^2 \cos\omega t$, $v = -A\omega \sin\omega t$, $x = A\cos\omega t$ and $T = \frac{1}{f} = \frac{2\pi}{\omega}$ as applied to a simple harmonic oscillator	2.2
121	obtain a displacement–time graph for an oscillating object and recognise that the gradient at a point gives the velocity at that point	2.2
122	recall that the total energy of an undamped simple harmonic system remains constant and recognise and use expressions for total energy of an oscillator	2.2
123	distinguish between free, damped and forced oscillations	2.1, 3.1
124	investigate and recall how the amplitude of a forced oscillation changes at and around the natural frequency of a system and describe, qualitatively, how damping affects resonance	2.1, 2.4, 3.1
125	explain how damping and the plastic deformation of ductile materials reduce the amplitude of oscillation	3.1, 3.2

Answers

1 Distance = 0.5 × circumference = πr

$$\text{time} = \frac{\text{distance}}{\text{speed}} = \frac{\pi \times 6.378 \times 10^6 \text{ m}}{3 \times 10^3 \text{ m s}^{-1}} = 6680 \text{ s}$$

= 111 minutes (to the nearest minute)

(In fact faster waves also travel through the Earth as well as around the surface, so they will arrive sooner than this. The waves in the crust will also gradually get smaller as energy is transferred to other forms, so in practice these surface waves will have little effect on the far side of the Earth.)

2 (a) stress, $\sigma = \frac{\text{force}}{\text{area}} = \frac{F}{A}$

strain, $\varepsilon = \frac{\text{extension}}{\text{original length}} = \frac{x}{l}$

$E = \frac{\sigma}{\varepsilon}$

and so $\varepsilon = \frac{(F/A)}{E}$, hence

$$x = \frac{Fl}{AE} = \frac{2 \times 10^{-3} \text{ kg} \times 9.81 \text{ N kg}^{-1} \times 10 \times 10^{-2} \text{ m}}{\pi \times (3 \times 10^{-6} \text{ m})^2 \times 73.1 \times 10^9 \text{ N m}^{-2}}$$

$= 0.95 \times 10^{-3}$ m = 0.95 mm

(b) maximum mass, m = tensile strength × $\frac{A}{g}$

$$= \frac{1000 \times 10^6 \text{ N m}^{-2} \times \pi \times (3 \times 10^{-6} \text{ m})^2}{9.81 \text{ N kg}^{-1}}$$

$= 2.9 \times 10^{-3}$ kg = 2.9 g

3 (a) Increasing the elastic modulus appears to increase the wave speed. You can deduce this from various numbers in Table 4.1; eg a shear modulus of zero corresponds to a wave speed of zero; shale and sandstone have similar densities, but the one with the larger bulk modulus has the larger P-wave speed.

(b) Increasing density appears to reduce the speed. For example, sandstone and salt have the same bulk modulus; salt is less dense and has a higher wave speed.

4 The phenomenon is called refraction. The rays bend away from the normal and the wavelength increases. See Figure 4.59.

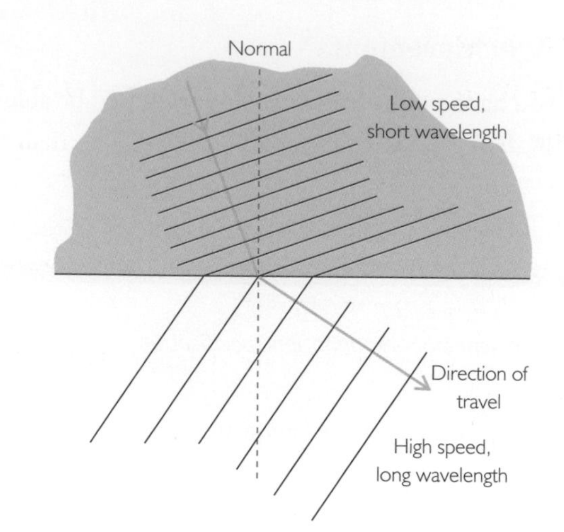

Figure 4.59 Answer to Question 4

5 (a) (i) has travelled directly; (ii) has met the interface at an angle greater than the critical angle and undergone total internal reflection; (iii) has been refracted into a deeper layer.

(b) The time of arrival of the waves will give information about the paths taken, and so the depth of the Moho (or other boundaries between layers) can be found.

6 refractive index, $\mu = \frac{v_1}{v_2}$

$= \frac{\text{velocity in lower mantle}}{\text{velocity in outer core}}$

From Figure 4.8, $\mu = \frac{13.5 \text{ km s}^{-1}}{7.5 \text{ km s}^{-1}} = 1.80$

Snell's Law: $\mu = \frac{\sin i}{\sin r}$

$\sin r = \frac{\sin i}{\mu} = \frac{\sin 30°}{1.80} = 0.278$

$r = 16.1°$

7 (a) The speed must increase with depth – the waves follow paths similar to that in Figure 4.12 only they curve gradually rather than changing direction abruptly.

(b) Density will increase with depth.

(c) On its own, an increase in density will produce a *reduction* in wave speed (see Equation 1). To bring about an increase in speed the elastic modulus must also increase with depth, and the increase must outstrip the increase in density.

8 From Equation 3:

$D = 0.001\ \text{mm} \times 10^{5.2} = 1.58 \times 10^2\ \text{mm} = 0.158\ \text{m}.$

This is a very large deflection – much larger than could in practice be recorded by a seismometer.

The actual ground movement would be much less. (People at this distance felt the quake, but the ground did not move visibly.)

9 (a) (i) From Equation 7, max. speed occurs when $\sin\omega t = 1$; ie:

$v = A\omega = 0.30\ \text{m} \times 5.0\ \text{s}^{-1} = 1.5\ \text{m s}^{-1}$

(ii) $f = \frac{1}{T} = \frac{\omega}{2\pi} = \frac{5.0\ \text{s}^{-1}}{2\pi} = 0.80\ \text{Hz}$ (2 sig. fig.).

(b) $\omega^2 = \frac{k}{m}$ so $k = m\omega^2 = 0.40\ \text{kg} \times (5.0\ \text{s}^{-1})^2$

$= 10\ \text{kg s}^{-2} = 10\ \text{N m}^{-1}.$

(c) When $t = 0.20$ s, $\omega t = 1$ rad

(i) $x = A\cos(1\ \text{rad}) = 0.16\ \text{m}$

(ii) $a = -\omega^2 x = -(5.0\ \text{s}^{-1})^2 \times 0.16\ \text{m} = 4.0\ \text{m s}^{-2}$

10 (a) (i) $v_{max} = A\omega$ (from Equation 8) and

$\omega = 2\pi f$ (from Equation 15) so:

$v_{max} = 2\pi fA = 2\pi \times 50 \times 10^3\ \text{s}^{-1} \times 10^{-6}\ \text{m}$

$= 0.6\ \text{m s}^{-1}$

(ii) From Equation 12, $a_{max} = \omega^2 A$

$= (2\pi \times 50 \times 10^3\ \text{s}^{-1})^2 \times 2 \times 10^{-6}\ \text{m}$

$= 2 \times 10^5\ \text{m s}^{-2}$

(b) $a \approx 2 \times 10^4\ g$. The peak force will be 20 000 times the weight of an oscillating part – so the material needs to withstand a high stress.

(c) Since $a_{max} = \omega^2 A = (2\pi f)^2 A$, the acceleration (and hence the internal stresses) are proportional to frequency squared.

11 (a) (i) From Figure 4.60, there is an unbalanced force of magnitude $F_x = mg\sin\theta$ acting on the mass along the direction marked x. If θ is small, $\sin\theta \approx \theta$ (in radians) $\approx \frac{x}{l}$. Hence $F_x \approx \frac{mgx}{l}$ when θ is small. There is therefore a restoring force that is proportional to displacement; ie the resulting motion is SHM. Since $a = \frac{F}{m}$ =, along the direction x we have $a_x = \frac{gx}{l}$. From Equation 13, $\omega^2 = \frac{a}{x}$ so we can identify $\omega^2 = \frac{g}{l}$.

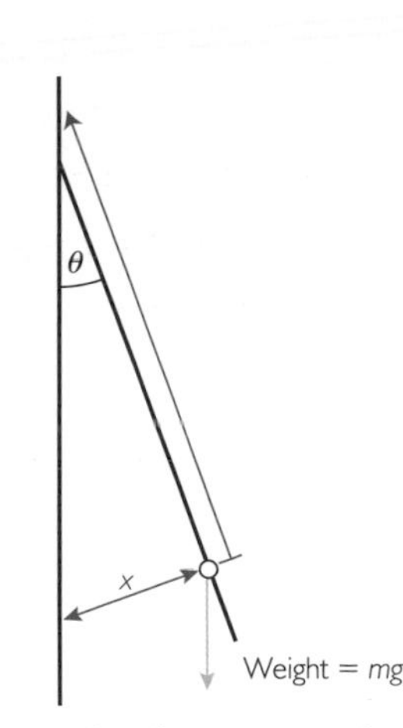

Figure 4.60 Diagram for the answer to Question 11

(ii) From Equation 14, $T = \frac{2\pi}{\omega} = 2\pi\sqrt{\frac{l}{g}}$

(b) For $f = 1$ Hz, $T = 1$ s. From the answer above:

$l = \frac{gT}{4\pi^2} = \frac{9.81\ \text{m s}^{-2} \times 1\ \text{s}}{4\pi^2} = 0.25\ \text{m}$

12 (a) From worked example,

$k = 50\ \text{N m}^{-1}$, $A = 3.0 \times 10^{-2}$ m.

$E_{tot} = \frac{1}{2}kA^2$

$= \frac{1}{2} \times 50\ \text{N m}^{-1} \times (3.0 \times 10^{-2}\ \text{m})^2$

$= 2.25 \times 10^{-2}\ \text{J}.$

$m = 0.1\ \text{kg}$, $\omega = 22.36\ \text{s}^{-1}$

$E_{tot} = \frac{1}{2}m\omega^2 A^2$

$= \frac{1}{2} \times 0.1\ \text{kg} \times (22.36\ \text{s})^2 \times (3.0 \times 10^{-2}\ \text{m})^2$

$= 2.25 \times 10^{-2}\ \text{J}$

(b) From Question 9, $k = 10\ \text{N m}^{-1}$, $A = 0.30$ m, $m = 0.40$ kg, $\omega = 5.0\ \text{s}^{-1}$.

$E_{tot} = \frac{1}{2}kA^2 = \frac{1}{2} \times 10\ \text{Nm}^{-1} \times (0.30\ \text{m})^2$

$= 0.45\ \text{J}$

$E_{tot} = \frac{1}{2}m\omega^2 A^2$

$= \frac{1}{2} \times 0.40\ \text{kg} \times (5.0\ \text{s}^{-1})^2 \times (0.30\ \text{m})^2$

$= 0.45\ \text{J}.$

13 (a) $E_p = \frac{1}{2}kx^2 = \frac{1}{2} \times 1.0 \times 10^2\text{N m}^{-11} \times (0.15\ \text{m})^2$

$= 1.1\ \text{J}$

(b) Assuming that the energy transfer to the missile is 100% efficient; ie that E_k gained equals E_p lost by catapult:

$\frac{1}{2}mv^2 = 1.1\ \text{J}$

$$v = \sqrt{\left(\frac{2 \times 1.1\ \text{J}}{2.0 \times 10^{-3}\ \text{kg}}\right)} = 33\ \text{m s}^{-1}$$

(c) $\frac{1}{2}mv^2 = \frac{1}{2}kx^2$ so $v = x\sqrt{\left(\frac{k}{m}\right)}$

(d) (i) Since $v \propto x$, the speed is one-third that in (b); ie 11 m s^{-1}

(ii) $v \propto 1/\sqrt{m}$ so multiplying m by a factor of two will lead to v being divided by $\sqrt{2}$; ie $v = \frac{33\ \text{m s}^{-1}}{\sqrt{2}} = 23\ \text{m s}^{-1}$

14 (a) Speed $= \frac{2 \times \text{length of bar}}{\text{time taken}}$

$$= \frac{1.40\ \text{m}}{512 \times 10^{-6}\ \text{s}} = 2.73 \times 10^3\ \text{m s}^{-1}$$

(b) It suggests that the circuit contains elements of resistance and capacitance – the decay resembles that of an *RC* discharge.

15 (a) The Young modulus of oak must be significantly greater than that of pine – despite oak having a higher density, it also has a higher wave speed. Similarly, metals in general have higher Young moduli than woods – their waves speeds are higher despite their higher densities.

(b) From Equation 1, $E = \rho v^2$.

(i) Pine: $E = 500\ \text{kg m}^{-3} \times (3313\ \text{m s}^{-1})^2 = 5.49 \times 10^9\ \text{N m}^{-2}$

(ii) Copper: $E = 8933\ \text{kg m}^{-3} \times (3650\ \text{m s}^{-1})^2 = 1.19 \times 10^{11}\ \text{N m}^{-2}$

16 (a) From Section 2.2, $f = \frac{1}{2\pi}\sqrt{\left(\frac{k}{m}\right)}$ so

$$k_{\text{horiz}} = m \times (2\pi f)^2$$
$$= 40 \times 10^3\ \text{kg} \times (2\pi \times 0.60\ \text{Hz})^2$$
$$= 5.7 \times 10^5\ \text{N m}^{-1}.$$

(b) (i) $k \propto f^2$, so $k_{\text{vert}} = k_{\text{horiz}} \times \left(\frac{10\ \text{Hz}}{0.60\ \text{Hz}}\right)^2$

$$= 1.6 \times 10^8\ \text{N m}^{-1}.$$

(ii) Compression $\Delta h = \frac{F}{k}$.

$F = mg$, so

$$\Delta h = \frac{40 \times 10^3\ \text{kg} \times 9.81\ \text{N kg}^{-1}}{1.6 \times 10^8\ \text{N m}^{-1}}$$

$$= 2.5 \times 10^{-3}\ \text{m (2.5 mm)}$$

(c) (i) Use Equation 13: $a = -\omega^2 x$, where $\omega = 2\pi f$.

$x = \frac{a}{(2\pi f)^2}$ (dropping the minus as we are only interested in size not direction

$$x = \frac{12\ \text{m s}^{-2}}{(2\pi \times 3.0\ \text{Hz})^2} = 3.4 \times 10^{-2}\ \text{m}$$

(ii) If the amplitude scales up by a factor 3, then it will be 0.1 m – which would be alarming.

17 (a) Linear dimensions are all multiplied by 3, so volume (and hence mass) will be multiplied by 27. Actual mass = 27 × 40 tonnes = 1080 t.

(b) As f depends on k/m, stiffness would also have to be multiplied by 27.

(c) Horizontal oscillations will increase in amplitude as you go up the building, but the vertical amplitude will remain more or less constant with height. And any building is under vertical compression because of its weight (largest stress at the bottom). With vertical oscillations the acceleration will increase and decrease this compressive stress, but will be unlikely to put materials under tension (when they are more likely to fail).

18 (a) We need to find $f = \frac{1}{T}$, where $T = 2\pi\sqrt{\frac{m}{k}}$. The suspended mass is unknown – but it is not needed. Let the mass be m and its weight mg.

Force extending spring, $F = mg$

Extension, $x = 0.20$ m.

So $k = \frac{F}{x} = \frac{mg}{x}$

Hence $\frac{m}{k} = \frac{x}{g}$ and we have $T = 2\pi\sqrt{\frac{x}{g}}$

$$f_{\text{spring}} = \frac{1}{2\pi} \times \sqrt{\frac{g}{x}} = \frac{1}{2\pi} \times \sqrt{\frac{9.81\ \text{m s}^{-2}}{0.20\ \text{m}}}$$

= 1.1 Hz.

(b) The overall length $l = 0.80$ m. From Question 11, $T = 2\pi\sqrt{\frac{l}{g}}$.

$$f_{\text{pend}} = \frac{1}{2\pi} \times \sqrt{\frac{g}{l}} = \frac{1}{2\pi} \times \sqrt{\frac{9.81\ \text{m s}^{-2}}{0.80\ \text{m}}}$$

= 0.56.

(c) $\frac{f_{spring}}{f_{pend}} = 2:1$

Both frequencies are proportional to $\sqrt{\frac{1}{\text{length}}}$. The lengths are in a ratio 4: 1 so the frequencies must be in the ratio $\sqrt{4}$: $\sqrt{1}$; ie 2:1.

(d) There are two spring oscillations to every pendulum oscillation. Starting off a spring oscillation can gradually set off a pendulum oscillation so that one complete spring cycle takes place in half a pendulum cycle. The system gradually switches from one kind of oscillation to the other – and then back again. It can be thought of as a kind of resonance, but where the frequencies are in the ratio 2:1 and it is not clear which is the driving system and which the driven.

19 (a) Use $E_p = \frac{1}{2}Fx$

$= \frac{1}{2} \times 1.0 \times 10^5 \text{ N} \times 50 \times 10^{-3} \text{ m} = 2.5 \times 10^3 \text{ J}.$

(b) Assuming that $E_p = E_k = \frac{1}{2}mv^2$:

$$v = \sqrt{\frac{2E_p}{m}} = \sqrt{\frac{2 \times 2.5 \times 10^3 \text{ J}}{5.5 \times 10^3 \text{ kg}}} = 0.95 \text{ m s}^{-1}$$

(c) The decay *is* exponential. You could use any of the following tests: constant half-life; equal fractions (of A) in equal times; a graph of log(A) against n is linear.

(d) See Table 4.6.

Cycle number, n	Amplitude A, $/10^{-3}$ m	Stored energy at end of cycle/J
0	50	2500
1	36	1300
2	26	680
3	19	360
4	13	170
5	10	100

Table 4.6 Answer to Question 19(d)

(e) Fractions lost are: energy, 0.48; amplitude, 0.28. The energy decreases by a larger fraction because energy is proportional to the square of the amplitude (eg if amplitude halves, energy falls to one-quarter of its initial value).

(f) (i) 1200 J

(ii) 300 J over $\frac{1}{4}$, cycle

(iii) The kinetic energy as the tower passes through its midpoint is thus 2.2×10^3 J. Using the same method as in (b), this gives $v = 0.89 \text{ m s}^{-1}$.

(g) (i) $v_{av} = \frac{0.89 \text{ m s}^{-1}}{2} = 0.45 \text{ m s}^{-1}$

(ii) Displacement, $x = 50 \times 10^{-3}$ m, so

$$\frac{T}{4} = \frac{x}{v_{av}} = \frac{50 \times 10^{-3} \text{ m}}{0.45 \text{ m s}^{-1}} = 0.11 \text{ s}$$

(iii) $T = 0.44$ s

(iv) $f = \frac{1}{T} = \frac{1}{0.44 \text{ s}} = 2.3 \text{ Hz}$

You could probably imagine a tower oscillating at around this frequency – it is not implausible and it is within the range of earthquake frequencies.

20 $\Delta E = 1 \text{ kg} \times 4200 \text{ J kg}^{-1} \text{ °C}^{-1} \times 20 \text{ °C} = 84000 \text{ J}$

$$\text{Power} = \frac{84000 \text{ J}}{120 \text{ s}} = 700 \text{ W}$$

21 (a) (i) The larger the radiator's area, the cooler the radiator.

(ii) The higher the sink temperature the higher the radiator temperature.

(b) Approximately 1 m^2.

22 (a) Use Equation 33 ($\Delta E = mc\Delta\theta$) and deal with changes that take place in one minute.

For each wide face, $\Delta\theta = 7$°C and $m = 2400$ kg:

$\Delta E = 2400 \text{ kg} \times 4179 \text{ J kg}^{-1} \text{ °C}^{-1} \times 7 \text{ °C}$
$= 70\,207\,200 \text{ J } (7.02 \times 10^7 \text{ J})$

For each narrow face $\Delta\theta = 8$°C and $m = 240$ kg

$\Delta E = 240 \text{ kg} \times 4179 \text{ J kg}^{-1} \text{ °C}^{-1} \times 8\text{°C}$
$= 8\,023\,680 \text{ J } (8.02 \times 10^6 \text{ J})$

So the total energy transferred in one minute is

$2 \times (70.2 + 8.02) \times 10^6 \text{ J} = 1.56 \times 10^8 \text{ J}$.

(b) Power $P = \frac{\Delta E}{\Delta t}$, $\Delta t = 1 \text{ min} = 60\text{s}$

$$P = \frac{1.56 \times 10^8 \text{ J}}{60 \text{ s}}$$

$= 2.61 \times 10^6 \text{ W} = 2.61 \text{ MW}$

23 From Equation 34 with $\Delta\theta = 10°C$ and $A = 6\ m^2$:

$P = 2.2\ W\ m^{-2}\ K^{-1} \times 6\ m^2 \times 10°C = 132\ W$

24 (a) $\Delta E_{in} = P\Delta t$
$= 3.4 \times 10^3\ W \times 3.0 \times 3600\ s$
$= 3.67 \times 10^7\ J$

From Equation 33:

$$\Delta\theta = \frac{\Delta E}{mc} = \frac{3.67 \times 10^7\ J}{152\ kg \times 950\ J\ kg°C^{-1}} = 254\ °C$$

(b) Brick will be much hotter by the heater. At the surface it will be much closer to room temperature.

(c) Using Equation 34:
output power $P_{out} = UA\Delta\theta$
$= 11.2\ W\ m^{-2}°C \times 0.23\ m^2 \times 254°C = 654\ W$

25 (a) The large reflector focuses the solar radiation onto the water pipes. The smaller reflector 'beams' radiation from the heated water back into the room. Without the smaller reflector most of the radiation not emitted directly into the room would be absorbed by the insulation.

(b) Insulation helps ensure that the absorbed energy is stored in the water and is not transferred directly into the room. Glass fibre or rock wool (such as used in loft insulation) could be used. These materials trap air within and prevent it from circulating. The materials themselves are also poor conductors, as is the air. They therefore reduce both convection and conduction.

(c) (i) Warm water is less dense than cold water, so during the day the water circulates within the system by convection: heated water moves upwards to the tank while cooler water descends to the blind.

(ii) At night, when warm water is required to move downwards from tank to blind, a pump is needed.

(d) Energy transfer will be by mainly by convection (warmed air near the blind rises, setting up convention currents in the room) and also by radiation. Transfer by conduction will be minimal as air is a poor conductor.

(e) Ideally the tubes should be good absorbers of solar radiation (mostly visible and infrared), and good emitters of infrared radiation – preferably their surfaces should be dull black. They should also be good thermal conductors in order to transfer energy to and from the water.

Copper is a better thermal conductor than glass so energy would transfer more easily through the tube walls. However, depending on its density and specific heat capacity, more energy might be wasted in heating copper tubes at the expense of the water.

(f) (i) The surface area $A = 4\ m^3$ so the total amount of solar energy intercepted must be $15\ MJ\ m^{-2} \times 4\ m^2 = 60\ MJ$. Energy transferred to water is 70% of 60 MJ; ie 42 MJ ($= 42 \times 10^6\ J$).

(ii) From Equation 33,

$\Delta E = mc\Delta\theta$ so $\Delta\theta = \frac{\Delta E}{mc}$:

$$\Delta\theta = \frac{42 \times 10^6\ J}{6000\ kg \times 4200\ J\ kg^{-1}\ °C^{-1}} = 1.7\ °C$$

26 (a) The resistance of the thermistors decreases with increasing temperature.

(b) (i) Power transferred from heater
$= 240\ V \times 15A = 3600\ W$.

(ii) In 1s, 0.1 kg of gas is heated through 13°C. Energy transferred in 1 s is found from $\Delta E = mc\Delta\theta$:

$\Delta E = 0.1\ kg \times 2200\ J\ kg^{-1}\ °C^{-1} \times 13°C$
$= 2860\ J$

So power transferred to gas $= 2860\ J\ s^{-1}$
$= 2860\ W$.

(iii) The 'missing' power is transferred to the thermistors and the rest of the electric circuit, and to the pipeline walls and through them to the surroundings. Using the law of conservation of energy, this 'missing' power must be
3600 W – 2860 W = 740W.

(c) (i) Since the 'missing' power is again 740 W, power transferred to gas
= 2500 W – 740W = 1760W.

(ii) In 1 s, 1760 J is transferred to the gas, producing a temperature rise of 13°C. $\Delta E = mc\Delta\theta$ so the mass, m, flowing past the heater in 1 s is given by:

$$m = \frac{\Delta E}{c\Delta\theta}$$

$$= \frac{1760\ J}{(2200\ J\ kg^{-1}\ °C^{-1} \times 13°C)} = 0.06\ kg$$

so the flow rate is 0.06 kg $^{-1}$

CHAPTER 5

Reach for the Stars

Why a chapter called *Reach for the Stars*?

Ever since humans with modern brains have walked the planet, they've been doing cosmology – asking themselves questions about their ultimate origins: how the world came into being and what will happen to it – but it's only since the 20th century that science has been able to tackle such questions; and the pace of discovery has been astounding. Halfway through the 20th century, the physicist Hermann Bondi said: 'there are only two-and-a-half facts in cosmology'. Now we have experimental and observational evidence (Figure 5.1) to support our own myths of creation. And we can even begin to speculate on almost unimaginable questions like the existence of other Universes.

This chapter will show how physics has helped modern-day cosmologists unlock some of the secrets of the Universe and its origins: What makes stars shine? Do black holes exist? Is the Universe infinite?

Figure 5.1 Some images obtained with the Hubble Space Telescope: (a) the Eagle nebula, a region where stars are forming;

Figure 5.1 (b) a supernova remnant, the exploded remains of a 'dead' star

Figure 5.1 (c) some of the most distant galaxies yet observed

Overview of physics principles and techniques

In this chapter, you will study several areas of physics that relate to astronomy and cosmology. You will learn about the behaviour of light that enables astronomers to interpret observations of distant objects, the nuclear reactions that power stars, the use of radioactivity to date rocks that make up our planet, the ideas about molecules that help us to understand how stars form, and about the gravitational forces that keep stars and planets in their orbits and will determine the ultimate fate of the Universe. This chapter will also give you several opportunities to look back over your work from previous chapters and to review your knowledge and understanding.

In this chapter you will extend your knowledge of:

- motion in a circle from *Probing the Heart of Matter*
- energy and work from *Higher, Faster, Stronger*, *Transport on Track*, *Probing the Heart of Matter*, and *Build or Bust?*
- radiation from *Technology in Space* and *The Medium is the Message;*
- radioactivity and nuclear reactions from *Probing the Heart of Matter*
- waves from *The Sound of Music, Technology in Space* and *Build or Bust?*
- inverse-square law fields from *Probing the Heart of Matter*.

1 In the beginning

1.1 Big questions

As the following two passages show, **cosmology** (the study of the Universe) is an ancient science, which today is still probing some of the deepest questions about the origins and future of the Universe.

AD 1200, India

As Vishnor watched from the rock beside the River Indus, the sky shifted through a kaleidoscope of colours, from blood red, to rich velvet, to inky black. Then, out of the darkness popped a single white dot. Soon the whole sky was studded with bright points of light, twinkling at him.

On such occasions, Vishnor's mind was full of mysterious questions: How did all this get here? His friends were no help, they just laughed at him. So, as in all matters of wisdom, he turned to one of the great Indian legends for an answer:

> *Some foolish men declare that a Creator made the world. The doctrine that the world was created is ill-advised, and should be rejected. If God created the world, where was He before creation? How could God have made the world without any raw material? If you say He made this first, and then the world, you are faced with an endless regression. Know that the world is uncreated, as time itself is, without beginning and end.*

Extract from *The Mahapurana,* a 9th century Hindu text

AD 1999 Yorkshire, England

One kilometre below the Earth's surface, and David Davidge was sweating more than ever. He was making his way through the mine tunnel, towards the experiment control room. The air was a roasting 50°C, and the bare rock felt hot to the touch. Not really an obvious place for a group of physicists to go looking for clues to the fate of the Universe.

Could we really find those strange, elusive particles, known only as **dark matter**?' he mused. 'The consequences were mind-blowing. We might actually know what would happen to the whole Universe in the distant future. Would it go on forever? Or collapse again in a titanic fireball, like the one that created it?

Certainly down here, the team had a good chance of detecting the particles. All other radiation should get absorbed by the rocks above him. But dark matter was so hard to pick up. The particles – whatever they were – hardly interacted with a damn thing. They passed through virtually anything, as though it weren't there. And instruments that were sensitive enough to detect dark matter were still beyond current technology. But as everyone said, just give it a few more years…

In 1999, David Davidge was a second-year PhD student at Imperial College, London. He talked about his work.

> *People think you have to be a genius to work in any area like mine. It's not true. You just gradually learn to think in certain ways. If you'd told me 10 years ago I would be doing this now, I wouldn't have believed you.*

What first got me excited about cosmology was reading books like Stephen Hawking's A Brief History of Time. *The Universe sounded such a bizarre, alien place. I wanted to understand it, to get closer to the truth. As I studied physics through school and university, I felt I was getting closer, towards the limit of what was known. Then I opted to do astronomy and cosmology in my last year. The concepts were making sense to me now.*

Getting onto the dark matter project happened by chance. It seems such an important area of research, and I thought I could help. I love the work. It can be frustrating working on such a long difficult project, but every so often there's a breakthrough, and then it makes me think we can do it.

You will find out more about the search for dark matter later in the chapter. You will also learn about other aspects of astronomy and cosmology, as illustrated in Figures 5.1(c) and 5.2. These pictures show how modern instruments are helping to probe the mysteries of the Universe. To the naked eye, the picture in Figure 5.1(c) would be a faint speck of light, probably too faint to see. The Hubble Space Telescope reveals that it is a cluster of galaxies, made of hundreds of galaxies, each containing perhaps hundreds of billions of stars. Figure 5.2 shows what might be a possible planet outside our solar system. The planet is the faint object in the lower left. As there are other planets, then perhaps life exists elsewhere in the Universe.

Study note

Note that in scientific usage a billion is defined to be 10^9 – an 'American' billion.

Figure 5.2 Hubble Space Telescope (HST) image of a possible planet

2 Our nearest star

2.1 The Sun

Question: Which is the star nearest the Earth?

Answer: No, it's not Proxima Centauri, a faint star in the constellation Centaurus and second closest to Earth. The answer is the Sun.

An old joke, but it is still easy to trick people with this question. Because the Sun looks so different from anything else in the sky we sometimes forget that it is really a star. In fact the Sun is a very ordinary, average-sized typical star. Compared with the billions of other stars in the Universe there is nothing special about it – but it is very much closer to Earth.

Proxima Centauri is a distance of 4.24 light years from Earth. A **light year** is the distance light travels in a vacuum in one year. The average distance between the Earth and the Sun is 1.50×10^{11} m.

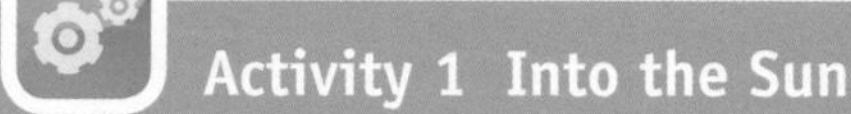

Activity 1 Into the Sun

Use the internet to research some information about the Sun. In particular, look for information about space missions and telescopes that are studying physics processes and conditions in the Sun. Go to **www.shaplinks.co.uk** for some useful links.

Study note

This activity should remind you about aspects of the AS chapter *Technology in Space* and help you to look ahead to later sections of this chapter.

Question

1 (a) How long does it take light to travel from the Sun to Earth, given that light travels at 3.00×10^8 ms^{-1}?

(b) Calculate the distance to the Sun in light years, given that 1 year $= 3.16 \times 10^7$ s.

(c) Calculate the ratio $\frac{d_{pc}}{d_{sun}}$

where d_{pc} is the distance from Earth to Proxima Centauri and d_{sun} is the distance from Earth to the Sun

Then express your answer as an order of magnitude – that is, state the nearest power of ten (10, … 10^4, 10^5, 10^6, …).

Because the Sun is so close to Earth, it is the star we are best placed to study, and so our detailed study of stars begins here. But first a word of caution before you start any activities involving the Sun. Many of the early scientists who studied the Sun lost their sight by staring directly at the Sun.

There are ways to image the Sun safely; these will be explained in the activities later in this chapter.

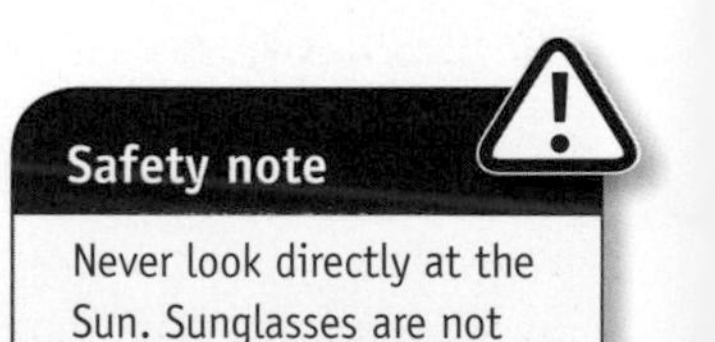

Safety note

Never look directly at the Sun. Sunglasses are not sufficient protection.

Brightness and distance

The reason that the Sun appears so much brighter than the other stars is because it is much closer. As you move further away from the source of any radiation, the radiation becomes less intense. This is because the radiation is spreading out in all directions, and so the amount that will land on you will become less as you move away. As you blow up a balloon the rubber gets thinner and thinner – the radiation gets weaker and weaker in the same way. Figure 5.3 shows that if you double the distance from a star, then the area over which the radiations is spread is multiplied by four; if you multiply the distance by three, then the area is multiplied by nine.

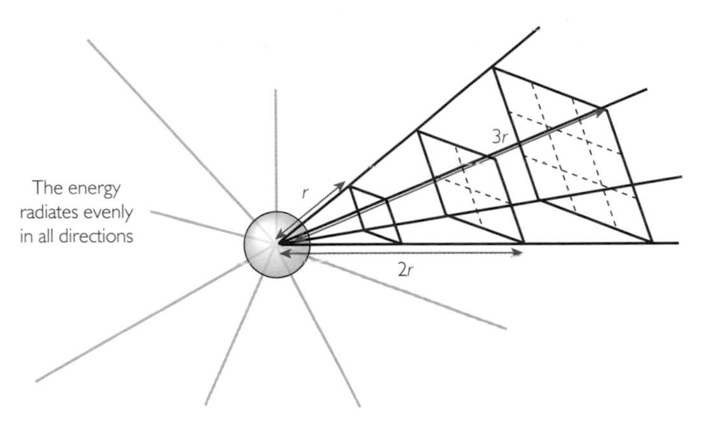

Figure 5.3 Radiation spreads out as it travels

Figure 5.3 shows that the **energy flux**, F, or **intensity**, I, of the radiation must obey an **inverse-square law**. An inverse-square law is one in which two quantities (say x and y) are related by an equation of the type:

$$y = \frac{\text{constant}}{x^2}$$

Study note

You met the idea of energy flux in the AS chapter *Technology in Space*.

Energy flux is the rate at which energy is transferred across a unit area perpendicular to the light beam. Energy flux is not quite the same as observed brightness, since the brightness we observe depends only on the visible component of the radiation, whereas energy flux refers to the whole electromagnetic spectrum. Think of this page. If you increase the radiation hitting the page it will appear to be more brightly lit, won't it? It will if the increased radiation is in the visible spectrum; but we have to remember that most of the electromagnetic spectrum is invisible to us. For example, if the page was illuminated with X-rays, we wouldn't detect them with our eyes, even though the actual radiant energy flux might be quite large.

The **luminosity**, L, of a radiation source (such as a star) is defined as the rate at which the source radiates energy – the total energy it loses every second; ie its radiated power. The SI unit of luminosity is the watt. You may like to think of luminosity as being a measure of how bright a star really is, as opposed to how bright it appears from Earth.

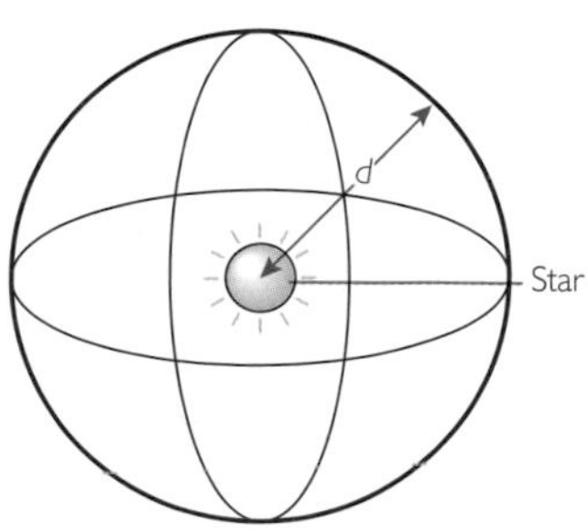

Figure 5.4 Radiation from a star

Energy flux is related to luminosity – the brighter a star, the greater the energy flux at a given distance. F is proportional to L. However, we must allow for the distance, d, to the star. Figure 5.4 shows a star at the centre of an imaginary sphere of radius d.

The radiation emitted from the star in one second will spread out so that it is shared evenly over the surface of the sphere. Travelling at a steady speed, if it is not absorbed or scattered en route, the radiation will pass through the imaginary sphere in one second in such a way that:

$$\text{Flux} = \frac{\text{total energy emitted from source per second}}{\text{surface area of sphere}}$$

$$F = \frac{L}{4\pi d^2} \qquad (1)$$

Equation 1 is important. It expresses the inverse-square law for energy flux, and enables us to calculate the luminosity of a star if we can measure its radiant energy flux at Earth and its distance.

Activity 2 Exploring the inverse-square law

Measure the conductance of a light-dependent resistor (LDR) when illuminated by a small light bulb at different distances. Comment on the extent to which the results are consistent with an inverse-square law for radiation flux.

Questions

2 The Sun is the main source of energy for all the planets and other bodies in the solar system. By calculating radiant flux energy at a planet we can begin to have some idea of the environmental conditions on that planet. Table 5.1 shows the distance to the Sun from various bodies in the solar system.

Body	Distance from Sun/10^9 m
Mercury	57.9
Venus	108.2
Earth	149.6
Mars	227.9
Pluto	5900

Table 5.1 Distances to bodies from the Sun

(a) For each body in Table 5.1, calculate the ratio F_{body}/F_{Earth}, where F_{Earth} is the energy flux at Earth and F_{body} in the energy flux at the planet in question. (Hint: use Equation 1. You don't need to know the Sun's luminosity, L, because when you calculate the ratio it cancels.)

(b) How will the radiant energy flux from the Sun affect the temperature on each planet?

(c) What other factors do you think might affect the temperature on the surface of a planet?

3 In the 17th century Christiaan Huygens (Figure 5.5) worked out the approximate distance to the star Sirius by the 'faintness means farness' principle. He made a small hole in the blackened window of a dark room and adjusted the opening until the beam of sunlight appeared to match the brightness of the star as he remembered it. He then measured the diameter of his hole, D_a, and the apparent diameter of the Sun, D_s, as it appeared at the window. He found that the hole had a diameter $D_s/26\,644$. Since the brightness of the beam depends on the area of the light beam, he then reasoned that:

$$\frac{F_{Sir}}{F_{Sun}} = \frac{\text{area of aperture}}{\text{apparent area of Sun}} = \left(\frac{1}{26644}\right)^2$$

(a) Huygens assumed that Sirius and the Sun were equally luminous. Use his assumption and Equation 1 to calculate, as he did, the ratio of distances from the Earth, d_{Sir}/d_{Sun}.

(b) In Question 1 you calculated the distance to the Sun as 1.58×10^{-5} light years. Given that Sirius's true distance is about 8 light years, calculate the real ratio d_{Sir}/d_{Sun}.

(c) Compare your answers from parts (a) and (b). At a time when no one had any idea how far it was to the stars, Huygens's work was brilliant, but not above criticism. Suggest an explanation for his result being so far out, and criticise his experimental technique.

Figure 5.5 Christiaan Huygens

Luminosity, flux and distance

Equation 1 describes how the radiant energy flux from a source depends on the distance to the source and its luminosity. If we know (or assume) the luminosity of a star and measure its energy flux, then we can calculate its distance – or, knowing the distance, we can calculate its luminosity. As you will see in Part 3, it is fairly easy to deduce a star's luminosity from its visual appearance, and so Equation 1 underlies most methods of determining stellar distances.

A very simple way to measure the radiant energy flux, using an oil spot on a piece of paper, was devised in the middle of the 19th century by Robert Bunsen (1811–1899). Ordinary paper becomes extremely translucent when it is made greasy, so it is easily lit from behind. If a piece of paper is held in front of a lamp, any oil spots show up brightly. However, if the paper is also being lit from the observer's side, the ordinary white paper is bright. Bunsen made the assumption that if the oil spot and the paper in such a situation appeared to be of equal brightness, then the radiant energy flux falling on the front and the back of the paper were equal.

In Activity 4 you will use Bunsen's method to find the Sun's luminosity. The essence of the experiment is very simple. You need a clear sunny day to carry it out. A piece of paper, dotted with oil, is held between a lamp and an observer. The lamp illuminates the oil from below while the Sun illuminates the paper from above (see Figure 5.6). The distance between the paper and the lamp is adjusted until the paper and the oil spot appear equally bright.

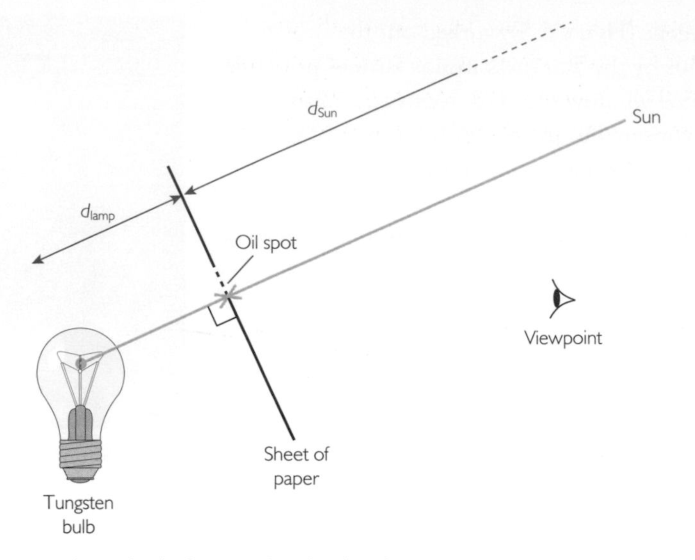

Figure 5.6 Bunsen's method of measuring luminosity

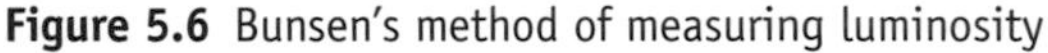

We know that:

$$F_{Sun} = F_{lamp}$$

where F refers to the radiant energy flux on the paper. From Equation 1:

$$\frac{L_{Sun}}{4\pi d^2_{Sun}} = \frac{L_{lamp}}{4\pi d^2_{lamp}}$$

which can be rearranged to find L_{Sun} if the other quantities are known.

Activity 3 The distance to the Sun

In Activity 4, you will need to use a known value of the distance to the Sun in order to measure its luminosity. With a partner, discuss how you think this distance might have been measured.

Activity 4 The Sun's luminosity

Using Bunsen's method, determine the Sun's luminosity. You will need to know the distance to the Sun: $d_{Sun} = 1.50 \times 10^{11}$ m.

Your measurement will be subject to experimental uncertainty, and your calculation of luminosity is based on several assumptions. Try to think what these assumptions are, and suggest how you might correct for them.

Size and distance

It is difficult to imagine the vastness of the Sun. We know that it is much further away than the Moon, but appears to be just about the same size as we observe it from Earth

– in other words it has a very similar **angular diameter**. The Moon is at a distance of about 3.85×10^8 m from Earth. The Sun is approximately 400 times further away than the Moon – but it is much bigger. Figure 5.7 shows how angular diameter is defined; the ratio of actual diameter to distance defines the angle α, measured in radians, that the object subtends at the observer:

$$\alpha = \frac{D}{d} \qquad (2)$$

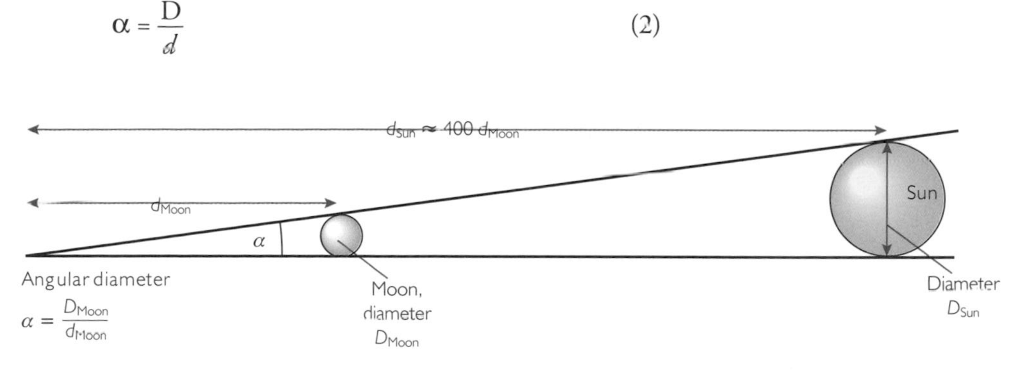

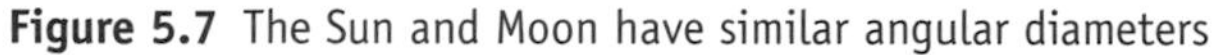

Figure 5.7 The Sun and Moon have similar angular diameters

Angular diameters and angular separations are used quite a lot by astronomers since they can be measured directly, unlike true sizes and separations.

Unlike the Earth, the Sun doesn't have a solid surface. When we talk about the 'surface' of the Sun we usually mean the edge of the ball of light that we see. This is the photosphere – a layer of extremely hot gases that emit most of the Sun's electromagnetic radiation. The photosphere has a diameter of 1.4×10^9 m. The Sun's outer atmosphere, the corona, extends out into space sometimes by more than twice the radius of the photosphere, and the flares that erupt from the surface of the Sun arch in loops bigger than the Earth (Figure 5.8). We can observe the solar corona only during a total eclipse, when light from the photosphere is blocked by the Moon.

Maths reference

Degree and radians
See Maths note 6.1

The small angle approximations
See Maths note 6.6

Angular units
See Maths note 6.7

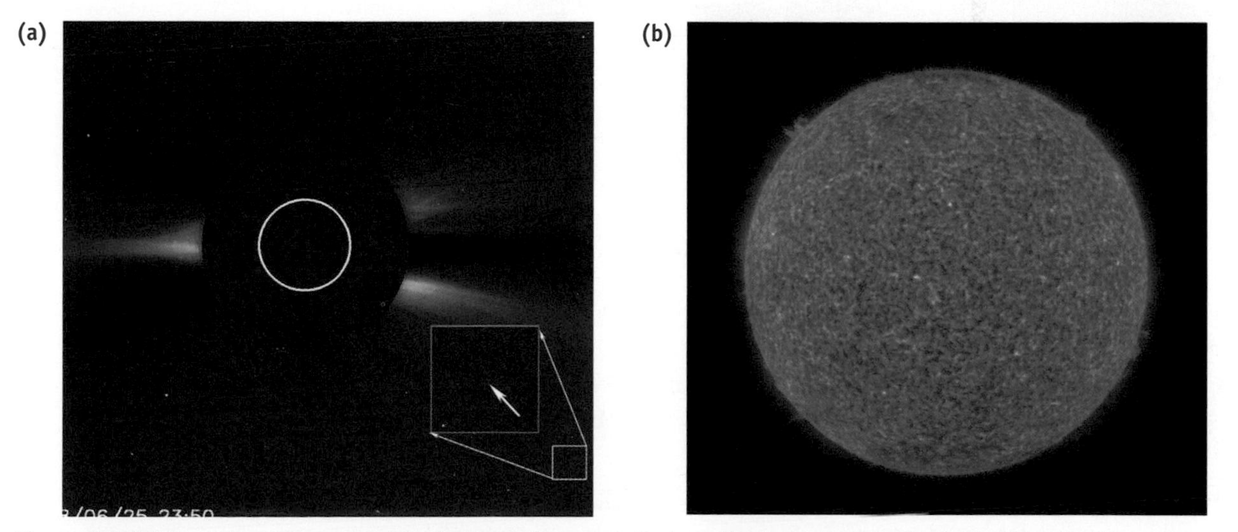

Figure 5.8 (a) The solar corona, showing the inner streamer belt along the Sun's equator; the field of view in this coronagraph encompasses 8.4 million kilometres; (b) a solar flare arches above the photosphere

It is relatively easy to measure the angular diameter of the Sun's photosphere. In Activity 5, you do this using a pinhole camera. It works best with a big camera so that the image of the Sun can be measured precisely.

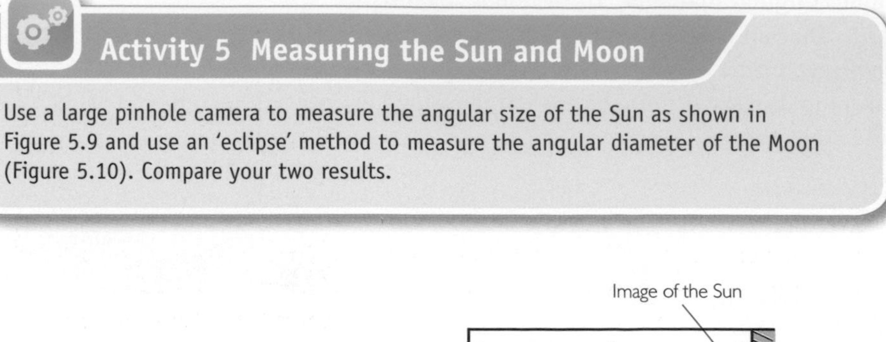

Activity 5 Measuring the Sun and Moon

Use a large pinhole camera to measure the angular size of the Sun as shown in Figure 5.9 and use an 'eclipse' method to measure the angular diameter of the Moon (Figure 5.10). Compare your two results.

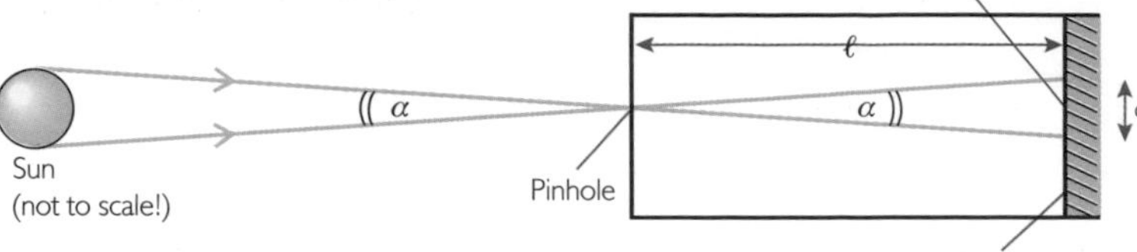

Figure 5.9 Measuring the angular diameter of the Sun

Figure 5.10 Measuring the angular diameter of the Moon

Questions

For these questions, you will need to use information from the previous paragraphs.

4 Calculate the approximate angular diameters, in radians, of the Sun and Moon, as observed from Earth.

5 The Earth has a radius of about 6.4×10^3 km. Calculate the volumes of the Earth and Sun. To the nearest order of magnitude, how many Earth volumes would fit into the Sun?

6 According to the *Windows to the Universe* website (see **www.shaplinks.co.uk**) each square centimetre of solar surface emits as much light as a 6000 W lamp. Check this claim by calculation. (Use $L_{Sun} = 3.84 \times 10^{26}$ W.)

2.2 How old is the solar system?

Figure 5.11 Mayan Sun temple

According to the Bible, and most other creation stories, the first thing that God made was the Sun. Without the Sun there can be no life as we know it. The Sun was worshipped by many civilisations (for example, see Figure 5.11). The Chinese appointed special astronomers to predict the times of eclipses. This meant that when the eclipse occurred the people were ready to make lots of noise and frighten off the dragon they believed was attacking the Sun. The job of forecasting eclipses was felt to be so crucial that when two astronomers failed in their duty, and an eclipse occurred without warning, the astronomers were executed. Feelings ran high about the Sun in Western Europe, too. When Galileo first observed spots on the 'immaculate' Sun the church leaders were so shocked that they questioned Galileo's sanity rather than entertain the possibility that the Sun could be 'blemished'.

Humans have always been interested in where we came from, and where we are going. All civilisations and cultures have stories that explain our place in the history of the world, and the Universe. Because the Sun is so obviously crucial to our life, there has always been a great deal of speculation and thought concerning the age of the Sun (for example, see Figure 5.12). We can't imagine that life began before the Sun shone on the Earth, and we know that if the Sun ceases to shine, life as we know it will come to an end.

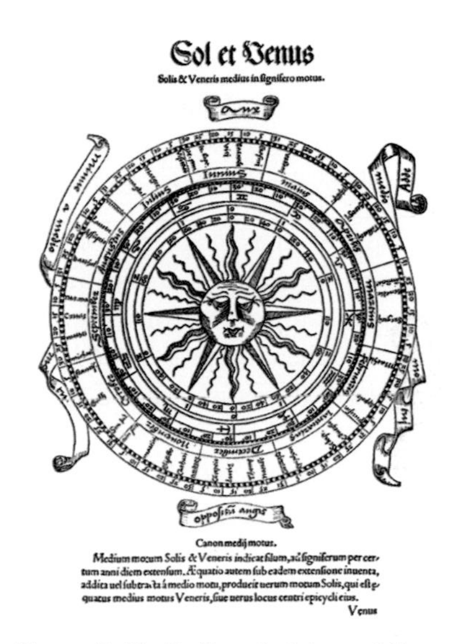

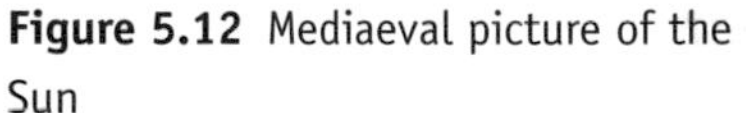

Figure 5.12 Mediaeval picture of the Sun

Activity 6 The Sun in history

Use the internet to find out what your ancestors may have been told about the Sun. The Stanford Solar Center website (see **www.shaplinks.co.uk**) contains a wealth of information about the Sun's folklore of many cultures.

Some 19th century theologians worked out the date of creation by counting back the generations to Adam and Eve (if the Bible is taken literally, Adam and Eve were created in the same week as the Sun). This made the Sun around 10 000 years old. However, during the 19th century more and more scientists became convinced that life on Earth developed from extremely simple organisms by the process of evolution. The idea of evolution is now well established – we can observe changes in populations of insects as environmental conditions change. However, these changes are very small, and they occur slowly over many generations. If we assume that in the past evolution changed species at the same slow pace, we can see that life on Earth must be much more than 10 000 years old. It must have taken millions of years to develop the complex organisms we observe on Earth today if life began as simple replicating molecules. If the Sun had to shine to begin life, the Sun must be older.

Geologists nowadays can determine the relative ages of rocks back to about 600 million years ago. (During this time many shelly fossils first appear in sedimentary rocks.) If the development of life depends on the Sun, then the Sun must be more than 600 million years old.

In 1796 the French scientist Pierre Simon Laplace suggested that the whole solar system formed from one large rotating cloud of gas and dust, now known as the solar nebula. Most of the material contracted to form the Sun, but other fragments condensed ultimately to form the planets (Figure 5.13). Since Laplace proposed his idea, powerful telescopes have allowed us to observe discs of matter surrounding young stars, making the solar nebula theory widely accepted. (Figure 5.2 showed the next stage on – a possible planet.) This gives us another possibility for dating the Sun – if we can find some of the rocks that condensed while the Sun was forming, and if we can date them, then we can date the Sun.

Figure 5.13 Artist's impression of the formation of the Sun and planets

Isotopes

If we want to find out where a rock has come from we might study the minerals, the fossils or the basic elements present in the rock. But we can do better than that: it is possible to date rocks by studying the **isotopes** they contain.

Isotopes of an element are atoms that are chemically identical, but have different masses. They contain exactly the same number of protons in the nucleus of the atom. Neutral isotopes must therefore contain the same number of electrons, so they have the same chemical properties. Isotopes differ only in the number of neutrons in the nucleus of the atom, hence they have very slightly different masses. Only using special equipment, which measures the mass of individual atoms (a mass spectrometer), can we tell isotopes apart. By analysing the atoms in a rock according to mass we can determine the relative proportions of each isotope in the rock. This gives us a clue to the origin of the rock.

When we refer to different isotopes of an element we add the **nucleon number** to the name of the element. The nucleon number is the total number of protons and neutrons in the nucleus of the atom. (Protons and neutrons are collectively known as **nucleons**.) For example, oxygen exists on the Earth as several stable isotopes: oxygen-16, oxygen-17 and oxygen-18. Chemically identical, all these isotopes have eight protons in their nuclei; they each have a **proton number** 8 (the proton number is also known as the **atomic number**). Oxygen-16 or ${}^{16}_{8}O$ as it is written, contains 16 nucleons of which eight are protons; this means that there are $16 - 8 = 8$ neutrons in the nucleus. Similarly, ${}^{17}_{8}O$ contains nine neutrons, and ${}^{18}_{8}O$ contains ten neutrons (see Figure 5.14). Table 5.2 lists the symbols for chemical elements and their proton number. Note that the hydrogen nucleus, ${}^{1}_{1}H$ is simply a proton, and so it is sometimes represented by the symbol p.

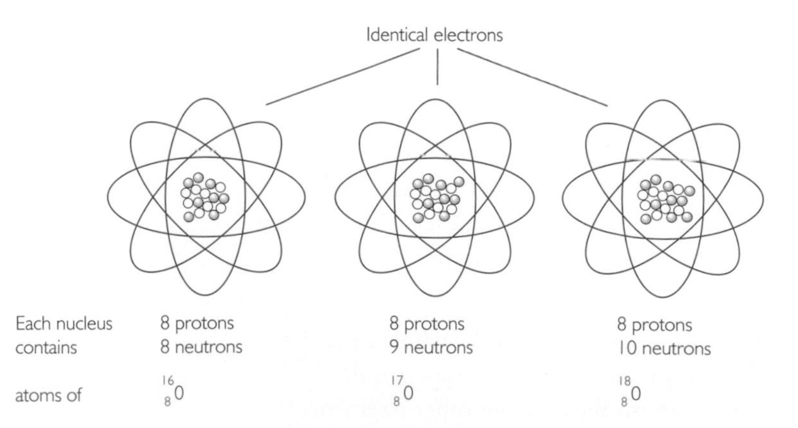

Figure 5.14 Schematic representations of oxygen isotopes

Proton number	Name	Chemical symbol
1	hydrogen	H
2	helium	He
3	lithium	Li
4	beryllium	Be
5	boron	B
6	carbon	C
7	nitrogen	N
8	oxygen	O
9	fluorine	F
10	neon	Ne
11	sodium	Na
12	magnesium	Mg
13	aluminium	Al
14	silicon	Si
15	phosphorus	P
16	sulfur	S
17	chlorine	Cl
18	argon	Ar
19	potassium	K
20	calcium	Ca
21	scandium	Sc
22	titanium	Ti
23	vanadium	V
24	chromium	Cr
25	manganese	Mn
26	iron	Fe
27	cobalt	Co
28	nickel	Ni
29	copper	Cu
30	zinc	Zn
31	gallium	Ga
32	germanium	Ge
33	arsenic	As
34	selenium	Se
35	bromine	Br
36	krypton	Kr
37	rubidium	Rb
38	strontium	Sr
39	yttrium	Y
40	zirconium	Zr
41	niobium	Nb
42	molybdenum	Mo
43	technetium	Tc
44	ruthenium	Ru
45	rhodium	Rh
46	palladium	Pd
47	silver	Ag
48	cadmium	Cd
49	indium	In
50	tin	Sn
51	antimony	Sb
52	tellurium	Te
53	iodine	I
54	xenon	Xe
55	caesium	Cs
56	barium	Ba
57	lanthanum	La
58	cerium	Ce
59	prascodymium	Pr
60	neodymium	Nd
61	promethium	Pm
62	samarium	Sm
63	europium	Eu
64	gadolinium	Gd
65	terbium	Tb
66	dysprosium	Dy
67	holmium	H
68	erbium	Er
69	thulium	Tm
70	ytterbium	Yb
71	lutetium	Lu
72	hafnium	Hf
73	tantalum	Ta
74	tungsten	W
75	rhenium	Re
76	osmium	Os
77	iridium	Ir
78	platinum	Pt
79	gold	Au
80	mercury	Hg
81	thallium	Tl
82	lead	Pb
83	bismuth	Bi
84	polonium	Po
85	astatine	At
86	radon	Rn
87	francium	Fr
88	radium	Ra
89	actinium	Ac
90	thorium	Th
91	protoactinium	Pa
92	uranium	U
93	neptunium	Np
94	plutonium	Pu
95	americium	Am
96	curium	Cm
97	berkelium	Bk
98	californium	Cf
99	einsteinium	Es
100	fermium	Fm
101	mendelevium	Md
102	nobelium	No
103	lawrencium	Lr
104	unnilquadium	Unq
105	unnilpentium	Unp

Table 5.2 Chemical elements, their symbols and their proton numbers

Since masses of atoms and nuclei are sometimes expressed as multiples of the proton mass, and since protons and neutrons have virtually the same mass, the nucleon number is also called the **mass number**. The mass of each nucleon is close to 1 atomic mass unit (1 u) and so, rounded to the nearest whole number of atomic mass units, the mass of a nucleus is equal to its mass number.

Study note

You met the atomic mass unit in the chapter *Probing the Heart of Matter*.

By measuring the oxygen isotope compositions in different rocks we find that rocks, the atmosphere and water on the Earth have oxygen-isotope compositions that tell us that they formed from the same kind of original material.

What about rocks that were not formed on the Earth? Meteorites are simply extraterrestrial lumps of rock that land on Earth. They may be made of rock, metal, or a mixture of both. For dating the solar system the most important meteorites to study are those known as carbonaceous chondrites (Figure 5.15). These meteorites have chemical compositions very similar to that of the Sun, leading us to suspect that they have the composition of the solar nebula. Carbonaceous chondrites are relatively unaltered by processes that have affected other planetary matter. Their oxygen isotope compositions are very different from those measured on materials from Earth. We can thus be certain that they formed in a different part of the solar nebula.

Figure 5.15 Layer from a larger carbonaceous chondrite meteorite seen under a microscope

Radioactive decay

Not all isotopes of an element are stable. Some undergo **radioactive decay,** which is a general name for processes in which nuclei rearrange themselves to become more stable by emitting radiation that has high enough energy to cause ionisation. There are three types of radioactive emission: **alpha α**, **beta β** and **gamma γ**.

Activity 7 α, β and γ radiation

Using your knowledge from earlier work, or using information from textbooks, draw up a table to summarise the properties of α, β and γ emission. Your table should include:

- the nature of the emission
- the sign of any electric charge
- the mass
- the typical speed.

What happens when an isotope undergoes radioactive decay? Take carbon, for example. Much of the material that makes up your body is based on chains of carbon atoms, the vast majority of which are stable, carbon-12, atoms. However, an unstable version of carbon, carbon-14, which is chemically identical to carbon-12, also forms a tiny percentage of the carbon in your body.

Both carbon isotopes contain six protons, but carbon-14 contains two extra neutrons compared with carbon-12. This makes its nucleus unstable. There are too many neutrons for the nucleus to stay together for ever, and it will arrange itself to become more stable. The nucleus converts a neutron into a proton plus an electron, flinging out the electron from the nucleus at high speed, accompanied by an antineutrino. This changes the nucleus into that of another element – one that has one more proton and one fewer neutrons than the original carbon-14 nucleus. This is the element with a proton number 6 + 1, ie it is nitrogen, ${}^{14}_{7}\mathrm{N}$ (Figure 5.16). Using the convention that an electron has a proton number of –1 (ie its charge is equal and opposite to the proton's), we can write this reaction using symbols:

$$ {}^{14}_{6}\mathrm{C} \rightarrow {}^{14}_{7}\mathrm{N} + {}^{0}_{-1}\mathrm{e}^{-} + \bar{\nu}_{\mathrm{e}} $$

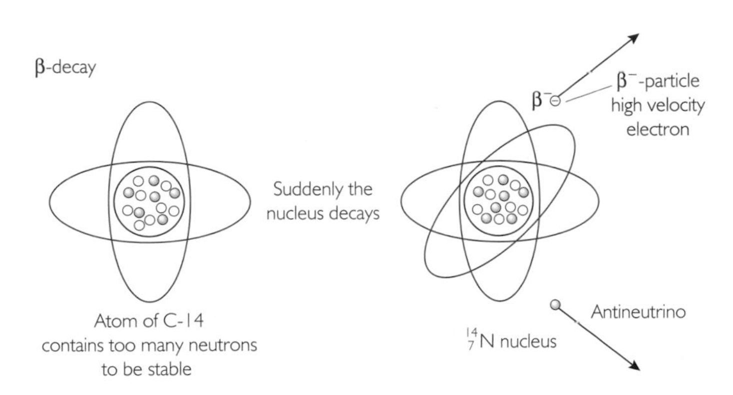

Figure 5.16 Radioactive β^{-} decay of carbon-14

The atom changes from one element to another. For example, when carbon-14 decays it becomes nitrogen, with all the usual chemical properties of nitrogen and none of the properties of carbon. Notice that the total electrical charge (proton number) is unchanged by the reaction (you can check this by adding up the subscripts on the right- and left-hand sides). The electrons that orbit the nucleus of the decaying atoms are disturbed during the decay, then settle back into orbit around the new **daughter nucleus.** The total nucleon number is also unchanged, as shown by the superscripts on the right- and left-hand sides.

Study note

You used charge conservation in the chapter *Probing the Heart of Matter*.

The high-speed electron ejected from the nucleus is known as a beta-minus particle β^{-}, and the whole reaction is one example of the process called **beta-minus decay** (or often simply beta decay). The high-speed electron travels at almost the speed of light, and has enough energy to ionise atoms it encounters, knocking out some of their electrons. The release of energy that accompanies radioactive decay is vast compared with energies involved in chemical reactions.

The decay of carbon-14 is of enormous scientific interest because it allows us to date plant- or animal-derived materials. Living plants absorb carbon-14 from the air. Once they die, the carbon-14 concentration falls as the carbon-14 decays. By measuring the carbon-14 concentrations we can estimate the age of organic matter.

Study note

You might have found out about carbon-14 dating in the AS chapter *Digging Up the Past*.

Beta-minus decay is one type of **radioactive decay**, a general name for processes in which nuclei rearrange themselves to become more stable by emitting a particle that has high enough energy to cause ionisation.

There is also a process known as **beta-plus decay**, which involves the emission of a positron and a neutrino while a proton in the nucleus turns into a neutron. The positron is symbolised ${}^{0}_{+1}e$; perhaps confusingly, it has a proton number of +1 because it has a positive charge, but the nucleon number of 0 shows that it is definitely not a proton – it is not a nucleon. The same result is produced if one of the orbiting electrons is captured by the nucleus and combines with a proton to produce a neutron.

Study note

You met the positron, neutrinos and antineutrinos, and their symbols, in the chapter *Probing the Heart of Matter*.

A carbon-14 nucleus is unstable because it contains too many neutrons. Many more massive nuclei are unstable because they contain too many protons. Such nuclei become more stable by alpha decay in which they emit an alpha particle (α), which consists of two neutrons and two protons (and is identical to a helium nucleus). Uranium-238 is one example of an alpha emitter. The equation for the alpha decay of uranium-238 is:

$${}^{238}_{92}U \rightarrow {}^{234}_{90}Th + {}^{4}_{2}He$$

See Figure 5.17. Again, notice that the total electric charge and nucleon number are each conserved.

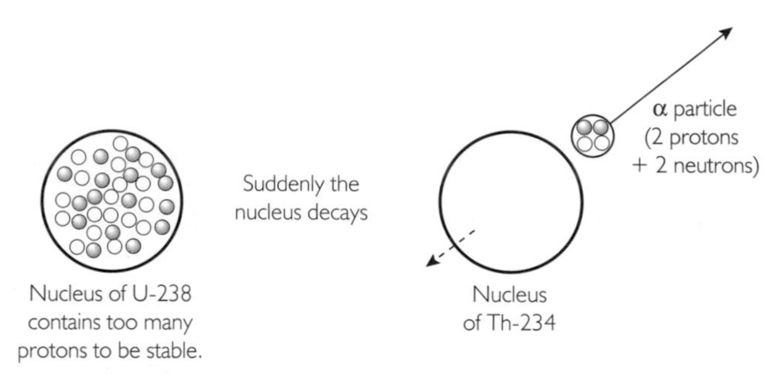

Figure 5.17 Radioactive alpha decay of uranium-238

The third main type of radioactive decay is **gamma decay**, in which a nucleus becomes more stable by emitting a photon of gamma radiation. Gamma radiation is very-high-energy electromagnetic radiation. In common with all electromagnetic radiation it travels at a speed of 3.00×10^{8} m s^{-1} in a vacuum. Gamma radiation has a very short wavelength (about 10^{-14} m or less); each photon carries a large amount of energy and is very penetrating. In gamma decay, there is no change to the proton number or neutron number, just a loss of energy from the nucleus. Gamma emissions usually accompany alpha or beta decay, because these decays often produce a new nucleus which is in an excited state.

Alpha, beta and gamma emissions all cause ionisation of air and other materials through which they pass. This ionisation can be very dangerous to life. If the atoms of a living organism become ionised, electrons are disturbed and chemical bonds are disrupted. Biochemistry, the chemistry of life, depends on precise chemical reactions. Ionisation can cause things to go badly wrong. Large doses of radiation cause death very quickly, but even tiny doses can damage the DNA in replicating cells so that the cells cannot replicate correctly and become cancerous. We are always surrounded by very-low-level **background radiation** from naturally occurring rocks and cosmic rays, but we should avoid any unnecessary extra exposure to ionising radiations. Pay particular attention to the safety instructions when carrying out experiments involving radioactivity.

Activity 8 Investigating α, β and γ radiation

Set up a Geiger-Müller (GM) tube with a data logger or scaler-ratemeter and measure the **background count rate** (the number of ionising event detected per second) in your laboratory.

Investigate the properties of α, β and γ radiation using sealed lab sources. They can all be detected with a suitably adjusted GM tube.

Use paper and aluminium foils of different thickness to investigate how easy it is to stop each type of radiation.

Send a stream of α, β and γ radiation through a strong magnetic field and observe any deflection.

Safety note

Before attempting this activity, make sure that you know the safety precautions you must follow when using radioactive materials.

Study note

You will have studied cloud chamber tracks in the chapter *Probing the Heart of Matter*.

Cloud chambers are designed to show the tracks of ionising radiation (Figure 5.18). Rather like miniature versions of the vapour trails we see in the sky when jets fly high overhead, the tracks form as droplets condense in a supersaturated vapour. Radioactive emissions ionise air and this causes the vapour to condense, and so the vapour trails show the particles' tracks.

Activity 9 Using a cloud chamber

Set up the cloud chamber according to the manufacturer's instructions. Using information from Activity 8, decide which type of radiation will produce the thickest tracks.

Look for tracks from the cloud chamber's own source and tracks due to background radiation.

Watch the tracks appear and answer the following questions.

- Can you predict where or when a track will form?
- Are all the tracks the same length?

Safety note

Before attempting this activity, make sure that you know the safety precautions you must follow when using radioactive materials.

Questions

7 Carbon dating, by measuring carbon-14 concentrations, is a valuable tool in archaeology. Why couldn't it be used to date the Earth or the Sun?

8 The core of the Earth is heated by the energy released by the radioactive decay of naturally occurring isotopes. The most significant of these are uranium-235, uranium-238, thorium-232 and potassium-40. For each of these isotopes write its abbreviated symbol, including the nucleon and proton numbers, and state the number of neutrons in the nucleus.

9 One of the radioactive decay processes commonly used for dating rocks is the decay ${}^{87}_{37}\text{Rb}$ to form ${}^{87}_{38}\text{Sr}$. What type of decay occurs? Write a full, balanced equation to show the decay process.

10 What is the difference between a helium atom and an alpha particle?

Figure 5.18 Tracks in a cloud chamber

Time to go

The minerals we find in rocks originally crystallised when the rock was molten. Like all crystals, minerals crystallise with a fairly precise chemical formula. However, if some of the atoms in the mineral are radioactive isotopes, they will, in time, decay. Chemical analysis of old rocks may reveal elements that could not have been present in the original crystal; they must have formed as daughter products by radioactive decay. By measuring the relative proportions of parent and daughter nuclei, we can find out how much time has passed since the mineral first crystallised. This method of dating rocks depends on our understanding of how rapidly a sample of radioactive material will decay.

Study note

You met exponential changes in the units *Transport on Track* and *The Medium is the Message*.

As you will see in Activity 10, the activity of a radioactive sample changes exponentially with time. If some quantity, N, decreases exponentially with time, t, then the change of N with time is described by equations of the form:

$$\frac{dN}{dt} = -\lambda N \qquad (3)$$

and (equivalently)

$$N = N_0 e^{-\lambda t} \qquad (4)$$

where λ is a constant – a **decay constant** in this case – and N_0 the value of N when $t = 0$. A graph of N against t shows N changing by equal fractions in equal time intervals, and a graph of $\log(N)$ against t is a straight line.

Maths reference

Exponential changes
See Maths note 9.1

Exponential functions
See Maths note 9.2

Exponentials and logs
See Maths note 9.3

Another way of looking at the rate of change of N, dN/dt, is to realise that it is the number of disintegrations per unit time (eg per second or per minute) and is hence also the number of alphas, betas or gammas emitted per unit time, which can quite easily be measured using a GM tube or similar instrument. The number of disintegrations per unit time is the **activity**, A, of a sample:

$$A = \frac{dN}{dt} \qquad (5)$$

The SI unit of activity is the bequerel, Bq. 1 Bq is one disintegration per second.

Since the number of unstable nuclei decays exponentially, so does the activity – with the same decay constant. If you are familiar with calculus, you can show this by differentiating Equation 4:

$$A = \frac{dN}{dt} = -\lambda N_0 e^{-\lambda t}$$

and then using Equation 3 to substitute an expression for A_0; ie the activity when $t = 0$:

$$A = A_0 e^{-\lambda t} \qquad (6)$$

Another way to see this is by analogy with the decay of charge and current when a capacitor discharges through a resistor. As you saw in the chapter *The Medium is the Message*, the stored charge, Q, decays exponentially, so does the current I where

$$I = \frac{dQ}{dt}.$$

The decay of a radioactive isotope is usually described in terms of its **half-life**, $t_{1/2}$ – that is, the time for the number of nuclei of that isotope to halve. Some isotopes are very unstable and have short half-lives (a few seconds or even less), whereas more stable isotopes have much longer half-lives – perhaps millions of years.

The half-life is related to the decay constant. From the definition of half-life, when $N = N_0/2$, $t = t_{1/2}$. Putting these values into Equation 4 we have:

$$\frac{N_0}{2} = N_0 \, e^{-\lambda t_{1/2}}$$

Dividing by N_0 and taking the reciprocal of both sides:

$$2 = e^{\lambda t_{1/2}}$$

Taking natural logs of both sides:

$$\log_e(2) = \lambda t_{1/2}$$

$$t_{1/2} = \frac{\log_e(2)}{\lambda} \qquad (7)$$

Note that $t_{1/2}$ can be expressed in any unit of time; for example, seconds, minutes or years, and the corresponding units for λ are then s^{-1}, min^{-1} or yr^{-1}.

Maths reference

Using log graphs
See Maths note 8.7

Exponential changes
See Maths note 9.1

Exponential functions
See Maths note 9.2

Exponentials and logs
See Maths note 9.3

Activity 10 Measuring half-life

Measure the half-life of a short-lived radioactive isotope.

Measure the background count and subtract it from your measurements.

Plot a suitable graph showing how the activity of the sample changes with time, and hence show that the decay is indeed exponential.

Safety note

Before attempting this activity, make sure that you know the safety precautions you must follow when using radioactive materials.

Why is radioactive decay exponential?

A radioactive nucleus is an unstable nucleus that could decay at any moment. The chance that it will decay at a given moment is governed by the laws of probability. If you toss a coin you know that each time there is a 50% chance of it coming up heads, and that within a few tosses it almost certainly will come up heads. If you roll a die there is a one in six chance of a given number coming up. Atoms of radioactive isotopes behave in rather the same way: in any one second there is a certain probability that the isotopes will decay. Some isotopes are extremely unstable, so there is a high probability that in one second a nucleus will decay. Some isotopes are relatively stable, so any given nucleus is very unlikely to decay in any one second. Activity 11 uses a model to demonstrate how the number of radioactive nuclei varies with time.

Activity 11 Modelling radioactive decay

Use dice, coins or a computer model to simulate the random decay of radioactive nuclei. Plot a graph showing how the number of remaining 'nuclei' changes with time.

In Activity 11 you will have seen that the number of 'nuclei', ΔN, decaying in a given time interval, Δt, is proportional to the number of nuclei, N, remaining in the sample. Expressing this mathematically for a very small time interval (effectively 'at an instant'):

$$\frac{dN}{dt} \propto N$$

This is one of the characteristic equations of exponential decay (Equation 3). The fact that radioactive decay is exponential indicates that the decay of individual nuclei really is governed by probability. This is quite a disturbing (and interesting!) idea. After all, in most situations that you meet in physics, something either definitely happens or definitely does not; on an atomic scale things look rather different. It is also interesting that, even though the decay of each nucleus happens by chance, we can still make very definite predictions and measurements of the behaviour of a large number of nuclei – the half-life of a sample, for example.

Dating

Radiometric dating of rocks (or of archaeological artefacts) essentially involves knowing, or assuming, the amount of an isotope of known half-life initially present in the sample, measuring the amount present now, and hence calculating the time elapsed since the sample was formed. The following examples illustrate the principle, and also show how the units of $t_{1/2}$ and λ are interrelated.

Worked example

Q Uranium-238 ($^{238}_{92}U$) decays by alpha emission to form thorium-234 ($^{234}_{90}Th$). The decay constant for this process is 1.54×10^{-10} yr^{-1}. If you had 1.00 tonne of pure uranium-238:

(a) how long will it take before the activity of the uranium-238 has halved?

(b) how much would you expect to remain 1.00 billion years later (ie after 1.00×10^9 yr)?

(c) how long would it take for the mass of uranium-238 to decrease to 0.80 tonne?

A (a) The time to halve (ie the half-life) does not depend on the initial activity. Using Equation 7:

$$t_{1/2} = \frac{\log_e(2)}{\lambda}$$

$$= \frac{\log_e(2)}{1.54 \times 10^{-10}\ \text{yr}^{-1}} = 4.5 \times 10^9\ \text{yr}$$

(b) Here we need to use Equation 4 with $t = 1.00 \times 10^9$ yr.

$$\lambda t = 1.54 \times 10^{-10}\ \text{yr}^{-1} \times 1.00 \times 10^9\ \text{yr} = 0.154$$

(Note that the quantity λt has no units.)

$e^{-\lambda t} = 0.857$, and so $N = 0.857\ N_0$, and there will be 0.857 tonne of uranium-238 remaining.

(c) Now we need Equation 4 with $N = 0.80N_0$:

$$0.8N_0 = N_0 e^{-\lambda t}$$

$$\text{so } 0.80 = e^{-\lambda t}$$

Taking natural logs of both sides:

$$\log_e(0.80) = -\lambda t$$

$$t = \frac{\log_e(0.80)}{-\lambda} = \frac{\log_e(0.80)}{-1.54 \times 10^{-10}\ \text{yr}^{-1}}$$

$$= 1.45 \times 10^9\ \text{yr}$$

Table 5.3 shows a slightly different way of relating the age of a sample to its half-life. The 'amount' could be either the mass or the number of nuclei.

Number of half-lives elapsed	Amount of parent isotope remaining		
0	100%		
1	50%	1/2	$1/2^1$
2	25%	1/4	$1/2^2$
3	12.5%	1/8	$1/2^3$

Table 5.3 Half-life and age

Question

11 Study Table 5.3 and write a mathematical expression relating the number of half-lives that have elapsed to the amount of parent isotope remaining.

When we are dating rocks we need to study isotopes with relatively long half-lives if they are still to be detectable today. Table 5.4 shows some of the isotopes commonly used for dating rocks by geologists. Note: the daughter products listed for ^{238}U and ^{235}U are not the immediate products, but are the stable products produced by a series of decay reactions. The stated half-lives and decay constants refer to the complete series of decays.

Parent isotope	Daughter isotope	Half-life, $t_{1/2}/10^6$ yr	decay constant, $\lambda/10^{-10}$ yr^{-1}
$^{238}_{92}$U	$^{206}_{82}$Pb		1.552
$^{235}_{92}$U	$^{207}_{82}$Pb	704	
$^{40}_{19}$K	See Q 22		5.810
$^{87}_{37}$Rb	$^{87}_{38}$Sr	48 800	

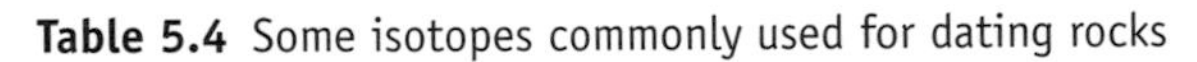

Table 5.4 Some isotopes commonly used for dating rocks

If a sample initially contained none of the daughter isotope, then dating is straightforward. Figure 5.19 shows how the amounts of ^{235}U and ^{207}Pb change with time in a rock that initially contained no ^{207}Pb. If we measure the current proportions of ^{235}U and ^{207}Pb, we can calculate the time since the sample crystallised. The proportions of ^{235}U and ^{207}Pb are equal after one half-life, and as time goes on the proportion of lead increases.

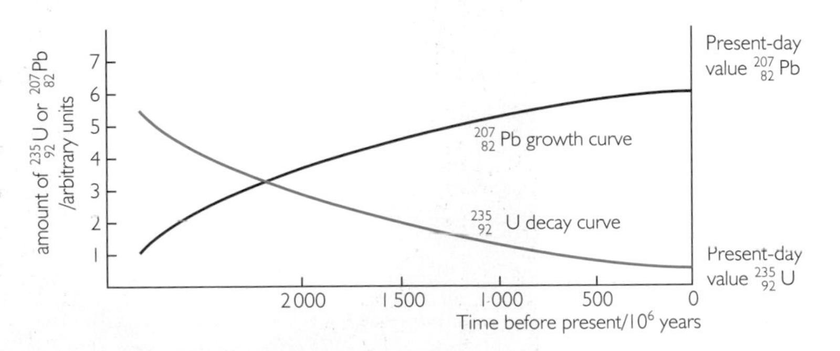

Figure 5.19 Proportions of ^{235}U and ^{207}Pb changing with time

Radiometric dating techniques typically assign ages of over 4 billion years to many Earth rocks. By comparing the ^{87}Rb and ^{87}Sr ratios in zircon crystals (Figure 5.20) found in ancient rocks in Australia, the rocks have been dated as 4200 million years old. Some deep-sea sediments give results that suggest they are even older. Using similar techniques, the oldest Moon rocks and meteorites have been shown to be about 4600 million years old. If we accept the solar-nebula theory of the formation of the solar system, it is fair to give the Sun a similar age.

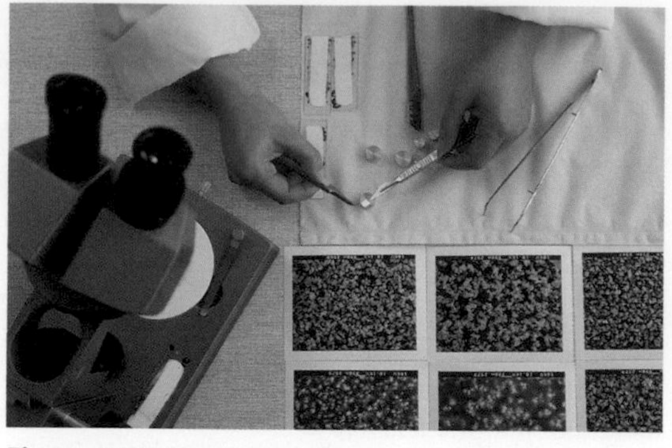

Figure 5.20 Zircon crystals

Questions

In answering these questions, you will need to refer to Table 5.4 and to information in the text.

12 Calculate the missing decay constants and half-lives needed to complete Table 5.4.

13 Using your answer to Question 12, calculate the proportion of ^{87}Rb you would expect to find remaining in a sample that is analysed 24 400 million years after it crystallised.

14 A grain of mineral, which contained uranium, but was assumed to have contained no lead when it crystallised, was found to have 15 times as much $^{207}_{82}$Pb as $^{235}_{92}$U. How old is it?

15 Some recently discovered meteorites are believed to have come from Mars – the pockets of gas trapped in the rock closely match that of the Martian atmosphere. The crystallisation ages of these rocks (the time since they crystallised) are found to be 0.2–1.3 billion years.

(a) If we assume that Mars formed from the solar nebula at the same time as Earth, what crystallisation age would we expect from these rocks?

(b) Mars has many extinct volcanoes, some very high. Suggest a way in which the meteorites got to Earth.

(c) If Mars formed at the same time as Earth, how might we explain the younger crystallisation ages of the Martian meteorites?

2.3 What fuels the Sun?

The active Sun

Until the invention of telescopes, people thought that the Sun was a globe of pure fire and light. Galileo shook the world when he discovered imperfections – dark spots – on the Sun (Figure 5.21). In the 18th and 19th centuries most scientists believed that the Sun had a cool, dark interior surrounded by a burning shell: sunspots were explained as either mountains of cool material poking out of the burning clouds, or holes in the burning clouds.

We now know that sunspots are relatively cool regions of the photosphere, and are related to large-scale upheavals in the photosphere that involve the Sun's magnetic field and prevent hot material reaching the photosphere. The

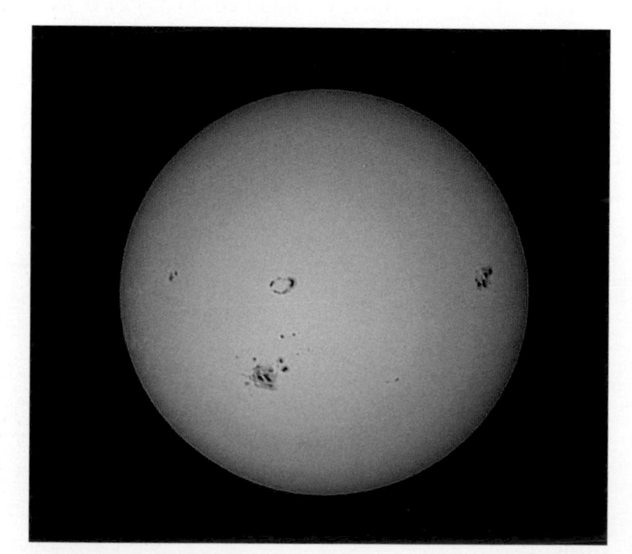

Figure 5.21 Sunspots

amount of this so-called solar activity, and the numbers of sunspots, vary regularly with time and follow an approximately 11-year cycle.

Activity 12 Sunspots

By projecting an image of the Sun on to a screen, study and record the positions of sunspots and hence demonstrate the Sun's rotation.

Use the internet to find images of the Sun produced at solar observatories.

Compare your images of the Sun with images obtained in quiet and active periods of the sunspot cycle. Where are we now in the cycle?

Safety note

Never look directly at the Sun. Sunglasses are not sufficient protection.

In the 19th century the technique of spectroscopy was developed. This allowed a chemical analysis of distant objects by studying their light. It was soon discovered that the Sun emitted the continuous spectrum of an intensely hot object, but that there were dark lines in the spectrum where light had been absorbed by elements in a cooler outer atmosphere beyond the photosphere (Figure 5.22). These are named Fraunhofer lines after the German physicist Joseph von Fraunhofer (1787–1826) who first recorded them.

Study note

You will learn more about the radiation from hot objects in Part 3 of this chapter.

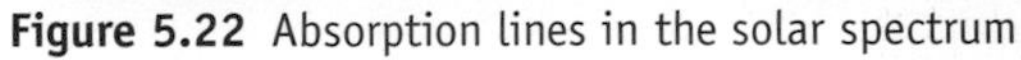

Figure 5.22 Absorption lines in the solar spectrum

By shining continuous spectra through vaporised materials in the laboratory and comparing them with the spectra of the Sun, the German scientist Gustav Kirchhoff (1824–1827) was able to detect many elements in the outer atmosphere of the Sun: sodium, calcium, magnesium and iron; he even discovered new elements that were only later detected on Earth. Kirchhoff worked with Robert Bunsen and Henry Roscoe (see Figure 5.23). Figure 5.24 shows a spectroscopes dating from 1861, similar to those made and used earlier by Bunsen.

Figure 5.23 Kirchhoff, Bunsen and Roscoe

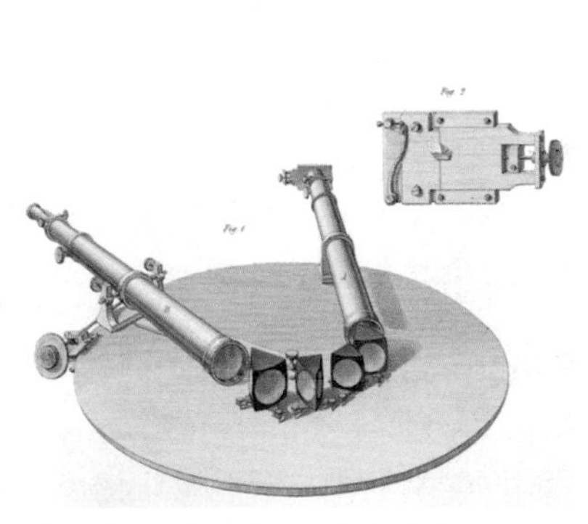

Figure 5.24 A spectroscope from 1861

Activity 13 The Sun's spectrum

Use a spectroscope to observe the solar spectrum in sunlight reflected from a matt white surface or from clouds. Do not point a spectroscope directly at the Sun. Alternatively use a spectrometer fitted with a diffraction grating. You should be able to see the full 'rainbow' of colours, plus the Fraunhofer lines.

Safety note

Be careful not to look directly at the Sun.

Solar power

In December 1838 an English astronomer, Sir John Herschel (1792–1871), was visiting South Africa. At noon, while the Sun was almost directly overhead (12° north), he used the Sun's rays to heat a carefully measured amount of water for 10 minutes. By measuring the rise in temperature of the water he calculated the energy received by the Earth from the Sun, and hence he was able to calculate the total power output of the Sun.

Question

16 We currently believe the luminosity of the Sun to be about 3.84×10^{26} W. In 1838 the unit of power was the horse: 1 horsepower (hp) = 746 W. A steam engine with a 1000 hp engine would have seemed incredibly powerful. What is the approximate luminosity of the Sun in horse power?

It was soon shown that such a vast energy output from the Sun was difficult to explain. Scientists began to wonder where such energy came from and in 1842 Julius Robert Mayer (1848–1878) proposed the law of conservation of energy. Scientists realised that the energy from the Sun must be being transformed from another form of energy. But what kind of fuel could sustain such a power output?

As far as the 19th-century scientists knew, the richest form of energy was coal, so the idea was proposed that the Sun was a huge piece of coal. However, if the Sun consisted of anthracite (a type of coal), a layer of 20 feet in thickness would have to be burned every hour to account for the energy released by the Sun. This would have consumed the whole of the Sun in less than 5000 years, and no one then believed the solar system to be younger than that.

Mayer's idea was that the kinetic energy of meteorites falling into the Sun would provide the energy for the observed output of radiation. But this would have increased the mass of the Sun and changed the orbit of the planets. In 1854 German physicist Hermann von Helmholtz (1821–1894), came up with a more successful theory. He proposed that the Sun contracted and condensed, and as it did so it became hotter – so hot that it could have been radiating at its current rate for 22 million years and still have another 17 million years before it cooled. In the mid-19th century this seemed plausible, but as geologists and palaeontologists became more precise in their estimates of the time for life on Earth (then 100–250 million years) the theory lost credibility.

Nuclear fusion

With the discover of radioactivity and the nuclear atom at the beginning of the 20th century, physicists and astronomers began to realise that the source of the Sun's energy could more realistically be explained by nuclear energy of some kind. In

1919 Rutherford showed that it was possible to change one element into another by bombarding it with very rapidly-moving particles and in a classic book of 1926 Sir Arthur Eddington proposed that the Sun's energy might be accounted for by the combination of hydrogen nuclei to form helium. We now believe that he was correct – the **fusion** of nuclei releases huge amounts of energy, and can fully explain why the Sun is still shining billions of years after it formed.

To understand why nuclear fusion can release energy, it can be helpful to carry out a 'thought experiment'. Imagine assembling a nucleus from its constituent protons and neutrons (which in practice would be extremely difficult to achieve). As these nucleons become very close together, they are attracted to one another by the strong nuclear force, and as they are drawn together and become more tightly bound they lose 'nuclear potential energy', rather like an object falling towards Earth loses gravitational potential energy. Making the lightest nuclei in this way only releases a relatively small amount of energy, but as more massive nuclei are formed, the amount of energy released per nucleon increases, up to the formation of iron and nickel (proton numbers 56 and 57), which are the most tightly bound of all nuclei. Figure 5.25 shows the energy that would be released per nucleon if each nucleus were to be created in this way. (Notice that the energy is measured in MeV per nucleon. 1 MeV = 1.6×10^{-13} J.)

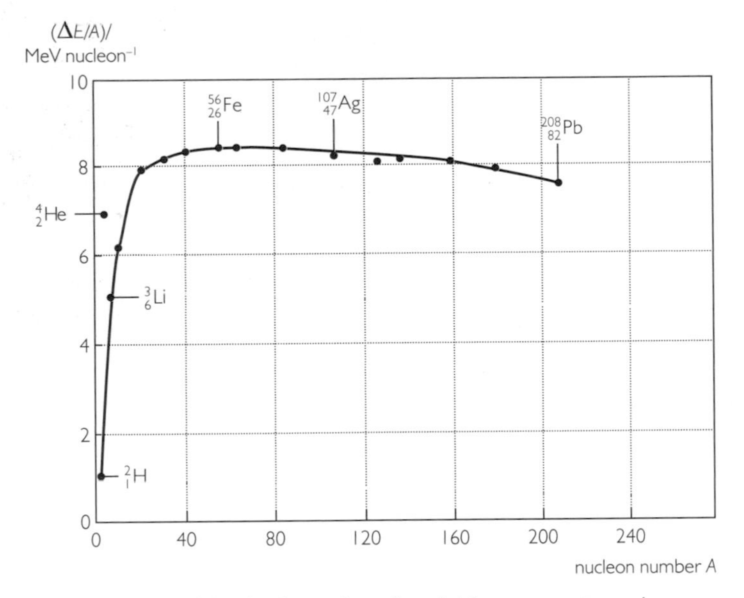

Figure 5.25 Energy released in the formation of nuclei from separate nucleons

The energy released when a nucleus is formed is called the nuclear **binding energy**. Remember that binding nucleons together involves a release of energy. One way to think of the binding energy is that it is the energy that you would need to supply in order to separate a nucleus into its nucleons – perhaps 'unbinding energy' would be a better term.

When two light nuclei react to produce a more massive nucleus, additional energy is released. In the thought experiment, you can picture this as pulling apart each reacting nucleus into its separate nucleons and then reassembling them to make a single, larger nucleus – this involves a net release of energy.

The Sun is composed mostly of hydrogen with some helium. Figure 5.25 shows that if hydrogen nuclei fuse to form helium, there is a release of several MeV per nucleon. The release of energy manifests itself mainly in the emission of high-energy (gamma-ray) photons and neutrinos. While a few MeV might not be a very large amount of energy on its own, remember that this is the energy released per nucleon. In just 1 g of hydrogen there would be about 6×10^{23} nuclei, and so the energy released by the fusion of 1 g of hydrogen would be many billions of joules – far more than the energy involved in any chemical reaction using a similar amount of material.

It would be nice if we could control a nuclear fusion reaction on the Earth to generate electricity, because hydrogen (in water, H_2O) is plentiful, and there would be no polluting waste products. Unfortunately, to get the hydrogen nuclei to fuse requires them to be moving close together at high speeds. This is because of their positive charge; unless they are moving very fast, the electrostatic repulsion ensures that they never become close enough to react. Also, to give the hydrogen nuclei a good chance of colliding with one another, the density, too, needs to be high.

It is very difficult to produce the high temperatures and densities required for fusion on Earth, but the core (centre) of a star (such as the Sun), with the enormous pressure due to the weight of overlying material, is able to reach a temperature of over 10^7 K, high enough to sustain hydrogen fusion. Such high temperatures can be produced in laboratories on Earth (Figure 5.26), but it is difficult to maintain them for very long and there is the additional problem of controlling a very hot hydrogen plasma (ionised gas), since it would vaporise the walls of any solid container.

Figure 5.26 Joint European Torus (JET), where research into nuclear fusion is carried out

The precise details of the fusion reactions that occur in a star depend on its temperature and pressure. It was only as recently as 1938 that the details of the nuclear reactions were worked out. Hans Bethe and Charles Critchfield showed that in the Sun it is a proton–proton chain reaction (known as the ppI chain) that releases energy. The chain has several stages, and the overall effect is:

$$4\,{}^{1}_{1}\text{H} \rightarrow {}^{4}_{2}\text{He} + 2\text{e}^{+} + 2\nu_{\text{e}} + 2\gamma$$

The two positrons annihilate with two electrons to produce more gamma photons:

$$2\text{e}^{+} + 2\text{e}^{-} \rightarrow 4\gamma$$

In stars like the Sun, the core temperature and pressure are just high enough to sustain the reaction, and so it actually proceeds quite slowly. Since its formation about 5 billion years ago, the Sun has converted only about half the hydrogen in its core, and so it will be able to go on shining for about another 5 billion years.

Nuclear fission

Very massive nuclei can also release energy. By breaking up in to smaller nuclei in a process called nuclear **fission**, they become more stable. Fission plays no part in supplying the energy output of stars, but it is worth mentioning for completeness. Some unstable massive nuclei can decay by splitting into two less massive nuclei, usually accompanied by a few neutrons. For example, an isotope of fermium can decay into xenon and palladium, like this:

$${}^{256}_{100}\text{Fm} \rightarrow {}^{140}_{54}\text{Xe} + {}^{112}_{46}\text{Pd} + 4\,{}^{1}_{0}\text{n}$$

As you can see from Figure 5.25, such a reaction involves a release of energy – the two lighter nuclei are more tightly bound than the initial nucleus. Spontaneous fission reactions such as this are rare, and tend to occur only in nuclei (such as $^{256}_{100}$Fm) that are produced artificially rather than occurring in nature. However, as with fusion, the energy released per reacting particle is vast, far outstripping that in chemical reactions, and since the mid 20th century people have sought to exploit this energy in nuclear reactors for generating electricity and in nuclear bombs. In both of these, a commonly used reaction is the fission of uranium-235. If a nucleus of ^{235}U absorbs an extra neutron, it becomes unstable and undergoes fission in reactions such as:

Figure 5.27 Nuclear power station

$$^{235}_{92}\mathrm{U} + ^{1}_{0}\mathrm{n} \rightarrow ^{236}_{92}\mathrm{U} \rightarrow ^{141}_{56}\mathrm{Ba} + ^{92}_{36}\mathrm{Kr} + 3^{1}_{0}\mathrm{n}$$

If the neutrons are absorbed by three further nuclei of ^{235}U, then they too will undergo fission, releasing further neutrons, and so on. Provided the neutrons neither escape from the sample nor get absorbed by other nuclei that cannot undergo fission, then the reaction is self-sustaining. In a nuclear reactor (Figure 5.27), the presence of other materials ensures that there are just enough surviving neutrons to sustain the reaction at a steady rate. But in a nuclear bomb, once the reaction is triggered to start, most of the neutrons go on to produce fission of another ^{235}U nucleus, leading to a runaway reaction – an explosion that causes huge damage not only because of the vast energy released, but also because of the ejection of highly radioactive materials into the surroundings (Figure 5.28).

Figure 5.28 Explosion of a nuclear bomb

Mass and energy

So much energy is released in a nuclear reaction that we can measure the associated loss of mass. Mass and energy are interrelated by Einstein's equation:

$$\Delta E = c^2 \Delta m \qquad (8)$$

where ΔE is the energy released, c is the speed of light in a vacuum (3.00×10^8 m s^{-1}) and Δm is the loss of mass, often called the **mass deficit** or **mass defect** when associated with nuclear reactions. In a spontaneous reaction, energy is always emitted and the total mass of the products is always less than that of the reactants. An example using nuclear fission illustrates this.

Study note

You met the relationship between mass and energy in the chapter *Probing the Heart of Matter*.

Worked example

Q Using data from Table 5.5, calculate the mass deficit when a nucleus of ^{235}U absorbs a neutron and undergoes fission into ^{141}Ba and ^{92}Kr, and hence calculate the energy released.

1u = 1.661×10^{-27} kg

A Initial mass = (1.008 665 + 235.043 93) u
= 236.052 6 u

Final mass = (91.926 25 + 140.914 34 + 3 × 1.008 665) u
= 235.866 6 u

Δm = (236.052 6 − 235.866 6) u
= 0.186 0 u
= $0.1860 \times 1.661 \times 10^{-27}$ kg = 3.090×10^{-28} kg

$\Delta E = c^2\Delta m = (3.00 \times 10^8 \text{ m s}^{-1})^2 \times 3.090 \times 10^{-28} \text{ kg} = 2.78 \times 10^{-11}$ J

Study note

We have calculated the actual values of initial and final mass, but you could reduce the arithmetic by doing the calculation in one stage, and noticing that, for this purpose, one neutron could be removed from each side.

Particle	Mass/u
$^{1}_{0}n$	1.008 665
$^{235}_{92}U$	235.043 93
$^{92}_{36}Kr$	91.926 25
$^{141}_{56}Ba$	140.914 34

Table 5.5 Particles involved in a fission reaction of uranium-235

Study note

We have left the conversion from u to kg until the final stage as the masses are quoted to a greater precision than the conversion factor 1.661×10^{-27} and we do not wish to introduce rounding errors at an intermediate stage.

Questions

17 The Sun has a mass of 1.99×10^{30} kg and is emitting energy at a rate of 3.84×10^{26} W.

(a) How much energy will the Sun emit in 1000 years? (1 yr $\approx 3 \times 10^7$ s)

(b) (i) If this energy were converted back into mass, how much mass would that be?

(ii) What percentage of the Sun's mass is this?

(c) What percentage of its mass will the Sun lose over its lifetime (roughly 10^{10} years)?

18 (a) Write an equation that shows the overall effect of the ppI chain including the electron–positron annihilations. Include sub- and superscripts to show that both charge and nucleon number are conserved.

(b) Use the data in Table 5.6 to calculate the mass deficit associated with the production of one nucleus of $^{4}_{2}He$ in the ppI fusion chain.

(c) How much energy is therefore released when one nucleus of $^{4}_{2}He$ is produced?

(d) How many hydrogen nuclei must be consumed per second in order to maintain the Sun's luminosity of 3.84×10^{26} W?

(e) The Sun's mass is 1.99×10^{30} kg. Given that when it formed from the solar nebula about 75% of its mass was hydrogen, and that about 15% of this hydrogen will eventually be converted to helium, estimate the Sun's lifetime. Express your answer in years. (1 year $\approx 3 \times 10^7$ s.)

Particle	Mass/10^{-27} kg
$^{4}_{2}He$	6.645
$^{1}_{1}H$	1.673
e^+	9.110×10^{-4}
e^-	9.110×10^{-4}
ν	0.000

Table 5.6 Particles involved in the ppI chain

19 Without doing any calculations, sketch a graph showing how the mass per nucleon will vary with atomic number (ie. the actual mass of the nucleus, not its mass number, divided by nucleon number).

2.4 Summing up Part 2

In this part of the chapter, there has been some geometry (angular size) and two main strands of physics:

- the inverse-square relationship between flux and distance for a point source
- nuclear physics – radioactive decay, fusion, fission and binding energy.

Activity 14 Understanding the Sun

Use the information from this part of the chapter to make a timeline showing how our knowledge and understanding of the Sun have developed. If you have time, use library resources and the internet to supplement the information given.

Activity 15 Summing up Part 2

Spend some time looking through your work on this part of the chapter, making sure you understand all the terms highlighted in bold.

Questions

20 The planet Jupiter orbits the Sun at a distance of 7.78×10^{11} m and has a radius of 7.14×10^{7} m. It is composed mainly of hydrogen and helium, but it is believed that its central core has a similar composition to the rocks of the Earth. The outer parts of Jupiter are at a temperature of about 130 K, which is slightly higher than that which is calculated on the basis of its absorption of solar energy alone, suggesting that there is some internal heat source. The Earth is 1.49×10^{11} m from the Sun.

(a) (i) Calculate the maximum and minimum angular diameter, in radians, of Jupiter when observed from the Earth.

(ii) Express the smaller value as a percentage of the larger one.

(b) Calculate the ratio $F_{Jupiter}/F_{Earth}$, where F is the solar energy flux at each planet. Express the ratio as a percentage.

(c) Suggest a possible mechanism by which Jupiter might be heated internally.

21 The radioactive sources used for demonstrations in schools and colleges have long half-lives. Suggest a reason why it is desirable to use such sources for this purpose.

22 To answer this question you will need to refer to Tables 5.2, 5.4 and 5.6.

(a) The isotope $^{40}_{19}K$ is used in dating rocks. It decays to form a stable isotope of argon gas, which becomes trapped in the rock. Write a balanced equation for the decay of ^{40}K to argon, and hence deduce the nucleon number of the daughter isotope and the type of decay that might be involved. (Hint: there are two possibilities.)

(b) A geophysicist finds that a certain rock sample contains 0.5 μg of ^{40}K and 3.3 μg of ^{40}Ar. Assuming that the rock initially contained no argon, and that all the argon produced has remained trapped in the rock, calculate the age of the sample.

(c) Calculate the approximate number of ^{40}K nuclei in the sample and hence calculate the activity due to their decay. Express your answer in suitable units. (1 yr ≈ 3×10^7 s.)

3 Stars

In Part 2 of this chapter you studied the nature of our own Sun, its structure and the way in which it generates the heat and light that keep us in comfort here on Earth. What we have seen is a snapshot view of our local star, which is how it appears to us over a very limited portion of its immense lifespan.

Now we will broaden the discussion of one star at a particular moment of its life to include the formation and evolution of stars in general. You will see how simple physical ideas and calculations can help us to understand the physical and evolutionary processes in stars.

For many centuries, the stars were seen as unchanging points of light, fixed in position for eternity, and astronomers had little concept of their nature. In the early part of the 19th century it was thought unlikely that it would be possible to find anything out about the structure and composition of stars. Within a few years, however, the science of spectroscopy had been developed and the first spectra of stars had been obtained. It turned out that the light from the stars contained a wealth of scientific information that would help astronomers to unravel their hidden secrets.

At the same time, astronomers such as Sir William Herschel, with access to new and powerful telescopes, were beginning to make new detailed maps of the heavens that would eventually result in the understanding of the structure of our galaxy; you will learn more about this in Part 4.

3.1 Studying the stars

How far are the stars?

As the Earth rotates on its axis we see the stars move across the sky, as is demonstrated in Activity 16. Figure 5.29 also reveals the stars' apparent motion.

Figure 5.29 Long-exposure photograph of the night sky

Activity 16 Star watch

On a clear night, find the major constellation known as the Plough. Figure 5.30 shows you how to find Polaris, the Pole Star, using two of the stars in the Plough.

Alternatively, use Stellarium, which is is a free open-source planetarium for your computer (see **www.shaplinks.co.uk**) It shows a realistic sky, just like you see with the naked eye, binoculars or a telescope. If you set your coordinates, you can see the night sky from your own location at any time now or in the future.

Draw a circle on a sheet of A4 paper and put the Pole Star in the centre. Observe the position of the Plough and draw it onto your diagram. Do the same thing every hour for, say, three hours.

Note how the Plough changes its position, relative to the Pole Star and the horizon over the three-hour period. Find the approximate angle through which the Plough rotates in 1 hour and relate this to the length of the day.

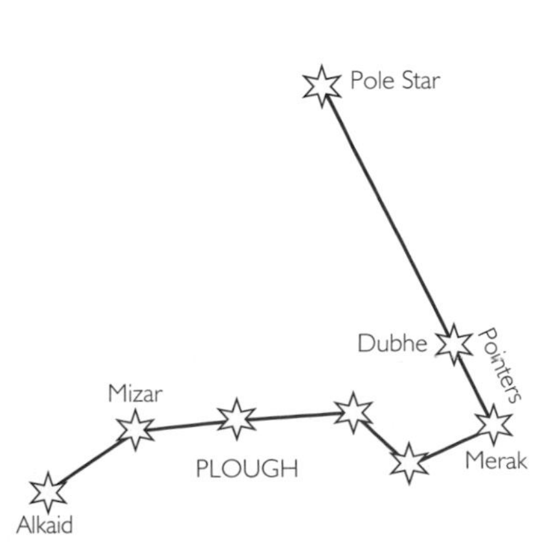

Figure 5.30 Finding Polaris

An important thing to note from Activity 16 is that although the stars appear to move across the sky as the Earth rotates, the relative motion of the stars, that is the motion of one star with respect to another, seems to be zero. Another way of saying this is that the star patterns of the constellations are fixed. One might expect that, as the Earth moves around its orbit, the nearer stars would move relative to the more distant background stars in much the same way nearby objects move more rapidly across the field of view compared to more distant objects, eg as one drives down a motorway. This effect is called **parallax**.

Since there was no observable stellar parallax, some of Newton's contemporaries in the 17th century assumed that stars were infinitely distant and the observation led some astronomers to doubt the **heliocentric** (Sun-centred) model of the Universe that had been proposed by Copernicus in the 16th century. However, as you saw in Part 2, Huygens used the same inverse-square law (Equation 1) to estimate the distance to Sirius; Isaac Newton, too, devised ways to estimate stellar distances based on the same law.

Astronomical instruments became accurate enough to detect the parallax of nearby stars in 1838, when Frederick Bessel detected the parallax motion of 61 Cygni, and F. Struve detected that of the star Vega (this star is mentioned in the film *Contact*). Figure 5.31 shows the idea.

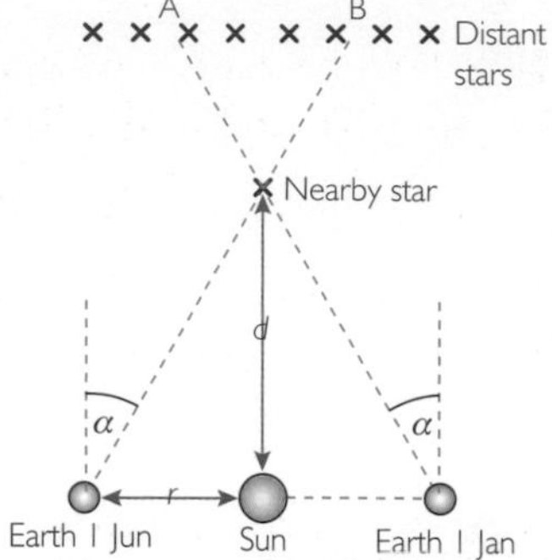

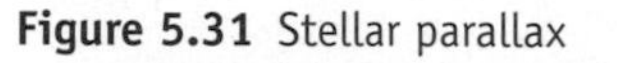

Figure 5.31 Stellar parallax

Study note

The angle α in Equation 9 must be in radians, as in Equation 2.

Study note

An angle of 1″ (1 arcsec) is $\frac{1^\circ}{3600}$.

Maths reference

Angular units
See Maths note 6.7

On 1st January, star X appears in position A against the background stars, but by 1st June, six months later, it appears in position B. The parallax angle, α, is defined as half the total angular motion, and so the star appears to have moved through an angle of 2α in six months (due only to the orbital motion of the Earth). Knowing the distance from Earth to Sun, r, we can calculate the distance, d, to the star:

$$d = \frac{r}{\alpha} \qquad (9)$$

where α, the parallax angle, is in radians. This equation relies on α being small. In practice, for stars, it always is; the nearest star to the Sun is Proxima Centauri, which has a parallax less than 1″.

Parallax measurements give rise to the unit of distance commonly used by professional astronomers, the **parsec**, pc, which is short for *par*allax *sec*ond. One parsec is defined as the distance of an object that would have a parallax of one arcsecond when observed from Earth. $1\,\text{pc} = 3.09 \times 10^{16}\,\text{m} = 3.26$ light years.

Measuring the distances to stars and galaxies may not seem the most interesting of activities for an astronomer, but its importance cannot be overestimated. The main reason for this is that it allows astronomers to determine the scale of the Universe, which in turn has a bearing on its past and future. Another important reason for knowing the distances to stars is that the fundamental physical properties of the stars can be found. This is essential for the study of stellar evolution.

So important is it to obtain accurate stellar parallaxes, the European Space Agency (ESA) designed and launched a special satellite called Hipparcos (High Precision Parallax Collecting Satellite), in August 1989, with the aim of measuring parallaxes to an unprecedented degree of accuracy (Figure 5.32). The name honours the Greek astronomer Hipparchus (190–120 BC), who first measured the parallax of the Moon. In its four years of operation, the satellite measured over 2.5 million stars, allowing astronomers to refine their estimates

Figure 5.32 Hipparcos satellite

of distances to many nearby familiar objects such as the Pleiades star cluster which, as you will see in Part 4, underpin measurements of the most distant objects in the Universe. You can learn more at the Hipparcos website (see **www.shaplinks.co.uk**).

Questions

Data: Earth–Sun distance, $r = 1.49 \times 10^{11}$ m; speed of light, $c = 3.00 \times 10^{8}$ ms^{-1}, 1 year = 3.1536×10^{7} s; 1 rad = 2.06×10^{5} arcsec; 1 parsec = 3.1×10^{16} m.

23 (a) The star Alpha Centauri is 4.351 light years from the solar system. Calculate its parallax in radians and in arcsec.

(b) Use the definition of the parsec to confirm that it has the value quoted above.

24 The smallest parallax Hipparcos could measure was 0.002″. Determine the distance of the furthest star that Hipparcos could measure accurately. Express your answer in metres, in light years and in parsecs.

How hot are the stars?

At first, you might think that measuring the temperature of a star is an impossible task – but it turns out that temperature is one of the easiest quantities to determine using remote observations. Photographs such as Figure 5.33 reveal that stars are coloured and, as you will see in Activity 17, temperature affects both the luminosity and the colour of a radiation source.

Figure 5.33 'True colour' photograph of part of the sky above the South Pole

Activity 17 Colour, luminosity and temperature

Switch on an electric heater or a laboratory radiant heater in a dimly lit room and observe the change in its appearance as it gets hotter. Notice in particular the changes in colour and in brightness.

What is happening? Discuss your observations with other students. There is a transfer of energy to the heating element, so it gets hotter. But why do we see what we do? And why does the element eventually stop getting hotter, even though energy is still being transferred to it at roughly the same rate?

Compare the colour and brightness of the electric heater with some hotter objects: eg a candle flame, a tungsten filament in a light bulb and the Sun. (Remember not to look directly at the Sun.)

Safety note

Never look directly at the Sun. Sunglasses are not sufficient protection.

In Activity 17 you saw that as objects get hotter they go from emitting invisible (infrared) radiation to emitting red, orange, yellow and white light; the hotter the whiter. Also as the object gets hotter the intensity of the radiation increases; the hotter the brighter. You might argue that red objects are not necessarily red hot – an object, such as a pillar box, will appear red if it is illuminated because it reflects red light. But a red-hot object glows red in the dark because of the light that it emits. This radiation given off from a body because of its temperature (which is determined by the thermal motion of its constituent atoms and molecules) is called **black-body radiation**. A body does not have to be black to emit black-body radiation – the term distinguishes bodies that are emitting radiation because they are hot from those that are coloured by dyes and paints. A perfectly black body will glow orange, red or white if you get it hot enough.

To the naked eye, stars mostly appear as faint pinpoints of white light. It is much easier to see their colours using binoculars or using a telescope to gather more light. However, most people can see the differences in colour when they look carefully at some of the brighter stars such as those in Orion (Figure 5.34). Even with the naked eye it is possible to see that Betelgeuse, one of the brightest stars in the constellation, has a red colour, whereas Sirius is blue-white. You can see more colour photographs of stars at the Science Photo Library website (see **www.shaplinks.co.uk**).

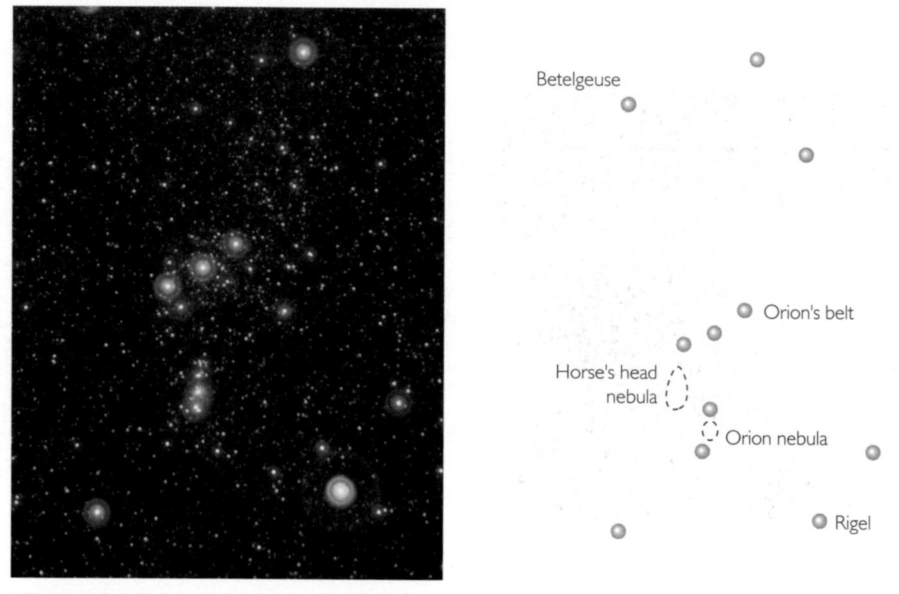

Figure 5.34 Orion: (a) colour photograph, (b) the brightest stars in the constellation

Activity 18 Colours of the stars

On a clear winter's night, find the constellation Orion and identify the stars Betelgeuse and Rigel as shown in Figure 5.34. (You could use Stellarium to help you find the right part of the sky. See Activity 16.) Give your eyes about 10 minutes to adapt to the darkness and then you should be able to see a distinct difference in the colours of these stars. Which of these two stars is the hotter?

When we view stars in the night sky, their light is weak and our eyes do not respond to colour well in these conditions. Some people will inevitably find if difficult to discern the colours of different stars. You will see the colours more clearly if you use binoculars. You could try to take a series of colour photographs of Orion with different exposure times, which should bring out the colour difference.

Question

25 Based on their names alone, put the following types of star in order of increasing temperature: red giant, blue supergiant, white dwarf, brown dwarf.

Black-body radiation

A better name for a **black body** would be an **ideal thermal radiator**. It is an object whose radiation spectrum depends only on its temperature and not on its composition. Such a body is also a 'perfect absorber' – it absorbs all radiation that falls on it, and no radiation is reflected. This second feature makes the term 'black body' seem more sensible. The radiation spectrum of a true black body is a smooth curve, like those shown in Figure 5.35.

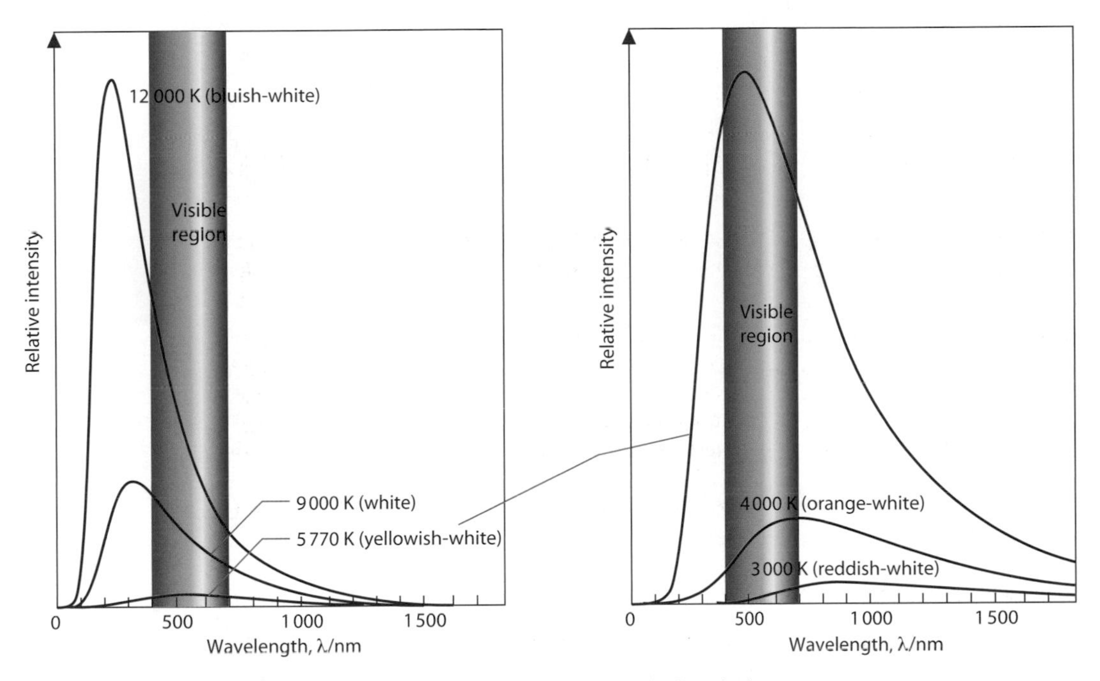

Figure 5.35 Black-body radiation curves (note the different vertical scales)

Figure 5.35 shows how the distribution of radiation (in other words, the spectrum) emitted from a black body varies as the body's temperature rises. The curves all have the same sort of shape but, as you observed in Activity 17, the intensity of the radiation increases dramatically as the temperature rises. Notice, too, that the position of the maximum shifts to shorter wavelengths.

To be a black body, an object must be at the same temperature throughout. In practice, very few objects behave exactly like true black bodies, but some are fairly close. Figure 5.36 shows that the Sun, for example, is not a true black body, particularly at wavelengths less than about 10^{-6} m, but the overall hump-shaped spectrum is quite similar to that of a black body.

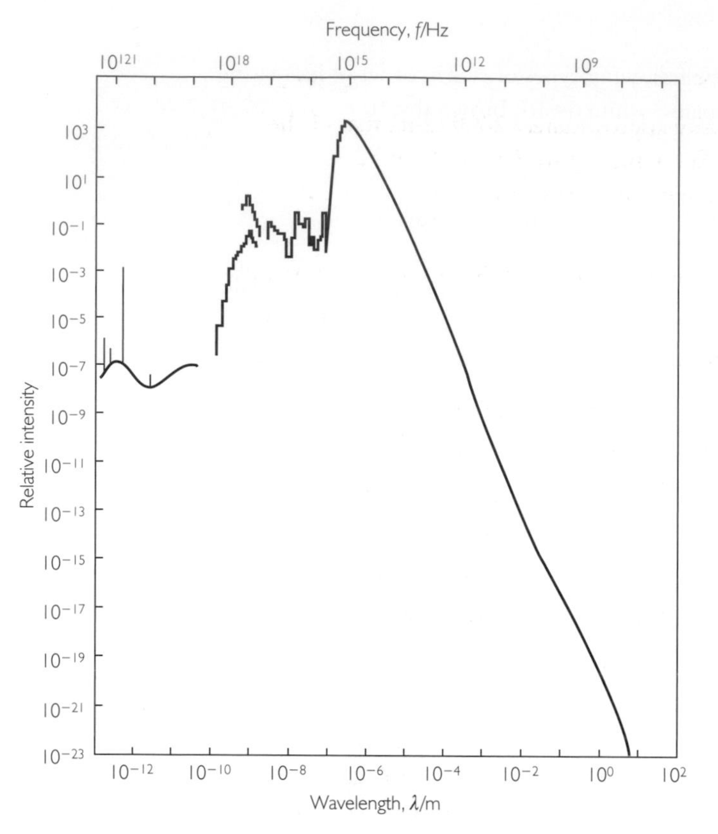

Figure 5.36 Radiation spectrum of the Sun

Study note

Notice that Figure 5.35 uses a linear scale on the wavelength axis whereas the scale in Figure 5.36 is logarithmic.

The radiation from a true black body is described by two laws, which can loosely be summarised as 'the hotter the whiter' and 'the hotter the brighter'.

As the temperature of a black body is raised, the peak of its radiation spectrum shifts to shorter wavelengths. This is described by **Wien's law** (sometimes called 'Wien's displacement law'):

$$\lambda_{max} T = 2.89 \times 10^{-3} \text{ m K} \qquad (10)$$

where T is the temperature on the absolute (Kelvin) scale and λ_{max} is the wavelength at which the brightness is maximum.

Study note

For more on the Kelvin temperature scale, see Section 3.3

Raising the temperature also increases the amount of radiation emitted from the object's surface. The radiation flux emitted from the surface is related to the absolute temperature, T, by **Stefan's law**:

$$F = \sigma T^4 \qquad (11)$$

where the constant, σ, is known as **Stefan's constant** or the **Stetan-Boltzmann constant**.

To find the luminosity of the object, multiply the emitted flux by the object's surface area. For a sphere of radius r, the surface area, A, is:

$$A = 4\pi r^2$$

and so:

$$L = 4\pi r^2 \sigma T^4 \qquad (12)$$

Questions

26 Deduce the SI units of Stefan's constant.

27 (a) The star Spica A has a luminosity approximately 2000 times that of the Sun (ie 8×10^{29} W) and a surface temperature 3×10^4 K. What, approximately, is its radius? Express your answer as a multiple of the Sun's radius ($r_{Sun} \approx 7 \times 10^8$ m).

(b) Repeat your calculations for Barnard's Star, which has $L \approx 4 \times 10^{23}$ W and $T \approx 2.7 \times 10^3$ K.

(Use $\sigma = 5.67 \times 10^{-8}$ W m^{-2} K^{-4}.)

28 Predict the colour of (a) a very hot star that has a surface temperature of 10 000 K; (b) a giant star with a surface temperature of about 2700 K.

29 When the Orion constellation is observed with binoculars, two stars are observed with distinct colours: Betelgeuse is red and Rigel is blue.

(a) Taking the frequency of red light to be 4.3×10^{14} Hz, and that of blue light to be 10×10^{14} Hz, estimate the surface temperature of each star.

(b) Explain why your calculations are only estimates.

Classifying stars

In Activity 17, 'Colour, luminosity and temperature', you saw that the radiation emitted by a black body depends on its temperature – the hotter the body, the brighter and whiter it appears, as summarised in Figure 5.35. The fact that the distribution of radiation over different wavelengths depends on the temperature enables astronomers to measure the surface temperatures of stars just by analysing the radiation they receive.

Astronomers classify stars according to their surface temperature, as shown in Table 5.7. (One interesting point to note about the colour of stars is that they are unlikely to appear to be green. Any star with a green peak wavelength is likely to have sufficient red and blue to appear white.) The spectral type indicates the temperature range, and was originally based on the appearance of absorption lines in stellar spectra. The way in which the spectra of stars can be systematically arranged was mainly developed by an American astronomer, Annie Jump Cannon (1863–1941) (Figure 5.37). She classified the spectra of about 200 000 stars, a monumental task requiring great skill and dedication.

Figure 5.37 Annie Jump Cannon

Surface temperature/K	Colour	Spectral type
>30 000	blue	O
11 000–30 000	bluish-white	B
7500–11 000	bluish-white	A
6000–7500	white	F
5000–6000	yellow-white	G
3500–6000	yellow-orange	K
<3500	red	M

Table 5.7 Spectral classification of stars

We will not go into detail about the spectral type of a star, but when reading other texts you may find it useful to know that when astronomers talk about an O-type star they mean a star with a temperature in excess of 30 000 K. As you will see in Part 4, such stars are much more massive and brighter than our Sun, but live short lives. On the other hand, the M-type stars glow a dull red because their surface temperature is low. The Sun is a type G star with a yellowish-white colour. Astronomers sometimes use a rather ridiculous (but memorable) mnemonic for the sequence of letters: Oh Be A Fine Girl (or Guy), Kiss Me.

How bright are the stars?

As we observe them in the sky, stars have a range of brightness. One reason why some stars appear brighter than others is simply that they are closer to us. Another reason is that some stars are more luminous than others. As you saw in Part 2, the energy flux, F, we receive is proportional to the luminosity, L, of a star and is inversely proportional to the square of its distance, d:

$$F = \frac{L}{4\pi d^2} \qquad \text{(Equation 1)}$$

There are other reasons too, for differences in observed brightness: the radiation from some stars might have been absorbed or scattered more than others en route to us, and the proportion of radiation in the visible part of the electromagnetic spectrum also varies between stars.

Questions

30 Table 5.8 lists the distances of some stars and the flux we receive from them. Assuming that the flux refers to the complete wavelength range, and that no radiation is absorbed or scattered en route, complete the table.

Star	Distance, d/m	Flux, F/W m^{-2}	Luminosity, L/W
Sirius	8.22×10^{16}	1.17×10^{-7}	
Rigel	2.38×10^{18}	5.37×10^{-7}	
Betelgeuse	6.18×10^{18}	1.11×10^{-8}	
Proxima Centauri	4.11×10^{16}	1.09×10^{-11}	
Sun	1.49×10^{11}		3.83×10^{26}

Table 5.8 Star data for Question 30

Of the stars listed in Table 5.8, the one that appears brightest to us is the one that produces the greatest energy flux at the Earth – the Sun, by a long way – and the one that seems faintest is Proxima Centauri. From the answer to Question 30 we can see that the Sun is actually much less luminous than some of the other stars listed, and that although Sirius appears brighter than Betelgeuse in the night sky it actually is much less luminous.

If astronomers wish to compare stars with one another, they can do one of two things. One is to measure their distances and thus calculate their luminosities as you did in Question 30. Alternatively they can study stars that they believe all to be at similar distances from Earth.

In the early years of the 20th century, American astronomer Henry Norris Russell (1877–1957) (Figure 5.38a) took the first approach. He studied stars that were close enough for their parallaxes to be measured, and looked for a relationship between stellar luminosity and colour (ie their surface temperature). Around the same time, Danish astronomer Ejnar Hertzsprung (1873–1967) (Figure 5.38b) looked at stars that were grouped closely together in clusters (such as those in Figure 5.39). He reasoned that stars found clustered closely together must be all at the same distance, hence their fluxes would be a good indication of the relative luminosities, and so he looked for a relationship between colour and flux.

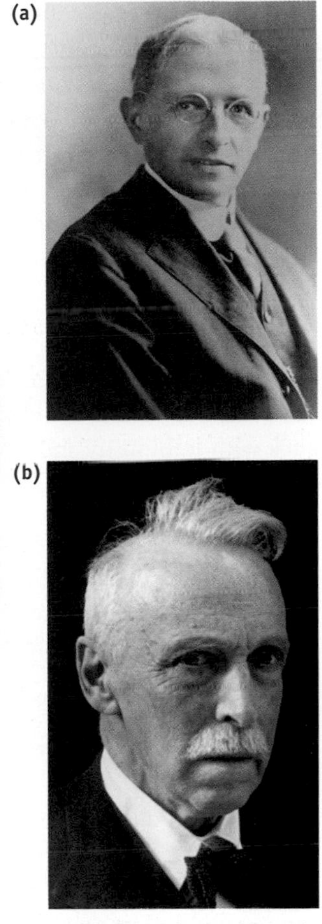

Figure 5.38 (a) Henry Norris Russell, (b) Ejnar Hertzsprung

Figure 5.39 Pleiades star cluster

Both Hertzsprung and Russell constructed graphs that were essentially plots of luminosity (or flux) against temperature for the stars they observed. This type of diagram is called a **Hertzsprung–Russell diagram** (or HR diagram) and was of tremendous importance in helping astronomers to understand stars; it is still very widely used as a way of displaying and analysing stellar data.

Figure 5.40 shows the axes for a Hertzsprung–Russell diagram. Notice that the luminosities are expressed in units of the Sun's luminosity, L_{Sun}, rather than watts, as this is more convenient than dealing with the huge numbers involved in stellar

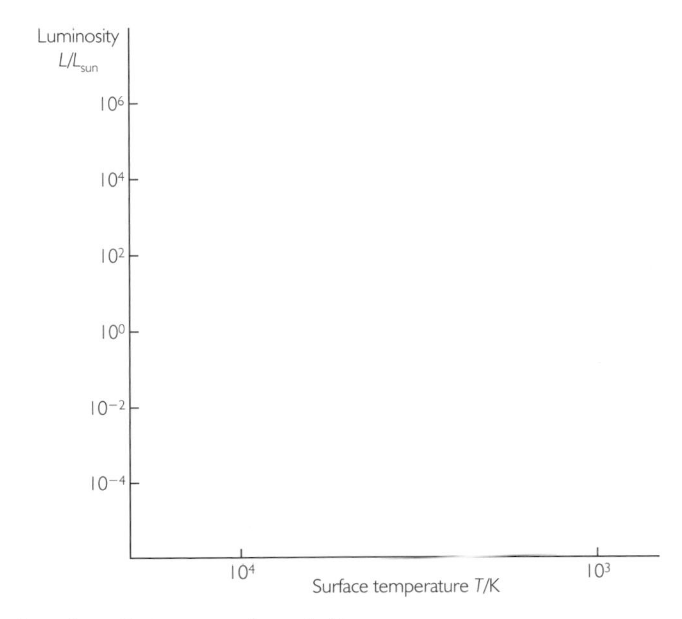

Figure 5.40 Axes for a Hertzsprung–Russell diagram

luminosity. The range of luminosities is large, from about $10^{-4}L_{Sun}$ to about $10^{6}L_{Sun}$. To cope with this range of values, a logarithmic scale is used. Although the range of surface temperatures is not as large it is still convenient to use a logarithmic scale. A very odd thing about the temperature axis (horizontal axis) is that it goes backwards. The high temperatures are plotted on the left of the axis and the low temperatures are on the right. As blue corresponds to short wavelength, it made sense to plot blue at the left; both Hertzsprung and Russell chose to put the blue-white stars on the left, and this convention remains today.

Maths reference

Using log scales
See Maths note 8.7

In other texts, you might come across HR diagrams plotted with other quantities. On the *y*-axis you might find 'absolute magnitude', which is related to the logarithm of flux. On the *x*-axis, you might find 'spectral type', which is, as you know, related to temperature, or 'colour index', which is also related to temperature. However, we will stick with luminosity and temperature.

Activity 19 A Hertzsprung–Russell diagram

Use logarithmic graph paper and a table of stellar luminosities and temperatures to construct a Hertzsprung-Russell diagram for a large number of stars. Include the Sun on your diagram. Comment on any grouping of stars that are apparent from your graph.

From the HR diagram you plotted in Activity 19, three distinct groups of stars can be identified. The main group, containing the greatest number of stars and running roughly diagonally across the diagram from top left to bottom right, is called the **main sequence**. Most of the stars we see in the night sky are in this band. A second group of stars inhabits the top right of the HR diagram. These stars are cool but also very luminous; they must be very large in order to emit such large amounts of radiation from their surface. These are the **red giant** stars. A third group is found at the bottom left of the HR diagram. These are the **white dwarf** stars. Despite being very hot they have low luminosity, so must be small. Because they are so faint, white dwarf stars are hard to find – there are no such stars visible with the naked eye from Earth – and it is likely that the data you used in Activity 19 included few, if any, white dwarfs.

Once the patterns apparent in the HR diagram became known, astronomers found that they had a powerful tool. For example, it enabled them to estimate stellar distances from measurements of flux and temperature, as you will see in Question 31.

In Sections 3.3 and 3.4 we shall see how the HR diagram can help us to understand the evolution of stars and we will see how these three regions of the HR diagram are linked together to give a coherent account of the birth, life and death of a star. But it is not easy to piece together the life history of a star from a 'snapshot' picture of a large number of stars at a particular point in time, as Question 32 indicates.

Questions

31 There are subtle differences in the line spectra of stars that have the same temperature but different luminosities, and so it is possible to distinguish a main-sequence star from, say, a red giant with the same surface temperature.

Suppose an astronomer observes a star at an unknown distance, whose spectrum indicates that it belongs on the main sequence. Outline how the distance to such a star might be measured – say what measurements need to be made, what quantities need to be deduced (and how) and explain how the distance would then be calculated.

32 (a) (i) Describe in words how luminosity varies with temperature for stars on the main sequence.

(ii) Suggest a possible mathematical relationship between L and T for stars on the main sequence.

(b) Is it possible to say anything about the size of stars on the main sequence? For example, how might a hot, highly luminous star at one end of the main sequence compare in size with a star at the other end?

(c) What, if anything, can be said about the masses of the stars on the main sequence?

(d) Does the HR diagram give any clues about the way stars form, or the way they change with age? If so, what?

3.2 Heavenly bodies in motion

In Section 3.1 you saw how astronomers can use the radiation they receive from stars to deduce the stars' temperatures and distances. Now we will turn our attention to another line of enquiry – how do stars and other bodies move, how do they influence one another's motion, and what does that enable us to deduce about their properties?

Observing orbits

In the early 17th century, Danish astronomer Tycho Brahe (1546–1601) and his German pupil Johannes Kepler (1571–1630) made many detailed observations of the planets' positions in the sky (see Figure 5.41). Kepler did a lot of work on analysing the data and looking for mathematical patterns, and was able to formulate three laws of planetary motion that summarised what he found. His first law stated that planets move in elliptical orbits, and the second that the line between a planet and the Sun sweeps out equal areas in equal times. The third law relates orbital radius to orbital period.

Figure 5.41 Tycho's Uraniborg observatory on the island of Hven, now in Sweden

The values in Table 5.9 are modern values of orbital radius and period. (Kepler had to deduce these from measurements of the planets' positions.) It is clear that orbital time is not proportional to the radius. If you assume that there is a power-law relationship:

$$T = Kr^p$$

it is possible to analyse the data using a log–log graph and find values for the constants K and p.

Maths reference

Using log graphs
See Maths note 8.7

Activity 20 Analysing orbits

Use the data from Table 5.9 to plot a graph of log (T) against log (r) and hence deduce the relationship known as Kepler's third law of planetary motion.

Planet	Mean radius of orbit, $r/10^{10}$ m	Planetary orbital period, T/Earth days
Mercury	5.785	87.97
Venus	10.81	224.7
Earth	14.95	365.3
Mars	22.78	687.1
Jupiter	77.76	4333
Saturn	142.58	10 760

Table 5.9 Orbital radius and period

Kepler's laws are empirical. That is, they describe *how* the planets move, based on observation, but they give no underlying explanation of *why* they move as they do. For an explanation, we need to call on Newton's work.

Newton and gravity

Before the discoveries of Isaac Newton (1642–1727), it must have seemed impossible to say anything meaningful about the masses of distant stars or even the Sun. However, Newton made fundamental breakthroughs in the science of gravitational physics that now enable astronomers to use these calculations as a matter of routine.

Arguably, Newton's greatest scientific work was *Philosophia Naturalis Principia Mathematica* or, in English, *The Mathematical Principles of Natural Philosophy* (Figure 5.42). Often referred to simply as the *Principia*, this book is regarded by many as the greatest feat of intellectual endeavour achieved. It was published in 1687 and contains a complete description of the laws of mechanics and gravitation, explaining the orbital motion of the planets and how the tides of the sea are generated – among many other things. These laws, discovered by Newton, are still used every day by scientists some 300 years after their formulation. They enable spacecraft to be launched to distant planets, and provide the means to measure the masses of stars and galaxies.

You probably know the famous tale of the falling apple. A descendant of the apple tree from which it fell can be seen today, in the garden of Woolsthorpe Manor in Lincolnshire, Newton's childhood home

PHILOSOPHIÆ
NATURALIS
PRINCIPIA
MATHEMATICA.

Autore *J S. NEWTON*, *Trin. Coll. Cantab. Soc.* Matheseos Professore *Lucasiano*, & Societatis Regalis Sodali.

IMPRIMATUR.
S. PEPYS, *Reg. Soc.* PRÆSES.
Julii 5. 1686.

LONDINI,
Jussu *Societatis Regiæ* ac Typis *Josephi Streater*. Prostat apud plures Bibliopolas. *Anno* MDCLXXXVII.

Figure 5.42 Newton's *Principia*

(Figure 5.43). The important thing about this story is that it inspired the concept of **universal gravitation**. The word 'universal' means what it says. Newton's great insight was to realise there is an attractive force of gravity between all masses, no matter where they are in the Universe – previously, it had been thought that the motion of objects on Earth was governed by rules quite different from those that applied to 'heavenly bodies'.

Figure 5.43 Woolsthorpe Manor

Newton argued that the force of gravity exerted by a body on another must be proportional to its mass. He also argued that the force exerted by a body of mass m_1 on one with mass m_2 must be equal in size to that exerted by the second on the first. From this it follows that the magnitude, F_{grav}, of the gravitational force must be proportional to the product m_1m_2:

> **Study note**
>
> This is an example of Newton's third law of motion, which you met in the AS chapter *Higher, Faster, Stronger*.

$$F_{grav} \propto m_1m_2$$

By studying the motion of the Moon in its orbit about Earth, Newton was also able to deduce how F_{grav} depends on the separation of the objects. He did this by realising that the gravitational force between the Earth and Moon must be responsible for the centripetal force needed to maintain the Moon's near circular motion. By an ingenious mathematical argument, he was able to show that the Earth–Moon force must diminish with distance according to an inverse-square law; that is:

> **Study note**
>
> You met centripetal force in *Probing the Heart of Matter*.

$$F_{grav} \propto \frac{m_1m_2}{r^2}$$

where r is the distance between the *centres* of the two masses (see Figure 5.44). To make this relationship into an equation, we need a constant of proportionality. This is called the **gravitational constant** and given the symbol G. The full mathematical expression of Newtons' law of universal gravitation is therefore:

$$F_{grav} = \frac{Gm_1m_2}{r^2} \qquad (13)$$

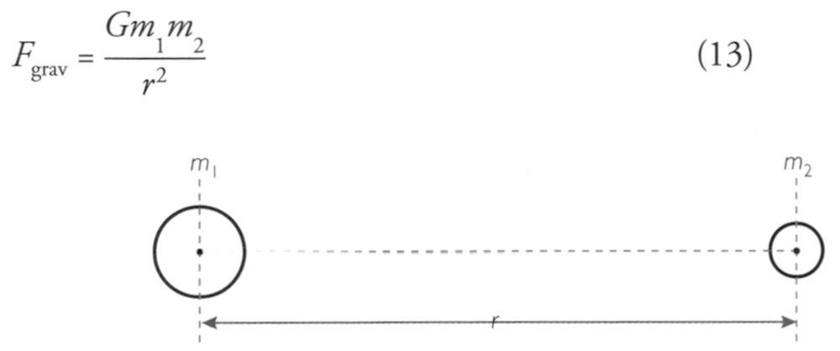

Figure 5.44 Two masses, m_1 and m_2, separated by a distance r

Question

33 Use Equation 13 to deduce the SI units of G.

Measuring G experimentally is difficult, mainly because the gravitational force is very small unless at least one of the masses is very large. In principle it is possible to deduce G from measuring the force between two large bodies such as the Earth and the Moon, but in practice this is difficult since determining the masses of such bodies relies on knowing G in the first place, as you will see shortly. However, it is possible to set up delicate laboratory experiments to measure the forces between quite small masses, and these yield a value of $G = 6.67 \times 10^{-11}$ N m^2 kg^{-2}. In Activity 21 you will get a feel for the sizes of some gravitational forces.

Activity 21 Universal gravity

Table 5.10 lists some objects, their masses and separations. Use a spreadsheet to calculate the magnitude of the gravitational force between each pair, and hence complete Table 5.10.

Objects	m_1/kg	m_2/kg	r/m	F_{grav}/N
Moon in orbit around Earth	7.35×10^{22}	5.98×10^{24}	3.85×10^{8}	
Earth in orbit around Sun	5.98×10^{24}	1.99×10^{30}	1.49×10^{11}	
Moon and Sun	7.35×10^{22}	1.99×10^{30}	1.49×10^{11}	
Person on surface of Earth	60.0	5.98×10^{24}	6.38×10^{6}	
Person on surface of Moon	60.0	7.35×10^{22}	1.74×10^{6}	
Person at Sun's photosphere(!)	60.0	1.99×10^{30}	6.96×10^{8}	
Two people standing together	60.0	60.0	1.00	
Electron in orbit around proton	9.11×10^{-31}	1.67×10^{-27}	5.3×10^{-12}	

Table 5.10 Data for Activity 21; the mass m_1 is that of the first object listed in each pair

Force and field

Gravitational field strength, g, is defined as the gravitational force per unit mass. Close to the Earth's surface, $g = 9.8$ N kg^{-1}. This gravitational force on an object, ie its weight, W, is related to its mass, m, via the gravitational field strength:

$$W = F_{grav} = mg \qquad (14)$$

Electric field is defined as a region in which a charged object experiences an electrostatic force. The direction of an electric field is defined to be that of the force acting on a positive charge, and the strength, E, of an electric field is defined as the force per unit charge. The electrostatic force, F_{elec}, on a charged particle is related to its charge, q, via the electric field strength:

$$F_{elec} = qE \qquad (15)$$

There is a marked similarity between Equations 14 and 15. You have probably also noticed the similarity between the mathematical expression of Newton's law of gravitational and Coulomb's law of electrostatic force:

$$F_{elec} = \frac{kQ_1Q_2}{r^2} \qquad (16)$$

From this relationship we can drive an expression for the field strength at a distance r from a point charge Q:

$$E = \frac{kQ}{r^2} \qquad (17)$$

Study note

You met gravitational field strength in the AS chapter *Higher, Faster, Stronger*.

Study note

You met electric field strength in the chapters *The Medium is the Message* and *Probing the Heart of Matter*.

Study note

Coulomb's law was introduced in the chapter *Probing the Heart of Matter*.

Similarly, we can use the definition of gravitational field strength to derive an expression for the gravitational field due to a point mass. Imagine placing a small mass m at a distance r from the centre of a mass M. The gravitational force acting on m is given by:

$$F_{grav} = \frac{GmM}{r^2} \qquad \text{(Equation 13)}$$

and the gravitational field strength is then:

$$g = \frac{GM}{r^2} \qquad (18)$$

Like force, gravitational field is a vector, and its direction is the same as that of the gravitational force. Just as we represented electric fields by drawing field lines, we can draw lines of gravitational field. Figure 5.45 shows the gravitational field around a planet (or any spherical mass). The field is represented by arrows, each arrow representing the direction and magnitude of the force on a 1 kg mass placed at the blunt end of the arrow. Notice that the field is spherically symmetric, which means that the field strength is dependent only on the radial distance from the mass and not on direction. Also notice that the length of the arrows decreases with increasing distance from the mass. This indicates that the field strength decreases with radial distance, according to the inverse-square law.

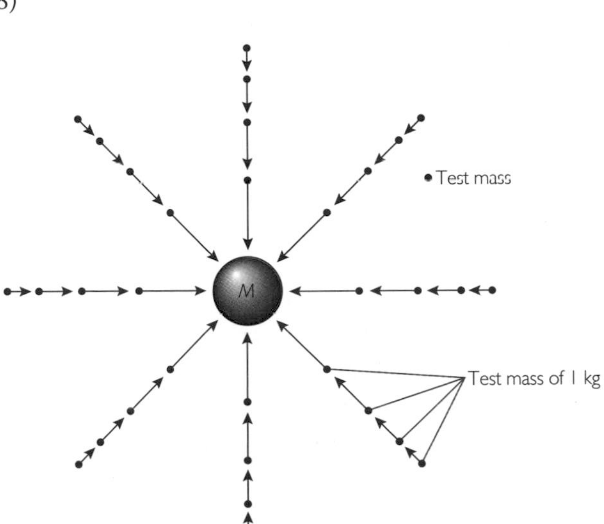

Figure 5.45 Gravitational field lines around a planet

Activity 22 Compare and contrast

As you have seen, there are many similarities between electrical and gravitational forces and fields, but there are also some important differences. Complete a copy of Table 5.11 to summarise key features of these two types of force. You may need to refer back to work from earlier parts of the course, and you may wish to add further rows to your table.

	Gravitational	Electrical
Force acts on ...	all matter	
Direction of force	always attractive	
Force law between two objects formulated by ...		Charles Coulomb 1736–1806
Equation for force law between two objects		$F = \frac{kQ_1Q_2}{r^2}$
SI units of constant in above equation		
Field strength	$g =$	$E = \frac{F}{q}$
SI units of field strength		
Direction of field around isolated object	towards object	
Change in potential energy in uniform field	$\Delta E_{grav} = mg\Delta h$	

Table 5.11 Electrical and gravitational forces

Questions

34 (a) Calculate the gravitational field strength at the surface of Mars which has a mass of 6.42×10^{23} kg and a radius of 3.39×10^{6} m.

(b) Calculate your own weight on the surface of Mars.

35 Assuming that a planet with radius r has uniform density, ρ, show that the gravitational field strength at its surface is given by $g = \frac{4\pi r\rho G}{3}$.

$\left(\text{Hint: the volume, } V\text{, of a sphere radius } r \text{ is } \frac{4\pi r^3}{3}.\right)$

Explaining orbits

Newton's work on gravitation provided an explanation for Kepler's laws in terms of more general ideas. While Kepler had successfully described what happened in the solar system, Newton's law of gravitation is, as we have already stated, universal. This means that it applies anywhere in the Universe and gives the gravitational force between any two masses, be they atoms or stars. The law of gravitation is particularly useful for determining the masses of stars and planets. This method is of great importance in astronomy, since it is the only means by which such masses can be directly measured.

Figure 5.46 shows two objects attracting one another gravitationally and moving in orbit. If the mass of one object is very much greater than the other, it remains essentially at rest while the less massive object orbits around it (Figure 5.46(a)). If the two are comparable in mass, they both move as shown in Figure 5.46(b), where point X (their centre of mass) remains at rest and the two masses orbit so that they are always on opposite sides of X. The two objects might be a planet and one of its moons, a star and a planet, or two stars – in which case they are called a binary pair of stars or, more often, the two together are called a **binary star**. It may surprise you to learn that many stars are either binary pairs or members of multiple-star systems whereas our Sun, as a single star, is in a minority. Imagine what it would be like to live in a binary star system. What would happen to night and day?

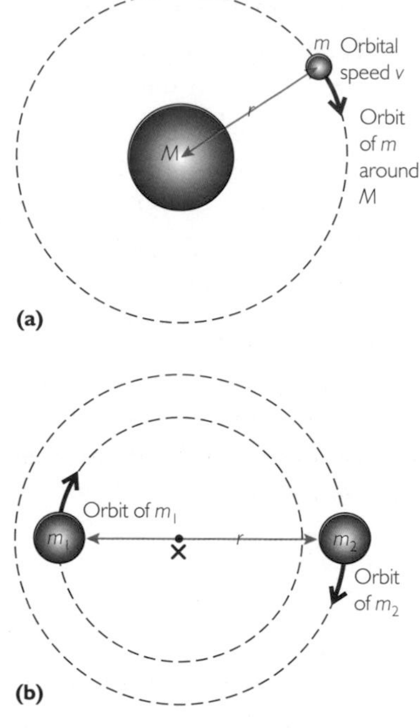

Figure 5.46 Two objects in orbit: (a) $M >> m$, (b) $M \sim m$

To see how mass can be measured, consider the situation in Figure 5.46(a), where the small mass, m, is moving in a circle, radius r, at speed v. There must be a centripetal force, F, acting to keep m moving in a circle, where:

$$F = \frac{mv^2}{r} \quad (19)$$

Study note

You met centripetal force and circular motion in the chapter *Probing the Heart of Matter*.

Here, the force must be provided by the gravitational attraction, F_{grav}, between the two masses:

$$F_{grav} = \frac{GMm}{r^2} \quad \text{(Equation 3)}$$

Identifying the two forces as one and the same:

$$\frac{GMm}{r^2} = \frac{mv^2}{r}$$

Cancelling m and r and rearranging gives:

$$M = \frac{v^2 r}{G} \quad (20)$$

It is often useful to rewrite Equation 20 in terms of the period T (the time for one orbit) where:

$$v = \frac{2\pi r}{T}$$

and so:

$$M = \left(\frac{2\pi r}{T}\right)^2 \times \frac{r}{G} = \frac{4\pi^2 r^3}{T^2 G} \qquad (21)$$

Rearranging Equation 21 to get a relationship between r and T yields:

$$T^2 = \frac{4\pi^2 r^3}{MG} \qquad (21a)$$

Comparing this with your results from Activity 20 shows how Kepler's third law is related to Newton's law of gravitation. Kepler's law is in fact an inevitable consequence of Newton's, but that's not the way in which it was discovered.

Equation 21 tells us that to measure the mass of the central object we need to know the radius of the orbit of its companion and the period of the orbit. It can be used to determine the masses of planets by observing the orbits of their moons, or the Sun's mass from measurements of the orbits of planets. Note that the mass, m, of the orbiting object does not feature in Equation 21 – we do not need to know m, nor can m be found using this method.

Analysis of the situation in Figure 5.46(b) is more complicated, but it yields the same expression where M is now the total mass of the pair of objects ($M = m_1 + m_2$). (This analysis also yields expressions that enable the two masses to be determined separately.) So to measure the mass of a star, it must have an observable gravitational effect on another object, usually another star. The masses of isolated stars must be deduced by other means.

Activity 23 In orbit

Use an applet (see **www.shaplinks.co.uk**) to explore the motion of an object under an inverse-square law force.

By trial and error, find out how to change the shape of the orbit so that it is either a circle or a highly eccentric (elongated) ellipse. Try to produce an orbit that is not closed - that is, the object disappears off the screen and does not return.

Experiment with changing the force law. For example, try $F = -kr^1$ and $F = -kr^3$.

Binary stars

Studying binary stars and measuring the periods and separations is a job for the observational astronomer. It is possible for us to observe some binaries directly and to see both stars in the eyepiece of the telescope, but mostly they are unresolved – that is, they appear as a single star, even with the most powerful telescopes available. So how do astronomers know that they are looking at a binary star?

> **Study note**
>
> You met the idea of resolution of an optical instrument in the AS chapter *Digging Up the Past*.

Unresolved binaries give themselves away in two ways. First, if the orbital plane of the binary system is in our line of sight, then one star will at some stage eclipse the other. When this happens there will be a change in the brightness of the star, just as when the

Moon eclipses the Sun. Figure 5.47 shows the effect. The graph (known as light curve) shows the variation in observed brightness as the small bright star passes behind, and later in front of, the larger star.

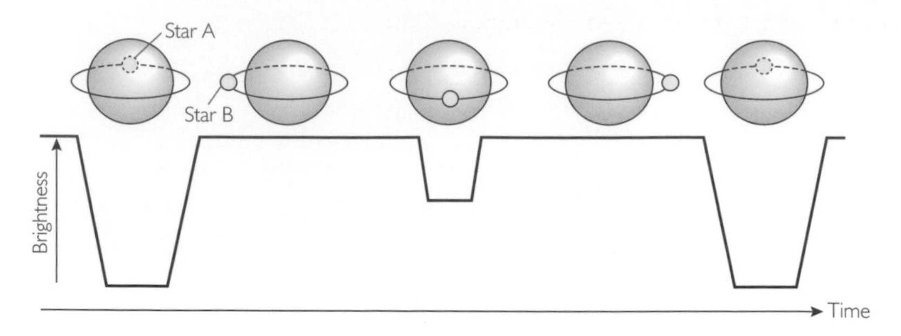

Figure 5.47 Light curve of an eclipsing binary star

The bright star Algol, in the constellation of Perseus, is an eclipsing binary (see Figure 5.48). It is most easily observed in the autumn night sky. For most of the time it is about as bright as Polaris, the pole star, but every $2\frac{1}{2}$ days it fades noticeably over about 4 hours. It stays dim for about 20 minutes, and then takes another four hours to return to its normal brightness. Algol's behaviour was first explained by a young astronomer, John Goodricke of York (1764–1786) (Figure 5.49). A sculpture representing the two stars of Algol stands outside the college named after Goodricke at the University of York (Figure 5.50).

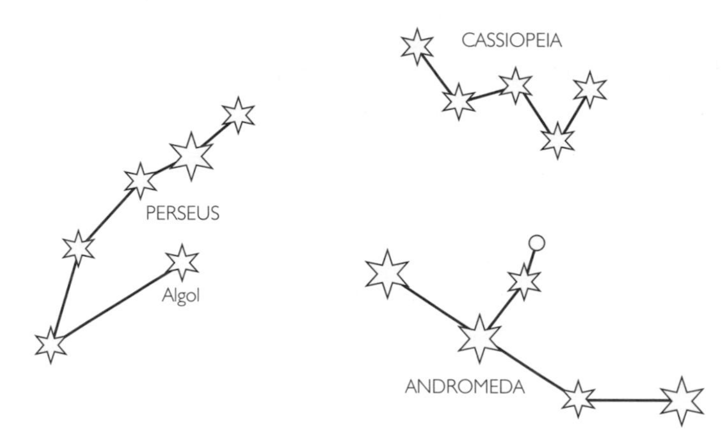

Figure 5.48 Algol in the constellation of Perseus, which is close to Andromeda and Cassiopeia

Figure 5.49 John Goodricke

Figure 5.50 Algol sculpture at Goodricke College

The second way that an astronomer knows that a star is a binary is to observe the spectrum of the star. Consider again the smaller star in Figure 5.47. As it moves towards or away from us, the lines in its observed spectrum will be Doppler shifted. As the star recedes the lines will be shifted to longer wavelengths, and as it approaches they will be shifted to shorter wavelengths. You met the Doppler effect in the AS chapter *Technology in Space*, and used an expression for the shift in frequency:

$$\frac{\Delta f}{f} \approx \frac{u}{v} \qquad (22)$$

where u is the speed of the source or detector, v is the wave speed and $\Delta f = f_{em} - f_{rec}$ where 'em' and 'rec' refer to the emitted and received frequencies. The expression applies only when $u << v$. In the case of binary stars, $v = c$ (the speed of light) and it is still true that $u << c$ (u is the speed of the star along our line of sight). When studying spectra it is more convenient to measure wavelength, λ, than frequency, and the appropriate expression is then:

$$\frac{\Delta\lambda}{\lambda_{em}} \approx \frac{u}{c} \qquad (23)$$

where:

$$\Delta\lambda = \lambda_{rec} - \lambda_{em} \qquad (24)$$

So, using Equation 23, it is quite straightforward to measure the orbital speed. Orbital period, T, can be deduced simply from seeing how the amount of Doppler shift changes with time, and thus all the information is available for determining the separation, r, and hence the mass, M.

Masses of stars

In the middle decades of the 20th century, measurement of large numbers of stellar masses provided astronomers with further clues to the nature and evolution of stars. One important finding was that the masses of main-sequence stars are strongly linked to their temperature and luminosity. Main-sequence stars with the same luminosity and temperature also have the same mass. The most luminous main-sequence stars are also the most massive, with masses up to about 20 times that of the Sun. The faintest main-sequence stars, on the other hand, have masses less than half the Sun's mass.

The measurement of mass provides very good reasons for believing that main-sequence stars do *not* change their luminosity and temperature, since to do so they would need to change their mass by a considerable amount – it would be very difficult to see how they could do this. The main sequence, then, does not represent a star's evolution, even though its name tends to suggest otherwise.

In Sections 3.3 to 3.5 you will see how the 'story' of star formation and evolution has been developed to explain the features of the HR diagram that you saw in Section 3.1.

Questions

36 (a) From the graph in Figure 5.47, decide which of the two stars has the brighter surface and explain how you reached your decision.

(b) Label a copy of Figure 5.47 to show the orbital period.

37 (a) Sketch the light curve of a binary pair of stars, of equal sizes and luminosities, in a circular orbit viewed edge-on. Mark the orbital period on your sketch.

(b) How would the light curve be affected if the orbit were elliptical, as in Figure 5.51?

38 Use the wave equation, $v = f\lambda$, to derive Equation 23 from Equation 22. (Hint: near the end you will need to use the fact that $\lambda_{rec} \approx \lambda_{em}$.)

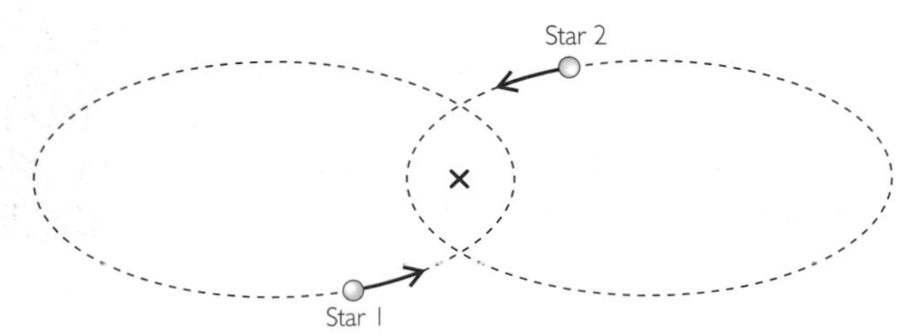

Figure 5.51 Binary pair of stars with elliptical orbits

39 In February 1987, astronomers observed a huge stellar explosion (a supernova) in a nearby galaxy. About six months later, they measured the wavelength of a spectral line of helium as being 1.070 μm, whereas in the laboratory this line has a wavelength 1.083 μm.

Were the gases from the explosion approaching or receding, and how fast were they moving? Express your answer as a fraction of the speed of light, c, and in km s^{-1} ($c = 3.00 \times 10^8$ m s^{-1}).

40 Analysis of the spectrum of an eclipsing, spectroscopic binary of period $T = 8.6$ years shows that the maximum Doppler shift of a hydrogen absorption line with $\lambda_{em} = 656.28$ nm is 0.041 nm.

Assume the system consists of a small star in orbit about a much more massive companion. Calculate the radius of the small star's orbit (assuming it to be circular) and hence determine the mass of the large star. ($c = 3.00 \times 10^8$ m s^{-1}; 1 year $= 3.16 \times 10^7$ s; $G = 6.67 \times 10^{-11}$ N m^2 kg^{-2}.)

3.3 Star formation

We have seen how some of the properties of stars are measured and calculated – we now turn to the evolution of stars: how they are born, live and die. This is a story of real scientific endeavour and achievement. Until the early part of the 20th century, it was thought to be impossible to find out anything about the nature of stars and how they evolved, but since then, astronomers, physicists and mathematicians have made great progress in this area, to the extent that we can now describe the life story of a star with some degree of confidence. In this section, we will follow some aspects of this story, from the gas clouds that are the stellar nurseries, and in Section 3.4 we will continue the story to the final stages of the lives of stars.

What lies between the stars? You might think 'empty space', but that would not be quite true. Within our galaxy, the space between the stars is filled with gas and dust (minute solid particles). By our own everyday standards, this **interstellar material** is very tenuous (ie has very low density), but we can still detect it. Figure 5.52 shows the Orion nebula – visible to the naked eye as a fuzzy patch of light in Orion (see Figure 5.34). This and photographs of other **nebulae** (a general term for 'fuzzy objects' or clouds) show that between the stars there are huge clouds of hot gas, many light years across, that emit their own light, and cool, darker regions that show their presence by blocking out the light from more distant stars. And in between the visible clouds, there is gas that is too cold, or too tenuous, for us to detect with visible light.

Figure 5.52 Orion nebula

The bright stars embedded in the Orion nebula are believed to have formed quite recently, and so the nebula is a site of star formation. But how can a tenuous cloud of gas give rise to the conditions of very high density and temperature needed for the nuclear fusion reactions that power stars? Understanding what is involved requires some knowledge of gravitation (from Section 3.2) and of the behaviour of gases.

In the 17th century, scientists such as Robert Boyle first investigated the physical properties of gases. They did experiments on the relationships between the pressure, volume and temperature of gases. We can repeat their pioneering experiments fairly easily in a modern physics laboratory.

Activity 24 Experiments on gases

Use a gas syringe or the apparatus shown in Figure 5.53 to measure how the volume of a fixed amount of gas changes as the pressure is *slowly* increased. (The pressure must be slowly increased, otherwise the temperature of the gas will also increase.)

Use a spreadsheet to plot pressure against volume. Suggest a simple mathematical form for the relationship between these two variables and use the spreadsheet to test your idea.

If apparatus is available, also investigate the relationships between pressure and temperature at constant volume, and between volume and temperature at constant pressure.

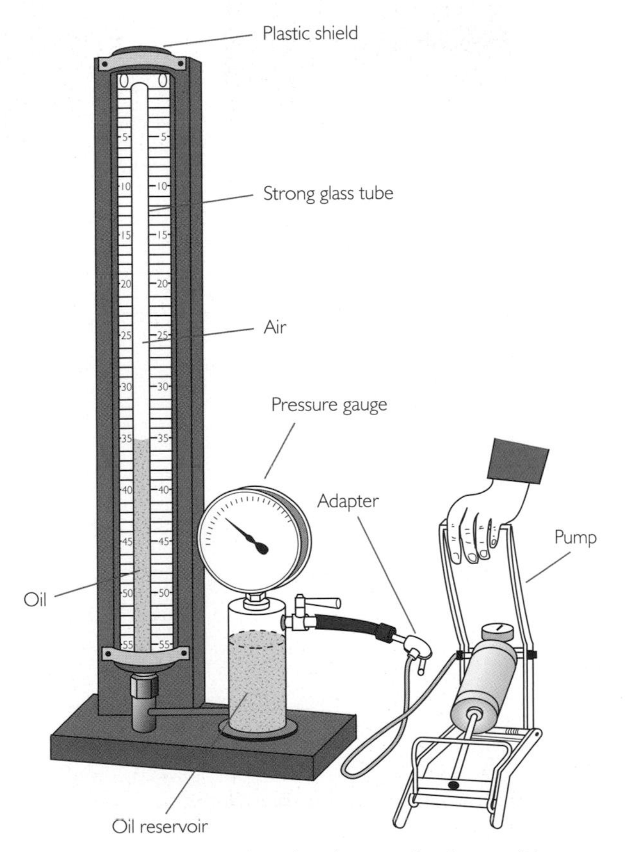

Figure 5.53 Apparatus for demonstrating the change of volume with pressure of a gas

The graphs in Figure 5.54 are based on experiments such as those in Activity 24, and show the relationships between pressure, p, volume, V, and temperature, T, for a fixed mass of gas as each variable in turn is held constant.

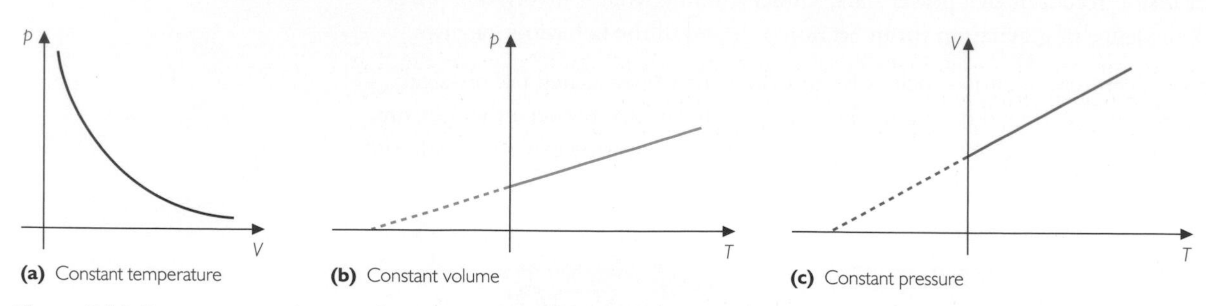

(a) Constant temperature **(b)** Constant volume **(c)** Constant pressure

Figure 5.54 Temperature, volume and pressure graphs for a fixed mass of gas

Temperature scales

Notice that the pressure–temperature graph and the volume–temperature graph are both straight lines. If extrapolated, both these lines should meet the horizontal axis at a temperature of –273°C. You should realise that this could never be achieved in practice, since any real gas will liquefy and stop exerting a pressure well before zero temperature is reached, and it is impossible to imagine a gas with zero volume (what would happen to the molecules?). The extrapolation shows the behaviour of an **ideal gas** – one that does not liquefy, among other things. However, the theoretical idea can be put to good use in the definition of an **absolute temperature scale**. The **absolute zero** of temperature, zero kelvins (0 K – notice no degree symbol), is defined as the temperature at which the pressure of an ideal gas becomes zero. Absolute zero, 0 K, is equivalent to –273°C.

The other fixed point on the absolute temperature scale is the 'triple point of water', which is the unique temperature at which pure water, pure ice and water vapour can coexist; this is 0°C. Thus we have:

$$\text{temperature in K} = \text{temperature in °C} + 273 \qquad (25)$$

We have emphasised the connection between the Celsius and the Kelvin scales as the gas laws use kelvins as a measure of temperature, but everyday life uses degrees Celsius – you will need to convert between the two. Note, though, that the kelvin is the same size as the degree Celsius, so a temperature *change* ΔT has the same numerical value expressed in either scale.

Gas laws

Provided temperature is measured in kelvins, then the results shown in Figure 5.54 for a fixed mass of gas can be expressed as follows:

- The **pressure law**: at constant volume, pressure is directly proportional to temperature:

  $$p \propto T \qquad (26)$$

- **Charles's law**: at constant pressure, volume is directly proportional to temperature:

  $$V \propto T \qquad (27)$$

- **Boyle's law**: at constant temperature (in any units!) pressure is inversely proportional to volume:

$$p \propto \frac{1}{V} \qquad (28)$$

Collectively, these three laws are known as the **gas laws**. These laws are empirical, that is, they are purely experimental — they tell us how ideal gases behave under a variety of conditions, but they do not tell us *why* they behave in this way. The gas laws can be combined in the form of a single equation:

$$\frac{pV}{T} = \text{constant} \qquad (29)$$

This is called the **ideal-gas equation** or the **equation of state for an ideal gas**. An **ideal gas** is defined as one that exactly obeys the gas laws. In reality, ideal gases do not exist – there is no gas that exactly obeys the gas laws, although if a gas is at a high enough temperature and low enough pressure to avoid liquefying then it will approximate to an ideal gas.

Experiment shows that the constant is proportional to the quantity of gas. This might give the impression that the 'constant' is not constant, but remember that the ideal-gas equation applies to a fixed mass of gas and the 'constant' is only constant if the quantity of gas is fixed. If the mass of gas is changed, or the type of gas, then the constant also changes. A complete statement of the ideal-gas equation is:

$$pV = NkT \qquad (30)$$

where N is the number of particles in the gas and k is the **Boltzmann constant**; $k = 1.38 \times 10^{-23}$ J K^{-1}.

Notice that the world 'particle' has been used rather than 'atom' or 'molecule'. In a monatomic material such as copper, sodium or argon, the particle would be an atom, whereas for a substance that exists in molecular form, such as hydrogen (H_2), the particle would be a molecule of hydrogen.

Study note

In chemistry the ideal-gas equation is usually written $pV = nRT$ where n is the number of moles and R the universal gas constant.

Questions

41 Show that pV has SI units of J and that k has SI units of J K^{-1}.

42 The Sun has a mass $M = 2 \times 10^{30}$ kg and a radius $r = 7 \times 10^{8}$ m. It can be modelled as a spherical ball of hydrogen ions, H^+ (each of mass $m = 1.67 \times 10^{-27}$ kg), and an equal number of electrons (whose mass is negligible compared to the ions). If the temperature inside the Sun is $T = 1 \times 10^{7}$ K (ie high enough for fusion) what is the pressure? ($k = 1.38 \times 10^{-23}$ J K^{-1}.)

Assume the temperature and pressure are uniform throughout the Sun.

Kinetic theory of gases

In the 19th century, around 200 years after the gas laws were established, scientists began to develop the **molecular kinetic theory of gases**. In this work, a theory was used to link the particle model of matter to the large-scale measurable properties of matter. Essentially, the ideal-gas equation was described in terms of the motion of invisible molecules. This was remarkable work at the time as many scientists disputed the existence of atoms.

The kinetic theory of gases is one of the best theories produced in the 19th century. It was developed principally by James Clerk Maxwell (Scottish physicist, 1831–1879) (Figure 5.55(a)) and subsequently by Ludwig Boltzmann (Austrian physicist, 1844–1906) (Figure 5.55(b)). Maxwell was a remarkable scientist. As well as the kinetic theory, he produced the theory of electromagnetism, which led to his prediction of electromagnetic waves.

Figure 5.55 (a) James Clerk Maxwell; (b) Ludwig Boltzmann

The motivation behind developing the theory was pure intellectual curiosity rather than any application (though it has come to be applied in many situations, including star formation as you will see later). The theory was remarkable because it set out to link **macroscopic properties** (properties we can measure with instruments) to the invisible **microscopic properties** of matter (randomly moving atoms and molecules) (see Table 5.12). What is also remarkable about the molecular kinetic theory of gases is that it uses nothing more sophisticated than Newton's laws of motion to produce an elegant and testable theory about invisible particles of matter.

Macroscopic	Microscopic
temperature	average kinetic energy of a particle
pressure	forces due to collisions of particle: rate of change of momentum

Table 5.12 Macroscopic and microscopic properties

The elegance referred to is part of the beauty of physics – the final equation is neat. Many scientists have argued that mathematical beauty is necessary for an equation to be correct. Einstein said: "The only physical theories that we are willing to accept are the beautiful ones". Any scientific theory must also be testable. One test of the kinetic theory of gases is that it predicts the average speed of particles in a gas, which can be measured.

The theory is based on the idea that all matter (solids, liquids and gases) is made of discrete particles (molecular or atoms) in a state of random motion (see Figure 5.56). One way to understand the theory is to approach it 'backwards'; that is, start with an experimental estimate of particle speed and see how that leads to a calculation of gas pressure.

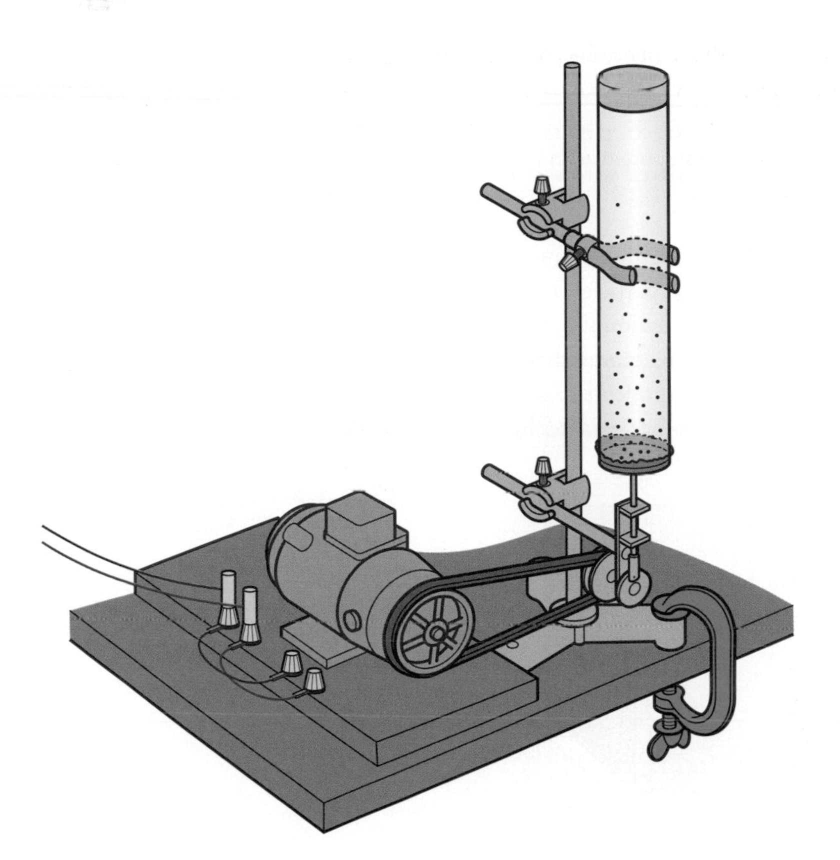

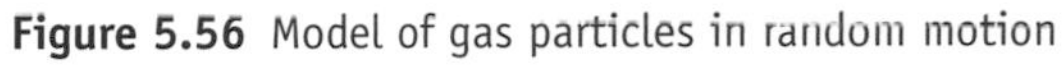

Figure 5.56 Model of gas particles in random motion

Activity 25 Particle speed and gas pressure

Observe the release of bromine vapour into an evacuated tube and hence estimate the speed of the bromine particles.

Then calculate the force exerted by air particles moving at this speed as they bounce against the walls of your lab. Use this force calculation to predict the pressure exerted by a room full of air particles, and compare the result with known atmospheric pressure.

In the calculations for Activity 25, you had to make some assumptions about the air particles and some simplifications. You may not like assumptions and simplifications, but without them, scientists would not be able to make much progress. The final result from the kinetic theory makes predictions that can be tested. If the predictions did not agree with what actually happened, scientists would refine the theory to get a better agreement with experiments.

The assumptions that you made in Activity 25 are the same as those made by Maxwell and Boltzmann, namely that:

- particles in gas collide frequently
- a very large number of particles are present
- the particles have negligible long-range forces between them
- the particles travel in straight lines between collisions and obey Newton's laws of motion

- the volume of the particles themselves is negligible compared with the volume of their container
- collisions between particles in a gas and walls of containers are perfectly elastic
- the time taken for collisions is negligible compared with the time between collisions.

Study note

You met elastic collisions in the chapter *Transport on Track*, only they involved much larger objects.

These assumptions, along with ideas about force and momentum, lead to an equation that relates the pressure of a gas to the motion of its particles. For an algebraic version of the calculation from Activity 25, consider the situation shown in Figure 5.57, which contains a gas of N particles, each of mass m, moving randomly. We will first fix attention on just one particle, moving at velocity $\mathbf{c}$, with components u, v and w in the x, y and z directions, respectively – and imagine that all the other particles have been removed.

Study note

Notice the use of vector notation to describe the particle's motion.

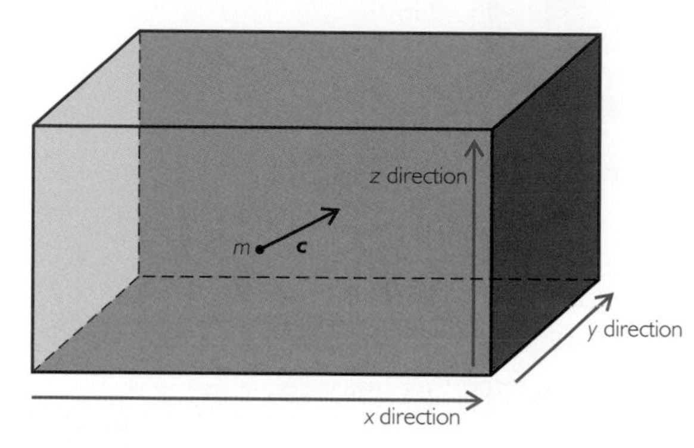

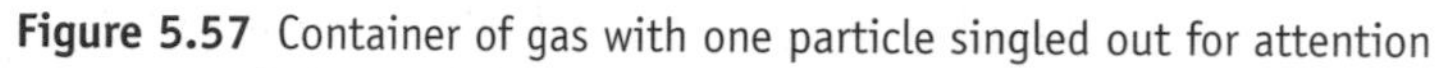

Figure 5.57 Container of gas with one particle singled out for attention

Suppose the particle collides elastically with the shaded face of the container. The x-component, u, of its velocity will be reversed, and the other components unchanged. It therefore undergoes a change of momentum in the x-direction, of magnitude ΔP_X, where:

$$\begin{aligned}\Delta P_x &= \text{final momentum} - \text{initial momentum}\\ &= -mu - mu\\ &= -2mu \qquad (31)\end{aligned}$$

Study note

Unfortunately lower-case letter p is conventionally used for both pressure and momentum. We will use upper case P for momentum here to reduce confusion.

The particle continues to bounce back and forth, colliding repeatedly with the shaded face of the box. Between successive collisions it travels a distance $2x$ (there and back), so the time interval, Δt, between collisions is:

$$\Delta t = \frac{2x}{u} \qquad (32)$$

The repeated bombardment of the face of the box produces a force on it. Since force is equal to the rate of change of momentum, the force exerted by the particle on the shaded face of the container has magnitude:

$$\frac{\Delta P_x}{\Delta t} = \frac{-2mu}{2x/u} = \frac{-mu^2}{x}$$

Then by Newton's third law, the (outward) force on the wall is:

$$F_x = +\frac{mu^2}{x} \qquad (33)$$

We add a suffix 'x' because the force acts in the x-direction.

The pressure exerted on the shaded face is:

$$p_x = \frac{F_x}{yz} \qquad (34)$$

because yz is the area of the face. Hence we have:

$$p_x = \frac{mu^2}{xyz} = \frac{mu^2}{V} \qquad (35)$$

where V is the volume of the container.

Now think of all N particles in the container, each with its own velocity which can be resolved into components in the x, y and z directions. Each particle contributes to the pressure exerted on the shaded face so that the total pressure on that face is:

$$p_x = \frac{m}{V} \times (u_1^2 + u_1^2 + \ldots + u_N^2) \qquad (36)$$

which we can write as:

$$p_x = \frac{Nm\langle u^2 \rangle}{V} \qquad (37a)$$

where $\langle u^2 \rangle$ is the mean value of u^2 for all the particles — see notes below.

However, there is nothing special about the x-direction, so we could equally well have derived similar expressions by considering the other directions:

$$p_y = \frac{Nm\langle v^2 \rangle}{V} \qquad (37b)$$

$$p_x = \frac{Nm\langle w^2 \rangle}{V} \qquad (37c)$$

But the pressure, p, of a gas is the same in all directions, so we can drop the subscripts and write:

$$3p = p_x + p_y + p_z = \frac{Nm}{V} \times (\langle u^2 \rangle + \langle v^2 \rangle + \langle w^2 \rangle) \qquad (38)$$

At this stage we can also note that the velocity components for any one particle are related by:

$$u^2 + v^2 + w^2 = c^2 \qquad (39)$$

and so:

$$(\langle u^2 \rangle + \langle v^2 \rangle + \langle w^2 \rangle) = \langle c^2 \rangle \qquad (40)$$

where $\langle c^2 \rangle$ is the **mean square speed** of the particles (see below). Equation 38 then becomes:

$$pV = \frac{Nm\langle c^2 \rangle}{3} \qquad (41)$$

This equation relates properties that we can measure (p and V) to the motion of the particles, which we cannot see. It is also remarkably similar to the ideal-gas equation:

$$pV = NkT \qquad \text{(Equation 30)}$$

Study note

Alternatively, you can say that the number of collisions per second on the shaded end is $\frac{u}{2x}$, and so the total change of momentum per second is $\frac{u}{2x} \times 2mu$, which gives the same result.

Study note

We use angle brackets ⟨ and ⟩ to denote a mean value. Some texts use a bar over the averaged quantity: $\overline{c^2}$.

which indicates that the kinetic theory can indeed explain the behaviour of gases. Using Equations 30 and 41 together, we can produce many useful relationships that describe the behaviour of gases, as you will see below.

Mean squares and rms speeds

In producing Equations 37 and 41, we used the idea of a mean square speed. This is the mean value of c^2, averaged over all the particles, which is *not* the same thing as finding the mean speed and squaring it. The square root of the mean square speed, called the **root-mean-square speed** (or **rms speed**), $\sqrt{\langle c^2 \rangle}$, is a useful quantity when dealing with particles in a gas, since it is an 'average' speed that is directly related to the pressure exerted by the particles.

To find $\sqrt{\langle c^2 \rangle}$, you need to:

- square all the individual speeds
- sum all the squares
- divide by the total number of particles
- take the square root of the result.

The rms speed is in general not the same as the 'ordinary' mean speed. For example, suppose we have three particles with speed 200 m s^{-1}, five with speed 300 m s^{-1} and one with speed 700 m s^{-1}:

$$\sqrt{\langle c^2 \rangle} = \sqrt{\left\{\frac{(3 \times 2^2 + 5 \times 3^2 + 1 \times 7^2)}{9}\right\}} \times 100 \text{ m s}^{-1}$$

$$\sqrt{\langle c^2 \rangle} = 343 \text{ m s}^{-1}.$$

But the 'ordinary' average speed is 311 m s^{-1}.

Note that the average *velocity* of particles in a gas is zero (there is equal movement in all positive and negative directions) unless the gas as a whole is drifting in a particular direction.

Question

43 At a certain moment, five particles have the following speeds: 300 ms^{-1}, 400 ms^{-1}, 500 ms^{-1}, 500 ms^{-1} and 600 ms^{-1}. Calculate (a) their mean speed (b) their rms speed.

Activity 26 Root mean squares

Use a spreadsheet to explore root-mean-square speed.

Use the RAND() function to generate 100 different speeds. In one of the cells put the formula ' = RAND()*1000'. This will generate 100 different speeds between 0 and 1000 m s^{-1}.

In the next column put 100 randomly chosen integers between 1 and 10. This represents the number of particles with that speed. To do this use the function ' = INT(RAND()*10)'.

Use these values to find the rms speed and the average speed of the molecules in the spreadsheet. Compare the two values.

Results based on the kinetic theory

The kinetic theory and the gas laws together provide a powerful tool in furthering our understanding of gases. By rewriting and combining Equations 30 and 41 in various ways, we can make many deductions about the behaviour of gases and the particles of which they are made. Here you will derive just some of the many useful relationships. The fact that these all produce a self-consistent picture of gases, which ties in with experimental results, provides a good reason to believe that particles in real gases do behave more or less in the ways assumed by the kinetic theory.

Look first at Equation 41. On the right-hand side the product Nm, the number of particles times the mass of a single particle, gives the total mass, M, of the gas in the container. Noting that:

$$\frac{M}{V} = \rho$$

where ρ is the gas density, we can write Equation 41 as:

$$p = \frac{\rho\langle c^2\rangle}{3} \qquad (42)$$

So, we now have an expression that relates the rms speed to properties of the gas itself, regardless of the volume present. Note that Equation 42 does not require a gas of identical particles – it applies equally well to a mixture.

Look again at Equations 30 and 41. Both have pV on the left, so we can eliminate pV and write:

$$\frac{Nm\langle c^2\rangle}{3} = NkT$$

Cancelling N and rearranging gives another expression for the particles' mean-square speed, this time in terms of temperature:

$$\langle c^2\rangle = \frac{3kT}{m} \qquad (43)$$

It is only a short step from here to kinetic energy. The average kinetic energy per particle is:

$$\langle E_\mathrm{k}\rangle = \frac{m\langle c^2\rangle}{2}$$

Using Equation 43 we can therefore write:

$$\langle E_\mathrm{k}\rangle = \frac{3kT}{2} \qquad (44)$$

Notice that the average kinetic energy depends *only* on the temperature, and not on the mass of the particles. This gives us a way to interpret absolute zero in terms of the behaviours of particles in an ideal gas: it is the temperature at which they cease to move.

It is worth thinking a bit more about the speed and kinetic energy of particles at any temperature above 0 K. The particles have a range of kinetic energies and collide frequently with one another. When they collide there is often a transfer of kinetic energy from one to another, but total momentum and total kinetic energy are conserved; ie the collisions are elastic. The way the total energy is shared between the particles depends only on the temperature, and the speeds found in a particular gas depend only on the temperature and the mass of the particles in the gas. At a particular instant some particles will have high kinetic energy and others will only have low kinetic energy. The majority will have kinetic energy close to the average for that particular temperature. As the temperature increases the average kinetic energy will increase, as will the average speed of the particles. Figure 5.58 shows the number of particles plotted on the vertical axis against the speed on the horizontal axis. A few particles have very small speeds and a few have very large speeds. The majority of the particles have intermediate speeds.

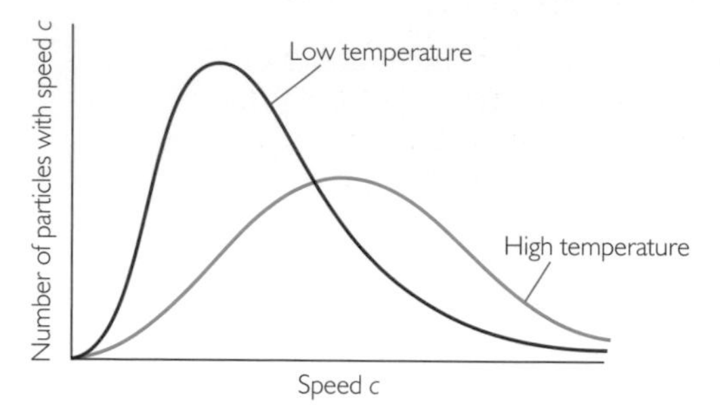

Figure 5.58 Distributions of particle speeds

Developing the kinetic theory

Science does not progress quickly, and it often involves a lot of argument and small fragments being pieced together. When condensed into a few paragraphs of a history book, it can seem that scientists push on remarkably quickly. But the kinetic theory of gases was developed over 200 years after the ideal-gas equation. Many scientists contributed to deriving Equation 41. J. J. Waterson worked out that the pressure of a gas would depend on the square of its molecular speed: doubling the speed would double the momentum change in collisions, but halve the time between collisions, thus producing a factor of four increase in the rate of change of momentum (force). R. Clausius then derived Equation 41. However, Clausius did not have a mathematical grasp of what fraction of molecules would have a particular speed. It was James Clerk Maxwell in 1860 who worked out the distribution of speeds for a gas in thermal equilibrium shown in Figure 5.58.

And don't forget Brownian motion and Einstein!

The first recorded observation of what we now call Brownian motion was made in 1785 by Jan Ingenhauz using charcoal dust. Then in 1827 the biologist Robert Brown published a scientific paper in which he described observing through a microscope the motion of small particles extracted from plant pollen jiggling about in water. He did not know what was causing it. It looked as if the particles had a life of their own, but he worked hard to prove that it was not due to some biological motion, observing many different particles immersed in water. His hard work was rewarded, as from then on the random jiggling motion was referred to as 'Brownian motion'.

In a paper based on his PhD thesis and published on 30 April 1905, Einstein mathematically described the collisions and motion of large particles buffeted by invisible molecules. This was the first of three brilliant papers he wrote in that year. This and the work of other scientists in the same field in the first few years of the 20th century convinced even hardened sceptics that atoms existed. Einstein derived an equation for diffusion, which describes how complex fluids such as milk spread into water, using the idea of atoms. Importantly, Einstein's paper also made predictions about the properties of atoms that could be tested. The French physicist Jean Perrin used Einstein's predictions to work out the diameter of atoms and remove any remaining doubts about their existence.

For showing that Einstein's mathematical prediction was right, Perrin later got a Nobel Prize. Einstein got his Nobel Prize in 1921, largely for his 1905 paper on the photoelectric effect; but he became most famous for his paper on relativity, most notable for $E = mc^2$; however, his most cited paper from 1905 is the one concerning Brownian motion. Research scientists have often referred to aspects of the paper in their own research. The paper has spawned a large number of related PhD theses over the last century. There is even an annual Brownian motion conference where scientists share research work related to this jiggling motion. The speed of the buffeting motion is important for working out the rate of chemical reactions; it is used by scientists working on such varied topics as aerosol particles in pollution research and the properties of paints.

Activity 27 A good theory

Discuss why some scientists might regard the molecular kinetic theory of matter as being an elegant and strong theory. What makes it so good?

And finally...

Returning at last to our astronomical starting point, we can use the results of the kinetic theory to gain some insight into the low-density interstellar material such as the Horsehead nebula (Figure 5.59) where we might look for star formation. This is the subject of Question 47.

Figure 5.59 Horsehead nebula

Questions

44 At atmospheric pressure (1×10^5 Pa) and room temperature, the density of air is about 1 kg m^{-3}. What is the approximate rms speed of the air molecules?

45 The mass of an oxygen molecule is 16 times that of a hydrogen molecule. At a given temperature, which molecules would have the greater rms speed, and by what factor?

46 Use Equation 43 to find the rms speed of hydrogen ions ($m = 1.67 \times 10^{-27}$ kg) at a temperature of 1×10^7 K.

47 Table 5.13 contains some data on various types of interstellar gas. (You can see pictures of the examples mentioned in many websites, astronomy books and CD-ROMs.)

(a) Find the rms speed of:

(i) hydrogen molecules ($m = 2 \times 1.67 \times 10^{-27}$ kg) in a typical molecular cloud

(ii) hydrogen ions in a supernova remnant which has $T = 10^6$ K. (Hint: there is a short cut to (ii) using your answer to Question 46.)

(b) The number density of particles, n, is the number per unit volume, equal to N/V.

(i) Use this information and the Equation 30 to calculate the values missing from the 'pressure' column of Table 5.13.

(ii) Comment on the range of values in this column.

Type of region	Typical diameter, d/light years	Typical number density of particles, n/m^{-3}	Typical temperature T/K	Composition	Typical pressure, $p/10^{-13}$Pa	Example(s)
Molecular clouds (also called dense clouds)	0.3–60	10^{10}	30	H_2, He, small and large molecules	40	Horsehead nebula
Diffuse clouds	10–300	10^{8}	100	H_2, H, He, other atoms, diatomic molecules	1.4	(not detected in visible light)
Supernova remnants	up to 3000	10^{5}	10^{6}	H^+ or H, other atoms and/or ions		Crab nebula Vela nebula
Regions of ionised hydrogen (also called HII regions)	3–60	10^{8}	8000	H^+, He^+, other singly-charged ions		Rosette nebula Carina nebula Lagoon nebula

Table 5.13 Interstellar gas regions

Internal energy and specific heat capacity

In an ideal gas, the particles do not interact with one another (except when they collide). But in a liquid or solid (and in a non-ideal gas) the particles exert forces on one another (in a solid rather as if they were joined by springs). So the energy of the particles cannot be described in terms of kinetic energy alone – their potential energy also varies – and so the **internal energy**, U, of a collection of particles is now the total of their kinetic and potential energies, which are distributed at random between the particles. However, it is still the case that the interactions between particles ensure that the overall distribution of kinetic-energy depends only on the temperature as described by Equation 44.

The molecular kinetic theory allows us to explain **specific heat capacity** in terms of microscopic properties of matter. Specific heat capacity, C, is related to the amount of energy, ΔE, that must be supplied to a mass M of a substance to produce a temperature rise of $\Delta\theta$:

$$\Delta E = MC\Delta\theta \text{ or } MC\Delta T \qquad (45)$$

In an ideal gas, particles have only kinetic energy, and so the internal energy is related to temperature by:

$$U = N\langle E_k\rangle = \frac{3NkT}{2} \qquad (46)$$

Study note

In the chapter *Build or Bust*, specific heat capacity is treated in terms of macrosopic properties of matter.

Study note

Normally c is used for both specific heat capacity and particle speed. Here we have used C for specific heat capacity to avoid confusion.

We have used M here for the mass of substance so as to avoid confusion with m that represents the mass of a particle.

The symbols $\Delta\theta$ and ΔT are both used for temperature change.

Provided the gas is not allowed to expand, all the energy supplied, ΔE, is accounted for by an increase in the internal energy of the gas:

$$\Delta E = \Delta U = \frac{3Nk\Delta T}{2} \qquad (47)$$

(If the gas is allowed to expand, it must do work pushing against the pressure of its surroundings. This requires energy. Some of the energy supplied is used to do work, so there is less energy available to increase the particles' kinetic energy and the temperature does not rise so much.)

Imagine two gases. Gas A has particles with twice the mass of those in gas B. What can you deduce about the temperature rise when equal masses of each gas are heated with the same energy? Sample B has twice as many particles as sample A. Therefore this same energy will be distributed amongst a smaller number of particles in A, giving gas A an average increase in particle kinetic energy that is twice that of B. The temperature rise in A will be twice as big: smaller N gives a larger ΔT for the same ΔE. The specific heat capacity of gas A is *half* that of gas B – it needs only half as much energy for a given rise in temperature.

More generally, we can express Equation 45 in terms of particles. If there are N particles each of mass m, then:

$$M = Nm$$

$$\Delta E = NmC\Delta T \qquad (48)$$

From Equations 47 and 48 we then have:

$$NmC\Delta T = \frac{3Nk\Delta T}{2}$$

$$C = \frac{3k}{2m} \qquad (49)$$

Molecules and real gases

All our discussion so far has been about ideal gases made from particles whose only motion is translational – that is, movement from place to place. But molecules made from two or more atoms also have energy associated with their vibration and rotation. At any given temperature, a gas made of such molecules has more internal energy than a monatomic gas because, like translational kinetic energy, the energy of vibration and rotation must also increase with temperature. This means that the specific heat capacity of such a gas is greater than that of a monatomic gas.

The situation is complicated still further in real gases, because there are long-range forces between particles. Now, if the gas expands, the potential energy of its particles increases so there is less energy available for increasing the kinetic energy, and the temperature rise is less than would be expected for an ideal gas.

Study note

A monatomic gas is one in which the particles are single atoms. The gases helium and argon are monatomic. Most gases, such as oxygen (O_2) and nitrogen (N_2), have molecules made up of more than one atom.

Question

48 (a) 1000 J is used to heat 1.00 kg of helium at constant volume. Assuming helium behaves as an ideal monatomic gas, what is the resulting temperature rise?

(Mass of helium atom $m_{\text{He}} = 6.68 \times 10^{-27}$ kg; Boltzmann constant $k = 1.38 \times 10^{-23}$ J K^{-1}.)

(b) Argon is a monatomic gas. The mass of an argon atom (^{40}Ar) is 10 times that of a helium atom. Assuming argon behaves as an ideal gas, what would be the temperature rise if 1000 J were used to heat 1.00 kg of argon at constant volume?

(c) The mass of a water molecule (H_2O) is 4.5 times that of a helium atom. Discuss how you would expect the specific heat capacity to compare with that of helium.

Making stars

Having established some important points about the behaviour of particles in gases, we now return the question of star formation. The stars we see in the sky are hot and dense, and yet they form from interstellar material (ISM) that is very tenuous. As you saw in Table 5.13, the temperature in the hottest parts of the ISM is not quite high enough for hydrogen fusion (which needs over 10^7 K), but the main problem is the density and pressure. You saw in Question 42 that the pressure inside the Sun is about 10^{14} Pa, while Question 47 showed that in the ISM it is no more than about 10^{-11} Pa – a difference of some 25 orders of magnitude. The following passage explains how star formation happens. Read it carefully, then answer Questions 49 to 51.

To form stars, an interstellar cloud must collapse under its own gravity to form a hot, dense concentration of matter. To see how this works, imagine a particle (eg a hydrogen atom or molecule) near the edge of an interstellar cloud (Figure 5.60). Newton's law of gravitation tells us that the particle will experience gravitational forces due to all the other matter around it. As there is a greater concentration of matter on one side of the particle, it accelerates towards the centre of the cloud – the particles are all drawn gradually inwards and so after many thousands of years all the material will eventually become concentrated within a small volume.

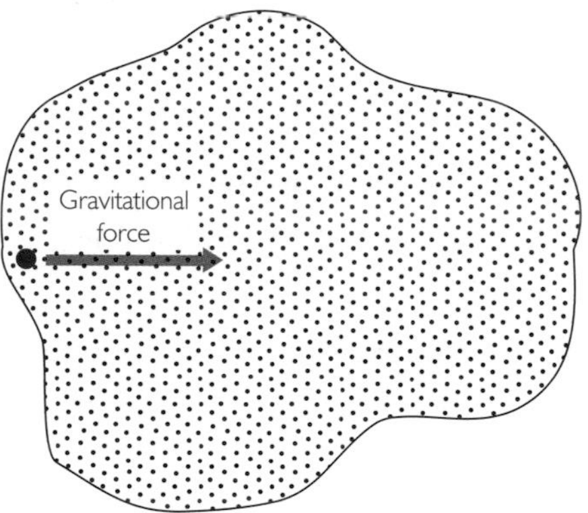

Figure 5.60 Particle near the edge of an interstellar cloud

But if that were all that happens, then all interstellar material would be clumped together in stars, rather than existing as tenuous gas. As we know from the success of the kinetic theory of gases, the particles in a gas are moving around at high speeds. If they are moving fast enough, they can escape from the cloud, rather as a rocket fired from Earth at high enough speed can travel out to space instead of falling back to the ground; and so the cloud will disperse rather than collapse.

Whether a cloud collapses under its own gravity to form stars, or disperses due to the motion of its particles, depends on the 'competition' between those two factors. A detailed analysis shows that only the coldest, densest regions of the ISM can collapse under their own gravity. But that then raises the problem of how such material can ever become hot enough for nuclear fusion to occur. It turns out that gravity and kinetic theory again play a part.

During the collapse of a cloud, particles can be thought of as gradually 'falling' towards the centre, rather like objects falling to earth. As they fall, they accelerate. Put another way, they gain kinetic energy at the expense of gravitational potential energy. The increasing concentration of matter is enough to maintain the collapse, despite the increasing speeds of the particles.

As you can see from pictures of interstellar clouds (such as the Orion nebula in Figure 5.52), they are not smooth and uniform. As a cloud collapses, it breaks up into fragments that continue to collapse separately.

By the time a fragment has collapsed into a small volume – several million years after the collapse first started – the particles are moving at very high speeds and are colliding with one another. As the temperature rises, the collisions become increasingly violent, so molecules are first disrupted into atoms, and then the electrons become dislodged from the atoms to make ions.

By the time the material has collapsed to the size of a star, it is a very dense plasma (an ionised gas) with a temperature of about 10^7 K. Now nuclear fusion begins to take place, and the release of energy halts the collapse. A star is born.

This is a nice story, but do we have any evidence to support it? The processes are all much too slow for us to observe them happening during our lifetime, but one key piece of evidence comes from infrared astronomy. We know that all objects emit radiation at wavelengths that depend on their surface temperature, and so we would expect to detect infrared radiation from fragments that were collapsing and getting hotter on the way to becoming stars. And this is what astronomers find when they turn infrared-sensitive telescopes to regions such as the Orion nebula – clusters of 'hotspots' that are just as we could expect to see if the cloud were in the process of collapsing, fragmenting and heating up (Figure 5.61).

Figure 5.61 Infrared image of the Orion nebula

Activity 28 First generation star

Sort a set of cards so that they correctly describe the formation of a star as described above.

Questions

In answering these questions, you will need to refer to Table 5.13 and to your work in Section 3.2 as well as to the passage above.

49 In the following paragraphs, choose the correct word or phrase from each pair to make a correct summary of star formation.

Stars are formed by the collapse of a molecular cloud/region of ionised hydrogen. The particles in the cloud are far apart/close together. The force of gravitational/electrostatic attraction pulls them all towards the edge/centre of the cloud. Each particle is falling inwards, like a ball falling towards the Earth. As the cloud collapses, the particles move more quickly/slowly, and they collide more frequently with one another, sharing their kinetic energy. The faster the particles move, the higher/lower the temperature of the gas. As the cloud collapses: its density increases/decreases; its pressure increases/decreases; the particles' kinetic energy increases/decreases; and their gravitational potential energy increases/decreases.

The particles collide frequently and energetically so that they break up to form a hot plasma of fast moving/slow moving positively/negatively charged nuclei and electrons. When the temperature is high/low enough, the nuclei can approach one another so closely that nuclear fission/fusion can start.

50 Estimate the mass of a typical molecular cloud, assuming that it is made entirely of hydrogen molecules, H_2, and hence estimate the number of Sun-like stars that could form when such a cloud collapses. Quote your answer to the nearest order of magnitude. (Mass of H_2, $m = 3.34 \times 10^{-27}$ kg; mass of Sun, $M_{sun} = 2 \times 10^{30}$ kg; 1 light year = 9.46×10^{15} m.)

51 (a) Estimate (i) the acceleration of a hydrogen molecule at the edge of a typical molecular cloud (see Question 50); (ii) the time it would take the molecule to reach the centre of the cloud if it started from rest, travelled in a straight line, and its acceleration did not change; (iii) the molecule's speed when it reached the centre; and (iv) the temperature of a gas or plasma where this was the rms speed of hydrogen molecules.

(b) Comment on your answer to (a)(iv).

($G = 6.67 \times 10^{-11}$ N m^2 kg^{-2}; 1 year $\approx 3 \times 10^7$ s, $k = 1.38 \times 10^{-23}$ J K^{-1}.)

Activity 29 Star formation

Scientists who wish to publicise their research findings often produce a poster for display and discussion at conferences. The most eye-catching of these is likely to have the most immediate impact.

Make a conference poster showing the main stages in star formation from an interstellar cloud to the arrival of the star onto the main sequence. Use sketches or photographs and add plenty of information, including details of density, temperature and pressure in the interstellar medium. Include any relevant equations and example calculations to make your poster as informative as possible.

3.4 Evolution and end points – Summing up Part 3

This section continues the story of stars with an account of what happens after they are formed and, in doing so, reminds you of the physics that you have studied in Sections 3.1 to 3.3 (and also some of the physics from Part 2 of this chapter). The questions and activities are intended to help you look back over your work.

Activity 30 Summing up Part 3

Before going on to read more about stars, look back through your work in this part of the chapter and make sure you understand the meaning of all the terms printed in bold.

A star on the main sequence

In Section 3.3 you saw how a fragment of interstellar cloud could collapse under its own weight to form a small, dense ball of matter that was hot enough for nuclear fusion to begin. Once the star has settled down to a steady rate of fusion, its size, luminosity and temperature remain more or less constant until its supply of hydrogen fuel is exhausted. It has become a main-sequence star. This stage of its life takes a long time, which is why we observe so many stars on the main sequence of a Hertzsprung–Russell diagram (see Figure 5.62), which is based on observations of a very large number of stars. The numbers give the masses of the stars in units of the Sun's mass, $M_{Sun} \approx 2 \times 10^{30}$ kg.

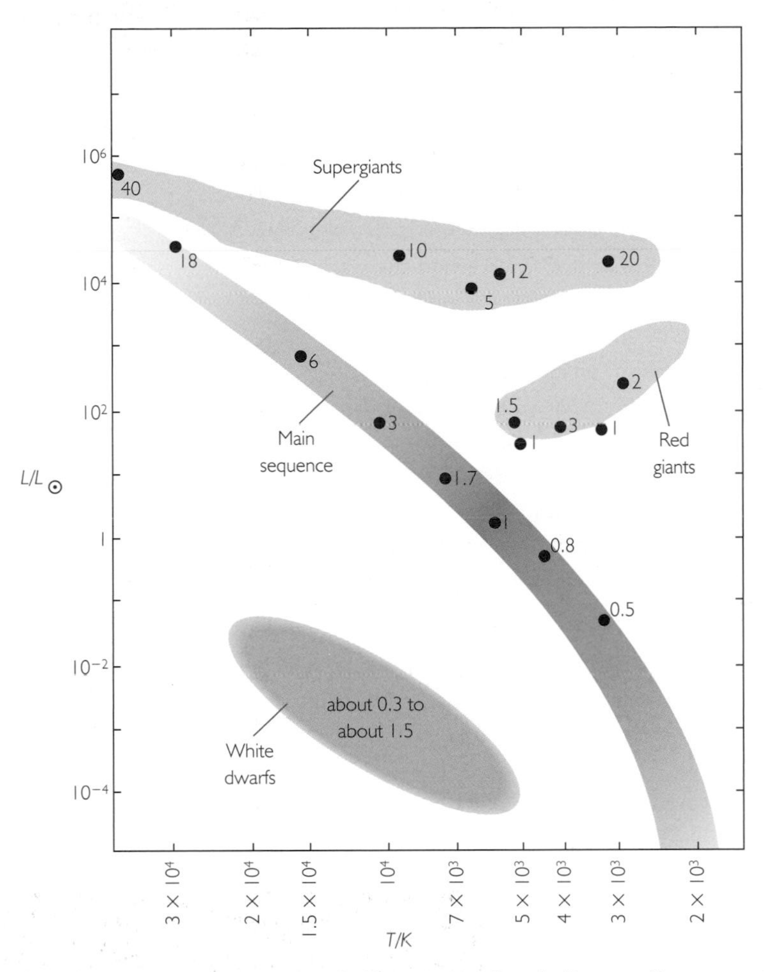

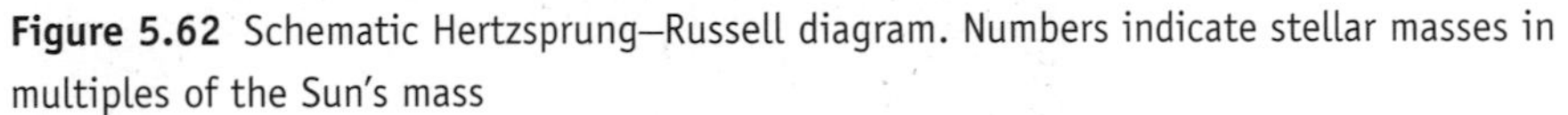
Figure 5.62 Schematic Hertzsprung–Russell diagram. Numbers indicate stellar masses in multiples of the Sun's mass

A star with the same mass as the Sun will remain on the main sequence for about 10^{10} years (see Question 18 in Section 2.3). The main-sequence lifetime of a star depends on its mass and the rate at which it converts hydrogen to helium. Perhaps surprisingly, low-mass stars live for much longer than the Sun. A star of $0.5M_{sun}$ has $L \approx 0.1\ L_{Sun}$. It has only half the amount of fuel but it is consuming it at one-tenth the rate found in the Sun, and so it will remain on the main sequence for five times as long – about 5×10^{10} years. The most massive stars are the shortest lived, because their nuclear reactions proceed more rapidly, giving them a very high luminosity, and so they exhaust their fuel supplies in a relatively short time. So if we observe any very luminous main-sequence stars, we know they must have formed quite recently – only a few million years ago.

Questions

52 In two or three sentences, outline how the stars' masses in Figure 5.62 might have been determined.

53 Use Figure 5.62 to find the approximate luminosity of a star of mass $18M_{sun}$. Assuming that all stars fuse the same fraction of their total mass while on the main sequence, estimate the main-sequence lifetime of this star.

After the main sequence

As with the 'story' of star formation, astronomers have drawn on observations and on physics in order to piece together an account of what happens to a star after the hydrogen in its core has all undergone fusion to helium.

When hydrogen fusion ceases, the inner parts of the star cool down, which allows the star to collapse again under its own gravity. Just as happened when the star was forming, there is a gain of kinetic energy at the expense of gravitational potential energy, which leads to heating. Now the material surrounding the helium core becomes hot enough for hydrogen fusion to helium to begin, so inside the star there is now a cool(ish) shrinking core surrounded by a layer where nuclear fusion is releasing energy. The material surrounding this fusion layer expands enormously, cooling as it does so. If the star's mass is less than that of the Sun, or only a few times greater, it is now a red giant – a star with a cool exterior, but very luminous because of its large size. A more massive star (say about five times the Sun's mass or more) becomes a **supergiant** – it is not only large but the exterior is still quite hot. Figure 5.62 shows where red giants and supergiants are located on an HR diagram.

As the core continues to shrink under its own gravity, the temperature rises. In stars at least as massive as the Sun, the core becomes hot enough for new fusion reactions to take place:

$${}^{4}_{2}\text{He} + {}^{4}_{2}\text{He} \rightarrow {}^{8}_{4}\text{Be}$$

$${}^{4}_{2}\text{He} + {}^{8}_{4}\text{Be} \rightarrow {}^{12}_{6}\text{C} + 2\gamma$$

(The beryllium nucleus is very unstable, and decays in less than a second if it does not immediately undergo fusion with another helium nucleus.)

In stars up to about twice the mass of the Sun, the onset of helium fusion causes great upheavals in the outer parts of the star. The upheavals can cause the outer parts of a star to pulsate in and out with periods of only a few days, and astronomers observe a variable star as its luminosity changes. The outermost layers are eventually thrown off to form an expanding shell of glowing gas called a **planetary nebula** (though it has absolutely nothing to do with the planets). Figure 5.63 shows an example of such a shell – you can also see the star in its centre.

Figure 5.63 Helix nebula in Aquarius

When the helium in the core has been converted to carbon, then the star again begins to collapse under its own weight and to heat up. In stars up to about the mass of the Sun, the temperature never gets high enough for any further fusion reactions that would stop the gravitational collapse, and so the star gradually shrinks to become a white dwarf – a small, hot, dense star about the size of the Earth that will eventually cool and fade from view.

Activity 31 End of the Sun

Figure 5.64 summarises the evolution of a Sun-like star by tracing the changes in its luminosity and surface temperature from just before it joins the main sequence right up to its end as a white dwarf. Table 5.14 summarises the physical processes that take place at each stage (though not in the right order). Use Figure 5.64 and the preceding text to help you complete the table – in the 'luminosity' column, state the nearest order of magnitude. In your answer, arrange the stages in the correct order.

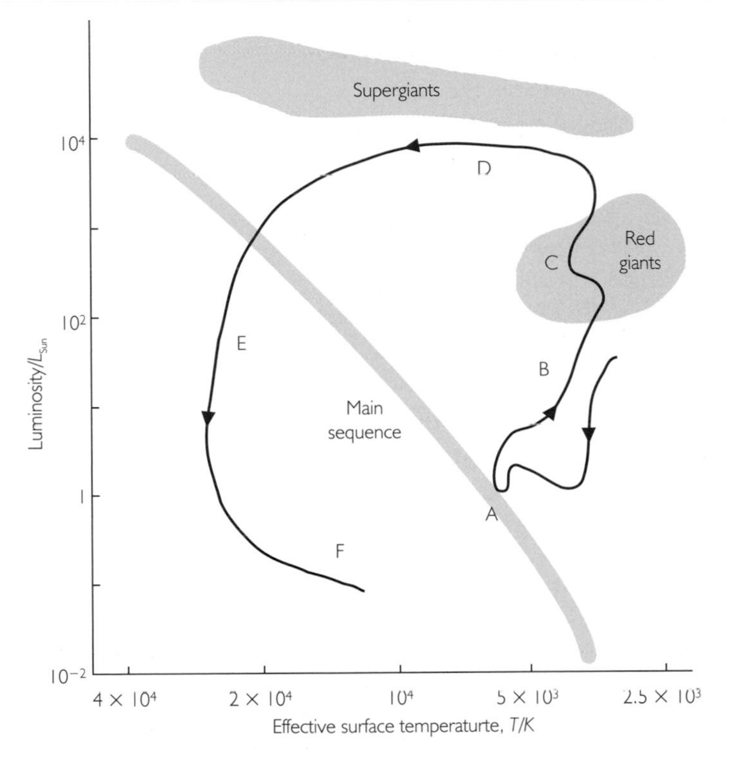

Figure 5.64 Evolution of a Sun-like star

Point on Figure 5.64	Stage	Physical process	Luminosity/L_{Sun}	Surface temperature T/K
	evolution away from main sequence	fusion of H continues		
		all fusion stops; slow cooling		
	red giant			
		outer layers thrown off leaving hot core		
	variable star	violent pulsations of outer layers		
	main sequence		1	

Table 5.14 The evolution of a Sun-like star

The fate of massive stars

In stars more massive than the Sun, there is now a pattern that repeats itself:

- the star contracts
- loss of gravitational energy leads to heating
- new fusion reactions produce more massive nuclei
- reactions cease
- the star contracts.

... and so it goes on. The successive stages of fusion each happen more and more rapidly, and so a star can get through many stages of fusion in a much shorter time than it spends on the main sequence. Each stage of fusion can give rise to changes in the external appearance of the star: it may become a variable star as the outer parts repeatedly expand and shrink before being ejected to form a planetary nebula.

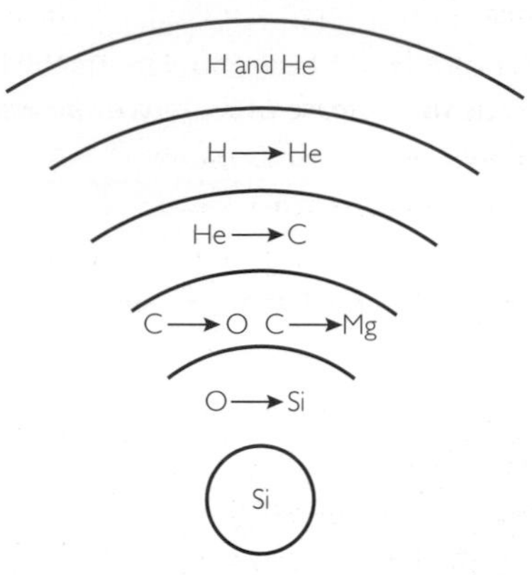

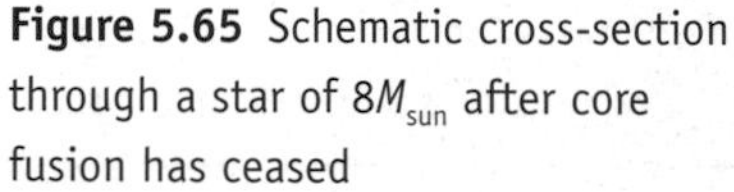

Figure 5.65 Schematic cross-section through a star of $8M_{sun}$ after core fusion has ceased

For stars with masses up to about eight times that of the Sun, the series of fusion reactions will eventually cease when the rise in temperature is not great enough to trigger the next stage. Figure 5.65 shows the interior of an eight solar mass star at the point when fusion stops.

More massive stars can continue with fusion until they produce iron nuclei in their cores (Figure 5.66), reaching a core temperature of about 4×10^9 K. The successive fusion episodes, leading to the creation of iron, have effectively kept the star stable. As the core has collapsed, the material in the core has started to fuse, releasing energy that has kept gravity in check by providing the pressure to support the weight of the star. This feedback mechanism fails when the star is composed of iron because the fusion of iron nuclei to heavier nuclei requires a net input of energy. At this point the star consists of an iron core about as big as the Earth and a hydrogen envelope that may be as big as the orbit of Jupiter.

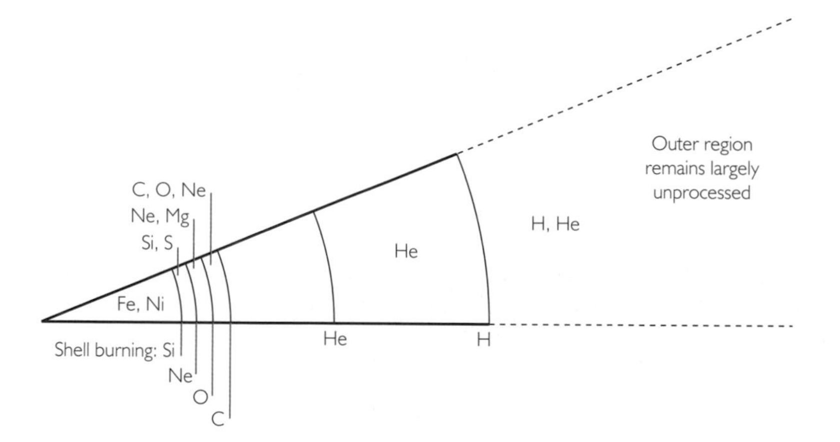

Figure 5.66 Schematic cross-section through a star greater than $8M_{sun}$

It has taken the star millions of years to reach this stage, but the next stage takes place with extraordinary rapidity. As fusion ceases, the core collapses. The temperature rises, but instead of the production of heavier elements by nuclear fusion, the opposite happens. The collapse causes the temperature of the core to exceed 5 billion K. The gamma rays associated with this high temperature cause the iron atoms to break down into alpha particles. This reaction takes energy away from the core and the collapse continues to gather pace. The alpha particles themselves break down into protons and neutrons. The protons combine with electrons and the core is now nuclear matter (neutrons) with a density of about 4×10^{17} kg m^{-3}. The core now ceases to collapse.

The hydrogen envelope falls onto the now-rigid core. This material rises in temperature and pressure, and creates a shock wave that travels outwards, blowing the star apart in just a few seconds in a tremendous explosion called a **supernova**. The explosion produces vast numbers of neutrons that can be absorbed by remaining heavy nuclei to form elements heavier than iron. (Note that this process of neutron capture is not quite nuclear fusion – it requires a source of energy, whereas fusion leads to a release of energy.) The entire star, apart from a tiny condensed core, is blown out into the interstellar medium as a rapidly expanding cloud of very hot gases called a **supernova remnant**. Figure 5.67 shows one of the most famous supernova remnants in our galaxy. The Crab nebula is the remains of a star that was seen to explode in AD 1054, and was so bright that it was visible in daytime despite being about 6000 light years away.

Figure 5.67 Crab nebula

Supernovae are the source of nearly all the elements heavier than hydrogen and helium. The Big Bang that began the Universe created only hydrogen, helium and traces of some other light elements. Heavier elements are created by nuclear fusion in stars, but most remain locked in stellar cores as they cool. It is only through the violent supernova explosions that elements are scattered into the ISM. Here they cool down and might eventually condense to form another star, perhaps with planets in orbit around it, where carbon-based life (or perhaps some other life forms) can evolve.

Activity 32 Fusion in stars

The more massive the star, the more massive the nuclei it can produce by fusion, but even the most massive stars cannot produce nuclei more massive than iron in this way.

Write a paragraph explaining the physics behind this statement. Include the following terms: gravitational force, electrostatic force, binding energy, kinetic energy. Also include any other key terms from Parts 2 and 3 of this chapter that you think are relevant.

Activity 33 SETI

Imagine you are an astronomer involved in the SETI (Search for Extraterrestrial Intelligence) programme. You are bidding for time to use the Arecibo Radio Astronomy facility in Puerto Rico (Figure 5.68). Your aim is to try and pick up radio transmissions that might have been emitted by an alien civilisation on a planet orbiting another star. Prepare a bid to be given to the time-allocation committee that explains which type of stars you intend to observe and why.

You can find out more about SETI, planets and the possibility of life elsewhere in the Universe from the websites listed at **www.shaplinks.co.uk**.

Figure 5.68 Giant radio telescope at Arecibo

Questions

54 Explain the physics behind the phrase 'gamma rays associated with this high temperature'.

55 (a) Making the rash assumption that the core of a highly evolved star behaves like an ideal gas calculate (i) the mean kinetic energy and (ii) the rms speed of ^{56}Fe nuclei at a temperature of 5×10^9 K. (Mass of ^{56}Fe nucleus, $m = 9.3 \times 10^{-26}$ kg; $k = 1.38 \times 10^{-23}$ J K^{-1}.)

(b) Repeat for an electron in the star's core, and comment on your answer. (Electron mass, m_e 9.11×10^{-31} kg.)

End points

After a massive star has blown itself up in a supernova explosion, most of its material is scattered into the interstellar medium leaving just the innermost core of the star. If this central object has less than about twice the mass of the Sun, it survives as a **neutron star**. This is an object made mainly of neutrons and with the density of nuclear matter. The entire mass is contained within a sphere about 10 km in diameter. Neutron stars are small and emit little visible light, but some can be detected using radio telescopes. In 1967 Jocelyn Bell Burnell, who at the time was a research student at Cambridge, was the first to detect a neutron star – completely unintentionally. The star produced a rapid periodic radio pulse, so regular that at first the Cambridge astronomers thought they had discovered extra-terrestrial intelligent life. The regular pulses led to this type of star being named a **pulsar**, and the regular pulses were explained as being due to a rapidly spinning neutron star.

If a star is rotating before it explodes (and all stars do rotate), then as the core collapses it will spin much more rapidly. This happens to any rotating object that becomes more compact. You can illustrate this by sitting on a swivel chair and spinning round slowly with your arms and legs outstretched; as you bring your limbs closer to your body, you can feel yourself spin faster. When a star's core shrinks to become a neutron star, it can be spinning with a period of less than a second. The collapse also concentrates the star's magnetic field, and the interaction of this very strong field and any charged particles in the vicinity gives rise to a narrow beam of radiation emerging from the star's magnetic poles. If the beam emerges at an angle to the rotation axis, then an observer in the path of the beam will detect a sharp 'flash' each time the beam sweeps past, rather like a lighthouse beam (see Figure 5.69). Many pulsars have now been observed, including one in the centre of the Crab nebula supernova remnant (see Figure 5.67).

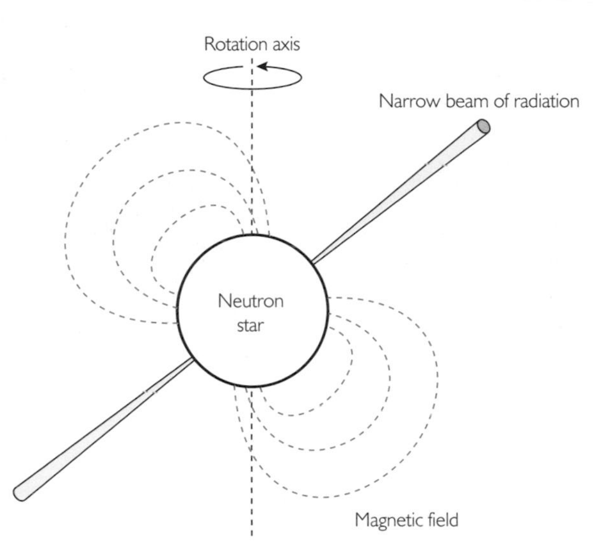

Figure 5.69 Schematic diagram of a pulsar

If the central core has more than about twice the mass of the Sun, then even neutrons are crushed by the strong gravitational forces. The object becomes unimaginably small and compact, and it has such a strong gravitational field that all nearby matter will inexorably be drawn into it; nothing can escape – not even light. It is a **black hole**. It may surprise you to learn that the notion of black holes has been around since the middle of the 18th century when they were speculated upon by the Rev. John Michell, vicar of Thornhill near Barnsley in Yorkshire. Like many clergymen of his day, he had studied mathematics at Cambridge and was aware of the discoveries of Sir Isaac Newton. In his paper to the Royal Society in 1783, he said 'Let us now suppose the particles of light to be attracted (to a body) in the same manner as all other bodies which we are acquainted... all light emitted from such a body would be made to return towards it, by its own proper gravity.' (It is interesting that he thought of light as being made of particles, as did Isaac Newton. The use of waves to explain the behaviour of light came rather later.)

By their very nature black holes cannot be seen directly. However, astronomers have detected several objects called X-ray binaries. These are believed to be binary stars consisting of a black hole and an ordinary star, in which the intense gravity of the black hole is dragging material away from its companion. As this material spirals towards the black hole like water down a plughole, it is accelerated to high speed and so becomes very hot and emits X-rays (Figure 5.70).

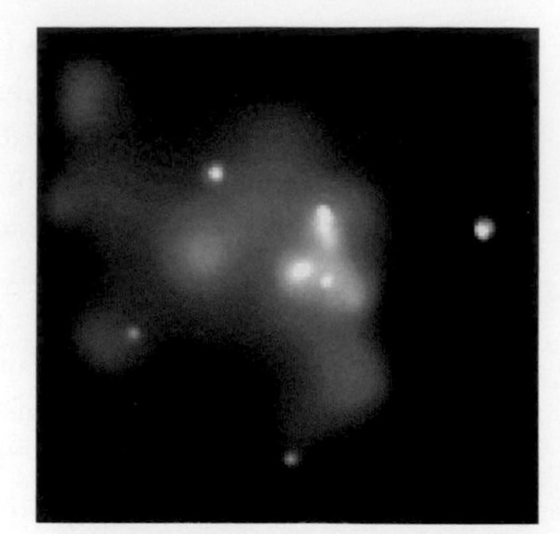

Figure 5.70 X-rays detected from a (probable) black hole in the Andromeda galaxy

Last word

The final words in this long section from astronaut Edgar Mitchell who flew the Apollo 14 spacecraft to the Moon:

> *The biggest joy was on the way home. In my cockpit window every 2 minutes I would see the Earth, the Moon and the Sun and the whole 360 degree panorama of the heavens. It was a powerful and overwhelming experience. And suddenly I realised that the molecules of my body, the molecules of the spacecraft, the molecules in the bodies of my partners were prototyped and manufactured in some ancient generation of star. And there was an overwhelming sense of oneness, of connectedness. It wasn't them and us. It was me – that was all of it – it's all one thing. And it was accompanied by an ecstasy or a sense of "Oh my God, wow, yes!" – an insight, an epiphany.*

The Apollo astronauts were amongst the most highly trained scientists and pilots. The most exciting aspect of the Apollo missions was not the science and technology, but the realisation of how connected we are to the natural Universe. It was an emotional, almost spiritual reaction. A cynic might say we are nuclear waste; a poet would say we are made of stardust.

Questions

56 (a) Calculate the densities of (i) a neutron star that has a radius of 5 km and a mass of 4×10^{30} kg (ie twice the mass of the Sun) and (ii) a helium nucleus that has a radius of approximately 10^{-15} m and a mass of 6.7×10^{-27} kg. (iii) Hence comment on the statement that a neutron star 'has the density of nuclear matter'.

(b) (i) Calculate the gravitational field strength at the surface of the neutron star in (a). (ii) To the nearest order of magnitude, express your answer as a multiple of the field strength at the Earth's surface. ($G = 6.67 \times 10^{-11}$ N m^2 kg^{-2}; $g_{\text{Earth}} = 9.8$ N kg^{-1}.)

57 Pulsars have been detected with periods of about a millisecond. The material at the surface of a spinning neutron star must experience a centripetal force strong enough to keep it in orbit, and the only force available is the star's own gravity. If the gravitational force is not strong enough, then the star will fly apart.

(a) Show that the minimum possible rotation period, T, of a star mass M and radius r is given by:

$$T \geq \sqrt{\frac{4\pi^2 r^3}{GM}}$$

(b) Find the minimum possible rotation period for the neutron star in Question 56 and hence say whether a 'millisecond pulsar' could reasonably be a rotating neutron star.

58 Theoretical physicists (and science-fiction writers) have speculated what might happen to space travellers who approached a black hole. One scenario is that as a person fell towards a black hole, they would suffer a process dubbed 'spaghettification', which this question illustrates.

Jo is very tall and she has a large head and large feet (see Figure 5.71). Her spacecraft has brought her close to a black hole of mass $M = 10^{26}$ kg.

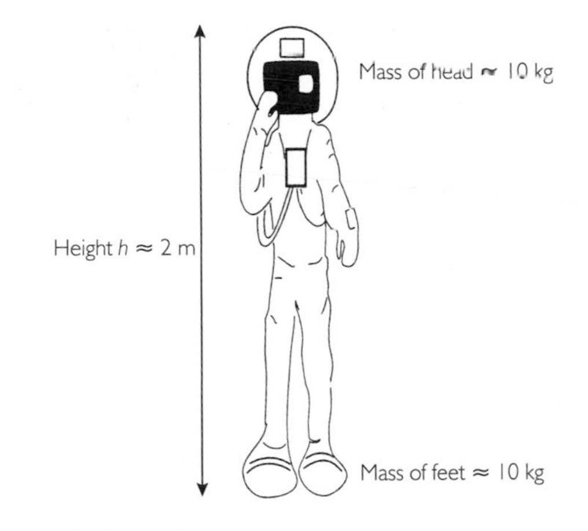

Figure 5.71 Diagram for Question 55

(a) Calculate (i) the gravitational force on Jo's feet, which are exactly 50 km from the black hole, and (ii) the gravitational force on her head, which is exactly 2.0 m further from the black hole. (Keep at least five figures in each answer, even though you might not feel that this is justified.)

(b) Explain how these forces will affect Jo.

Activity 34 Going out with a bang

Produce a sequence of annotated cartoon sketches, based on the information above, to summarise the post-main-sequence evolution of a massive star. Present your work as a series of overhead projection transparencies or use a presentation package such as PowerPoint or Flash.

4 The story of our Universe

In this final part of the chapter, you will see how, since the early 20th century, scientists have been looking for answers to such incredible questions as: 'How did the Universe begin?' and 'How will it end?' You will see how some of these questions can be tackled using physics that you have been studying in Parts 2 and 3 of the chapter, and extend your knowledge and understanding of those areas of physics.

4.1 Taking on the biggest question in science

A telescope is a time machine.

J. Silk, *A Short History of the Universe*

The birth of modern cosmology

The greatest scientist of the 20th century (Figure 5.72) hated school. But he was fascinated by science and maths, and read up on them at home. Disappointed about not getting a post as a lecturer, he left Germany, moved to Switzerland, and became an examiner of patents for scientific inventions. In his spare time, he came up with two major new theories that changed the whole face of physics in the 20th century. First, 'special relativity', which revolutionised the way we think about time; and then, general relativity, which changed the way we think about space.

Figure 5.72 The young Albert Einstein

But Albert Einstein (1879–1955) was a complex and tragic character. He was deeply concerned about human rights and freedom on the one hand, yet completely detached from everyday life on the other. He desperately needed affection, yet could not maintain close relationships with his family. Despite his giant steps in physics, he died feeling a failure, because he hadn't managed to work out a complete theory of physics.

In 1917 he turned his attention to cosmology and tried to devise equations for how the whole Universe behaves. He didn't know much about astronomy, and had no experimental evidence on which to base his equations – truly pioneering work. One of his ideas produced a foundation for cosmology in the future...

Einstein made the bold assumption that, on average, the Universe looks the same from any point and the same in all directions. Matter, galaxies and clusters of galaxies are evenly scattered on a large scale. The Universe is 'homogenous'. This is called the **cosmological principle**. Without the cosmological principle, the equations of cosmology would be just too hard to solve, and scientists would not have made the astonishing progress they achieved in the 20th century. Mercifully Einstein's assumption turns out to be true – the Universe does seem to be pretty much the same everywhere.

Study note

You will see some evidence for homogeneity in Section 4.2

Like virtually everyone in his day, Einstein believed that the Universe was unchanging, that it was the same now as it was when it was created. But he realised that the gravitational attraction of all the matter in the Universe would cause it to be drawn together (rather like the collapsing cloud idea that you met in Section 3.3). He therefore included in his equations a term describing an unknown repulsive force that would ensure that the Universe would indeed continue to look the same for ever. He later called this 'the biggest blunder in my life', as Edwin Hubble made some observations that indicated that the Universe would not always stay the same.

Hubble and the galaxies

I see beyond this island Universe,
Beyond our sun, and all those other suns
That throng the Milky Way, far, far beyond,
A thousand little wisps, faint nebulae.

Alfred Noyes, *The Torch Bearers*

Edwin Hubble (1889–1953, Figure 5.73), a tall, athletic young American, had a tricky choice to make. He excelled in maths, astronomy and boxing. Would he take the chance of fighting the then world heavyweight champion he had been offered? Thankfully for the future of cosmology, he chose instead to come and study at Oxford University, England. But his discoveries had to wait until he'd served in the First World War. (He began as a private in the US army and ended the war a major.)

Figure 5.73 Edwin Hubble

Eventually, at the age of 30, he began his famous work at the new 100-inch telescope at Mount Wilson, California. His first discovery was a special type of star within a faint misty patch of light known as the Andromeda nebulae (Figure 5.74). Astronomers had puzzled over what these nebulae were for centuries. Hubble was able to measure the distance to the nebula accurately, and managed to show that it was simply a congregation of stars – just like our Milky Way galaxy, but far beyond it. For the first time, we knew something about the real scale of the Universe. The basic units of matter in the Universe are galaxies, not stars. And if that was not enough, Hubble was soon to make sense of a phenomenon that was to change our whole conception of the Universe even more. This concerned the line spectra of distant galaxies.

As mentioned in Part 2 of this chapter, the spectrum of light from the Sun and other stars contains many narrow absorption lines, produced when atoms in the cool outer parts of the star absorb radiation at their own characteristic wavelengths. When astronomers began looking at the spectra of whole galaxies beyond our own, they found the same general pattern of absorption lines in the light from all the stars superimposed, but they also noticed something rather surprising: for all but the closest galaxies, the dark lines seemed to be shifted from their normal positions, towards longer wavelengths. Wouldn't you be surprised if you looked outside and saw a green sky, orange grass, and a red midday sun? This phenomenon became known as **redshift**, because the lines were shifted towards the red (low-frequency, long-wavelength) end of the spectrum.

Figure 5.74 Andromeda nebula

Redshift is defined as the ratio of the change in wavelength $\Delta\lambda$ as a fraction of the wavelength, λ, measured in the laboratory and is usually given the symbol z. Provided the wavelength change is small, this is very nearly the same as ratio of the change in frequency Δf to the 'lab' frequency f, and so:

$$z = \frac{\Delta\lambda}{\lambda} \approx \frac{\Delta f}{f} \qquad (50)$$

Figure 5.75 shows the spectra of three galaxies with different redshifts.

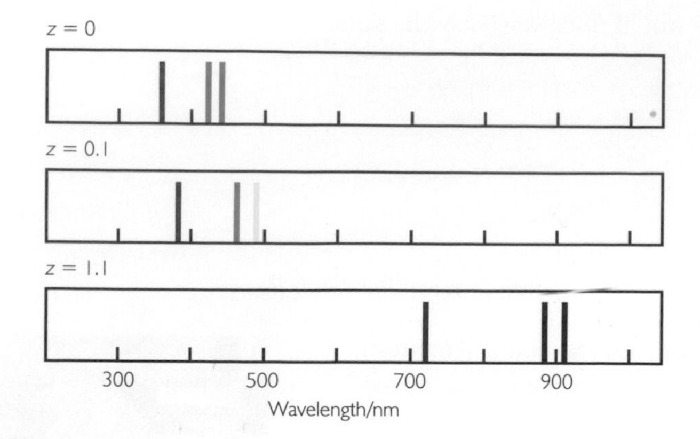

Figure 5.75 Effect of redshift on wavelength

You have met something like Equation 50 before – in the AS chapter *Technology in Space* and in Part 3 of this chapter. There you saw that the Doppler shift was given by:

$$\frac{\Delta\lambda}{\lambda} \approx \frac{\Delta f}{f} \approx \frac{v}{c} \qquad \text{(Equations 22 and 23)}$$

provided $v << c$, where v is the speed of the source of the radiation and c the speed of light in a vacuum. In other words, the redshift is related to the speed of the source of radiation. We can combine these expressions and write:

$$z = \frac{\Delta\lambda}{\lambda} \approx \frac{\Delta f}{f} \approx \frac{v}{c} \qquad (51)$$

Study note

Note that we, and other texts, also use u to represent the speed of the source.

Using the largest telescope in the world, Hubble was able to measure the redshifts and distances of 24 galaxies. He noticed that the redshift was greater for more distant galaxies. He plotted his results on a graph of distance against 'recessional' speed. Figure 5.76 shows his results, taken from his paper published in 1929. Notice that the unit of distance here is the parsec, pc (see Section 3.1). Each black dot is an individual galaxy, and each circle is a group of galaxies. Figure 5.77 shows more recent data. You can see the linear relationship extends to very distant galaxies – note the huge range of

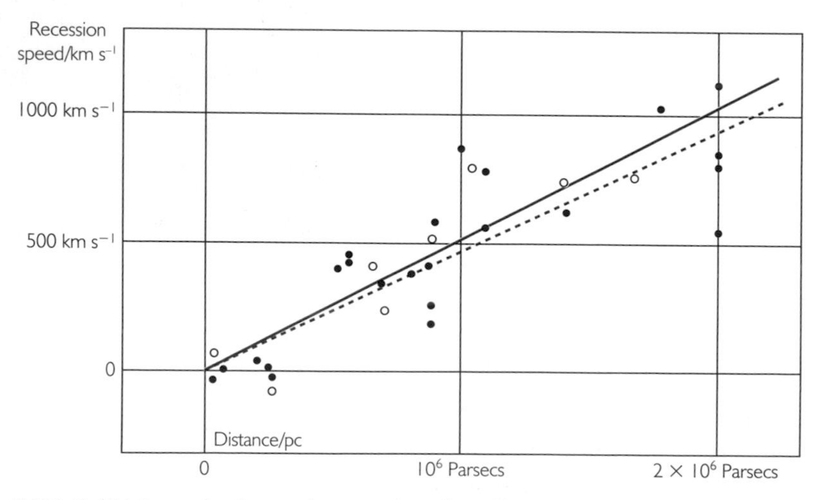

Figure 5.76 Hubble's graph of recession speed against distance

values plotted, making it necessary to use a logarithmic scale. The distance unit here is the megaparsec, Mpc. 1Mpc = 10^6 pc.

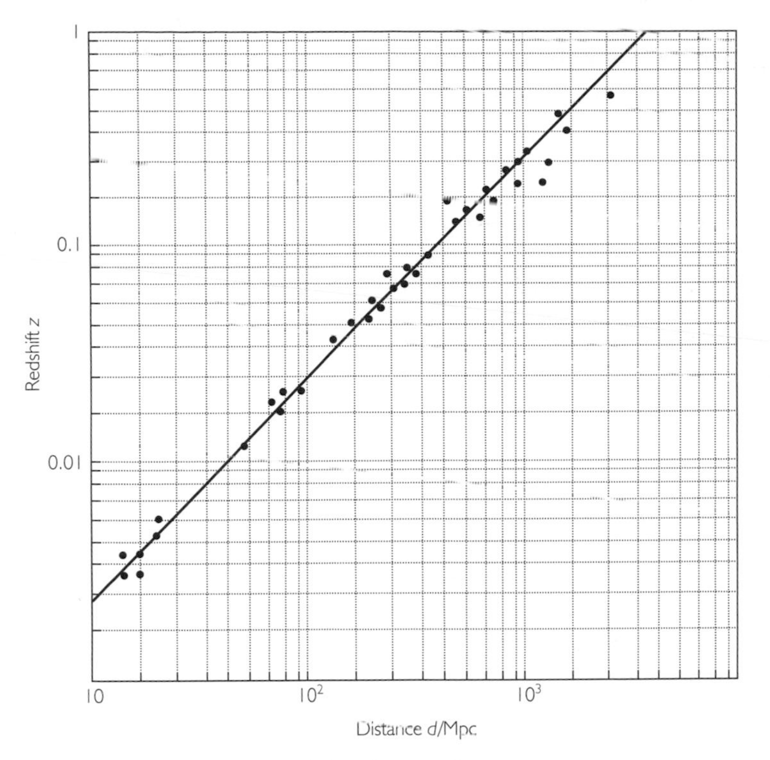

Figure 5.77 Graph of more recent redshift–distance data

Study note

Notice the log scales in Figure 5.77. Zero does not feature on the graph, since that would require an infinite length of graph paper.

Activity 35 Redshift

The CLEA software lets you experience some aspects of modern observational astronomy and provides you with some authentic data. Use the software to measure the redshifts of some galaxies and to relate the redshifts to their distances.

Hubble had found a simple link between distance, d, and redshift:

$$z \propto d$$

This is known as **Hubble's law**. If z is small, we can use Equation 51 to write:

$$v \approx H_0 d \qquad (52)$$

where H_0 is called the **Hubble constant**. For large redshifts, it is no longer true that $z \approx \frac{v}{c}$. The correct version of Equation 52 is:

$$z = \frac{H_0 d}{c} \qquad (53)$$

The Hubble constant can be found from the gradients of graphs such as those in Figures 5.76 and 5.77. It is usually quoted in units of km s^{-1} Mpc^{-1} (think of Equation 52 with v in km s^{-1} and d in Mpc). It is difficult to determine accurately, because it relies on the accurate measurement of d for very distant galaxies, and values range from about 50 km s^{-1} Mpc^{-1} to about 100 km s^{-1} Mpc^{-1} with the range 70–80 km s^{-1} Mpc^{-1} being the most favoured.

Study note

The current 'favourite' value for the Hubble constant is 70.1 ±1.3 $kms^{-1}Mpc^{-1}$. This is based upon the WMAP five-year data. You can find out more at the NASA website via **www.shaplinks.so.uk/**.

The expanding Universe

The most surprising thing about Hubble's discovery was its interpretation. It implied that the Universe was expanding! Activity 36 shows how Hubble's law implies an expanding Universe.

Activity 36 The expanding rubber Universe

Cut a wide elastic band to make a strip of rubber. With a pen, make three marks at 1 cm intervals, labelled 1, 2, and 3. These represent galaxies.

Draw an observer (a small bug?) on one of the galaxies.

Stretch the rubber band to simulate the expansion of the Universe (Figure 5.78).

Measure how far the other galaxies have moved away from the observer, and so compare the recession speeds of the two galaxies.

Repeat with the observer on one of the other galaxies.

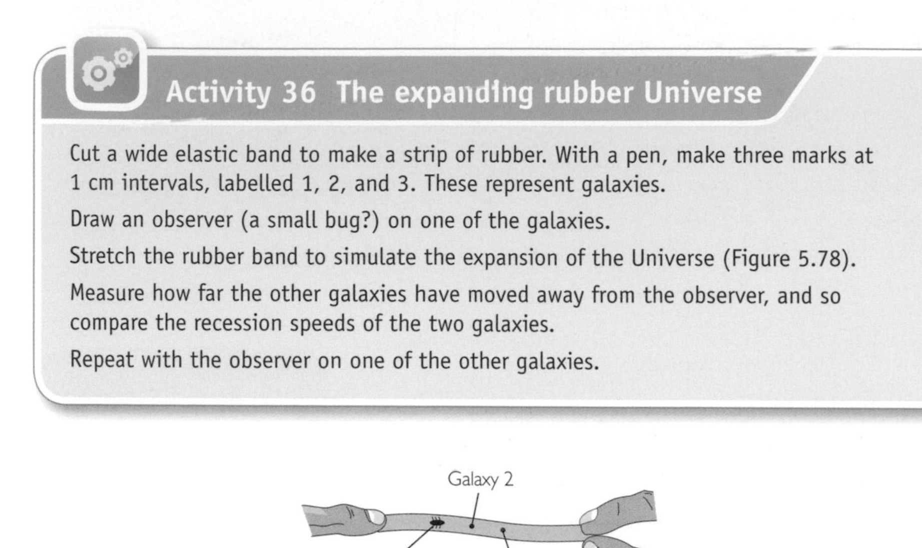

Figure 5.78 Stretching the Universe

Stretching the rubber band produces Hubble's law: speed is proportional to distance. Any uniformly stretching material would do the same. And the converse is true: if galaxies obey Hubble's law, you know that the 'material' that holds them – in this case the Universe – is uniformly stretching.

Hubble's observational evidence wasn't completely convincing. He had only been able to measure relatively nearby galaxies. It was a leap of faith to believe that Hubble's law applied to the rest of the galaxies in the Universe. But the leap turned out to be correct, as you can see from Figure 5.77. Hubble's law is now so well established that astronomers use it to estimate distances by measuring redshifts from galaxies that are too far away to have their distances measured by other methods.

Figure 5.79 Georges Lemaître

In fact, Hubble's law had already been predicted. A Belgian physicist and priest, Georges Lemaître (Figure 5.79), had just written a theoretical paper that claimed that the Universe was expanding. The link that Hubble discovered followed directly from this. So, it seemed that Einstein's static view of the Universe had to be thrown away. The evidence pointed to a Universe that was expanding equally in all directions. Georges Lemaître is sometimes called the father of the Big Bang, as he predicted the expanding Universe before Hubble discovered it.

Thinking about Activity 36 may help you to understand a fundamental fact about the Universe. The Universe is expanding not because the things in it (galaxies) are moving outwards, but because space itself is expanding. The rubber band is like the space in the Universe. When it expands, it carries the dots (representing galaxies) with it.

Try not to imagine the Big Bang as being like an ordinary explosion, where matter flies outwards into space in all directions. Instead, think of the Universe as a balloon, and imagine that the surface of the balloon contains every single thing in the Universe within it. So when the Universe expands, it's like blowing up the balloon. The Universe (or balloon) is not expanding into anything – it is everything. It's just that the surface of the balloon grows bigger. Or you can imagine that the expanding Universe is like raisin bread rising. In the edible Universe it is the expansion of the dough that moves the raisins apart (Figure 5.80). In the real Universe, it is the expansion of space that moves galaxies apart.

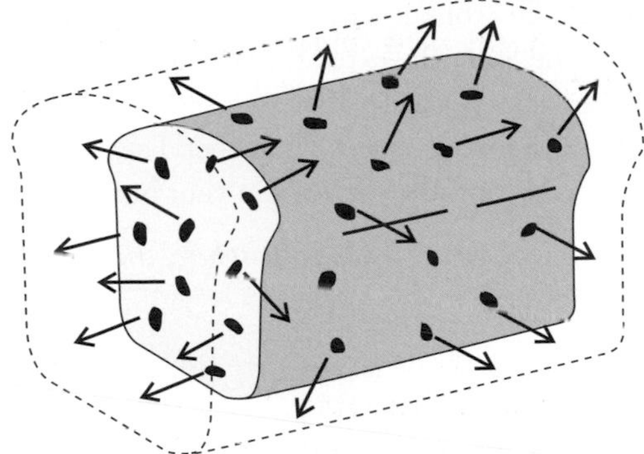

Figure 5.80 Raisin bread rising

When astronomers explain why galaxies appear redshifted, they are careful not to say that it's because of the Doppler effect. Why? Because the Doppler effect is a change in observed wavelength owing to a moving source. In the case of galaxies, it is the expansion of space itself that causes the shift in wavelength; Figure 5.81 illustrates the difference. To mark the difference, the redshift of receding galaxies is properly called

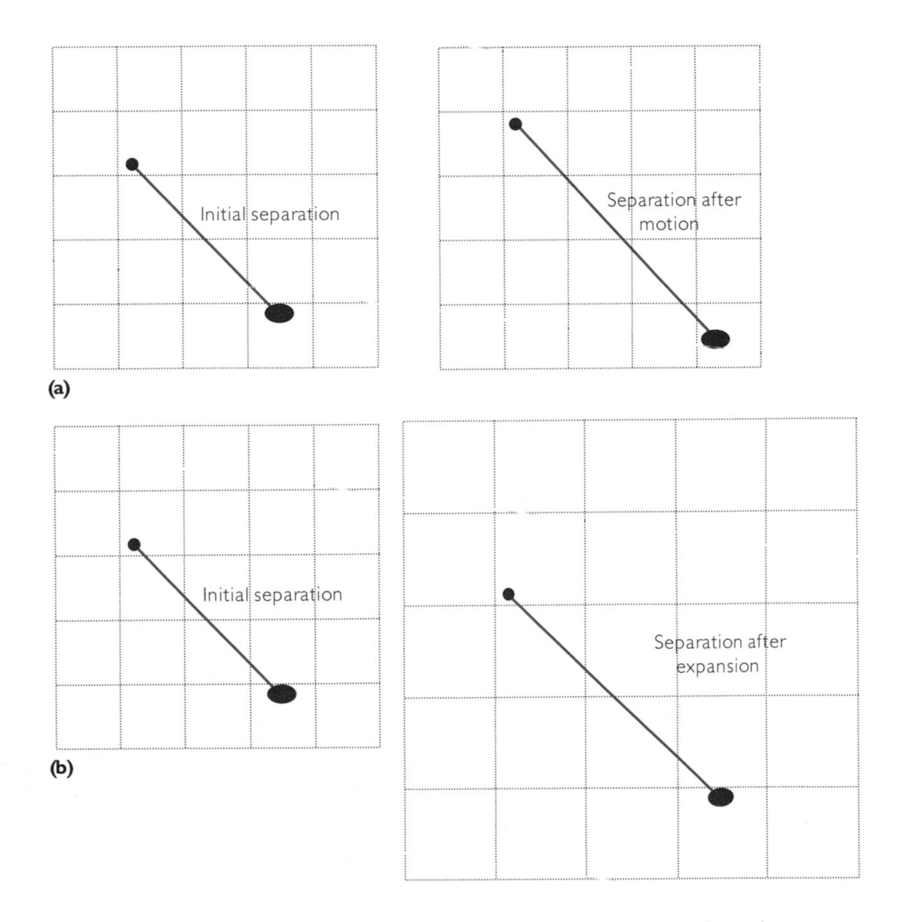

Figure 5.81 Galaxies increasing their separation because of (a) motion through space; (b) the expansion of the Universe

cosmological redshift. However, you can use the formula for the Doppler effect to calculate the cosmological redshift for receding galaxies, because the result of expanding space is that galaxies are speeding away from us.

You might wonder why, if the Universe is expanding, the Earth is not getting further from the Sun. The expansion of the Universe only applies to sufficiently distant objects like galaxies. Particles that are relatively close together are much more affected by attractive forces, and stay together. For instance, the Sun and Earth are held together by gravity, and the particles in your body are kept in place by electrical forces.

> *The Universe is an infinite sphere, the centre of which is everywhere, the circumference nowhere.*
>
> Pascal, *Pensées*

Activity 36 helps answer the question: is our galaxy at the centre of the expanding Universe? Whichever ink mark you measure from, you will see that speed is proportional to distance. Likewise, an intelligent being on a planet in another galaxy would see all other galaxies speeding away from them too.

It started with a bang

If we imagine travelling backwards in time, then galaxies would all be speeding towards each other, making the Universe a denser place in the past. If we continue going backwards, then we should reach a moment in time when the Universe was crammed together into a state of almost infinite density.

This reasoning suggests that the Universe began in a sort of explosion. It's hard to conceive of such a colossal event. The Universe began at infinite density and temperature, and all the matter and energy in the Universe today was released. Ever since, the matter and energy have been travelling outwards, thinning out and cooling. This is basically the **Big Bang** model of the origin of the Universe.

People sometimes ask what happened before the Big Bang. Strange as it may seem, there is no before the Big Bang. Time itself came into being at the moment of the Big Bang.

The UCLA (University of California at Los Angeles) has a useful website that addresses some frequently asked questions (FAQs) about cosmology that you an access from **www.shaplinks.co.uk**.

The Big Bang in fact follows logically from Einstein's theory of general relativity (and a few other assumptions). But it wasn't that long before an alternative theory appeared, as you will see in Section 4.2.

Questions

59 Hubble's law holds from any viewpoint in the Universe. Explain why this supports the cosmological principle.

60 Explain the difference between Doppler redshift and cosmological redshift.

4.2 Was there a Big Bang?

In the 1990s, *Focus* magazine published an article about an imagined punch-up between two well-known cosmologists (Figure 5.82).

Figure 5.82 Adapted from *Focus* magazine, November 1996

Big Bang punch up

The Big Bang is a load of old balderdash

Vs

The Big Bang is the way it all started

In the red corner: that old bruiser Professor Sir Fred Hoyle, arch-enemy of the Big Bang theory of the origin of the Universe, who describes it as 'a form of religious fundamentalism'.

In the blue corner: the urbane Astronomer Royal, Professor Sir Martin Rees, and just about every other astronomer in the world.

It may seem like an uneven contest, but that is failing to take into account the legendary prowess of Sir Fred, the greatest living scientist never to win a Nobel Prize. Together with some redoubtable companions around the world, he has consistently argued that the Big Bang theory is total and utter scientific codswallop and presented a welter of evidence to back his claims, from stars seemingly older than the Universe to galaxies separated by unfeasibly huge distances being joined by 'impossible' arcs of gas and dust.

And then there are all those riddles such as what came 'before the Big Bang'. Sir Fred's answer is simple; the Big Bang never happened. Instead, he and his followers claim that the Universe exists in a 'steady state', which never had a beginning, but is kept topped up with fresh matter bursting into our Universe at the centres of galaxies.

For those in the blue corner with Sir Martin Rees, the knock-out punch is obvious...

Big Bang or steady state?

Fred Hoyle and colleagues devised their steady-state theory in the 1950s. Oddly enough, the idea for the steady-state theory came to Hoyle after seeing a film. The fact the film ended with everything returning to how it was at the beginning started him wondering whether the Universe could be like that too (with the expansion of the Universe compensated by extra matter forming within it).

For a while, the steady-state theory presented a real challenge to the Big Bang explanation. It explained the expanding Universe by proposing that new matter is continually created in space, compensating for the spreading apart of individual galaxies. In this way, the Universe maintained its steady state.

So what makes us think that Universe did begin with a bang? The two theories make different predictions about what the Universe looked like a long time ago. One way to find this out is simply to look far out into space. As light travels with a finite speed, we observe distant galaxies as they were a long time ago, not as they are now.

The steady-state theory predicts that distant places in the Universe should look just the same as nearby places. However, the Big Bang theory predicts that distant places should be denser, since in the early Universe the same amount of matter occupied a smaller space.

In the mid-1950s, the British astronomer Sir Martin Ryle and colleagues tried to settle the debate using a radio telescope to look at nearby and at distant galaxies. They found that galaxies further away were much more concentrated than those near to us. This observation did not match the prediction of a steady-state Universe.

Fred Hoyle was on the ropes, struck by a theory that he had originally named the 'Big Bang', as a term of abuse. But despite the evidence, the controversy continued. It was in the late 1960s that the Big Bang struck the knock-out punch.

Knock-out punch

The Big Bang theory made another specific prediction. The birth of the Universe must have produced colossal numbers of extremely high-energy photons from energetic collisions between subatomic particles. Some radiation would have been absorbed, but not all. What happened to it? Shouldn't we be able to see it today?

In the 1950s Russian–American physicist George Gamow and colleagues realised that today's Universe should be bathed in **cosmic background radiation**, the remnant of the heat of creation, greatly cooled by the expansion of the Universe. (Remember, it was the whole Universe itself which emitted it, not just one part of it.) The radiation from the Big Bang has been travelling for about 15 billion years. The expansion of the Universe has literally stretched the waves, increasing the wavelength and reducing the energy of the photons (Figure 5.83). Gamow predicted that the peak wavelength of that radiation should now be in the microwave part of the spectrum, corresponding to the energy emitted by an object at approximately 3 K – just three degrees above absolute zero. Scientists thought the radiation would be very difficult to detect – it would require extremely sensitive microwave equipment.

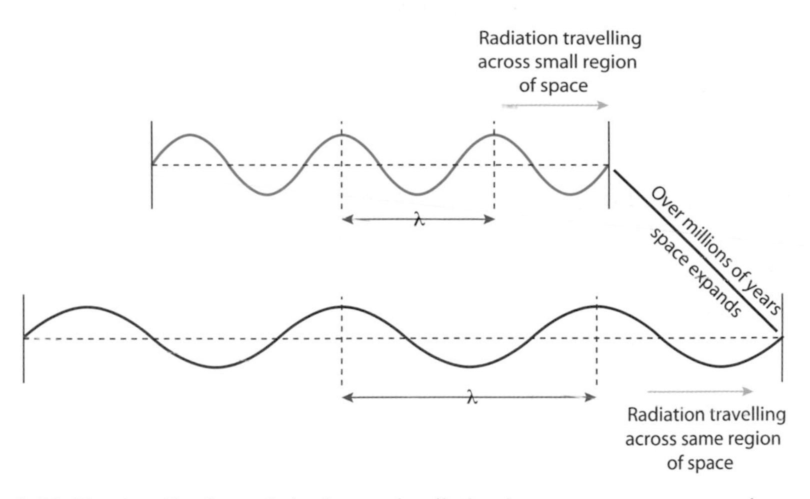

Figure 5.83 Wavelength of cosmic background radiation increases as space expands

However, in 1965, two American radio engineers, Arno Penzias and Robert Wilson (Figure 5.84), were testing a new horn-shaped microwave antenna. They had found an annoying constant hiss of static, which for months they thought was a black-body radiation from pigeon droppings. But try as they might, they couldn't get rid of it. Later they realised they had discovered the cosmic background radiation. It was like finding a smoking gun – direct evidence that the Universe started off incredibly hot and dense. The steady-state theory practically vanished overnight. The discovery of cosmic background radiation also supported the cosmological principle. The intensity of the microwaves was the same in all directions.

Figure 5.84 Arno Penzias (left) and Robert Wilson

Later measurements (including those by the Cosmic Background Explorer satellite in the 1980s) have confirmed that the spectrum closely matches that of a black body at a temperature of 2.7 K (see Figure 5.85 and compare it with Figure 5.35 in Part 3).

A third line of evidence for the Big Bang is its successful prediction of the amount of helium in the Universe. Large amounts were produced in the first few minutes of the Big Bang (see Section 4.3). The model predicts that about 23% of the matter in the early Universe was helium. Indeed, astronomers have been able to measure the proportion of helium in some of the oldest objects in the Universe – and they've found that it lies between 23% and 24%.

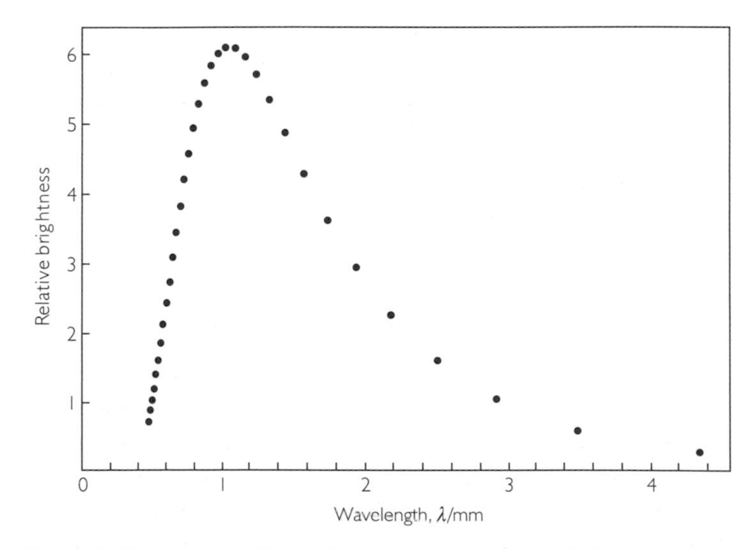

Figure 5.85 Spectrum of the microwave background radiation

Questions

61 Explain why the finding of the cosmic background radiation proved fatal to the steady-state theory.

62 What will happen to the spectrum of the cosmic background radiation if the Universe expands forever?

Activity 37 Cosmology chat show

Imagine a chat show where the two guests are Sir Martin Rees and Sir Fred Hoyle. Each comes on in turn, answers questions from the host about their theory, and then there is a chance for them to 'discuss' together. In a small group write and act out the script for the show – or improvise what you think they might say.

How old is the Universe?

The Big Bang may have neatly disposed of its rival theory, but it still had to contend with problems of its own (Figure 5.86).

Figure 5.86 The Big Bang theory has some problems

The Big Bang theory allows scientists to find the age of the Universe. By knowing how fast the Universe has been expanding, they can work out how long it must have taken to get to its current state. The Universe is believed to be between about 12 and 14 billion years old – scientists are still not completely in agreement. Current estimates put the 'best' age of the Universe at about 13.7 billion years.

You can take a virtual tour to the edge of the Universe at the NASA website and you can also view the simulated evolution of the Universe, as studied by the Wilkinson Microwave Anisotropy Probe (WMAP) (see **www.shaplinks.co.uk**).

A Universe that's at least 13.7 billion years old is fine. The problem arises if the age as predicted by the Big Bang turns out to be nearer 12 billion years, as some measurements by the Hubble Space Telescope have suggested – because astronomers have detected galaxies whose ages have been reliably measured to be 15–17 billion years old (by another method). It doesn't make any sense to say that a galaxy is older than the Universe. So is there something wrong with the Big Bang? Or is it simply difficult to get an accurate result of the age of the Universe? Let's now look at how cosmologists estimate the Universe's age.

The low-redshift approximation of Hubble's law gives us a simple way to estimate the age of the Universe:

$$v = H_0 d \qquad \text{(Equation 52)}$$

Suppose a galaxy has been travelling at speed v for a time t since the Big Bang. The distance it has travelled is given by:

$$d = v\,t$$

where t is the age of the Universe. Substituting for d from Equation 52 gives:

$$\frac{v}{H_0} = vt$$

so:

$$t = \frac{1}{H_0} \qquad (54)$$

With H_0 expressed in its normal units of km s^{-1} Mpc^{-1}, its relation to the age of the Universe is not obvious. The following example shows how to deal with the units.

Worked example

Q If H_0 = 70 km s^{-1} Mpc^{-1}, what does Equation 54 imply is the age of the Universe?

A 1 Mpc = 3.09×10^{22} m, and so 1 Mpc^{-1} = $(3.09 \times 10^{22}\text{ m})^{-1}$

In SI units:

$$H_0 = \frac{70 \times 10^3 \text{ m s}^{-1}}{(3.09 \times 10^{22}\text{ m})} = 2.27 \times 10^{-18}\text{ s}^{-1}$$

Then:

$$t = \frac{1}{H_0} = \left(\frac{1}{2.27 \times 10^{-18}}\right)\text{ s} = 4.41 \times 10^{17}\text{ s}$$

1 year = 3.16×10^7 s, so $t = 1.4 \times 10^{10}$ years (ie about 14 billion years)

One cause of the controversy over the age of the Universe is the uncertainty in the Hubble constant. Why is it difficult to measure H_0?

Determining H_0 relies on accurate measurements of distances to galaxies. Although astronomers can measure the distance to relatively nearby galaxies accurately, it's hard to get accurate measurements of distances of galaxies tens of hundreds of millions of light years away.

Figure 5.87 Standard candles

Most techniques for finding distances to galaxies and stars are based on the principle (covered in Section 2.1) that when light radiates from a source, its flux decreases with distance. If you know the intrinsic luminosity of a star or galaxy and its flux, you can work out how far away it is. To do this, we simply need types of object, known as **standard candles**, whose luminosity we already know or can reliably estimate (Figure 5.87). Easier said than done...

In the early 20th century, American astronomer Henrietta Swan Leavitt (Figure 5.88) found that certain variable stars, called Cepheid variables, which grow bright and then dim in a regular cycle (Figure 5.89), appear to be standard candles. Before Cepheids could be used as standard candles, their distance and hence their luminosity had to be determined using an independent method. Based on such independent distance measurements, astronomers now have a formula for determining the Cepheids' luminosity from their periods and can thus find their distance.

Figure 5.88 Henrietta Swan Leavitt

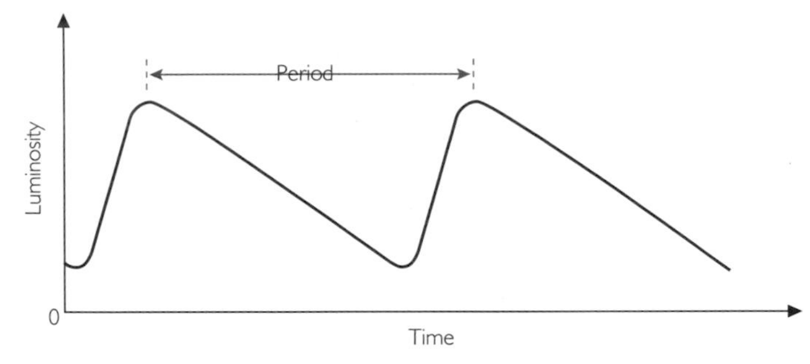

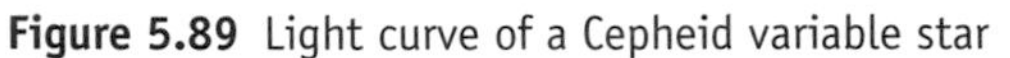

Figure 5.89 Light curve of a Cepheid variable star

Cepheid variables are no use beyond about 30 million light years away because their flux becomes too low and they cannot be seen. For such vast distances, astronomers use supernova explosions. One particular type, known as Type Ia, are very bright events that occur in binary systems containing a white dwarf star and a companion. Crucially, they always happen in the same way, producing the same luminosity which, at its peak, is about 50 million times that of the Sun. Other methods are based on assuming that all galaxies of the same type are equally luminous. The greater the distance, the more difficult it is to find standard candles, hence the uncertainties in the Hubble constant and the age of the Universe.

How big is the Universe?

The Universe is, presumably, infinitely big. If the Universe had an 'edge' then it would not be the same throughout – parts near the edge would appear different from parts near the centre, and the cosmological principle would not apply. Yet astronomers sometimes quote a size for the Universe. What can they mean? We can only see a small fraction of the Universe. If light has only been travelling for up to 14 billion years (since the Big Bang) we can only see parts of the Universe up to 14 billion light years away. Light from more distant parts hasn't reached us yet. When people talk about the 'size' of the Universe they mean the size of the part we can observe, which is related to its age.

The cosmic microwave background radiation we can detect is from about 300 000 years after the Big Bang, when the Universe became transparent to photons. This would correspond to a redshift of $z \approx 1000$. The highest confirmed redshift for a galaxy at the time of writing (autumn 2008) $z \approx 7$, with unconfirmed redshifts up to $z \approx 10$ for the furthest galaxies we can detect.

The big picture

Recent surveys, such as the 2dF Sky Survey have allowed us to measure the redshift of distant galaxies accurately and plot 3D maps of their distance from us, producing the largest scale maps of the Universe (Figure 5.90). These reveal some very unusual distributions, resulting in 'structures' such as enormous voids and the Sloan Great Wall, the largest known structure in the Universe (nearly 1.4 billion light years long). You can read more about the large-scale distribution of galaxies using the Harvard weblink (see **www.shaplinks.co.uk**).

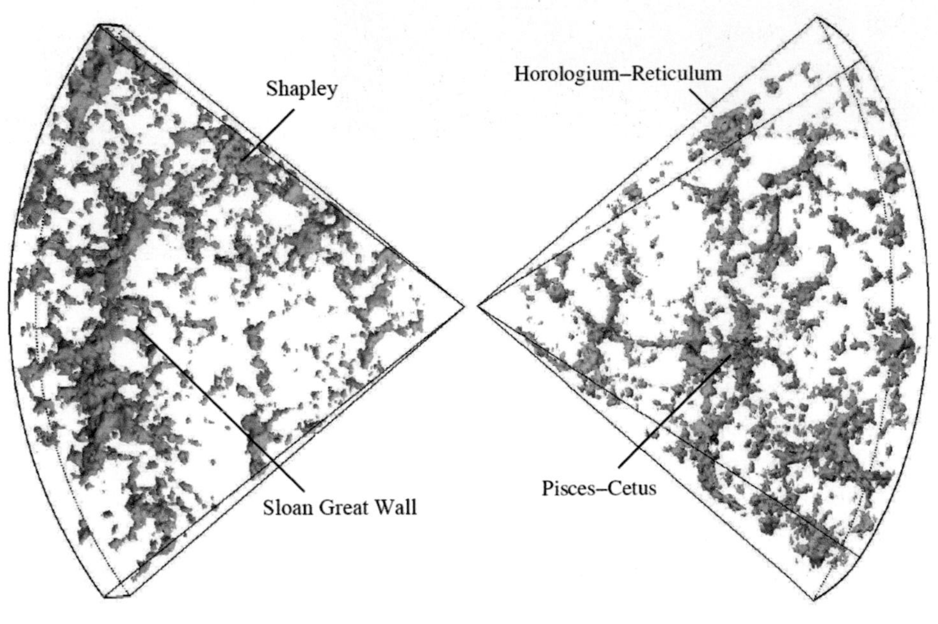

Figure 5.90 2dF sky survey map

The Shapley Supercluster is the largest concentration of galaxies yet seen (tens of thousands that gravitationally interact, remaining a single unit). This was rediscovered by Dr Somak Raychaudhury at the University of Birmingham, and named after its original observer, the American astronomer Harlow Shapley (1907–1972).

On slightly smaller scales we can now clearly see some of these voids and superclusters (Figure 5.91). These superclusters are themselves composed of thousands of clusters of galaxies. Understanding the distribution of these features will help us to better explain how the Big Bang happened.

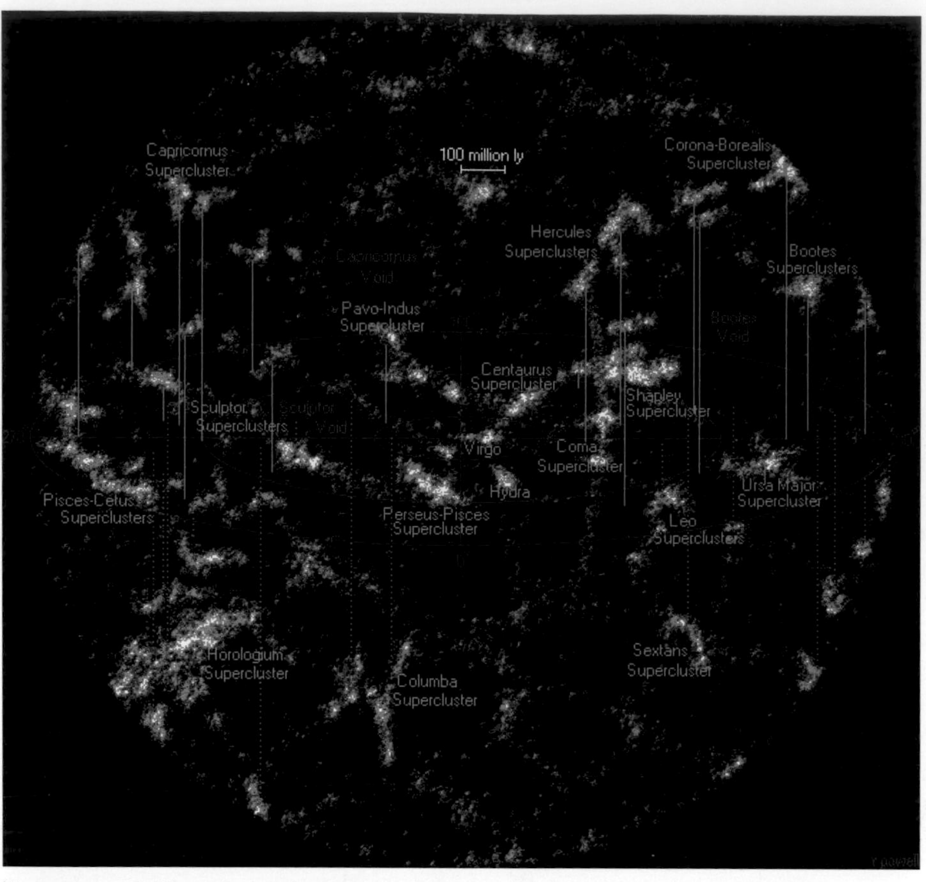

Figure 5.91 Voids and superclusters

Questions

63 If $H_0 = 100$ km s^{-1} Mpc^{-1}, how old, approximately, is the Universe?

64 A low-redshift quasar is believed to be 350 Mpc from our galaxy. If $H_0 = 70$ km s^{-1} Mpc^{-1}, what are (i) the quasar's recessional speed in km s^{-1} (ii) its redshift, z?

65 A Cepheid variable star in the large Magellanic Cloud (a small galaxy close to the Milky Way) has a radiant energy flux of $F_{LMC} = 9.5 \times 10^{-15}$ W m^{-2}. The Hubble telescope observed a Cepheid with the same period in the galaxy known as M100 and found its flux to be $F_{M100} = 2.7 \times 10^{-19}$ W m^{-2}. This is close to the smallest flux that can be measured with current instruments.

The distance to the LMC has been determined by other methods; it is 65 kpc.

(a) According to the Hubble Cepheid measurements, what is the distance to the galaxy M100?

(b) What, roughly, would be the furthest distance at which Hubble (or a similar instrument) could measure the peak flux from a Type Ia supernova ($L \approx 5 \times 10^7 L_{Sun}$)? Express your answer in pc and in light years.

(c) Suggest how the supernova luminosity might originally have been measured.

(d) Comment on your answer to (b) in relation to the apparent age of the Universe.

(Sun's luminosity, $L_{Sun} = 2.9 \times 10^{26}$ W; 1 pc $= 3.09 \times 10^{16}$ m;
1 ly $= 9.46 \times 10^{15}$ m.)

4.3 Riding through time

A newspaper once conducted a poll of its readers, asking 'Scientists have managed to work out the physics for how the Universe behaves from 1 s after the Big Bang, until now. True or false?'

The majority of people thought 'false'. But in fact we do have experimental evidence that backs up theories of the early Universe to within a second of the Big Bang. Not from telescopes, but from particle physics. A particle accelerator, by accelerating particles to close to the speed of light, can recreate the incredible temperatures of the birth of the Universe.

With the Big Bang model, and drawing on the results of particle-physics experiments, we can take a ride through time, charting the history of the Universe from the first instant of creation. So put on your virtual reality headset for a ride through the early history of the Universe – Figure 5.92 shows what you would see.

Display
Temp: 10^{11} K dropping

Time: 0 yrs
0 hr 00 min 00.01 sec

Goggles tuned to: gamma rays

Objects in focus: quarks

Commentary:
'A myriad of tiny particles swarm around in the murky haze. Some are much bigger than they exist today. But almost as soon as they appear, they are gone, never to be seen again. The temperature is dropping rapidly, and the particles are losing speed a little.

Here are quarks, the building blocks of matter. As they collide with one another, they fuse together to make two kinds of bigger particles, called protons and neutrons. There are about equal numbers of each.'

(a)

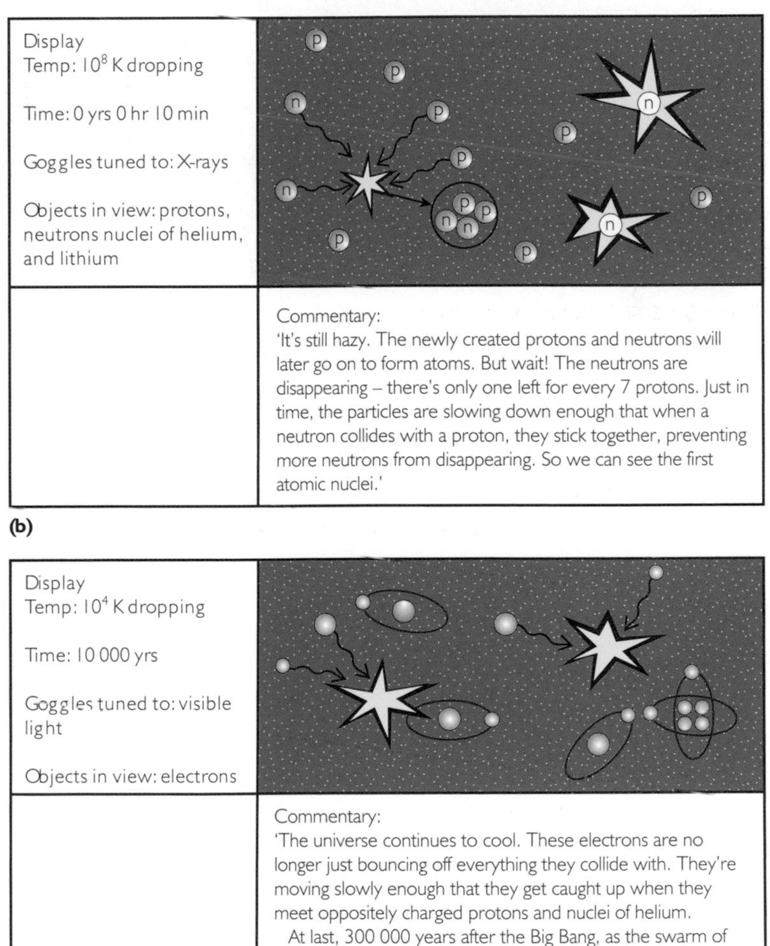

Figure 5.92 (a)-(f) Riding through time

Question

66 (a) Explain why looking at the light from distant galaxies is like looking back in time.

(b) We can't see through telescopes anything that happened earlier than 300 000 years after the Big Bang. Why not?

Activity 38 Matter in the Universe

Draw a diagram to show the stages by which fundamental particles in the early Universe became the large structures we see in the Universe today. Use the information in the 'ride' (Figure 5.92), and label each main stage with its temperature and its time from the Big Bang.

Display Temp: 10^4 K dropping Time: 10 000 yrs Goggles tuned to: microwave Objects in view: pink areas ñ hotter than average blue areas ñ colder than average	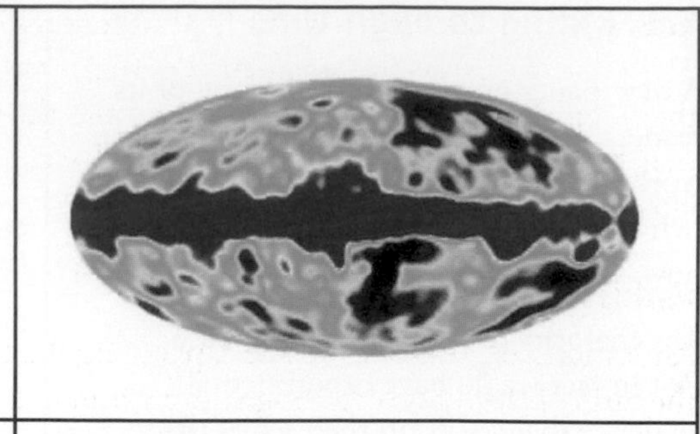
	Commentary: 'Without moving forward in time, we switch to a microwave view. This is no animation. This image came from a microwave detector sent into space in the 1980s. The radiation you can see began its journey nearly 15 billion years before it was detected. Most of the image is just "noise" from the detector, but careful analysis shows slight variations in the temperature and density. The denser parts are the seeds of galaxies that will form millions of years later.'

(d)

Display Temp: 1000 K dropping Time: 1 million yrs Goggles tuned to: visible Objects in view: dust, gas, clouds	
	Commentary: 'The Universe continues to cool and expand. Bits of dust and gas clump together, pulled by their gravity. The first structures start to appear in the Universe – big and small. As time goes on they grow.'

(e)

Display Temp: 100 K dropping Time: 30 million yrs Goggles tuned to: infrared Objects in view: galaxies, stars	
	Commentary: 'Now the Universe is a very cold place. You can see vast galaxies have formed. Inside them things are heating up dramatically. The blazing pinpoints of light that are breaking out all over the sky are the first stars beginning to shine.'

(f)

Figure 5.92 (a)-(f) Riding through time continued

The 'ride through time' is also a 'ride' through some important areas of physics. The reactions between fundamental particles are like some of those that you met in the chapter *Probing the Heart of Matter*, and the other key areas of physics are those that you studied in Parts 2 and 3 of this chapter: nuclear physics, kinetic theory and gravity. We will take a brief look at the role of each of these in turn.

Making nuclei

An important piece of evidence supporting the Big Bang theory is that it predicts the ratio of the mass of hydrogen to helium in the Universe exactly as scientists have measured it: 75% hydrogen, 25% helium. How were scientists able to make such a prediction?

According to the Big Bang model, the very early Universe was extremely hot and dense. Did you notice how in the virtual-reality ride, as the temperature dropped, particles began to move more slowly? This illustrates the relationship between the temperature of a substance, and the kinetic energy of its particles, which you met in Section 3.3.

Figure 5.93 shows how the temperature changed with time as the Universe expanded, as calculated using the Big Bang model. Particle physicists can work out what reactions took place, and which particles could survive at each stage. At first the Universe would consist of fundamental particles (quarks and leptons) and high-energy photons. As the temperature dropped, quarks could become bound into hadrons (eg protons and

Study note

You met and analysed Figure 5.93 in the chapter *Probing the Heart of Matter*.

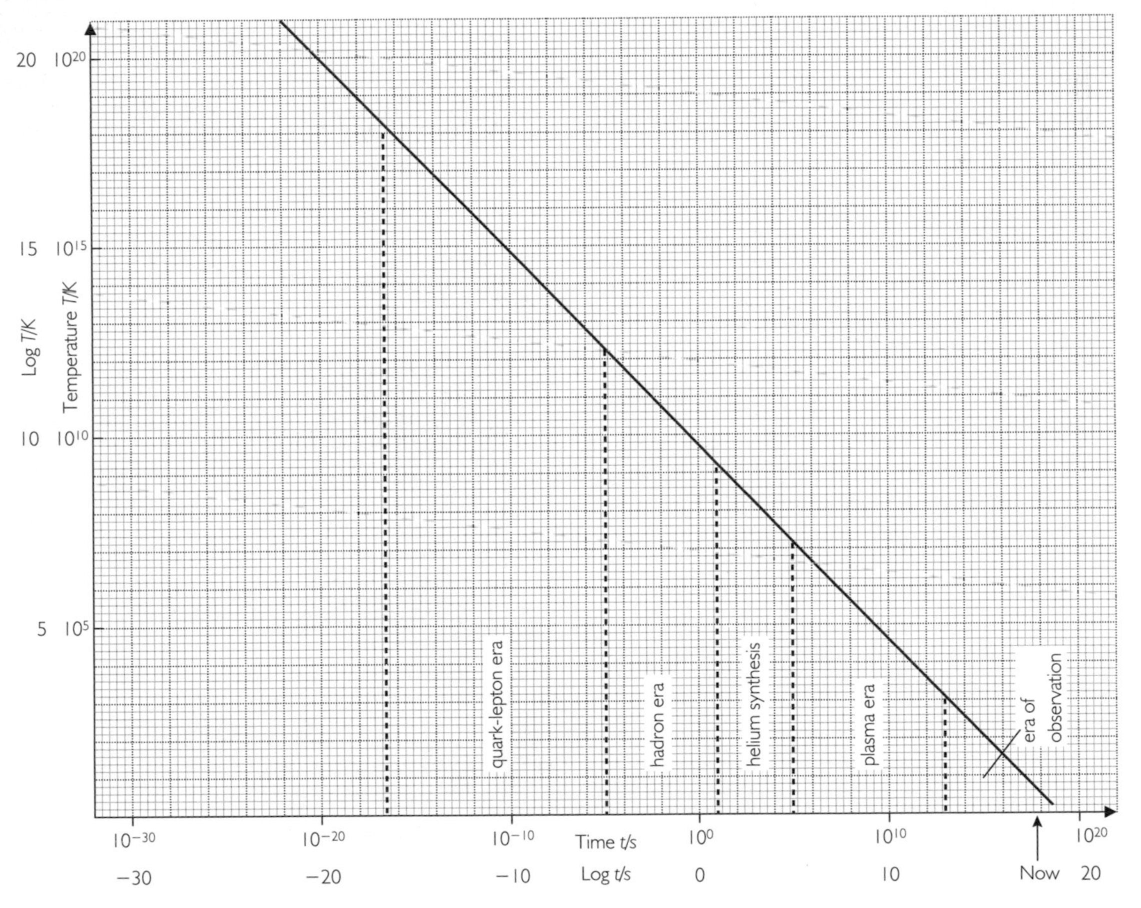

Figure 5.93 Temperature of the Universe falls with time

neutrons) without being shaken apart in violent collisions. Then, as the Universe cooled further, larger particles could begin to form in reactions like this:

$$^{1}_{0}\text{n} + ^{1}_{1}\text{H} \rightarrow ^{2}_{1}\text{H} + \gamma$$

forming the isotope $^{2}_{1}$H known as 'heavy hydrogen' or deuterium; its nucleus $^{2}_{1}$H is sometimes known as a deuteron. Fusion of $^{2}_{1}$H and $^{1}_{1}$H produces another hydrogen isotope, $^{3}_{1}$H, called tritium, and then a tritium and deuterium nucleus can fuse to make a nucleus of helium $^{4}_{2}$He.

Detailed calculations show that there is time only to convert less than one-twelfth of the protons into helium, and to produce minute amounts of lithium, before the Universe becomes too cold for further fusion to take place.

Questions

67 Write a balanced nuclear equation showing the fusion of $^{3}_{1}$H and $^{2}_{1}$H to make $^{4}_{2}$He. Which other particle must be produced in this reaction?

68 Calculate the mean kinetic energy and rms speed of helium nuclei when the temperature is 1.0×10^{8}K. (Mass of ^{4}He nucleus, $m = 6.68 \times 10^{-27}$ kg; $k = 1.38 \times 10^{-23}$ J K^{-1}.)

69 Given that the mass of a helium nucleus is four times that of a hydrogen nucleus, show that a ratio of 12 hydrogen atoms to one helium atom is consistent with a mass ratio of about 75% H to 25% He.

70 In stars, the production of helium is followed by further fusion to produce carbon. Explain why no carbon would have been produced in the early Universe.

Gravity takes over

Electric forces gave rise to atoms in the early Universe. Then, over large distances, gravity holds sway, and is the driving force of the evolution of the Universe. In the early Universe, the differences in density and temperature were tiny. But if there weren't differences, gravity would not have been able to clump matter together. Gravity pulls vast galaxies together. Gravity keeps planets orbiting their suns (Figure 5.94(a)). But with one finger, you can overcome the gravitational pull from the whole Earth (Figure 5.94(b)).

Gravity is an incredibly weak force. Compare it with another familiar force: the electrostatic force, which is responsible for the force when any two objects are in contact, for friction, and for surface tension.

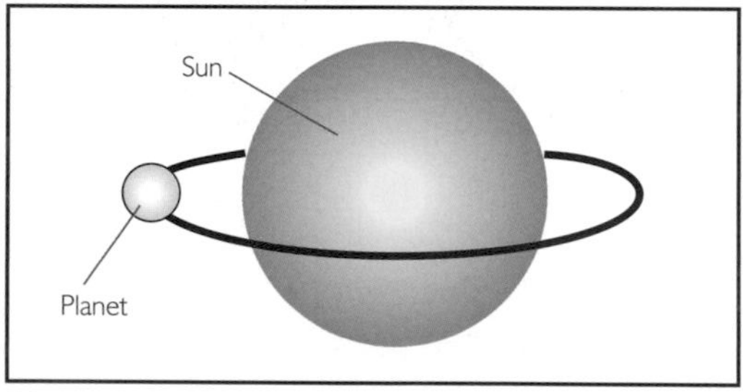

(a)

(b)

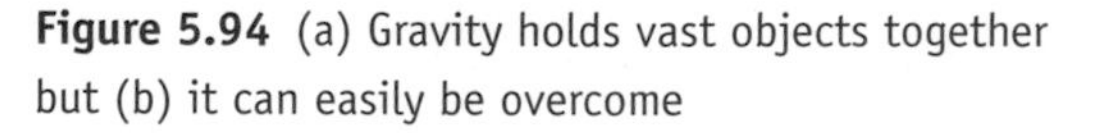
Figure 5.94 (a) Gravity holds vast objects together but (b) it can easily be overcome

Question

71 Compare the magnitudes of the electrostatic force and the gravitational force between two electrons.

(Data: electron mass, $m = 9.11 \times 10^{-31}$ kg; electron charge $e = -1.60 \times 10^{-19}$ C; $G = 6.67 \times 10^{-11}$ N m^2 kg^{-1}; $\varepsilon_0 = 8.85 \times 10^{-12}$ F m^{-1};
$k = \dfrac{1}{4\pi\varepsilon_0} = 8.99 \times 10^9$ N m^2 C^{-2}.)

Your answer to Question 71 should show that the electric force between electrons is an incredible 10^{42} times stronger than gravity, regardless of the separation of the particles.

So why is gravity so important on a big scale? Matter tends to be made of pairs of oppositely-charged particles whose charge cancels leaving no net electrostatic force. And the amounts of matter in big structures like stars and galaxies is huge, and so their gravitational effect is noticeable, causing atoms and molecules to gather together to form stars, holding stars together in galaxies and binding whole galaxies together into clusters (Figure 5.95).

Figure 5.95 Gravity causes clustering of matter

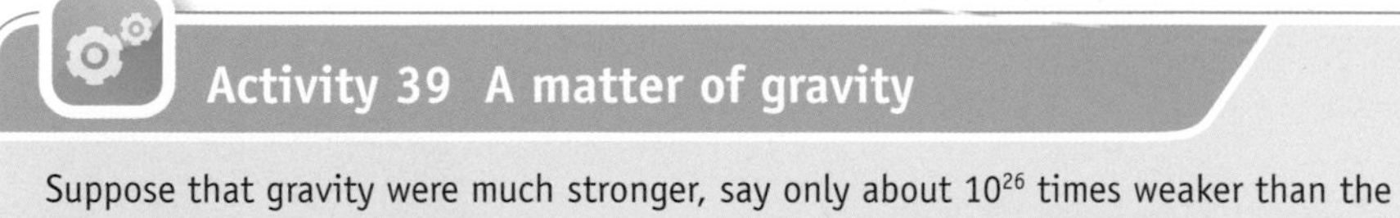

Activity 39 A matter of gravity

Suppose that gravity were much stronger, say only about 10^{26} times weaker than the forces between charged particles. Discuss what consequences this might have for star formation and the evolution of life.

We've seen how the Big Bang can give us a glimpse of the ancient past. But it can also take us on a trip to the future, as you will see in Section 4.4.

4.4 Into the future

One of the most embarrassing admissions which cosmologists and astronomers have to make is that they don't know what most of the mass of the Universe is made of.

Malcolm Longair, British astronomer

An open and shut case

What do cosmologists put bets on? Not horses or football matches. They're much more likely to be putting money on a number known as omega Ω. Why? Because, wrapped up in this single constant, is the fate of the Universe.

If $\Omega < 1$ the Universe will carry on expanding for ever, a so-called **open Universe** (a fate sometimes described as a 'big chill' or a 'big yawn'). There is a possibility that the expansion might accelerate rather than slow down.

If $\Omega > 1$ the Universe will eventually stop expanding, contract, and end up in a **Big Crunch**. This is a **closed Universe**.

You can see these possible futures illustrated in Figure 5.96. If the Universe is closed, then instead of ending after its collapse, it could begin a new expansion – a 'bounce'. In which case, maybe we are living in one of the bounces, and our Universe is actually much older than 10–20 billion years. However, scientists have found no plausible mechanism to produce these bounces and the theory is currently out of fashion.

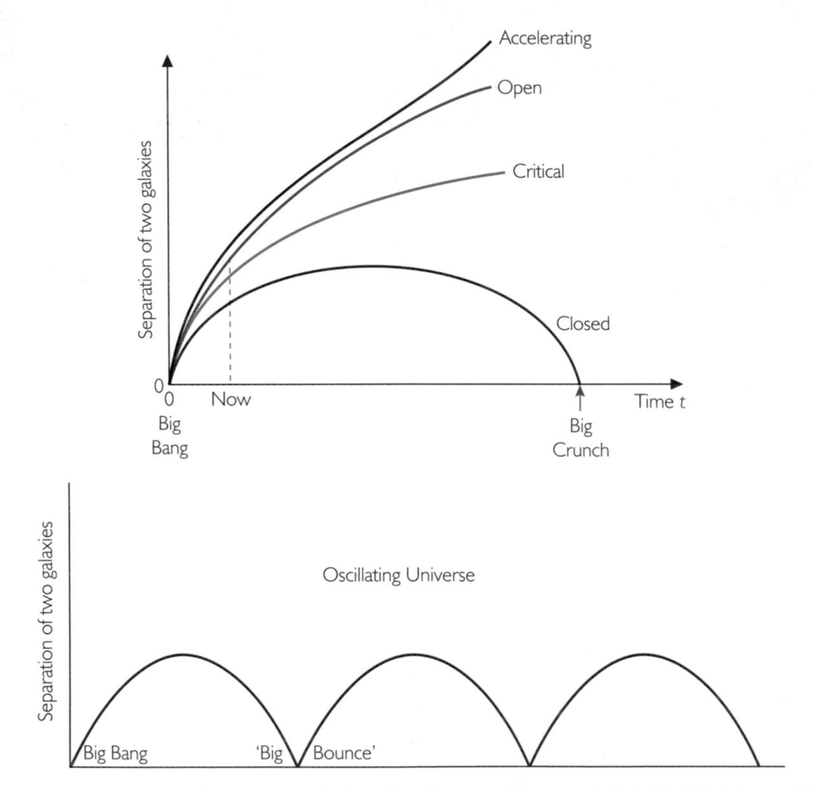

Figure 5.96 Graphs showing open and closed Universes (Source: NASA/WMAP Science Team)

Study note

The critical Universe and the accelerating Universe are discussed later in this section.

Why should the Universe ever stop expanding? Because the gravity between all the matter is trying to pull it together, and slow the expansion. In fact the expansion of the Universe has been slowing down ever since it started. Gravity and expansion are in direct competition. The winner dictates the fate of the Universe.

We can help explain the closed and open Universe by imagining what happens to a rocket fired away from the Earth (Figure 5.97). The rocket in Figure 5.97(a) has been fired at fairly low velocity. As it climbs, it loses kinetic energy as it gains gravitational potential energy. Eventually, its kinetic energy will decrease to zero, and it then falls back to Earth. This is very much like what happens in a closed Universe, where gravity wins (Figure 5.98). The rocket in Figure 5.97(b) has a higher velocity at launch. Even when it has reached a very great distance from Earth, it still has some kinetic energy and will keep on travelling out into space. This is like what happens in an open Universe, where the expansion wins over gravity – or, more correctly, kinetic energy wins over gravitational potential energy.

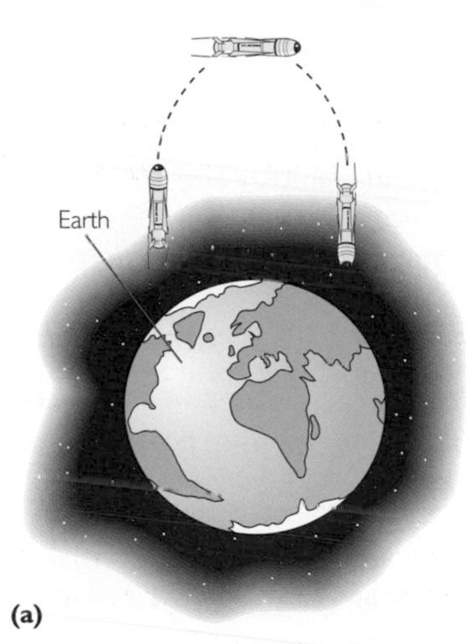

(a)

(b)

Figure 5.97 Rocket launched from Earth (a) at low velocity and (b) at high velocity

Figure 5.98 As the Universe continues to contract...

The argument is very like the one that determines whether a cloud will collapse and form stars or disperse owing to the motion of its particles. The main difference is that particles in a gas cloud are moving randomly, whereas galaxies are all moving apart from one another.

What is the difference between an open and closed Universe?

An open Universe will carry on expanding forever. A closed Universe will one day (a long time in the future) stop expanding and start contracting again.

That's the future. Is there any difference between the two now?

Yes. They have a different shape. A closed Universe has a limited size. That doesn't mean, as you might think, that there is anything outside it. It sort of bends around on itself, like the surface of a sphere does, so if you could walk across the Universe, you'd eventually end up where you started. But it's very hard to picture the shape of the Universe; you need mathematics.

An open Universe, on the other hand, has unlimited size. It extends infinitely in all directions. It might be a bit confusing to think something that's already infinite expanding even more. But all it means is that the distance between any two galaxies is increasing.

Do scientists know what would happen if the Universe did collapse?

Most of the time during the collapse, things will be pretty similar to conditions during the expansion phase we're in now. Planets and stars will be born, live and die. Similarly, galaxies will evolve and black holes will form from big dying stars.

The main difference is that spectra from distant galaxies will be blue-shifted (a cosmological blue shift), and the cosmic background radiation will get slowly warmer and warmer (though too slowly for people to notice).

Some people think that time will run backwards during collapse, but we don't know that.

Later on, clusters of galaxies will start to merge, and the galaxies they're made of will also combine. Stars in the galaxies will collide, forming black holes, and other stars and planets will evaporate as the temperature of space starts really hotting up.

Ultimately, the Universe may just end in the state it probably began in, as a 'quantum vacuum' (where particles pop in and out of existence). Or maybe it will bounce back into life again, reincarnated. If this happens, it's likely that the familiar forces of nature will be quite different, along with the particles that make up atoms. So who knows what will happen.

What if the Universe goes on expanding for ever? Isn't it going to get a bit lonely?

Quite true. One day (in about 120 000 billion years), all the stars will eventually run out of fuel, and they will stop shining. Then the Universe will become dark, populated by planets and black dwarfs. Its temperature will be just over absolute zero.

The Universe will shine its final brief dying light when remnant galaxies get sucked into the black holes at their centres. Then galaxies will lose their stars and planets, which will start wandering around in almost complete loneliness. They will be further apart than the whole of the observable Universe extends today. All they will have for company will be some very old particles that have been wandering around since the Big Bang.

We are not too sure what happens after that. But frankly, whatever it is, it won't be very interesting.

Questions

72 Another way of saying that the expansion of the Universe is slowing down, is to say that *the rate of change of expansion is decreasing.*

(a) Copy the graph for the open Universe (Figure 5.96). By drawing tangents to the line, explain the statement in italics.

(b) Copy the graph for the accelerating Universe. Draw tangents to this line and explain how the graph shows an acceleration.

73 In Section 4.1, the Hubble constant was used to estimate the age of the Universe, assuming a constant expansion. If the expansion is slowing down, will the values estimated in Section 4.2 be overestimates or underestimates?

Activity 40 Back to the future?

Plan a virtual-reality 'ride into the future'. You could write your plan as a story, or produce a series of cartoons showing what the user will see at each stage, called a 'storyboard', to describe what the user will experience. You could add commentary and sound effects, too. Think about whether the 'rider' will be able to choose to experience an open or a closed Universe, and how they will make that choice.

Big chill or big crunch?

This is the way the world ends,
This is the way the world ends,
This is the way the world ends:
Not with a bang but a whimper.

T. S. Eliot, *The Hollow Men*

You might think that cosmologists are wasting their time betting on the fate of the Universe. Surely we cannot know the result in advance. Wrong. The value of omega, which determines the outcome, is already fixed. It is as though it were in a sealed envelope waiting to be opened.

Omega is a ratio of densities:

$$\Omega = \frac{\rho}{\rho_c} \qquad (55)$$

where ρ is the actual average density of the Universe and ρ_c is called the **critical density**. If ρ is greater than this critical value (ie $\Omega > 1$), then there is enough mass in the Universe for gravity to defeat expansion, and the Universe to be closed. However, if $\rho < \rho_c$, the Universe is less dense than the critical value, expansion will win and the Universe will be open.

The way to calculate critical density is from the Hubble constant, as critical density is a measure of how much mass would be needed to halt the current expansion. Estimates of the critical density lie around 10^{-26} kg m^{-3}. This is an extremely small value, and it is uncertain because of the difficulties of measuring the Hubble constant. From this value, the Universe must on average contain at least the mass of one poppy seed in each volume the size of the Earth, for gravity to win, collapsing the Universe in on itself one day. Does the Universe contain this density of matter?

Scientists have estimated ρ, the actual density of the Universe, by examining huge volumes of space and calculating the mass of stars and galaxies needed to produce the observed amount of light. They have also calculated the density by estimating how the motions of galaxies are affected by local concentrations of mass. Unfortunately, all these estimates depend on knowing the distances to the galaxies, and we have seen that this is hard to find accurately. But the highest estimates turn out to be only about 10^{-27} kg m^{-3}, in which case:

$$\Omega \approx \frac{10^{-27}\ \text{kg m}^{-3}}{10^{-26}\ \text{kg m}^{-3}} = 0.1$$

Many estimates of actual density and omega are ten times lower. From these results, it looks as if the Universe is open, destined to end its days with matter spread ever wider, in a cold, lonely emptiness. But despite this, many cosmologists are betting on what seems an unlikely third way – that the density is exactly the critical value, making omega precisely one. This third possibility is called a **critical Universe** (or, sometimes, a flat Universe), as shown in Figure 5.96.

Why do many cosmologists think that gravity and expansion are so perfectly balanced? The evidence for omega being one is partly that we know it is close to one. It sounds absurd, but imagine the rocket again. This time it is launched with exactly the right vertical velocity just to keep on travelling away for ever. This critical velocity is called the escape velocity. If it launched with only slightly lower velocity, gravity would soon bring it back. Similarly, with slightly greater than escape velocity, you would soon know that it was going to escape easily. The longer time goes on the more the speed of the rocket would diverge from the critical value. What this means is, if you saw the rocket a *long* time later, still travelling with a speed just high enough to continue going for ever, it must have been launched with almost exactly the escape velocity.

Swapping the Universe for our rocket, seeing the value of Ω close to one means that it must have been almost exactly one at the Big Bang. Physicists have calculated that omega would have to be only one part in 10^{59} away from the value of one.

There is also reason to believe that a Universe with $\Omega = 1$ is 'easy' to create, on the grounds of energy. Finding the Universe so close to critical today, means the kinetic energy and gravitational potential energies are closely balanced. The potential energies of all the matter in the Universe (due to being in a gravitational field) are negative. The positive kinetic energy balances exactly with the negative potential energy, making the total energy in the Universe zero. Some scientists have suggested that creating a Universe with zero total energy is easier.

In the chapter *Probing the Heart of Matter*, you read that physicist Alan Guth had a spectacular realisation that led him to propose that the Universe had undergone a rapid inflation about 10^{-35} seconds after the Big Bang. One consequence of this inflation is that the Universe would achieve a density exactly equal to the critical density.

But as far as observational evidence is concerned for $\Omega = 1$, the jury is still out.

The riddle of dark matter

Nobody ever said all matter radiated.

Vera Rubin, American astronomer

I hope the missing matter isn't there.

Jesse Greenstein, American astronomer

If the Universe is flat, there is a big problem. We can only see a few per cent of the matter needed to make $\Omega = 1$. The rest is lurking unseen, invisible to our telescopes. The hidden stuff is called **dark matter** and there is also evidence for something called dark energy which we will discuss later. According to theoretical calculations, the relative amounts of dark matter and dark energy have changed over time (Figure 5.99).

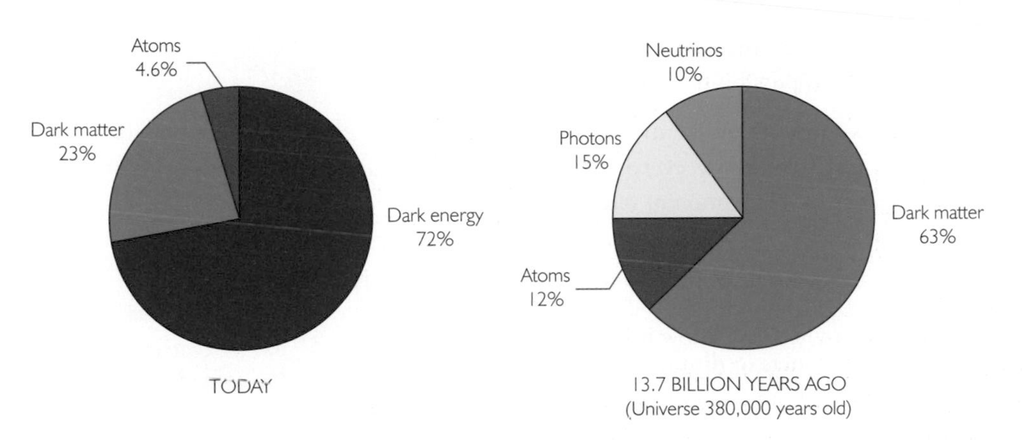

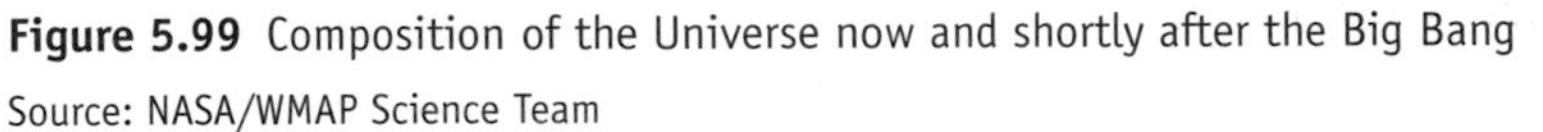
Figure 5.99 Composition of the Universe now and shortly after the Big Bang
Source: NASA/WMAP Science Team

Just because we cannot see dark matter does not mean it is not there. When you turn out the light in a room most objects disappear – only the luminous objects remain visible.

In fact we already know that dark matter exists. Though we cannot see it, astronomers have measured its gravitational effect on galaxies and stars, and believe there must be at least ten times as much matter as that we can actually see.

Some of the strongest evidence comes from the rotation of galaxies. Just as you saw in Part 3 that astronomers can determine the masses of stars from observing their effect on the orbits of planets or on one another, they can also determine the mass of a galaxy from the motions of stars near its edge. In the early 1970s, American astronomer Vera Rubin did just this for the Andromeda galaxy and some other spiral galaxies, and was surprised to find that the mass thus calculated was about ten times the mass of the stars and interstellar material needed to account for the output of radiation.

Other evidence comes from whole clusters of galaxies, in which the galaxies are seen to move around at random – rather like the kinetic-theory model of particles in a gas. Some years before Rubin's work, Swiss–American astronomer Fritz Zwicky measured the random speeds of galaxies in clusters and calculated the mass that would need to be present in order to stop the clusters dispersing. (This is another example of the sort of argument that we used in Part 3 when considering whether a gas cloud would disperse or collapse.) Here, too, there seemed to be about ten times as much mass in the clusters as could be accounted for by the radiating material.

Activity 41 Mass and motion

Use computer simulations to explore:

- how the rotation of a galaxy is affected by the distribution of its mass
- how the mass in a cluster of galaxies influences the galaxies' random speeds.

The website **www.shaplinks.co.uk** provides links to some websites that provide suitable applets.

The baffling question is: 'What is dark matter?' Simply, it is matter that we cannot detect by either its emission or its absorption of radiation (the dust that shows up as dark streaks on photographs of nebulae is not 'dark matter', as we detect its absorption of starlight). We often refer to this as 'cold dark matter' as it may interact very weakly (if at all) with normal matter. There are two possibilities:

- baryonic matter, made from baryons (protons and neutrons) and their accompanying electrons – in other words, ordinary, everyday matter in a form that renders it undetectable by the emission or absorption of radiation
- non-baryonic matter – which could include all manner of exotic particles not yet detected, probably at high GeV energies; we often refer to these as WIMPS (weakly interacting massive particles).

Study note

Baryons were discussed in *Probing the Heart of Matter.*

Study note

See the chapter *Probing the Heart of Matter* for more information about CERN and the hunt for new particles.

Baryonic matter could be in the form of planets or dim stars. Earth-like planets are not going to contribute a lot of mass to the Universe, but perhaps Jupiter-sized planets could. Other candidates include brown dwarfs – these are failed stars that were not quite massive enough to start fusion and become luminous. But we do not know. For all we know some dark matter could turn out to be bricks in space – they would be very hard to detect.

However, we know that not all the dark matter can be baryonic. The Big Bang theory makes a clear prediction for the total amount of ordinary, atom-forming matter in the Universe. It says there should only be enough baryonic matter to make $\Omega = 0.1$. If Ω really is one, then the remaining 90% of the Universe has to be made of stranger forms of matter.

One possible candidate for non-baryonic dark matter is the neutrino. It is not yet established whether neutrinos have any mass at all, but there are so many neutrinos in the Universe that with even a very small mass they could together make a significant contribution to the total mass of the Universe.

Current research

One technique that might help solve the riddle of dark matter goes by the name of gravitational lensing. It is based on Einstein's theory of gravity (general relativity). Einstein realised that, as well as matter, light is affected by gravity. So a beam of light would curve as it passes a large concentration of matter.

The technique is based on looking at the image of a very distant object, and seeing how it is altered by gravity. When light from a very distant galaxy (billions of light years away) makes the long journey to Earth, it is sometimes deflected on its journey to Earth by matter en route. The matter acts like a lens, distorting and splitting the image into two. About a dozen gravitational lenses have been discovered so far.

Figure 5.100 shows some examples of gravitational lensing. In Figure 5.100(a) the four images around the central one are all of the same galaxy, and in 5.100(b) the arc of light is an image of a galaxy that has been magnified and distorted. Figure 5.101 shows how a cluster of galaxies can produce gravitational lensing of light from a very distant object. Notice how the wavefronts and the rays behave as if they were passing through a converging lens. By studying images such as those in Figure 5.100, physicists can then reconstruct what the lens would be, its distribution in space and, importantly, its total mass.

Figure 5.100 Images produced by gravitational lensing

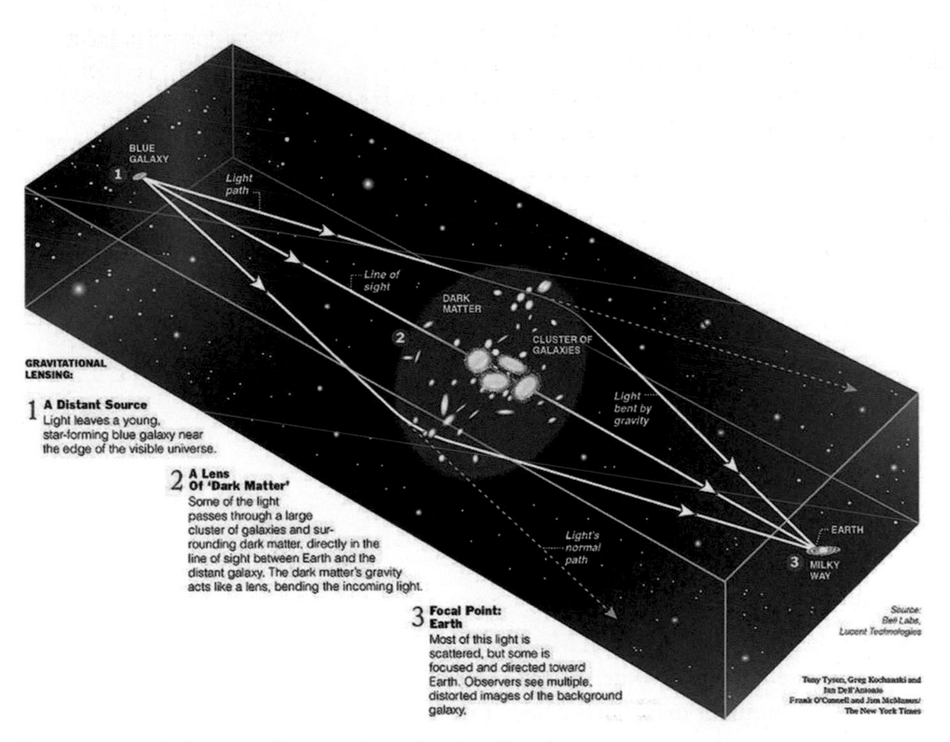

Figure 5.101 Schematic diagram of gravitational lensing

David Davidge and colleagues pursued a very different approach, searching for the missing matter a kilometre underground in a Yorkshire potash mine, for a week at a time (Figure 5.102). They set up an incredibly sensitive detector to pick up signals from one of the best candidates for dark matter. They were searching for WIMPs. Weakly interacting massive particles are generally heavy particles (thousands of times the mass of a proton) that only interact weakly with other matter. Scientists know that WIMPs must exist, and some of the proposed types of WIMP have been given exotic names like neutralinos, axions and others. But, right now, scientists do not know what WIMPs there actually are. That is because they do not interact readily with anything. WIMPs can glide through the gaps between electrons and nuclei with ease, and could be streaming right through you as you read this. So this makes them nearly impossible to detect.

Figure 5.102 You'll find physicists in some odd places

However, David Davidge's team expected that there ought to be one WIMP passing through their experiment (called WIMP event) every few days. You might think an event this rare would be easy to spot. But unfortunately, there are a million other events to confuse them, which are nothing to do with WIMPs, which is why they went down to the bottom of the mine (Figure 5.103), to shield the detector from all the unwanted signals.

David's team is one of several around the world, using different methods, but all in the race to find dark matter. And what a discovery it would make.

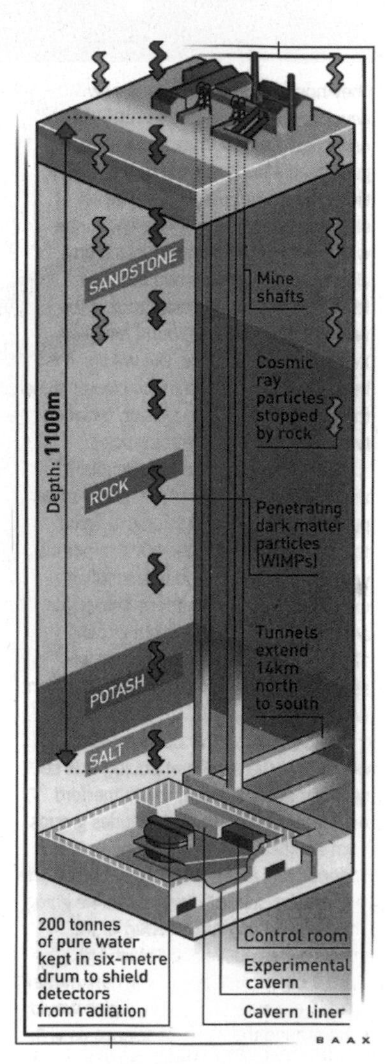

Figure 5.103 WIMP detector

Dark energy – a repulsive idea

Maybe even finding dark matter will not be the end of the story. Many physicists now believe the fate of the Universe is not wholly dependent on the value of Ω. It is possible that there is another number that affects whether the Universe expands forever or not. This is called the 'cosmological constant'. Einstein introduced it into his equations because he thought it was necessary to stop a static Universe collapsing under its own gravity. Then when Hubble discovered the Universe was expanding, the constant seemed unnecessary and Einstein called it his 'greatest blunder'. But maybe Albert Einstein will have the last laugh after all...

In the late 1990s, two separate groups of scientists studied the explosions of Type Ia supernovae in several very distant galaxies whose distances were known from their redshifts. Their aim was to determine the rate at which the expansion of the Universe was slowing down. They found that the supernovae were much fainter than expected, and their surprising interpretation of the results was that the Universe is actually expanding faster, and must have been speeding up for several billion years.

Study note

Type Ia supernovae were discussed in Section 4.2.

This has since been confirmed by other observations, including two detailed surveys of the sky – the *2 degree field* (2dF survey) and the *Sloan Digital Sky Survey* (SDSS). It means that the Universe isn't just open, it is accelerating (Figure 5.104).

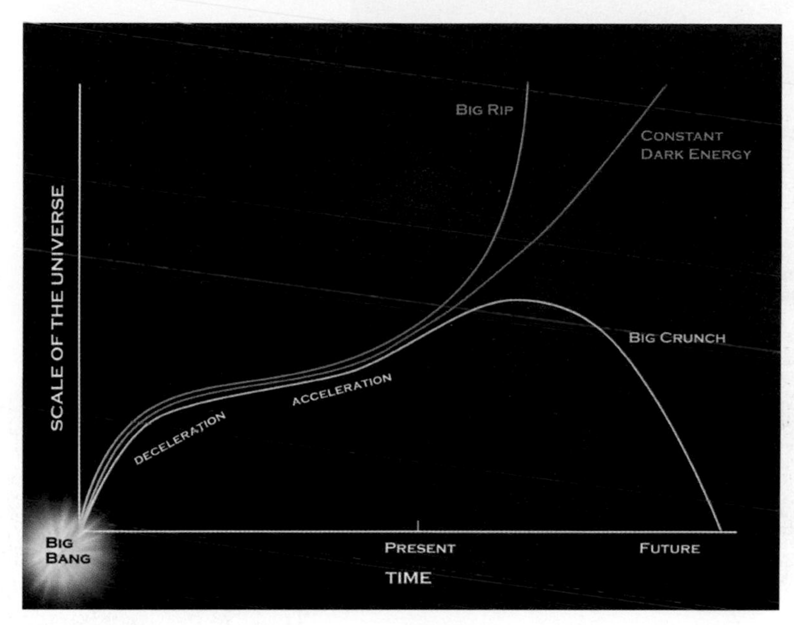

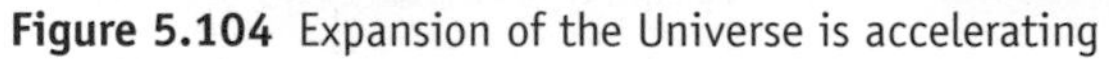
Figure 5.104 Expansion of the Universe is accelerating

To produce this effect, it is believed that three-quarters of the Universe must be composed of gravitationally repulsive **dark energy**, a radical and controversial idea in the world of physics. Several have suggested that the idea of dark energy is remarkably similar to Einstein's cosmological constant, and indeed, it may be that we need to revive the idea (although it would have a major impact upon our current theories about particle physics).

As the Universe expanded, it is believed that the balance between the attractive pull of dark matter and the repulsive push of dark energy changed, with the latter becoming more dominant 6 to 8 billion years ago (Figure 5.105).

It is difficult to explain dark energy using conventional physics. It produces effects similar to a negative pressure, hence its repulsion. It may be that dark matter and dark energy are connected in some way we don't yet understand. It appears that dark matter may be more 'clumpy' in its distribution than dark energy.

One alarming consequence of dark energy is the possibility of a *Big Rip*, tens of billions of years in the future, as suggested by Robert Caldwell of Dartmouth University. The accelerated expansion of the fabric of space–time by dark energy could cause matter to lose its attractive grip, and structures such as galaxies, then stars, planets and eventually atoms themselves would break apart. Could this then be the end of the Universe?

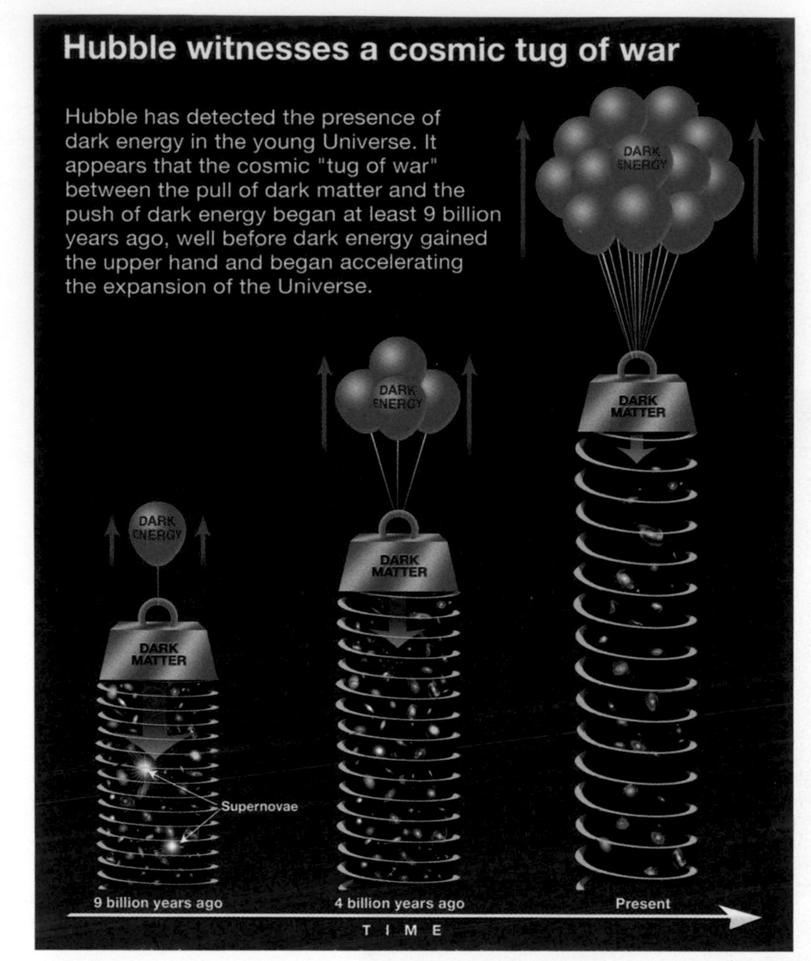

Figure 5.105 Cosmic tug of war

At the time of writing (2008) NASA is proposing a *Joint Dark Energy Mission* to measure the expansion rate of the Universe more accurately and try to pin down the evidence for, and effects of, dark energy.

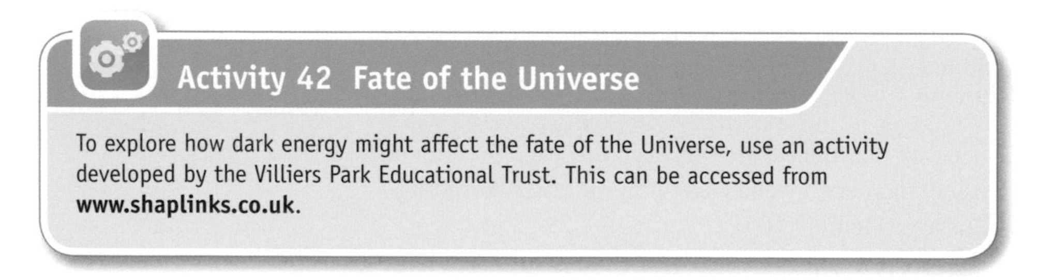

Activity 42 Fate of the Universe

To explore how dark energy might affect the fate of the Universe, use an activity developed by the Villiers Park Educational Trust. This can be accessed from **www.shaplinks.co.uk**.

4.5 Summing up Part 4

In this part of the chapter you have read about some of the biggest questions currently being addressed by astronomers and cosmologists, and seen how they relate to the physics that you have studied in the whole of this chapter and elsewhere in the course.

Activities 43 to 46 are intended to help you look back over this part of the chapter. You will find further questions in Part 5.

Activity 43 The cosmologists

Put yourself in the position of one of the great astronomers or cosmologists of the 20th century, giving a five-minute presentation about your ideas at a scientific conference. You could choose someone featured in this chapter or someone that you have researched yourself.

Do the presentation as a group. One person in the group could introduce the scientists, explaining to the audience, eg when the scientists worked, and a few details about their lives.

Include a brief account of your theory or experiment, and say why it was important. To help you communicate your message, use props or audiovisuals (eg a clear diagram on an overhead projector). You can find further information from books or websites – some useful websites can be accessed from **www.shaplinks.co.uk**.

Activity 44 In the beginning

Read the poem by Primo Levi printed below, and answer the following questions:

- Why do you think Levi calls the fireball 'solitary and eternal'?
- What does Primo Levi believe the fate of the Universe to be? How do you know?
- Explain (in terms a non-scientist would understand) how the 'thin echo' of the Big Bang 'resounds from the furthest reaches'.

Ten billion years before now,
Brilliant, soaring in space and time.
There was a bath of flame, solitary and eternal,
Our common father and our executioner.
It exploded, and every change began.
Even now the thin echo of this one reverse catastrophe
Resounds from the furthest reaches.

Primo Levi, 'In the Beginning'. From *Shema*, translated by Ruth Feldman and Brian Swann, Menard Press, 1976

Activity 45 Cosmology today

Use the internet to find out more about modern cosmology research. Go to **www.shaplinks.co.uk** for some useful websites.

Imagine you are applying for a research job in cosmology. Prepare an application and then take part in an 'interview' with other students acting as the interviewer.

Activity 46 Origins

Cosmology has come a long way since the ancient beliefs of civilisations from Greece, Rome, China, Maya and India. Summarise what you think are the main achievements of cosmology that have furthered our understanding of the Universe, and its origins and future.

5 Synthesis

5.1 Universal physics

In this chapter, you have studied four main areas of physics: black-body radiation, nuclear physics, gravitation and kinetic theory. You have also touched on several areas of physics from elsewhere in the course, including radiant energy flux, the Doppler effect, electrostatic force, and energy conservation. You have used many of the key ideas in more than one part of the chapter, and seen how they have helped, and are still helping, astronomers and cosmologists to find out more about stars, galaxies and the Universe itself.

Activity 47 Universal physics

For each of the major areas of physics covered in this chapter (black-body radiation, nuclear physics, gravitation and kinetic theory) make a summary chart in the form of a concept map.

Use a separate 'post-it' sticker for each key term, diagram or equation, and arrange the stickers on a large piece of paper, moving them around so as to make the connections between them as clear as possible. Then stick them down firmly, and add extra notes to emphasise the links between the points.

5.2 Questions on the whole chapter

74 A keen UFO spotter claims to have seen a 'mysterious object, about as bright as the brightest stars – maybe a bit brighter – and moving'. A cynic remarks that it was probably an aeroplane.

An Airbus landing-light has a power of 600 W. The brightest star in the night sky, Sirius, has a luminosity of 1.2×10^{28} W and is 8.3×10^{16} m from Earth. By carrying out a suitable order-of-magnitude calculation, decide whether an aeroplane light would typically be seen from a distance at which it could look about as bright as Sirius. Show the steps in your working, and state clearly any assumptions and approximations you have made.

75 The Sun can be approximated to a black body with a temperature of 6000 K, whose luminosity is 4×10^{26} W and whose spectrum peaks at 500 nm. If its temperature fell to 3000 K, what would be the new luminosity and peak wavelength?

76 Refer to Table 5.2 (Section 2.2) to help you answer this question.

(a) Complete the following reactions that take place in some giant stars:

(i) ${}^{4}_{2}\text{H} + {}^{12}_{6}\text{C} \rightarrow .... + \gamma$

(ii) ${}^{1}_{0}\text{n} + {}^{38}_{16}\text{S} \rightarrow {}^{39}_{16}\text{S} \rightarrow {}^{39}_{17}\text{Cl} + ...$

(b) In a supernova explosion, neutron absorption produces a long-lived isotope of indium, ${}^{115}_{49}\text{In}$. This can absorb another neutron before decaying to give ${}^{116}\text{In}$, which is itself unstable, decaying by beta-minus decay into an isotope of tin (Sn). Write equations for these two reactions, starting with ${}^{115}\text{In}$.

77 Nuclear power stations use nuclear fission to provide energy for electricity generation. When a nucleus of ${}^{235}\text{U}$ absorbs a neutron, it becomes unstable and undergoes fission, producing two lighter nuclei and some more neutrons. A typical reaction is:

$${}^{235}_{92}\text{U} + {}^{1}_{0}\text{n} \rightarrow {}^{141}_{56}\text{Ba} + {}^{92}_{36}\text{Kr} + \text{neutrons}$$

The neutrons released can produce more fission reactions if they are captured by further ${}^{235}\text{U}$ nuclei. A reactor in a power station contains neutron-absorbing materials designed so that the chain reaction proceeds at a steady rate. The energy released per unit mass of nuclear fuel is several orders of magnitude greater than that produced by burning chemical fuels such as oil or gas. After use, the spent fuel can be chemically treated to extract any useful materials. The remaining waste is highly radioactive.

(a) (i) How many neutrons are produced in the reaction above? Explain your reasoning.

(ii) Explain why a nuclear fission reaction releases energy.

(b) The uranium fuel used in reactors consists of a mixture of isotopes. ${}^{238}_{92}\text{U}$, the most common isotope, does not undergo fission when it absorbs a neutron. Instead, it continues to absorb neutrons and heavier nuclei are produced. These nuclei are radioactive, and contribute to the hazards associated with nuclear waste.

For example,

$${}^{238}_{92}\text{U} + {}^{1}_{0}\text{n} \rightarrow {}^{239}_{93}\text{Np} + \beta^{-} + \bar{\nu}_{e}$$

Further absorption reactions produce ${}^{239}_{94}\text{Pu}$ (plutonium), one of several nuclei that are particularly hazardous both because of the radiation they emit and because of their complex biological effects if they enter the body. ${}^{239}_{94}\text{Pu}$ has a half-life of 24 400 years and decays by alpha emission.

(i) What fraction of the ${}^{239}_{94}\text{Pu}$ nuclei in a sample of waste will remain after 1000 years?

(ii) By what percentage will the activity due to ${}^{239}_{94}\text{Pu}$ have decreased in this time?

(iii) If you started with a sample of pure $^{239}_{94}Pu$, the activity of the sample after 1000 years would in fact be greater than indicated by your answer above. Suggest a reason for this.

78 Geophysicists study volcanoes in order to understand how they work and to predict eruptions. In some types of volcano, magma (molten rock) from deep inside the Earth moves upwards through underground fractures (large cracks forming gaps in the solid rock) and emerges at the Earth's surface. As the fractures fill with magma, the gravitational field at the surface changes, giving warning of an eruption.

To measure small changes in gravitational field strength, geophysicists use a gravimeter, which is essentially a very sensitive spring balance. The gravimeter is so sensitive that precise measurements of height need to be made at the same time as the gravity measurements.

(Data: near Earth's surface $g = 9.81$ N kg^{-1}; $G = 6.67 \times 10^{-11}$ N m^2 kg^{-2}.)

(a) (i) What is meant by the term *gravitational field strength*?

(ii) If a piece of magma moves upwards towards the Earth's surface, filling a gap in the rocks, what would happen to the gravitational field strength at the surface?

(iii) One type of gravimeter used in volcanic studies can detect changes in gravitational field of one part in 100 million at the Earth's surface. What is the order of magnitude of the smallest change in gravitational field that it could detect?

(b) Calculate the strength of the gravitational field due to a cube of magma, density 4000 kg m^{-3}, measuring 100 m along each side, with its centre of mass 1.5 km below the surface.

(c) Explain why it is necessary to measure the height when monitoring changes in gravitational field due to underground magma movements.

79 During 1996 and 1997, the Galileo spacecraft investigated the planet Jupiter and its moons. One such moon is Io. Measurements of Io's orbital motion enable Jupiter's mass to be determined.

(a) The radius of Io's orbit around Jupiter, r, is 4.22×10^8 m and its period of rotation, T, is 1.53×10^5 s. What is Io's orbital speed v?

(b) (i) Write down an expression in terms of Io's mass, m_{Io}, for the centripetal force, F_{cent}, needed to keep it in orbit.

(ii) Using m_{Ju} to denote Jupiter's mass, write down an expression for the gravitational force of attraction, F_{grav}, between Jupiter and Io.

(c) Using your answer to (b), obtain an expression for m_{Ju} in terms of v, G and r.

(d) Calculate a value for the mass of Jupiter ($G = 6.67 \times 10^{-11}$ N m^2 kg^{-2}).

80 (a) A galaxy is moving away from Earth at 4.58×10^5 m s^{-1}. It emits light at a frequency 6.00×10^{14} Hz. Calculate the observed frequency of this light on Earth. ($c = 3.00 \times 10^8$ m s^{-1})

(b) Hydrogen atoms emit and absorb electromagnetic radiation with a wavelength of 21.2 cm. This '21cm line' is used by radio astronomers to study the Doppler shifts of radiation from the ISM rotating galaxies. The receivers must be tuned to precisely the right frequency.

(i) What is the emitted frequency of this radiation?

(ii) If the hydrogen in a galaxy is moving at 7370 km s^{-1} away from Earth, what is the difference between the observed frequency and the emitted frequency?

81 Our galaxy, the Milky Way, is one of a group of galaxies. The light from our neighbour in the group, called Andromeda, is slightly blue shifted. Suggest a reason why it is not redshifted as other galaxies are.

82 Creationists claim that the Universe was created 6000 years ago. If this were true, calculate the value that the Hubble constant would have (i) in $years^{-1}$, (ii) in s^{-1}, (iii) in km s^{-1} Mpc^{-1} (assuming uniform expansion). (1 year = 3.16×10^7 s; 1 Mpc = 3.09×10^{22} m.)

83 Figure 5.106 shows the spectrum of light received from the quasar known as 3C273 (a quasar is an extremely luminous type of galaxy). The four peaks marked are hydrogen lines that have been shifted in wavelength. In a laboratory, these lines have wavelengths 410 nm, 434 nm, 486 nm and 656 nm.

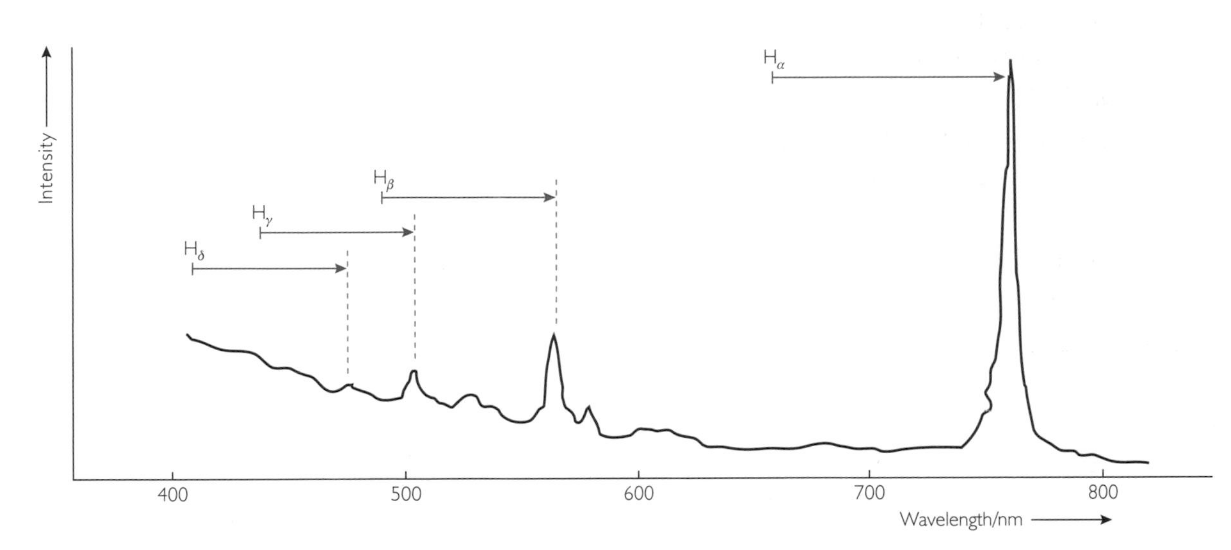

Figure 5.106 Spectrum of 3C273

(a) (i) What is the name given to this shift in wavelength?

(ii) Without doing any calculations, what does the spectrum of 3C273 allow you to deduce about its motion?

(b) Using any one of the lines marked in the diagram, calculate the speed of 3C273 relative to the Earth.

(c) Assuming that the Hubble constant H_0 = 75 km s^{-1} Mpc^{-1}, calculate the distance to 3C273. Express your answer in Mpc.

5.3 Achievements

Now you have studied this chapter you should be able to achieve the outcomes listed in Table 5.15.

Table 5.15 Achievements for the chapter *Reach for the Stars*

	Statement from examination specification	*Section(s) in this chapter*
109	investigate, recognise and use the expression $\Delta E = mc\Delta\theta$	3.3 and see BLD
110	explain the concept of internal energy as the random distribution of potential and kinetic energy amongst molecules	3.3
111	explain the concept of absolute zero and how the average kinetic energy of molecules is related to the absolute temperature	3.3
112	recognise and use the expression $\frac{1}{2}m\langle c^2\rangle = \frac{3}{2}kT$	3.3
113	use the expression $pV = NkT$ as the equation of state for an ideal gas	3.3
114	show an awareness of the existence and origin of background radiation, past and present	2.2
115	investigate and recognise nuclear radiations (alpha, beta and gamma) from their penetrating power and ionising ability	2.2
116	describe the spontaneous and random nature of nuclear decay	2.2
117	determine the half lives of radioactive isotopes graphically and recognise and use the expressions for radioactive decay: $\frac{dN}{dt} = -\lambda N$, $\lambda = \frac{\ln 2}{t_{\frac{1}{2}}}$ and $N = N_0 e^{-\lambda t}$	2.2
118	discuss the applications of radioactive materials, including ethical and environmental issues	2.2, 2.3
126	use the expression $F = \frac{Gm_1m_2}{r^2}$	3.2
127	derive and use the expression $g = -\frac{Gm}{r^2}$ for the gravitational field due to a point mass	3.2

CONTINUED ▶

Statement from examination specification		Section(s) in this chapter
128	recall similarities and differences between electric and gravitational fields	3.2
129	recognise and use the expression relating flux, luminosity and distance $F = \frac{L}{4\pi d^2}$ application to standard candles	2.1, 4.2
130	explain how distances can be determined using trigonometric parallax and by measurements on radiation flux received from objects of known luminosity (standard candles)	2.1, 3.1, 4.2
131	recognise and use a simple Hertzsprung–Russell diagram to relate luminosity and temperature. Use this diagram to explain the life cycle of stars	3.1, 3.3, 3.4
132	recognise and use the expression $L = \sigma T^4 \times$ surface area, (for a sphere $L = 4\pi r^2 \sigma T^4$) (Stefan–Boltzmann law) for black-body radiators	3.1
133	recognise and use the expression: $\lambda_{max} T = 2.898 \times 10^{-3}$ m K (Wien's law) for black-body radiators	3.1
134	recognise and use the expressions $z = \frac{\Delta\lambda}{\lambda} \approx \frac{\Delta f}{f} \approx \frac{v}{c}$ for a source of electromagnetic radiation moving relative to an observer and $v = H_0 d$ for objects at cosmological distances	3.2, 4.1
135	be aware of the controversy over the age and ultimate fate of the Universe associated with the value of the Hubble constant and the possible existence of dark matter	4.1, 4.2, 4.3, 4.4
136	explain the concept of nuclear binding energy, and recognise and use the expression $\Delta E = c^2 \Delta m$ and use the non SI atomic mass unit (u) in calculations of nuclear mass (including mass deficit) and energy	2.3 and see PRO
137	describe the processes of nuclear fusion and fission	2.3, 4.3
138	explain the mechanism of nuclear fusion and the need for high densities of matter and high temperatures to bring it about and maintain it	2.3

Answers

1 (a) time, t = distance/speed so:

$$t = \frac{1.50 \times 10^{11}\ \text{m}}{3.00 \times 10^{8}\ \text{m s}^{-1}} = 500\ \text{s}.$$

(b) Distance to Sun = 500 light seconds =

$$\frac{500\ \text{s}}{3.16 \times 10^{7}\ \text{s yr}^{-1}} = 1.58 \times 10^{-5}\ \text{light years}.$$

(c) $$\frac{d_{pc}}{d_{sun}} = \frac{4.24\ \text{light years}}{1.58 \times 10^{-5}\ \text{light years}}$$

$= 2.68 \times 10^{5}$

To the nearest order of magnitude, the ratio is 10^5.

2 (a) Using Equation 1:

$$\frac{F_{body}}{F_{Earth}} = \left(\frac{L}{4\pi d^2_{body}}\right) \div \left(\frac{L}{4\pi d^2_{Earth}}\right)$$

$$= \left(\frac{d_{Earth}}{d_{body}}\right)^2$$

For Mercury:

$$\frac{F_{Mercury}}{F_{Earth}} = \left(\frac{1.496 \times 10^{9}\ \text{m}}{57.9 \times 10^{9}\ \text{m}}\right)^2$$

$$= \left(\frac{149.6}{57.9}\right)^2 = 6.7$$

so $F_{Mercury} = 6.7F_{Earth}$

Notice that the units of m and the factor 10^9 cancel in the final calculation. Using a similar approach for the other bodies:

$$\frac{F_{Venus}}{F_{Earth}} = 1.91 \quad \frac{F_{Mars}}{F_{Earth}} = 0.43 \quad \frac{F_{Pluto}}{F_{Earth}} = 6.4 \times 10^{-4}$$

(b) In general we can expect that the bodies with the higher radiant energy flux (Mercury and Venus) will be hotter than Earth and have brighter sunlight. Mars is cooler, and Pluto is only very dimly lit by the Sun and very cold.

(c) You might have thought of some of the following effects. The curvature of the body reduces the incident energy flux near the poles. The body's surface may reflect radiation back into space rather than absorbing it. It may reflect or absorb radiation so the surface does not get so warm. Any atmosphere may trap heat radiation (the greenhouse effect). (This causes Venus to be hotter than Mercury, even though it is further from the Sun.)

Geothermal effects may warm up parts of the body (volcanic activity, hot springs etc.).

3 (a) From Equation 1, $d^2 = \dfrac{L}{4\pi F}$

Assuming that Sirius and the Sun have the same luminosity, L, then

$$\frac{d^2_{Sir}}{d^2_{Sun}} = \left(\frac{L}{4\pi F_{Sir}}\right) \div \left(\frac{L}{4\pi F_{Sun}}\right) = \frac{F_{Sun}}{F_{Sir}}$$

Using Huygens's measurements:

$$\frac{d^2_{Sir}}{d^2_{Sun}} = 26644^2$$

$$\frac{d_{Sir}}{d_{Sun}} = 26644 \approx 2.7 \times 10^4$$

(b) True $$\frac{d_{Sir}}{d_{Sun}} = \frac{8\ \text{light years}}{1.58 \times 10^{-5}\ \text{light years}} = 5.06 \times 10^5$$

(c) The ratio calculated by Huygens is too small by a factor of about 20. He was wrong to assume that Sirius has the same luminosity as the Sun; we now know that it is about 26 times more luminous. Also, his memory of the brightness of Sirius could not be very reliable, so the five significant figures he quotes for the ratio of diameters is incredibly optimistic.

4 We are told that the Sun is about 400 times further away than the Moon, and that the Sun and Moon have very similar angular diameters.

We can therefore deduce that

$d_{Sun} \approx 400\ d_{Moon} \approx 400 \times 3.85 \times 10^8$ m.

From Equation 2, $$\alpha_{Sun} = \frac{D_{Sun}}{d_{Sun}}$$

$$\approx \frac{1.4 \times 10^{9}\ \text{m}}{400 \times 3.85 \times 10^{8}\ \text{m}}$$

$$= 9 \times 10^{-3}\ \text{radians}$$

We have already said (and used the fact) that $\alpha_{Sun} = \alpha_{Moon}$, so, with no further calculation, $\alpha_{Moon} = 9 \times 10^{-3}$ radians.

(Alternatively, we could have started by saying $D_{Moon} = \dfrac{D_{Sun}}{400}$. This would give exactly the same answers.)

5 Volume of sphere, $V = \frac{4\pi r^3}{3}$

Volume of Earth, $V_{\text{Earth}} = \frac{4\pi \times (6.4 \times 10^6 \text{ m})^3}{3}$

$= 1.1 \times 10^{21} \text{ m}^3$

$V_{\text{Sun}} = \frac{4\pi \times (7.0 \times 10^8 \text{ m})^3}{3} = 1.4 \times 10^{27} \text{ m}^3$

$\frac{V_{\text{Sun}}}{V_{\text{Earth}}} = \frac{1.4 \times 10^{27} \text{ m}^3}{1.1 \times 10^{21} \text{ m}^3} = 1.3 \times 10^6.$

So in terms of volume, the Sun is about a million times bigger than the Earth.

6 Surface area of sphere, $A = 4\pi r^2$.

$A_{\text{Sun}} = 4\pi \times (7.0 \times 10^8 \text{ m})^2 = 6.16 \times 10^{18} \text{ m}^2$

Power emitted from each m^2 of surface $= \frac{L_{\text{Sun}}}{A_{\text{Sun}}}$

$= \frac{3.84 \times 10^{26} \text{ W}}{6.16 \times 10^{18} \text{ m}^2} = 6.23 \times 10^7 \text{ W m}^{-2}$

$1 \text{ m}^2 = 1 \times 10^4 \text{ cm}^2$, so each cm^2 emits a power of 6.23×10^3 W, ie very close to the figure stated in the question.

7 We don't have any material that was alive when the Earth formed. Also, while carbon-dating is good for dating objects from thousands up to millions of years old, in very ancient objects there would be too little carbon-14 to detect.

8 Uranium-235: ${}^{235}_{92}\text{U}$, contains 143 neutrons.

Uranium-238: ${}^{238}_{92}\text{U}$, 146 neutrons.

Thorium-232: ${}^{232}_{90}\text{Th}$, 142 neutrons.

Potassium-40: ${}^{40}_{19}\text{K}$, 21 neutrons.

9 This is β-decay. ${}^{87}_{37}\text{Rb} \rightarrow {}^{87}_{38}\text{Sr} + {}^{0}_{-1}\text{e} + \bar{\nu}_e$

10 An α particle is like a helium nucleus, with 2 protons and 2 neutrons. A helium atom also has 2 electrons in orbit about the nucleus.

11 If n half-lives have elapsed, then the amount of parent isotope is $\left(\frac{1}{2^n}\right)$ × the initial amount.

(Table 5.3 listed only whole-number values of n, but this rule works for n being any number.)

12 Using Equation 7:

For ${}^{238}\text{U}$, $t_{\frac{1}{2}} = \frac{\log_e(2)}{\lambda} = \frac{\log_e(2)}{1.552 \times 10^{-10} \text{ yr}^{-1}}$

$= 4.466 \times 10^9 \text{ yr} = 4466 \times 10^6 \text{ yr}$

Similarly ${}^{40}\text{K}$ has $t_{\frac{1}{2}} = 1193 \times 10^6$ yr

For ${}^{235}\text{U}$, $\lambda = \frac{\log_e(2)}{t_{\frac{1}{2}}} = \frac{\log_e(2)}{(704 \times 10^6 \text{ yr})}$

$= 9.846 \times 10^{-10} \text{ yr}^{-1}$

Similarly ${}^{87}\text{Rb}$ has $\lambda = 1.420 \times 10^{-11} \text{ yr}^{-1}$

$= 0.1420 \times 10^{-10} \text{ yr}^{-1}$

13 244000 million years is half of the half-life, so $n = 0.5$, and the fraction remaining is $\frac{1}{2^{0.5}} = 0.71$ or 71%.

14 $\frac{1}{16}$ of the parent nuclei must have survived (and $\frac{15}{16}$ changed into lead), so the rock must be 4 half-lives old; ie its age must be 4 × 704 million years = 2816 million years.

15 (a) 4600 million years

(b) The meteorites could have formed in a volcanic explosion in which they were hurled at very high speed from the Martian surface (exceeding the escape velocity) eventually landing on Earth.

(c) The rocks from Mars are much younger than those formed on Earth, suggesting that they crystallised from molten rock long after Mars formed. This is consistent with their crystallising from the molten rock ejected from a volcano.

16 $\frac{3.84 \times 10^{26} \text{ W}}{746 \text{ W hp}^{-1}} \approx 5 \times 10^{23}$ hp.

17 (a) Total energy emitted in 1000 years:

$\Delta E = 3.84 \times 10^{26}\ \text{J s}^{-1} \times 3 \times 10^{10}\ \text{s}$
$= 1.15 \times 10^{37}\ \text{J}$

(b) (i) $\Delta m = \dfrac{\Delta E}{c^2} = \dfrac{1.15 \times 10^{37}\ \text{J}}{(3.00 \times 10^{8}\ \text{m s}^{-1})^2}$

$= 1.28 \times 10^{20}\ \text{kg}$

(ii) Using m to represent mass of Sun,

$\dfrac{\Delta m}{m} = \dfrac{1.28 \times 10^{20}\ \text{kg}}{1.99 \times 10^{30}\ \text{kg}} = 6.43 \times 10^{-11}$

$= 6.43 \times 10^{-9}\ \%$

(c) In 10^{10} years, the percentage mass loss will be $10^{7} \times 6.43 \times 10^{-9}\% = 6.43 \times 10^{-2}\ \%$. (This is a very small fraction of the Sun's mass.)

18 (a) $4\,^{1}_{1}\text{H} + 2\,^{0}_{-1}\text{e}^{-} \rightarrow {}^{4}_{2}\text{He} + 2\nu_e + 6\gamma$

(b) Initial mass $= 4 \times 1.673 \times 10^{-27}\ \text{kg} + 2 \times 9.11 \times 10^{-31}\ \text{kg}$

$= 6.694 \times 10^{-27}\ \text{kg}$

Final mass $= 6.645 \times 10^{-27}\ \text{kg}$

Mass deficit $\Delta m = 6.694 \times 10^{-27}\ \text{kg} - 6.645 \times 10^{-27}\ \text{kg}$

$= 4.9 \times 10^{-29}\ \text{kg}$

(c) $\Delta E = c^2\Delta m = (3.00 \times 10^{8}\ \text{m s}^{-1})^2 \times 4.9 \times 10^{-29}\ \text{kg}$

$= 4.41 \times 10^{-12}\ \text{J}$

(d) The Sun emits energy at a rate of $3.84 \times 10^{26}\ \text{J s}^{-1}$. To maintain this output, n nuclei of ${}^{4}_{2}\text{He}$ produced must be produced per second, where:

$n = \dfrac{3.84 \times 10^{26}\ \text{J s}^{-1}}{4.41 \times 10^{-12}\ \text{J}} = 8.71 \times 10^{37}\ \text{s}^{-1}$

Each ${}^{4}_{2}\text{He}$ is made from 4 ${}^{1}_{1}\text{H}$, so the number of hydrogen nuclei consumed per second is $4n = 4 \times 8.71 \times 10^{37} = 3.48 \times 10^{38}\ \text{s}^{-1}$

(e) There are various ways to tackle this part, which all give the same answer. Here is one way.

Initial mass of hydrogen in Sun is
$m_0 = 0.75 \times 1.99 \times 10^{30}\ \text{kg}$
$= 1.49 \times 10^{30}\ \text{kg}$.

Mass of hydrogen that will be converted to helium over the Sun's lifetime is
$m = 0.15 \times 1.49 \times 10^{29}\ \text{kg}$
$= 2.24 \times 10^{29}\ \text{kg}$

No. of hydrogen nuclei, n_0 contained in this mass is

$n_0 = \dfrac{m}{m_{\text{H}}} = \dfrac{2.24 \times 10^{29}\ \text{kg}}{1.673 \times 10^{-27}\ \text{kg}} = 1.34 \times 10^{56}.$

If hydrogen nuclei are consumed at a rate of $3.48 \times 10^{38}\ \text{s}^{-1}$, then the time, t, for which the Sun can continue to emit energy is:

$t = \dfrac{1.34 \times 10^{56}}{3.48 \times 10^{38}\ \text{s}^{-1}} = 3.85 \times 10^{17}\ \text{s}$

$\approx 1.2 \times 10^{10}$ years

19 See Figure 5.107. The curve is like an upside-down binding-energy curve (Figure 5.25). Think of the thought experiment in which nuclei are assembled from their constituent nucleons. In forming the most tightly bound nuclei, the nucleons lose the greatest amount of 'nuclear potential energy' – they lose energy therefore they also lose mass. You have also seen, in the worked example in the text and in Question 18, that nuclei formed in fission and fusion reactions have less mass than the nuclei that initially took part in the reactions.

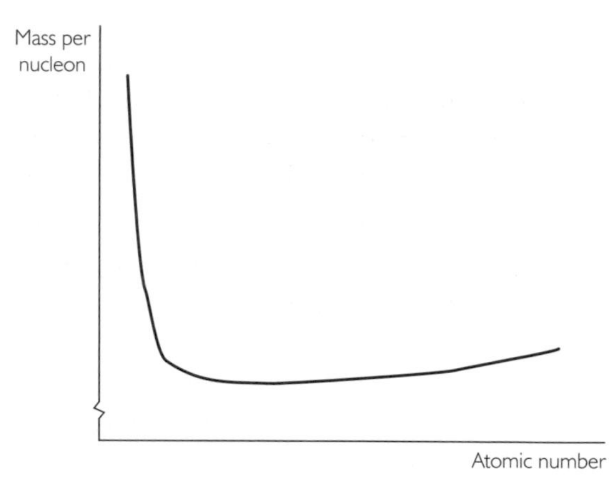

Figure 5.107 Answer to Question 19

20 (a) (i) Diameter of Jupiter, $D = 2 \times 7.14 \times 10^{7}\ \text{m}$

Maximum and minimum Earth–Jupiter separations:

$d_{\text{max}} = 7.78 \times 10^{11}\ \text{m} + 1.49 \times 10^{11}\ \text{m}$
$= 9.27 \times 10^{11}\ \text{m}$

$d_{\text{min}} = 7.78 \times 10^{11}\ \text{m} - 1.49 \times 10^{11}\ \text{m}$
$= 6.29 \times 10^{11}\ \text{m}$

Using Equation 2:

$$\alpha_{min} = \frac{D}{d_{max}} = \frac{2 \times 7.14 \times 10^7 \text{ m}}{9.27 \times 10^{11} \text{ m}}$$

$= 1.54 \times 10^{-4}$ radians

$$\alpha_{max} = \frac{D}{d_{min}} = \frac{2 \times 7.14 \times 10^7 \text{ m}}{6.29 \times 10^{11} \text{ m}}$$

$= 2.27 \times 10^{-4}$ radians

(ii) $\frac{\alpha_{min}}{\alpha_{max}} = \frac{1.54}{2.27} = 0.678 = 67.8\%$

(b) From Equation 1:

$$\frac{F_{Jupiter}}{F_{Earth}} = \frac{L}{4\pi d^2_{Jupiter}} \div \frac{L}{4\pi d^2_{Earth}}$$

where d now represents distance from the Sun, and L is the Sun's luminosity. Therefore:

$$\frac{F_{Jupiter}}{F_{Earth}} = \left(\frac{d_{Earth}}{d_{Jupiter}}\right)^2$$

$$= \left(\frac{1.49 \times 10^{11} \text{ m}}{7.78 \times 10^{11} \text{ m}}\right)^2 = 0.0367 = 3.67\%$$

(c) Jupiter is probably heated by the energy released in radioactive decay processes similar to those that occur in the Earth's rocks. (You might have suggested fusion of hydrogen – but this would only be possible at high temperatures so is unlikely because the outer parts of Jupiter, which contain hydrogen, are cold.)

21 The long half-life ensures that the activity of the source will not change much with time, and so the source will not need to be replaced. Also, a long half-life means that the activity is relatively low, and so the safety hazard is not so great as that associated with a more active, short-lived source.

22 (a) Argon has atomic number 18. To conserve charge, the emitted particle must have a proton number of +1. The particle must therefore be a positron – this is an example of beta-plus decay. Since a positron has a nucleon number of zero, the argon isotope must have a nucleon number 40 – the same as the parent isotope. And just as beta-minus emission is accompanied by the emission of an antineutrino, beta-plus is accompanied by a neutrino. The reaction is thus:

$${}^{40}_{19}\text{K} \rightarrow {}^{40}_{18}\text{Ar} + {}^{0}_{+1}\text{e} + \nu_e$$

Alternatively a captured electron could have joined with a proton in the nucleus to make a neutron. (This is in fact what happens.)

(b) One way to do this calculation is to use the same method as the worked example in Section 2.2, using Equation 4. The two isotopes have the same mass, so the initial mass of potassium is $m_0 = 3.8$ μg and the numbers of nuclei of each isotope are in proportion to their masses; ie

$$\frac{N}{N_0} = \frac{m}{m_0} = \frac{3.3 \text{ μg}}{3.8 \text{ μg}} = 0.868$$

From Table 5.4, $\lambda = 5.810 \times 10^{-10}$ yr^{-1}

$0.868N = N_0 e^{-\lambda t}$ so $0.868 = e^{-\lambda t}$

$\log_e(0.868) = -\lambda t$

$$t = \frac{\log_e(0.868)}{(-\lambda)} = \frac{\log_e(0.868)}{-5.810 \times 10^{-10} \text{ yr}^{-1}}$$

$= 3.5 \times 10^9$ yr

(Alternatively, using the method of Question 11 and the half-life you calculated in Question 12 gives the same answer.)

(c) Mass of ^{40}K nucleus ≈ 40 × mass of proton (from Table 5.6)

$= 40 \times 1.673 \times 10^{-27}$ kg $= 6.692 \times 10^{-26}$ kg

$$\text{no. of } ^{40}\text{K nuclei, } N = \frac{5.0 \times 10^{-10} \text{ kg}}{6.692 \times 10^{-26} \text{ kg}}$$

$= 7.5 \times 10^{15}$

From Equations 3 and 5, activity $A = \frac{dN}{dt} = -\lambda N$

so $A = 7.5 \times 10^{15} \times 5.810 \times 10^{-10}$ yr^{-1}

$$= 4.3 \times 10^6 \text{ yr}^{-1} = \frac{4.3 \times 10^6 \text{ yr}^{-1}}{3 \times 10^7 \text{ s yr}^{-1}}$$

≈ 0.14 s^{-1} = 0.14 Bq

23 (a) 1 light year $= 3.00 \times 10^8$ m s^{-1} $\times 3.1530 \times 10^7$ s
$= 9.46 \times 10^{15}$ m

$d = 4.351 \times 9.46 \times 10^{15}$ m

$$\alpha = \frac{r}{d} = \frac{1.49 \times 10^{11} \text{ m}}{4.351 \times 9.46 \times 10^{15} \text{ m}}$$

$= 3.62 \times 10^{-6}$ rad

$= 3.62 \times 10^{-6}$ rad $\times 2.06 \times 10^5$ arcsec rad^{-1}

$= 0.746$ arcsec

(b) $\alpha = 1$ arcsec $= \dfrac{1}{2.06 \times 10^5 \text{ rad}}$, so

$$d = \frac{r}{\alpha} = 1.49 \times 10^{11} \text{ m} \times 2.06 \times 10^5$$

$= 3.1 \times 10^{16}$ m, which is as stated.

24 $\alpha = 0.002 \text{ arcsec} = \dfrac{0.002 \text{ arcsec}}{2.06 \times 10^5 \text{ arcsec rad}^{-1}}$

$= 9.71 \times 10^{-9}$ rad

From Equation 9, $d = \dfrac{r}{\alpha/\text{rad}}$

$$= \frac{1.49 \times 10^{11} \text{ m}}{9.71 \times 10^{-9} \text{ rad}}$$

$= 1.53 \times 10^{19}$ m

From Question 23, 1 light year $= 9.46 \times 10^{15}$ m, so:

$$r = \frac{1.53 \times 10^{19} \text{ m}}{9.46 \times 10^{15} \text{ m lt yr}^{-1}}$$

$= 1.62 \times 10^3$ light years

1 pc $= 3.1 \times 10^{16}$ m so

$$r = \frac{1.53 \times 10^{19} \text{ m}}{3.1 \times 10^{16} \text{ m pc}^{-1}} = 494 \text{ pc}$$

25 Brown dwarf (coolest), red giant, white dwarf, blue supergiant (hottest).

26 From Equation 12, $\sigma = \dfrac{L}{4\pi r^2 T^4}$, so its SI units are W m^{-2} K^{-4}

27 From Equation 12, $r = \sqrt{\left(\dfrac{L}{4\pi\sigma T^4}\right)}$

(a) For Spica A:

$$r = \sqrt{\left\{\frac{8 \times 10^{29} \text{ W}}{(4\pi \times 5.67 \times 10^{-8} \text{ W m}^{-2} \text{ K}^{-4} \times (3 \times 10^4 \text{ K})^4)}\right\}}$$

$= 1.2 \times 10^9$ m

$\approx 1.7 \times r_{Sun}$

(b) For Barnard's Star,

$$r = \sqrt{\left\{\frac{4 \times 10^{23} \text{ W}}{(4\pi \times 5.67 \times 10^{-8} \text{ W m}^{-2} \text{ K}^{-4} \times (2.7 \times 10^3 \text{ K})^4)}\right\}}$$

$= 1.0 \times 10^8$ m

$\approx 0.15 \times r_{Sun}$

28 (a) Blue-white

(b) Orange-red

29 (a) Use Wien's law (Equation 10):

Betelgeuse: $\lambda_{max} = \dfrac{c}{f_{max}}$

$$= \frac{3.00 \times 10^8 \text{ ms}^{-1}}{4.3 \times 10^{14} \text{ Hz}}$$

$= 7.0 \times 10^{-7}$ m

$$T = \frac{2.89 \times 10^{-3} \text{ m K}}{\lambda_{max}}$$

$$\approx \frac{2.89 \times 10^{-3} \text{ m K}}{7.0 \times 10^{-7} \text{ m}}$$

$\approx 4.1 \times 10^3$K

Rigel: $\lambda_{max} = \dfrac{c}{f_{max}}$

$$= \frac{3.00 \times 10^8 \text{ ms}^{-1}}{10 \times 10^{14} \text{ Hz}}$$

$= 3.0 \times 10^{-7}$ m

$$T = \frac{2.89 \times 10^{-3} \text{ m K}}{\lambda_{max}}$$

$$\approx \frac{2.89 \times 10^{-3} \text{ m K}}{3.0 \times 10^{-7} \text{ m}}$$

$\approx 9.6 \times 10^3$ K

(b) Any hot object will emit radiation across a broad range of wavelengths, so even though Rigel appears blue, it will emit over a range of blue wavelengths. The quoted frequencies might not correspond to the peak wavelengths. A detailed spectral analysis would be needed. Also, stars are not perfect black-body emitters (there are absorption lines due to the cooler outer atmospheres).

30 Rearranging Equation 1, $L = 4\pi d^2 F$.

For Sirius:

$L = 4\pi \times (8.22 \times 10^{16} \text{ m})^2 \times 1.17 \times 10^{-7} \text{ W m}^{-2}$
$= 9.9 \times 10^{27}$ W

Similarly for the other stars – see Table 5.16.

Star	Distance, d/m	Flux, F/W m^{-2}	Luminosity, L/W
Sirius	8.22×10^{16}	1.17×10^{-7}	9.9×10^{27}
Rigel	2.38×10^{18}	5.37×10^{-7}	3.8×10^{31}
Betelgeuse	6.18×10^{18}	1.11×10^{-8}	5.3×10^{30}
Proxima Centauri	4.11×10^{16}	1.09×10^{-11}	2.3×10^{23}
Sun	1.49×10^{11}	1.37×10^{3}	3.83×10^{26}

Table 5.16 Answers to Question 30

31 You would need to determine the surface temperature from the star's spectrum. This would let you locate its position on the HR diagram (on the main sequence) from which you could read its luminosity, L. You would also need to measure its radiant energy flux, F, and then calculate its distance, d, using Equation 1: $d = \sqrt{\left(\frac{L}{4\pi F}\right)}$.

32 (a) (i) The hotter the star, the greater its luminosity.

(ii) The main sequence is not quite a straight line, so L and T cannot be related by a simple power law (see *Maths Note 8.7: Using log graphs*). But if you draw a straight line going approximately through the main sequence, then you can see that an increase of a factor 10 in temperature (eg from 2500 K to 25 000 K) corresponds to an increase in luminosity of about a factor 10^6 (from $10^{-2}L_{\text{Sun}}$ to $10^4 L_{\text{Sun}}$), which suggest a relationship something like $L \propto T^6$.

(b) It is difficult to be conclusive without more information. As you saw in Activity 17, increasing the surface temperature of an object increases its luminosity. We can also suppose that a large object is more luminous than a small one at the same temperature, as it has a greater surface area from which to radiate. All we can really say is that the most luminous main-sequence stars cannot be significantly smaller than the least luminous ones; otherwise the effect of increasing temperature would be outweighed by the effect of decreasing size. (In fact, the hotter main-sequence stars are larger than the cooler ones, and the differences in luminosity arise from the differences in temperature *and* in size.)

(c) As with the question about size, it is very difficult to say anything at all. Actually it is even more difficult, because size *can* be deduced from measurements of luminosity and temperature – but to deduce mass you would need to know a star's density, or else determine its mass by some completely independent means (which you will meet in Section 3.3).

(d) Yet again, it is very difficult to say anything conclusive. One might be tempted to speculate that stars are cool and faint as they are formed, and gradually heat up and become more luminous as a result of nuclear fusion in their interiors. Or perhaps they start off hot and luminous and gradually cool down. Similar suggestions were made when the HR diagram was first produced, but they are entirely speculative and, we now believe, totally wrong.

Assuming that the stars are all at different stages in their evolution, the most we can say is that some combinations of luminosity and temperature must be more stable than others, indicating long-lasting phases in a star's life, since there are large numbers of stars on the main sequence, quite a few red giants, and some parts of the HR diagram that contain no stars.

33 Rearranging Equation 13: $G = \frac{Fr^2}{m_1 m_2}$, so it has SI units N m^2 kg^{-2}.

34 (a) Using Equation 18 for Mars:

$$g = \frac{GM}{r^2}$$

$$= \frac{6.67 \times 10^{-11}\ \text{N m}^2\ \text{kg}^{-2} \times 6.42 \times 10^{23}\ \text{kg}}{(3.39 \times 10^6\ \text{m})^2}$$

$= 3.73$ N kg^{-1}.

(b) You need to multiply your mass (in kg) by the field strength; eg if your mass is 60.0 kg:

$W = mg = 3.73\ \text{N kg}^{-1} \times 60.0\ \text{kg} = 224\ \text{N}$.

35 Use Equation 18, $g = \frac{GM}{r^2}$, and substitute for the mass M:

$$M = \rho V = \frac{4\pi r^3 \rho}{3} \text{ so}$$

$$g = \frac{4\pi r^3 \rho G}{3r^2} = \frac{4\pi r \rho G}{3}$$

36 (a) The smaller star must have a brighter surface, since when it passes behind its companion there is a large dip in the light curve, whereas when it blocks off light from the larger star the dip is smaller.

(b) The period is the time for one complete orbit, which could be shown as the time from the middle of one large dip in the light curve to the middle of the next large dip.

37 (a) See Figure 5.108(a). The received flux halves each time one star passes in front of the other.

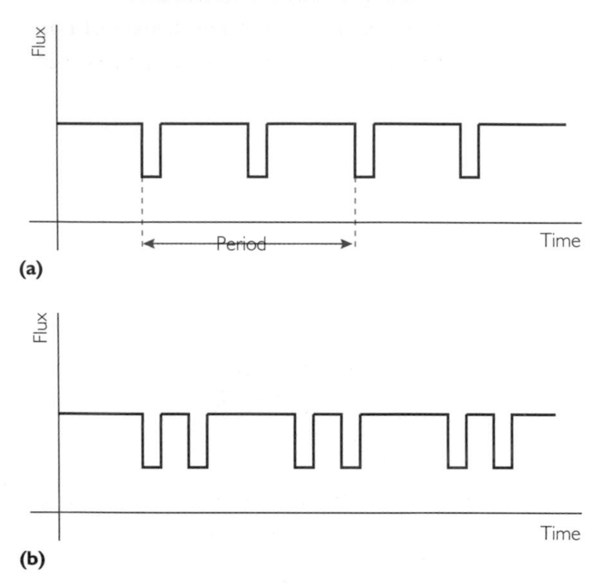

Figure 5.108 Answer to Question 37

(b) See Figure 5.108(b)

38 Rewriting Equation 22: $\frac{u}{c} \approx \frac{(f_{em} - f_{rec})}{f_{em}}$.

Substituting $f = \frac{c}{\lambda}$:

$$\frac{u}{c} \approx \left(\frac{c}{\lambda_{em}} - \frac{c}{\lambda_{rec}}\right) \times \left(\frac{\lambda_{em}}{c}\right)$$

On right-hand side, cancel c and combine the terms inside the bracket:

$$\frac{u}{c} \approx \left\{\frac{(\lambda_{rec} - \lambda_{em})}{\lambda_{em}\lambda_{rec}}\right\} \times \lambda_{em}$$

Then cancel λ_{em} to get:

$$\frac{u}{c} \approx \frac{(\lambda_{rec} - \lambda_{em})}{\lambda_{rec}}$$

and use $\lambda_{rec} \approx \lambda_{em}$ to get

$$\frac{u}{c} \approx \frac{(\lambda_{rec} - \lambda_{em})}{\lambda_{em}}$$

39 The received wavelength is shorter than the emitted wavelength, so the gases were approaching. Using Equations 23 and 24:

$$\frac{v}{c} \approx \frac{\Delta\lambda}{\lambda_{em}} = \frac{(\lambda_{rec} - \lambda_{em})}{\lambda_{em}}$$

$$= \frac{(1.070\ \mu\text{m} - 1.083\ \mu\text{m})}{1.083\ \mu\text{m}}$$

$$= -0.012$$

(The negative sign means the gases are approaching.)

$v = 0.012\ c = 3.6 \times 10^3\ \text{km s}^{-1}$

40 Using Equation 23,

$$u = \frac{c\Delta\lambda}{\lambda_{em}}$$

$$= \frac{3.00 \times 10^8\ \text{m s}^{-1} \times 0.041\ \text{nm}}{656.28\ \text{nm}}$$

$$= 1.87 \times 10^4\ \text{m s}^{-1}.$$

Assuming circular motion, $u = \frac{2\pi r}{T}$ so:

$$r = \frac{uT}{2\pi}$$

$$= \frac{1.87 \times 10^4\ \text{m s}^{-1} \times 8.6 \times 3.16 \times 10^7\ \text{s}}{2\pi}$$

$$= 8.1 \times 10^{11}\ \text{m}$$

Then using Equation 20:

$$M = \frac{v^2 r}{G}$$

$$= \frac{(1.87 \times 10^4\ \text{m s}^{-1})^2 \times 8.1 \times 10^{11}\ \text{m}}{6.67 \times 10^{-11}\ \text{N m}^2\ \text{kg}^{-2}}$$

$$= 4.2 \times 10^{30}\ \text{kg}$$

(You would get the same answer if you used Equation 21.)

41 pV has SI units N m^{-2} × m^3 = N m, and 1 N m = 1 J.

From Equation 30, $k = \frac{pV}{NT}$. N is a number (no units), T is in kelvin (K) so k has units J K^{-1}.

42 From Equation 30, $p = \frac{NkT}{V}$.

For a sphere, $V = \frac{4\pi r^3}{3}$, so $p = \frac{3NkT}{4\pi r^3}$

No. of ions $= \frac{M}{m}$

$$= \frac{2 \times 10^{30}\ \text{kg}}{1.67 \times 10^{-27}\ \text{kg}} \approx 1.2 \times 10^{57}$$

Total number of particles (ions + electrons)
$N = 2 \times$ no. of ions $\approx 2.4 \times 10^{57}$

$$p = \frac{3 \times 2.4 \times 10^{57} \times 1.38 \times 10^{-23}\ \text{J K}^{-1} \times 1 \times 10^{7}\ \text{K}}{4\pi \times (7 \times 10^{8}\ \text{m})^3}$$

$\approx 2 \times 10^{14}$ Pa.

43 (a) Mean speed = 460 m s^{-1}

(b) rms speed = 471 m s^{-1}

44 From Equation 42, $\langle c^2 \rangle = \frac{3p}{\rho}$, so $\sqrt{\langle c^2 \rangle} = \sqrt{\left(\frac{3p}{\rho}\right)}$

$$= \sqrt{\left(\frac{3 \times 10^5\ \text{Pa}}{1\ \text{kg m}^{-3}}\right)}$$

≈ 500 m s^{-1}

45 From Equation 43, at any given temperature $\sqrt{\langle c^2 \rangle} \propto \sqrt{\left(\frac{1}{m}\right)}$, so the less massive molecules will be moving faster. Since $\frac{m_{\text{ox}}}{m_{\text{hyd}}} = 16$, $\sqrt{\left(\frac{m_{\text{ox}}}{m_{\text{hyd}}}\right)} = 4$, and so the rms speed of hydrogen molecules will be four times that of oxygen molecules at the same temperature.

46 $\sqrt{\langle c^2 \rangle} = \sqrt{\left(\frac{3kT}{m}\right)}$

$$= \sqrt{\left\{\frac{(3 \times 1.38 \times 10^{-23}\ \text{J K}^{-1} \times 1 \times 10^{7}\ \text{K})}{(1.67 \times 10^{-27}\ \text{kg})}\right\}}$$

$= 5 \times 10^5$ m s^{-1}.

47 (a) (i) $\sqrt{\langle c^2 \rangle} = \sqrt{\left(\frac{3kT}{m}\right)}$

$$= \sqrt{\left\{\frac{3 \times 1.38 \times 10^{-23}\ \text{J K}^{-1} \times 30\ \text{K})}{(2 \times 1.67 \times 10^{-27}\ \text{kg})}\right\}}$$

≈ 600 m s^{-1}.

(ii) Since $c \propto \sqrt{T}$, and T is one-tenth that in Question 46,

$$c = \frac{5 \times 10^5\ \text{m s}^{-1}}{\sqrt{10}} = 1.6 \times 10^5\ \text{m s}^{-1}.$$

(b) (i) Supernova remnant:

$$p = \left(\frac{N}{V}\right) \times kT$$

$= 10^5\ \text{m}^{-3} \times 1.38 \times 10^{-23}\ \text{J K}^{-1} \times 10^6\ \text{K}$

$\approx 1.4 \times 10^{-12}\ \text{Pa} = 14 \times 10^{-13}\ \text{Pa}$

Similarly for ionised hydrogen region:

$p = 10^8\ \text{m}^{-3} \times 1.38 \times 10^{-23}\ \text{J K}^{-1} \times 8000\ \text{K}$

$\approx 1.1 \times 10^{-11}\ \text{Pa} = 110 \times 10^{-13}\ \text{Pa}$

(ii) Despite the huge differences in the temperatures and number densities in the intersellar gas, it is remarkable that the pressure varies only by about two orders of magnitude.

48 (a) From Equation 47, $\Delta T = \frac{2\Delta E}{3Nk}$:

$$N = \frac{M}{m} = \frac{1.00\ \text{kg}}{6.68 \times 10^{-27}\ \text{kg}}$$

$$\Delta T = \frac{2m\Delta E}{3Mk}$$

$$= \frac{2 \times 1000\ \text{J} \times 6.68 \times 10^{-27}\ \text{kg}}{3 \times 1.00\ \text{kg} \times 1.38 \times 10^{-23}\ \text{J K}^{-1}}$$

$= 0.32$ K

Equivalently, use Equations 48 and 49 to reach the same answer. (Equation 49 gives $C = 3.10 \times 10^3$ J kg K^{-1}.)

(b) As an argon atom has ten times the mass of a helium atom, there will be only one-tenth the number of atoms in the 1 kg sample. Each atom will receive (on average) ten times as much energy, and so the temperature rise will be ten times that for helium; ie 3.2 K.

Another way to look at this is to notice that, from Equation 49, the specific heat capacity is inversely proportional to atomic mass. Argon therefore has a specific heat capacity one-tenth that of helium, so it requires only one-tenth the amount of energy for a given temperature rise.

(c) If we were dealing with a monatomic gas, the reasoning would be the same as in (b); ie the specific heat capacity would be 1/4.5 that of helium. But water molecules have rotational and vibrational energy, and so a sample of water vapour needs additional energy in order to increase its temperature. Its specific heat capacity is therefore greater than that of a monatomic gas made from particles of the same mass.

49 Stars are formed by the collapse of a molecular cloud/~~region of ionised hydrogen~~. The particles in the cloud are far apart/~~close together~~. The force of gravitational/ ~~electrostatic~~ attraction pulls them all towards the ~~edge~~/centre of the cloud. Each particle is falling inwards, like a ball falling towards the Earth. As the cloud collapses, the particles move more quickly/~~slowly~~, and they collide more frequently with one another, sharing their kinetic energy. The faster the particles move, the higher/ ~~lower~~ the temperature of the gas. As the cloud collapses: its density increases/~~decreases~~; its pressure increases/~~decreases~~; the particles' kinetic energy increases/~~decreases~~; and their gravitational potential energy ~~increases~~/decreases.

The particles collide frequently and energetically so that they break up to form a hot plasma of fast moving/~~slow moving~~ positively/~~negatively~~ charged nuclei and electrons. When the temperature is high/ ~~low~~ enough, the nuclei can approach one another so closely nuclear ~~fission~~/fusion can start.

50 Your answer will depend on what you have assumed for the size of the cloud. For a rough estimate, volume, $V \approx d^3$ where d is the diameter. If $d \approx 10$ light years ~ 1×10^{17} m, then $V \sim 10^{51}$ m^3. Such a cloud would contain roughly 10^{61} molecules of H_2, so its mass would be:

$$M_{\text{cloud}} \approx 10^{61} \times 3.34 \times 10^{-27}\ \text{kg} \approx 3 \times 10^{34}\ \text{kg}$$

$$\frac{M_{\text{cloud}}}{M_{\text{Sun}}} \approx \frac{3 \times 10^{34}\ \text{kg}}{2 \times 10^{30}\ \text{kg}} \sim 10^4 \text{ stars}$$

(Some star clusters are found that do contain many thousand stars so this is reasonable.)

51 (a) (i) Since gravitational acceleration is equal to the field strength, use Equation 18

$\left(g = \dfrac{GM}{r^2}\right)$ and your values from Question 50.

With our values, $M = M_{\text{cloud}} = 3 \times 10^{34}$ kg, $r = \dfrac{d}{2} = 5 \times 10^{16}$ m, we get:

$$g = \frac{6.67 \times 10^{-11}\ \text{N m}^2\ \text{kg}^{-2} \times 3 \times 10^{34}\ \text{kg}}{(5 \times 10^{16}\ \text{m})^2}$$

$$\approx 8 \times 10^{-10}\ \text{m s}^{-2}.$$

(ii) $s = ut + \frac{1}{2}at^2$, with $u = 0$ and $s = r$,

$$t = \sqrt{\left(\frac{2r}{g}\right)} = \sqrt{\left(\frac{10^{17}\ \text{m}}{8 \times 10^{-10}\ \text{m s}^{-2}}\right)} \approx 1 \times 10^{13}\ \text{s}$$

$$\approx 3 \times 10^5 \text{ years}$$

(iii) $v = u + at$, $u = 0$ so

$$v = gt \approx 8 \times 10^{-10}\ \text{m s}^{-2} \times 10^{13}\ \text{s} \approx 8000\ \text{m s}^{-1}$$

(iv) From Equation 43, $T = \dfrac{m\langle c^2 \rangle}{3k}$

$$\approx \frac{3.34 \times 10^{-27}\ \text{kg} \times (8000\ \text{m s}^{-1})^2}{3 \times 1.38 \times 10^{-23}\ \text{J K}^{-1}}$$

$$\approx 5000\ \text{K}$$

(b) The temperature in (a) (iv) is not high enough for fusion. However, it does illustrate that gravitational collapse does lead to a substantial rise in temperature. In practice, the gravitational acceleration does *not* remain constant throughout the collapse – as the fragment shrinks to a smaller radius, the field strength at its surface increases, and so particles will 'fall' with an ever-increasing acceleration, leading to a much greater temperature rise than estimated here.

52 Many stars are found in binary systems. Doppler shifts in the stars' spectral lines enable their orbital speeds and periods, and hence the sizes of the orbits, to be determined. The stellar mass can then be calculated using equations of gravitation.

53 Luminosity $\approx 7 \times 10^4\ L_{\text{Sun}}$ (note that the axis has a log scale). So time t on the main sequence would be:

$$t \approx \frac{18 \times 10^{10} \text{ years}}{7 \times 10^4} \approx 2.5 \times 10^6 \text{ years}$$

54 All bodies emit electromagnetic radiation that depends on their temperature. The higher the temperature, the shorter the wavelength at which most of the radiation is emitted. Gamma radiation is electromagnetic radiation at very short wavelengths, so is emitted by very hot bodies.

55 (a) From Equation 44, $= \langle E_k \rangle = \frac{3kT}{2} = \frac{m\langle c^2 \rangle}{2}$

(i) $\langle E_k \rangle = \frac{3 \times 1.38 \times 10^{-23}\ \text{J K}^{-1} \times 5 \times 10^{9}\ \text{K}}{2}$

$= 1 \times 10^{-13}\ \text{J}$

(ii) $\sqrt{\langle c^2 \rangle} = \sqrt{\left(\frac{3kT}{m}\right)}$

$= \sqrt{\left(\frac{3 \times 1.38 \times 10^{-23}\ \text{J K}^{-1} \times 5 \times 10^{9}\ \text{K}}{9.3 \times 10^{-26}\ \text{kg}}\right)}$

$= 1.5 \times 10^{6}\ \text{m s}^{-1}$.

(b) (i) $\langle E_k \rangle = 1 \times 10^{-13}\ \text{J}$

(ie the same as the iron nuclei, as the mean kinetic energy is independent of mass.)

(ii) $\sqrt{\langle c^2 \rangle} = \sqrt{\left(\frac{3kT}{m}\right)}$

$= \sqrt{\left(\frac{3 \times 1.38 \times 10^{-23}\ \text{J K}^{-1} \times 5 \times 10^{9}\ \text{K}}{9.11 \times 10^{-31}\ \text{kg}}\right)}$

$= 4.8 \times 10^{8}\ \text{m s}^{-1}$

This is greater than the speed of light, but no particle can travel this fast. (The material is so hot that electrons do travel close to the speed of light. To describe their behaviour properly we need the equations of special relativity instead of the 'classical' equations used here.)

56 (a) $\rho = \frac{M}{V}$, $V = \frac{4\pi r^3}{3}$, so $\rho = \frac{3M}{4\pi r^3}$.

(i) Neutron star: $\rho = \frac{3 \times 4 \times 10^{30}\ \text{kg}}{4\pi \times (5 \times 10^{3}\ \text{m})^3}$

$= 7.6 \times 10^{18}\ \text{kg m}^{-3}$.

(ii) He nucleus: $\rho = \frac{3 \times 6.7 \times 10^{-27}\ \text{kg}}{4\pi \times (10^{-15}\ \text{m})^3}$

$= 1.6 \times 10^{18}\ \text{kg m}^{-3}$.

(iii) The two densities are the same order of magnitude so the statement is justified.

(b) (i) From Equation 18, $g = \frac{GM}{r^2}$

$= \frac{6.67 \times 10^{-11}\ \text{N m}^2\ \text{kg}^{-2} \times 4 \times 10^{30}\ \text{kg}}{(5 \times 10^{3}\ \text{m})^2}$

(ii) $= 1.1 \times 10^{13}\ \text{N kg}^{-1} \approx 10^{12} g_{\text{Earth}}$

57 (a) Consider a particle mass m that is part of the star's surface at its equator, which moves in a circle of radius r. The gravitational force must be at least strong enough to provide the centripetal force so, from Equations 13 and 19:

$\frac{GMm}{r^2} \geq \frac{mv^2}{r}$.

We can cancel m and multiply by r, and since

$v = \frac{2\pi r}{T}$, we can write:

$\frac{GM}{r} \geq \left(\frac{2\pi r}{T}\right)^2$

ie $\frac{GM}{r} \geq \frac{4\pi^2 r^2}{T^2}$ and hence

$T \geq \sqrt{\frac{4\pi^2 r^3}{GM}}$

(b) $T \geq \sqrt{\left(\frac{4\pi^2 \times (5 \times 10^{3}\ \text{m})^3}{6.67 \times 10^{-11}\ \text{N m}^2\ \text{kg}^{-2} \times 4 \times 10^{30}\ \text{kg}}\right)}$

$= 1.4 \times 10^{-4}\ \text{s} \approx 0.1\ \text{ms}$.

So a period of a few ms would be reasonable, but a neutron star of this size and mass with a period less than 0.1 ms would have to spin so fast that it would fly apart.

58 (a) (i) Using Equation 13, $F_{\text{feet}} = \frac{GMm}{r^2}$

$= \frac{6.67 \times 10^{-11}\ \text{N m}^2\ \text{kg}^{-2} \times 10^{26}\ \text{kg} \times 10.0\ \text{kg}}{(5.0 \times 10^{4}\ \text{m})^2}$

$= 2.6680 \times 10^{7}\ \text{N}$

(ii) Similarly for the head with

$r = 5.0002 \times 10^{4}\ \text{m}$

$F_{\text{head}} = 2.6678 \times 10^{7}\ \text{N}$.

(b) The difference between the forces is $0.0002 \times 10^{7}\ \text{N} = 2 \times 10^{3}\ \text{N}$. A difference of 2 kN will cause some serious stretching – imagine hanging by your hands with 200 kg suspended from your feet. Now you know why Jo is so tall.

59 The cosmological principle states that all large volumes of space have the same properties. They should then be expanding at the same rate, which implies Hubble's law holds everywhere.

60 Doppler shift is a change in wavelength (and frequency) resulting from movement of the source of waves away from the observer. Cosmological redshift is a similar change, but occurs as a result of space itself expanding.

61 The existence of the cosmic background radiation suggests the Universe was a hotter, denser place than it is now, whereas the steady-state theory predicts the Universe has always been as it is now.

62 As the Universe expands, the peak wavelength of the radiation will become longer and longer and the temperature will move ever closer to absolute zero.

63 If H_0 has about 1.5 times the value used in the worked example, then the age, t, will be about two-thirds that calculated, ie. about 10 billion years

64 (i) From Equation 52,

$$v = dH_0 = 350 \text{ Mpc} \times 70 \text{ km s}^{-1} \text{ Mpc}^{-1}$$

$$= 24 \text{ km s}^{-1}$$

(ii) As $v << c$ we can use Equation 51:

$$z = \frac{v}{c} = \frac{24 \times 10^3 \text{ m s}^{-1}}{3.00 \times 10^8 \text{ m s}^{-1}} = 8.0 \times 10^{-5}$$

65 (a) Using Equation 1, a Cepheid of this period has:

$$L = 4\pi d^2_{\text{LMC}} F_{\text{LMC}} = 4\pi d^2_{\text{M100}} F_{\text{M100}}$$

Hence

$$d^2_{\text{M100}} F_{\text{M100}} = d^2_{\text{LMC}} F_{\text{LMC}}$$

$$d_{\text{M100}} = d_{\text{LMC}} \sqrt{\frac{F_{\text{LMC}}}{F_{\text{M100}}}}$$

$$= 65 \text{ kpc} \times \sqrt{\frac{9.5 \times 10^{-15} \text{ W m}^{-2}}{2.7 \times 10^{-19} \text{ W m}^{-2}}}$$

$$= 1.2 \times 10^7 \text{ pc} = 12 \text{ Mpc}$$

As an intermediate step you could use Equation 1 to find the star's luminosity (4.5×10^{29} W) but this cancels out in the final calculation.

(b) From the text, the supernova has peak luminosity $L \approx 5 \times 10^7 \times 2.9 \times 10^{26}$ W.

The smallest flux that can be measured is close to $F = 2.7 \times 10^{-17} \text{ W m}^{-2}$:

Using Equation 1, the maximum distance at which the supernova can be measured is given approximately by:

$$d = \sqrt{\frac{L}{4\pi F}} = \sqrt{\frac{5 \times 10^7 \times 2.9 \times 10^{26} \text{ W}}{4\pi \times 2.7 \times 10^{-19} \text{ W m}^{-2}}}$$

$$\approx 6.5 \times 10^{25} \text{ m} \approx 2 \times 10^9 \text{ pc } (= 2 \text{ Gpc})$$

$$\approx 7 \times 10^9 \text{ ly}$$

(c) Supernovae of this type would need to be observed in galaxies whose distances were already known; eg from observations of Cepheids. Measurements of their flux would then allow their luminosity to be found using Equation 1.

(d) Light observed from such a distant object has been en route for 7 billion years. It must have been emitted when the Universe was only about half its present age.

66 (a) Light travels at finite speed. So light reaching us from a large distance has taken time to get here. We therefore see galaxies as they were when the light was emitted, not as they are right now (see the answer to Question 65(d).)

(b) In the first 300 000 years after the Big Bang, light could not travel freely through the Universe. Photons were continuously colliding with other particles.

67 ${}^3_1\text{H} + {}^2_1\text{H} \rightarrow {}^4_2\text{He} + {}^1_0\text{n}$

The neutron must be produced in order to conserve charge (proton number) and nucleon number.

68 Using Equation 44, $\langle E_k \rangle = \frac{3kT}{2} = \frac{m\langle c^2 \rangle}{2}$

$$\langle E_k \rangle = \frac{3 \times 1.38 \times 10^{-23} \text{ J K}^{-1} \times 1.0 \times 10^8 \text{ K}}{2}$$

$$= 2.1 \times 10^{-15} \text{ J.}$$

$$\sqrt{\langle c^2 \rangle} = \sqrt{\frac{3kT}{m}} = \sqrt{\frac{2\langle E_k \rangle}{m}}$$

$$= \sqrt{\frac{2 \times 2.1 \times 10^{-15} \text{ J}}{6.68 \times 10^{-27} \text{ kg}}}$$

$$= 7.9 \times 10^5 \text{ m s}^{-1}.$$

69 If there are $12n$ hydrogen atoms (^1_1H) to n helium atoms (^4_2He), then:

mass of H: mass of He = 12:4 = 75%:25%.

70 He nuclei carry more charge than H nuclei, so their mutual electrostatic repulsion is greater and hence they need to be moving with more kinetic energy in order to get close enough to fuse. Kinetic energy depends on temperature, so producing carbon requires a higher temperature than producing helium. In the expanding universe, unlike inside stars, temperature is falling so the next stage of fusion cannot take place.

71 From Equations 13 and 16:

$$\frac{F_e}{F_g} = \frac{ke^2}{Gm^2} = 4.16 \times 10^{42}$$

(Note that you do not need to know the separation r, since the magnitudes of both forces are proportional to $1/r^2$.)

72 (a) See Figure 5.109(a). The tangents to the curve show that the slope, which represents the rate of expansion, is decreasing.

(b) See Figure 5.109(b). Here the tangents to the curve show that the slope is increasing ie. the rate of expansion is increasing.

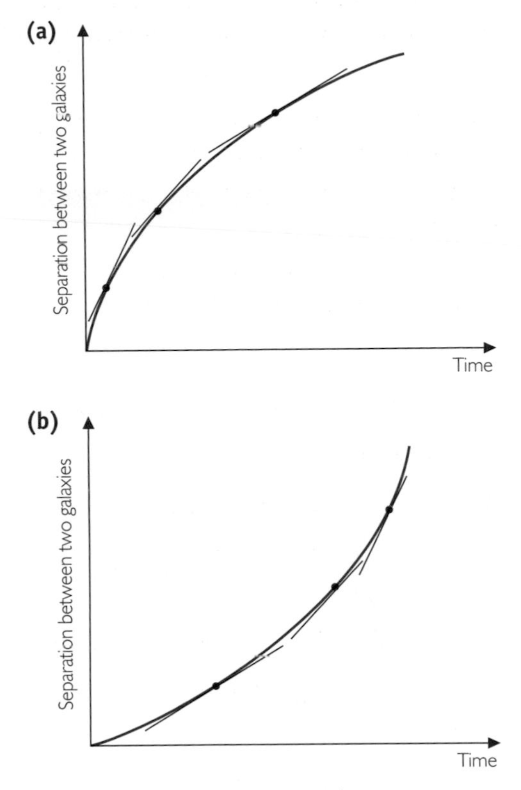

Figure 5.109 Answers to Question 72

73 If the expansion is slowing down, the actual age of the Universe will be less than estimated in Section 4.2. As the Universe must have been expanding more rapidly in the past, it will have reached its present size in a shorter time than if the expansion had always been at the rate which is deduced from the present value of the Hubble constant. (In the very distant past, H_0 would have been larger, as the expansion was more rapid.)

Maths Notes

0 Signs and symbols

0.1 Equations and comparisons

In physics, we are often interested in whether two quantities are exactly equal, or almost equal, or whether one is greater than the other. Table 1 lists the signs used for expressing such relationships.

Symbol	Meaning	Notes
$=$	is equal to	
$\equiv$	is exactly the same as	used to emphasis the point that two expressions are two ways of writing exactly the same thing (as opposed to two different things being the same size)
$\neq$	is not equal to	
$\approx$	is approximately equal to	
$\sim$	is the same order of magnitude as	
$<$	is less than	the smaller quantity is written at the narrow end of the symbol
$>$	is greater than	
$\leqslant$	is less than or equal to	
$\geqslant$	is greater than or equal to	
$\ll$	is much less than	
$\gg$	is much greater than	

Table 1 Signs for equations and comparisons

Maths reference

Order of magnitude

See Maths note 7.4

0.2 The delta symbol

The symbol Δ (the capital Greek letter delta) is used to mean 'a small amount of' or 'a change in'. Notice that Δ does *not* represent a number, so resist the temptation to cancel Δ, e.g. if it appears on the top and bottom of an expression.

For example, the symbol Δt represents a time interval and is often used when describing rates of flow or rates of change. For example, if an amount of charge ΔQ flows past a point in a time interval Δt, then the current I can be written

$$I = \frac{\Delta Q}{\Delta t}$$

The delta symbol is also used to denote an experimental uncertainty. For example, if a distance x is measured as 23 mm but could be out by 1 mm in either direction, then the uncertainty in the measurement is $\Delta x = 1$ mm. The measurement is written as $x \pm \Delta x$, i.e. 23 mm ± 1 mm.

0.3 Summation

The symbol $\sum$ (the capital Greek letter sigma) means 'the sum of'. Note that $\sum$ does *not* represent a number so resist the temptation to cancel it, e.g. if it appears on both sides of an equation.

The symbol $\sum$ is usually used with other symbols respresenting the items to be added together. For example, the mass of a sample of gas is $\sum m$, where m represents the mass of an individual gas molecule and the sum includes all the molecules in the sample.

When dealing with vectors, the sum is a vector sum i.e. it takes account of direction as well as magnitude. $\sum F$, meaning the sum of forces in a given situation, can be represented by a vetor diagram.

1 Index notation

An **index** (plural **indices**) or **power** is the superscript number which, when a positive whole number, means squared, cubed, etc. For example

$5^2 = 5 \times 5 = 25$

$7^3 = 7 \times 7 \times 7 = 343$

$0.6^4 = 0.6 \times 0.6 \times 0.6 \times 0.6 = 0.1296$

1.1 Index notation and powers of 10

Table 2 shows 'powers of 10'. The number in any row is found by dividing the number in the row above by 10.

100 000 =	$10 \times 10 \times 10 \times 10 \times 10 =$	10^5
10 000 =	$10 \times 10 \times 10 \times 10 =$	10^4
1 000 =	$10 \times 10 \times 10 =$	10^3
100 =	$10 \times 10 =$	10^2
10 =	$10 =$	10^1
1 =	$1 =$	10^0
0.1 =	$\frac{1}{10} =$	10^{-1}
0.01 =	$\frac{1}{10 \times 10} = \frac{1}{10^2} =$	10^{-2}
0.001 =	$\frac{1}{10 \times 10 \times 10} = \frac{1}{10^3} =$	10^{-3}

Table 2 Positive and negative powers of 10

Extending the pattern gives a meaning to zero and negative indices. If you replace all the 10s in Table 2 by any other number that you choose, you should be able to convince yourself that

$x^0 = 1$ for *any* value of x.

Maths reference

Units and physical quantitites

See Maths note 2.1

1.2 Standard form

To represent very large and very small numbers, we generally use **standard form**, also called **scientific notation**.

A number written in standard form consists of a number with a single digit (not zero) before the decimal point, multiplied by a power of 10.

Large numbers

5 620 000 (five million six hundred and twenty thousand) becomes 5.62×10^6

407 300 (four hundred and seven thousand, three hundred) becomes 4.073×10^5.

Small numbers

$0.5680 = 5.680 \times 0.1 = 5.68 \times 10^{-1}$

$0.000\,702\,3 = 7.023 \times 0.0001 = 7.023 \times 10^{-4}$

1.3 Combining powers

Powers of the same number

When multiplying two numbers expressed as 'powers' of the same number, the powers add:

$$10^2 \times 10^3 = (10 \times 10) \times (10 \times 10 \times 10) = 10^5$$

i.c. $\quad 10^2 \times 10^3 = 10^{(2+3)}$

$$6^2 \times 6^2 = (6 \times 6) \times (6 \times 6) = 6^4$$

When dividing, the powers subtract

$$10^6 \div 10^2 = (10 \times 10 \times 10 \times 10 \times 10 \times 10) \div (10 \times 10) = 10^4$$

i.e. $\quad 10^6 \div 10^2 = 10^{(6-2)}$

The rules still work when negative powers are involved:

$$10^5 \times 10^{-2} = 10^5 \times \left(\frac{1}{10^2}\right) = 10^5 \div 10^2 = 10^3$$

i.e. $\quad 10^5 \times 10^{-2} = 10^{(5-2)}$

$$x^4 \times x^{-3} = x^{(4-3)} = x$$

$$4^3 \div 4^{-2} = 4^3 \div \left(\frac{1}{4^2}\right) = 4^3 \times 4^2 = 4^5$$

i.e. $\quad 4^3 \div 4^{-2} = 4^{(3--2)} = 4^{(3+2)}$

Maths reference

Reciprocals

See Maths note 3.3

Powers of different numbers

When dealing with a mixture of numbers of different type, collect together all numbers of the same type and combine their powers by adding or subtracting:

$$2 \times 10^4 \times 3 \times 10^5 = (2 \times 3) \times (10^4 \times 10^5) = 6 \times 10^9$$

$$1.38 \times 10^{-23} \times 2.3 \times 10^3 = 1.38 \times 2.3 \times 10^{(-23+3)}$$

$$= 3.174 \times 10^{-20}$$

$$3y^2 \times 7y^5 = 21y^7$$

$$5z^2 \times 3z^{-2} = 15z^0 = 15$$

1.4 Manipulating powers on a calculator

Powers of 10

Think of the EXP or EE key as 'times 10 to the power of'.

To enter 7.54×10^9: enter 7.54, press EXP and enter 9. (Notice that you do *not* type in 10 – if you do, you will multiply your number by 10, making it 10 times too big.)

Your calculator might use its own shorthand to display this number as 7.54 09, or 7.54^9, or 7.54 EE 9 (or similar). But you should always *write* it as 7.54×10^9.

Negative powers of 10

To enter a negative index, use the ± or +/– key (*not* the 'minus' key, because that will subtract the next number from the one you have just entered).

To enter 1.38×10^{-23}: enter 1.38, press EXP, enter 23 and press ±.

Squares, etc.

To square a number, use the x^2 key. For example, to work out 1.3^2, enter 1.3 and press x^2 to get 1.69.

Pressing x^2 again squares the answer, i.e. calculates your original number to the power of 4. Pressing x^2 three times altogether gives you your original number to the power of 8, and so on – each time you press x^2, you double the power.

Other powers

Use the y^x key to raise one number to the power of a second number. y is the first number you enter, and x the second.

To calculate 2.5^3: enter 2.5, press y^x, enter 3, press =.

Other negative powers

As with powers of 10, use the ± or +/– key to enter negative numbers.

To calculate 2.5^{-3}: enter 2.5, press y^x, enter 3, press ±, press =.

1.5 Powers that are not whole numbers

The square root of a number x can be written as $x^{\frac{1}{2}}$ or $x^{1/2}$:

$$x^{\frac{1}{2}} \times x^{\frac{1}{2}} = x^{\left(\frac{1}{2}+\frac{1}{2}\right)} = x^1 = x$$

so $x^{\frac{1}{2}} = \sqrt{x}$.

Similarly, $x^{\frac{1}{3}} = \sqrt[3]{x}$ (the cube root of x); $x^{\frac{1}{4}} = \sqrt[4]{x}$ and so on.

Other fractional powers can also be interpreted in terms of roots, for example:

$$x^{\frac{3}{2}} = \sqrt{(x^3)} \text{ (the square root of } x\text{-cubed)}$$
$$= (\sqrt{x})^3 \text{ (the cube of the square root of } x\text{)}$$

and

$$x^{-\frac{1}{2}} = \frac{1}{x^{\frac{1}{2}}} = \frac{1}{\sqrt{x}}$$

Fractional powers can also be written using decimal numbers, for example:

$x^{\frac{1}{2}} = x^{0.5}$

$x^{\frac{3}{2}} = x^{1.5}$

Powers that are neither simple fractions nor whole numbers are less easy to interpret, but they still exist and can be calculated (e.g. using the y^x key of a calculator. For example:

$10^{0.333} = 2.153$

$10^{0.6021} = 4.000$

$5.6^{\pi} = 224.1$

$9.34^{-0.83} = 0.1565$

(All these answers are given to four significant figures.)

Maths reference

Significant figures

See Maths note 7.2

2 Units

The SI system of units (Système Internationale d'Unités) has been established by international agreement. In your study of physics you will use mainly SI units. The basic SI units are listed in Table 3. Notice that, when a unit is named after a person, the unit symbol has a capital but the *name* of the unit does not.

Quantity	SI unit	Notes
mass	kilogram, kg	
time	second, s	
length	metre, m	
electric current	ampere, A	used to define the unit of charge, the coulomb
temperature	kelvin, K	
luminous intensity	candela, cd	not used in this course, but included here for completeness
amount of substance	mole, mol	

Table 3 The basic SI units

2.1 Units and physical quantities; graphs and tables

A physical quantity consists of a number and a unit. Without the unit, the quantity is incomplete. When a symbol represents a physical quantity, it represents the *complete* quantity – units and all. For example, suppose v represents speed, and a particular speed is found to be 5 m s^{-1}. You should write

$v = 5 \text{ m s}^{-1}$

(*not* just $v = 5$ and *not* v (m s^{-1}) = 5).

Units can be manipulated just like numbers and other symbols. When labelling axes of graphs, and when listing physical quantities in tables, it is conventional to divide each quantity by its unit to get a pure number.

For example, you can divide both sides of the expression for v above by m s^{-1} and write

$v/(\text{m s}^{-1}) = 5$

If you are plotting values of v on a graph, or listing them in a table, you should label the graph axis, or the table column, as v/m s^{-1}.

Large and small numbers

Suppose you were dealing with speeds that were all several million metres per second:

$$v = 2 \times 10^6 \text{ m s}^{-1}, \qquad v = 7 \times 10^6 \text{ m s}^{-1}, \text{ etc.}$$

To make the numbers more manageable, you could use the same rule as above to write $v/(10^6 \text{ m s}^{-1}) = 2$, etc., and label your graph and table as shown in Figure 1.

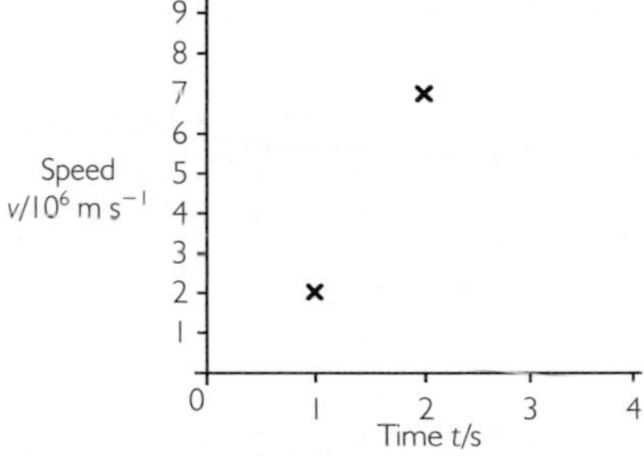

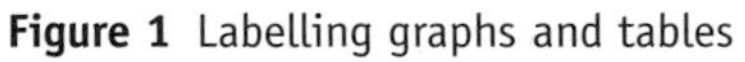

Figure 1 Labelling graphs and tables

2.2 Manipulating units; index notation and units

In calculations, the units should be manipulated as well as the numbers. This can help you keep track of what you are doing as well as being correct – so it is a good habit to get into.

Indices can be used with units and with algebraic symbols. For example,

$$4^{-1} = \frac{1}{4} = 0.25, \qquad x^{-2} = \frac{1}{x^2}$$

Units such as coulombs per second, or joules per coulomb, can be written either as C/s and J/C or using index notation: C s^{-1} and J C^{-1}. Similarly, metres per second, in calculations of unit of speed, can be written as m/s or m s^{-1}. For example,

$$70 \text{ m} \div 20 \text{ s} = 3.5 \text{ m s}^{-1}$$

Using the index notation helps prevent table headings and graph labels having too many oblique strokes.

When multiplying numbers, units or symbols, collect together all those of the same type and add their indices. For example:

$$2 \text{ C s}^{-1} \times 4 \text{ s} = 8 \text{ C}$$
$$10 \text{ m s}^{-1} \div 5 \text{ s} = 2 \text{ m s}^{-2}$$

Maths reference

Index notation and powers of 10
See Maths note 1.1

Units and physical quantities; graphs and tables
See Maths note 2.1

2.3 Derived units

Table 4 shows how SI units are combined to give units of various quantities. Some common combinations are given 'shorthand' names.

Quantity	Unit name	Symbol	Equivalent
speed			m s^{-1}
acceleration			m s^{-2}
force	newton	N	1 N = 1 kg m s^{-2}
gravitational field strength			1 N kg^{-1} = 1 m s^{-2}
energy, work	joule	J	1 J = 1 N m = 1 kg m^2 s^{-2}
power	watt	W	1 W = 1 J s^{-1} (= 1 kg m^2 s^{-3})
frequency	hertz	Hz	1 Hz = 1 s^{-1}
electric charge	coulomb	C	1 C = 1 A s 1 A = 1 C s^{-1}
potential difference, emf	volt	V	1 V = 1 J C^{-1} (= 1 kg m^2 C^{-1} s^{-2})
electrical resistance	ohm	Ω	1 Ω = 1 V A^{-1} (= 1 kg m^2 C^{-2} s^{-1})

Table 4 Some common derived SI units

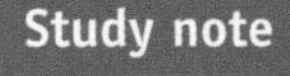

Study note

In writing units, the coulomb is often treated as if it were the basic unit rather that the ampere.

2.4 SI prefixes

When dealing with quantities that are large or small, we often use prefixes as an alternative to standard form. For example, a distance of 1.3×10^4 m could be written as 13 km, and a distance of 0.0037 m could be written as 3.7 mm. The official SI prefixes go up and down in steps of 10^3. Table 5 lists the SI prefixes that you are likely to encounter in your study of physics.

Prefix	Symbol	Equivalent in powers of 10
tera	T	10^{12}
giga	G	10^{9}
mega	M	10^{6}
kilo	k	10^{3}
centi	c	10^{-2}
milli	m	10^{-3}
micro	μ	10^{-6}
nano	n	10^{-9}
pico	p	10^{-12}
femto	f	10^{-15}

Table 5 SI prefixes

Study note

The centimetre is not officially an SI unit [because 'centi' (10^{-2}) does not fit the pattern] but it is widely used.

When dealing with conversions involving prefixes, it is wise to write down each step using appropriate powers of 10, *and include the units at each stage*. For example, suppose light of a certain colour has a wavelength of 468 nm and you want to use standard form to write the wavelength in metres:

$$\begin{aligned} 468 \text{ nm} &= 468 \times 10^{-9} \text{ m} \\ &= 4.68 \times 10^{2} \times 10^{-9} \text{ m} \\ &= 4.68 \times 10^{-7} \text{ m} \end{aligned}$$

Suppose the tension in a rope is 1.35×10^5 N and you want to express it in kN:

$$1 \text{ kN} = 10^3 \text{ N, so } 1 \text{ N} = \frac{1}{10^3} \text{ kN} = 10^{-3} \text{ kN}$$

$$\begin{aligned} 1.35 \times 10^5 \text{ N} &= 1.35 \times 10^5 \times 10^{-3} \text{ kN} \\ &= 1.35 \times 10^2 \text{ kN} \\ &= 135 \text{ kN} \end{aligned}$$

Suppose an electric current is 4.56×10^{-4} A and you want to express it in μA:

$$1\ \mu\text{A} = 10^{-6} \text{ A, so } 1 \text{ A} = \frac{1}{10^{-6}}\ \mu\text{A} = 10^{6}\ \mu\text{A}$$

$$\begin{aligned} 4.56 \times 10^{-4} \text{ A} &= 4.56 \times 10^{-4} \times 10^{6}\ \mu\text{A} \\ &= 4.56 \times 10^{2}\ \mu\text{A} \\ &= 456\ \mu\text{A} \end{aligned}$$

2.5 Dimensions

The **dimensions** of a quantity show how it is related to the basic quantities listed in Table 3. Symbols M, L and T are used to represent the dimensions of mass, length and time.

For example, volume is calculated from length × breadth × height so has dimension of length^3 or L^3; speed is found from distance ÷ time so has dimensions of L/T or LT^{-1}. The dimensions of force are those of mass × acceleration: MLT^{-2}.

Square brackets are used to denote the dimensions of a quantity. For example

$$[\text{velocity}] = LT^{-1}$$

$$[\text{force}] = [\text{mass}] \times [\text{acceleration}] = MLT^{-2}$$

Dimensions are more fundamental than units. You might, for example, choose to express a speed in miles per hour rather than SI units of m s^{-1}, but the dimensions are still LT^{-1}, i.e. length (miles) ÷ time (hours).

Any equation must be dimensionally consistent, that is, the dimensions of the left-hand side must be the same as those of the right-hand side. This can help you check whether a particular equation is correct, and can also enable you to derive relationships between quantities.

3 Arithmetic and algebra

3.1 Fractions, decimals and percentages

A fraction is really a division sum, e.g.

$$\frac{4}{5} = 4 \div 5; \qquad \frac{7}{3} = 7 \div 3.$$

You can express a fraction as a decimal number by doing the division on a calculator.

When fractions are multiplied together, you can often simplify the arithmetic by using the fact that the multiplication and division can be carried out in any order, e.g.

$$\frac{7}{5} \times \frac{3}{14} = \frac{7 \times 3}{5 \times 14}$$

and cancelling any common factors, e.g.

$$\frac{7 \times 3}{5 \times 14} = \frac{3}{5 \times 2} = \frac{3}{10} = 0.3.$$

You can think of the **percentage** sign, %, as being made up of a 1, 0, 0 to remind you that it is a fraction of 100 parts. To calculate a percentage from a number expressed as a fraction or a decimal, you multiply by 100:

$$\frac{1}{2} = 0.5 \text{ and } 100 \times 0.5 = 50 \text{ so } \frac{1}{2} = 50\% \text{ (or 50/100)}$$

$$\frac{1}{4} = 0.25 \text{ and } 100 \times 0.25 = 25 \text{ so } \frac{1}{4} = 25\% \text{ (or 25/100)}$$

$$\frac{7}{8} = 0.875 \text{ and } 100 \times 0.875 = 87.5 \text{ so } \frac{7}{8} = 87.5\%$$

For example, if a solar array produces an output power of 600 W from an input power of 4 kW (4000 W), its efficiency is

$$\frac{600\text{W}}{4000\text{W}} = 0.15 = 15\%.$$

To find a percentage of a quantity, you *multiply* the quantity by the percentage expressed as an ordinary fraction or decimal number. For example, to find 15% of 60 multiply 60 by 15/100 (or by 0.15)

$$\frac{15}{100} \times 60 = \frac{90}{10} = 9$$

or

$$0.15 \times 60 = 9.$$

3.2 Brackets and common factors

To evaluate an expression such as

$$6(2 + 3 - 4 + 5),\ \frac{12 + 8}{4} \text{ or } I(R_1 + R_2 + R_3)$$

you usually first deal with the additions and subtractions inside the bracket and then multiply or divide the result by the number or symbol outside. Alternatively you can carry out several separate multiplications or divisions on each number or symbol inside the bracket in turn, then do the additions or subtractions. For example

either $6(2 + 3 - 4 + 5) = 6 \times 6 = 36$

or $6(2 + 3 - 4 + 5) = 12 + 18 - 24 + 30 = 36$

either $\frac{12 + 8}{4} = \frac{20}{4} = 5$

or $\frac{12 + 8}{4} = \frac{12}{4} + \frac{8}{4} = 3 + 2 = 5$

A calculation that involves several multiplications or divisions using the same number and then adding or subtracting the results can be simplified if it is rewritten using brackets with the **common factor** outside. For example

$$25 + 30 + 35 = 5(5 + 6 + 7)$$

$$3x + 3y + 3z = 3(x + y + z)$$

$$IR_1 + IR_2 + IR_3 = I(R_1 + R_2 + R_3)$$

$$\frac{7}{2} + \frac{3}{2} + \frac{6}{2} = \frac{(7 + 3 + 6)}{2}$$

$$\frac{a}{x} + \frac{b}{x} + \frac{c}{x} = \frac{a + b + c}{x}$$

3.3 Reciprocals

The value obtained by dividing 1 by a number is called the **reciprocal** of the number (reciprocals can be found using the $1/x$ key of a calculator). Finding the reciprocal of a reciprocal gets you back to the original number. For example:

$$\frac{1}{2} = 0.5, \qquad \frac{1}{0.5} = 2.$$

For a wave or oscillation:

$$\text{period } T = \frac{1}{f}, \qquad \text{frequency } f = \frac{1}{T}.$$

Reciprocals are sometimes written using a negative index:

$$x^{-1} = \frac{1}{x}.$$

To find the reciprocal of a fraction, simply turn it the other way up. For example:

$$\frac{1}{\frac{1}{2}} = \frac{2}{1} = 2$$

$$\frac{1}{\frac{2}{3}} = \frac{3}{2} = 1\tfrac{1}{2}$$

$$\left(\frac{3}{7}\right)^{-1} = \frac{7}{3}$$

This is not just an arbitrary rule. It makes sense if you think in terms of division sums. Consider the second example above. Question: 'How many times does $\frac{2}{3}$ go into 1?' Answer: 'one-and-a-half times.'

Adding and subtracting

One place where you need to add and subtract reciprocals is in calculations of resistors in parallel. To find the net resistance R of several resistors connected in parallel, you must first find the reciprocal of each resistor, then add the reciprocals together (to get $1/R$), then find the reciprocal of $1/R$ to get R.

For example, if $R_1 = 2.0\ \Omega$, $R_2 = 5.0\ \Omega$, $R_3 = 1.0\ \Omega$, then

$$\frac{1}{R_1} = \frac{1}{2}\ \Omega^{-1} = 0.50\ \Omega^{-1} \text{ (notice the unit of } 1/R\text{)}$$

$$\frac{1}{R_2} = 0.20\ \Omega^{-1},$$

$$\frac{1}{R_3} = 1.00\ \Omega^{-1}$$

(notice that $1/1 = 1$ – the number stays the same but the unit still changes). So

$$\frac{1}{R} = (0.50 + 0.20 + 1.00)\ \Omega^{-1} = 1.70\ \Omega^{-1}$$

$$R = \frac{1}{1.70}\ \Omega = 0.59\ \Omega$$

Notice that adding the reciprocals of two numbers is *not* the same as adding the two numbers and then finding the reciprocal of their sum.

Multiplying and dividing

Multiplying by the reciprocal of a number is the same as dividing by that number. For example

$$7 \times \frac{1}{2} = 7 \div 2 = 3.50$$

Dividing by the reciprocal of a number is the same as multiplying by that number. For example

$$4 \div \frac{1}{3} = 4 \times 3 = 12$$

$$9 \div \frac{3}{4} = 9 \times \frac{4}{3} = \frac{9 \times 4}{3} = 12$$

For a wave,

$$f = \frac{v}{\lambda}, \qquad \text{time period } T = \frac{1}{f} = \frac{1}{(v/\lambda)} = \frac{\lambda}{v}$$

We can simplify divisions involving fractions. For example:

$$\frac{3}{4} \div \frac{5}{4} = \frac{3}{4} \times \frac{4}{5} = \frac{3 \times 4}{4 \times 5} = \frac{3}{5} = 0.6.$$

3.4 Algebra and elimination

If we have two different relationships that both involve some of the same things, we can combine them to produce a new equation. This allows us to avoid measuring, or calculating, something that is not already known – we can eliminate it (remove it) from the equations. For example, we can take an expression for electrical power

$$P = IV$$

and use the resistance equation

$$V = IR$$

to write IR instead of V:

$$P = I \times IR = I^2R$$

This enables us to relate P directly to I and R without needing to know or calculate V. Similarly, if we want to eliminate I:

$$P = \frac{V}{R} \times V = \frac{V^2}{R}$$

3.5 Adding and subtracting fractions

You can of course add and subtract fractions on a calculator – you carry out several division sums and add or subtract the results. But for simple fractions it can often be quicker to do the sums 'by hand'.

The trick is to write the fractions so that they have the same **denominator** (the number underneath the fraction). Sometimes it is quite easy to spot how to do this. For example:

$$\frac{3}{4} + \frac{5}{6} = \frac{3 \times 3}{3 \times 4} + \frac{2 \times 5}{2 \times 6}$$

$$= \frac{9}{12} + \frac{10}{12} = \frac{9 + 10}{12} = \frac{19}{12}$$

Otherwise, make a common denominator by multiplying the original denominators together:

$$\frac{2}{17} + \frac{4}{3} = \frac{2 \times 3}{17 \times 3} + \frac{4 \times 17}{3 \times 17}$$

$$= \frac{6}{51} + \frac{68}{51} = \frac{6 + 68}{51} = \frac{74}{51}$$

Another example:

$$\frac{1}{2} + \frac{1}{3} = \frac{3}{6} + \frac{2}{6} = \frac{5}{6}$$

4 Solving equations

It may sound obvious, but the main thing to understand about equations is that the '=' sign means that the two things on either side are *equal* to one another. So whatever you do to one side, you must also do to the other, otherwise they would no longer be equal. (Beware of getting into the bad habit of writing '=' when you really mean 'and so the next step is...'.)

One way to think of an equation is as a 'recipe' for calculating. For example, $F = ma$ tells you how to calculate the net force F if you know the acceleration a that it gives to a mass m. In this example, F is the **subject** of the equation – it is written on its own (usually on the left).

4.1 Rearranging an equation

Quite often, the quantity you want to calculate is wrapped up in the right-hand side of an equation, and you need to make it the subject. When doing this, it helps if you try to understand what you are doing rather than blindly trying to apply a set of rules. It is also wise to write down each step, justifying each one to yourself as you do so. This might sound time-consuming, but it isn't really because it helps you to keep track of what you are doing and, if you do make a slip, it is quite easy to go back and check.

Look at the part of the equation that contains the quantity that you want to know. Think what you need to do to get that quantity on its own, and do the same thing(s) to both sides.

For example, suppose you want to know the acceleration that a force F gives to a mass m:

$$F = ma$$

To get a on its own, you need to divide the right-hand side by m ($ma \div m = a$), so do the same to the left-hand side:

$$\frac{F}{m} = a, \qquad \text{or} \qquad a = \frac{F}{m}.$$

Another example: suppose you want to calculate internal resistance r from

$$V = \mathscr{E} - Ir$$

It is a good idea first to arrange that the thing you are interested in has a positive sign. You can do this by adding Ir to both sides:

$$V + Ir = \mathscr{E}$$

then to get r on its own you subtract V from both sides:

$$Ir = \mathscr{E} - V$$

and then divide by I

$$r = \frac{\mathscr{E} - V}{I} \qquad \text{or} \qquad r = (\mathscr{E} - V)/I$$

(Notice that you have to divide the *whole* of the right-hand side by I – hence the brackets.)

Maths reference

Brackets and common factors

See Maths note 3.2

4.2 Simultaneous equations

Simultaneous equations arise if we have two (or more) different ways of writing a relationship between quantities. If we have two unknown quantities, then they can both be found if we have two simultaneous equations. For three unknown quantities, we'd need three separate equations, and so on.

The trick in solving simultaneous equations is to carry out some algebra and arithmetic to get an expression that involves just *one* of the unknown things, and then use that value to calculate the other one.

For example, the equation $\mathscr{E} = V + Ir$ involves two things that can be measured (V and I). If neither $\mathscr{E}$ nor r is known, then they cannot be found from a single pair of values of V and I. However, if you obtain two *different* pairs of readings (V_1 and I_1, and V_2 and I_2) for the same power supply (using two different external loads), then you can write down two simultaneous equations – two different equations that both describe a relationship between the two unknown things $\mathscr{E}$ and r. These equations let you find both $\mathscr{E}$ and r. So

$$\mathscr{E} = V_1 + I_1 r$$

$$\mathscr{E} = V_2 + I_2 r$$

Since the right-hand side of each equation is equal to $\mathscr{E}$, then they must also be equal to each other:

$$V_1 + I_1 r = V_2 + I_2 r$$

Subtracting V_1 from each side

$$I_1 r = V_2 - V_1 + I_2 r$$

Subtracting $I_2 r$ from both sides (and being careful with signs and with the subscripts 1 and 2)

$$I_1 r - I_2 r = V_2 - V_1$$

Now r is a common factor on the left-hand side, so

$$r(I_1 - I_2) = V_2 - V_1$$

Dividing both sides by $(I_1 - I_2)$ (and using brackets to keep the subtracted things together)

$$r = \frac{(V_2 - V_1)}{(I_2 - I_1)} \qquad \text{or} \qquad r = (V_2 - V_1)/(I_1 - I_2)$$

This value of r can then be used in one of the original equations to find $\mathcal{E}$.

For example: a power supply gives readings of $V_1 = 3\,\text{V}$, $I_1 = 7\,\text{A}$, and $V_2 = 8\,\text{V}$, $I_2 = 2\,\text{A}$. So

$$r = \frac{8\,\text{V} - 3\,\text{V}}{7\,\text{A} - 2\,\text{A}} = \frac{5\,\text{V}}{5\,\text{A}} = 1\,\Omega$$

and

$$\mathcal{E} = V_1 + I_1 r = 3\,\text{V} + 7\,\text{A} \times 1\,\Omega - 3\,\text{V} + 7\,\text{V} = 10\,\text{V}$$

(you would find the same value using V_2 and I_2).

5 Relationships and graphs

Graphs are extremely useful in physics for giving us a pictorial representation of how one quantity is related to another. Trends in data are not always clear from a table of results, but become immediately evident when viewing a plot of the two quantities involved.

5.1 Graphs and proportionality

Many important relationships in physics involve the idea of direct proportion.

For example, if a conductor obeys Ohm's law, doubling the potential difference produces double the current, tripling the pd triples the current ... and so on. Mathematically, we say that the potential difference is **directly proportional** to the current. In symbols

$$V \propto I \qquad \text{or} \qquad V = kI$$

The symbol $\propto$ means 'is directly proportional to' and k is called a **constant of proportionality** and has a fixed value for a particular set of values of V and I. (The constant k in this example is the same thing as the electrical resistance R.)

If one quantity is directly proportional to another, then a graph of one plotted against the other is a straight line through the origin.

5.2 Linear relationships

The equation $V = \mathcal{E} - Ir$ is an example of a **linear relationship** between two variables, V and I in this case. A graph of V (on the vertical axis, the y-axis) against I (on the horizontal axis, the x-axis) gives a straight line. Linear relationships and graphs are often said to be of the type $y = mx + c$, where y stands for whatever is plotted on the y-axis and x for whatever is plotted on the x-axis, and m and c are constants (they remain fixed when x and y change). This type of graph has two properties that are often useful for doing calculations using experimental results. We can illustrate these with a graph of $y = 2x + 1$, i.e. $m = 2$, $c = 1$ (Figure 2).

On Figure 2, the line cuts the y-axis at $y = 1$ (using the equation, when $x = 0$, $y = c$). The line of such a graph always cuts the y-axis where $y = c$.

If y is directly proportional to x, then the line goes through the origin and $c = 0$.

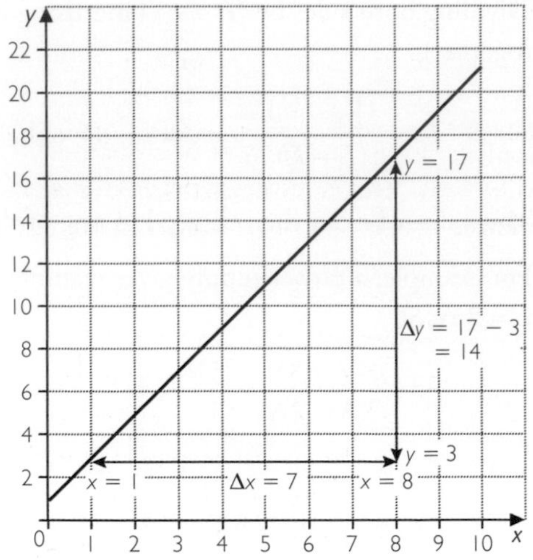

Figure 2 A graph of $y = 2x + 1$

5.3 Gradient of a linear graph

Figure 2 is a graph of the linear relationship $y = 2x + 1$.

The **gradient** (or slope) of the graph is defined as the rise of the graph (the increase in y, Δy) divided by the run (the corresponding increase in x, Δx) found by drawing a right angled triangle as shown in Figure 2. On Figure 2,

$$\Delta y = 14, \qquad \Delta x = 7,$$

$$\text{gradient} = \frac{\Delta y}{\Delta x} = 2.$$

Notice that Δy and Δx are numbers read from the graph scales, (*not* lengths measured with a ruler) and that any similar triangle drawn on the graph will give the same value of the gradient.

The gradient of a linear graph of y against x is always equal to the value m in the relationship $y = mx + c$.

The graph in Figure 2 has a positive gradient. If m is negative, then the graph slopes down from left to right.

If two variables measured in an experiment are related by a linear equation, then plotting them on a graph enables you to find the values of the constants relating them. It is helpful if you arrange the relationship so that it looks as much like $y = mx + c$ as possible. For example, by subtracting Ir from both sides you can write $\mathscr{E} = V + Ir$ as

$$V = (-r)I + \mathscr{E}$$

which can be compared directly with

$$y = mx + c$$

If you plot measured values of V on the y-axis against corresponding values of I on the x-axis, the graph will be a straight line that cuts the y-axis at $\mathscr{E}$, and with a gradient $m = -r$.

Maths reference

Error bars and error boxes

See Maths note 7.5

5.4 Inverse proportionality

If one quantity is **inversely proportional** to another, then as one increases, the other will decrease. For example, the acceleration produced by a given net force is inversely proportional to the mass on which it acts: doubling the mass halves the acceleration, tripling the mass divides the acceleration by three and so on – and vice versa.

Maths reference

Reciprocals

See Maths note 3.3

Such a relationship is written using reciprocals and the symbol for direct proportion:

$$a \propto \frac{1}{m} \qquad a = \frac{k}{m}$$

or

$$m \propto \frac{1}{a} \qquad m = \frac{k}{a}$$

(In this case, the constant of proportionality is the same as the net force F.)

If one quantity is inversely proportional to the other (Table 6), the graph of one plotted against the other is curved as in Figure 3.

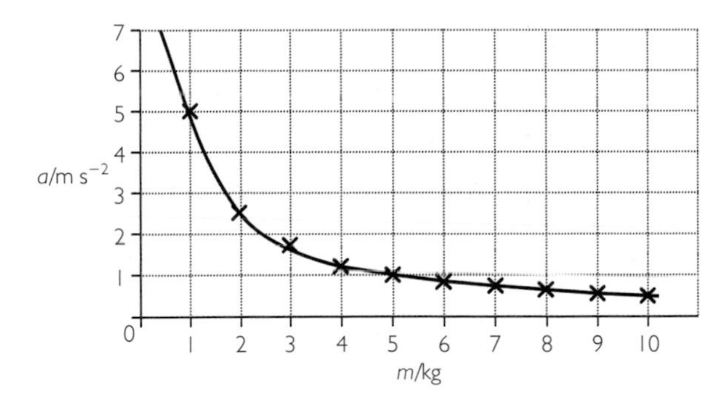

Figure 3 A graph showing how the accelertion a produced by a constant force F (= 5N) depends on mass m (data from Table 6)

m/kg	$(1/m)$/kg^{-1}	a/ms^{-2}
1	1.000	5.00
2	0.500	2.50
3	0.333	1.67
4	0.250	1.25
5	0.200	1.00
6	0.167	0.83
7	0.143	0.71
8	0.125	0.63
9	0.111	0.55
10	0.100	0.50

Table 6 Data for Figures 3 and 4

But if one quantity is plotted against the *reciprocal* of the other, then the graph is a straight line through the origin, as shown in Figure 4.

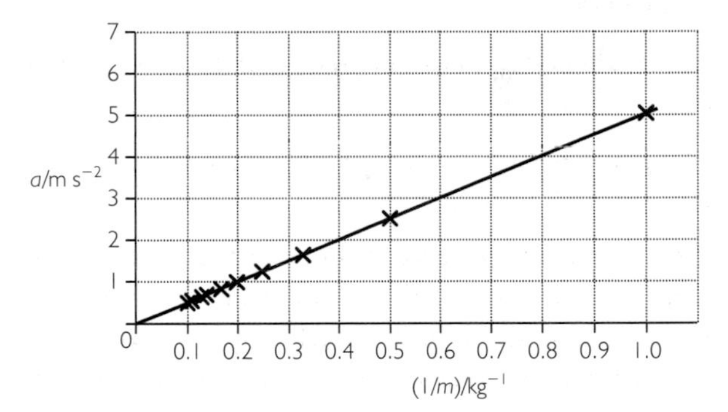

Figure 4 The data from Figure 3 plotted as a against $1/m$

5.5 Testing mathematical relationships

Sometimes we are interested in finding a mathematical relationship between two measured quantities. This usually involves some educated guesswork, based on ideas about the underlying physics and/or from looking at the numbers. Plotting graphs provides a way of testing the guesses.

Direct proportion

For example, if both quantities increase together, you might guess that one is directly proportional to the other. Plot a graph of one against the other and see whether you can draw a straight line through all the error boxes.

Examples that give straight-line graphs include:

$s \propto t$ for motion at constant speed
$I \propto V$ for an ohmic conductor.

If the plot does not give a straight line, try something else. For example, motion from rest at constant acceleration is described by the equation

$$s = \tfrac{1}{2}at^2$$
$$s \propto t^2$$

Maths reference

Experimental uncertainty
See Maths note 7.1

Error bars and error boxes
See Maths note 7.5

A graph of distance s against time t is a curve, but a graph of s against t^2 is a straight line with gradient $a/2$ or $\frac{1}{2}a$.

Sometimes you need to use the square root of a quantity to get a straight line. For example, for a simple pendulum a plot of its period T against the square root of its length l gives a straight line:

$$T \propto \sqrt{\ell}$$

Inverse proportion

If one quantity increases as the other decreases, you might guess that you are looking at inverse proportionality, so try plotting a graph using the reciprocal of one quantity.

If this does not give a straight line, try plotting the square, or the square root, of the reciprocal.

For example, suppose you measure the frequency f of the note from a plucked string of mass per unit length μ. Frequency f decreases as you increase μ, but suppose you find that a graph of f against $1/\mu$ is not a straight line.

If a graph of f against $\frac{1}{\mu^2}$ is a straight line, then $f \propto \frac{1}{\mu^2}$

If you need to plot f against $\frac{1}{\sqrt{\mu}}$ to get a straight line, then $f \propto \frac{1}{\sqrt{\mu}}$

6 Trigonometry and angular measurements

6.1 Degrees and radians

A **radian**, or **rad** for short, is a unit for measuring angles commonly used in physics instead of degrees. Figure 5 shows how the size of an angle, in radians, is defined.

For a full circle, length of arc = length of circumference = $2\pi r$.

Size of angle = $\frac{2\pi r}{r} = 2\pi$ radians, i.e. approximately 6.28 rad.

Table 7 lists some useful conversions between radians and degrees.

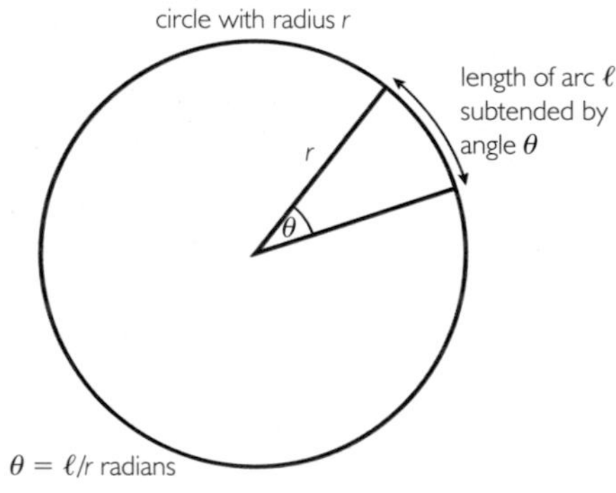

Figure 5 The size of an angle measured in radians

Angle	Size in degrees	Size in radians
full circle	360°	2π rad = 6.28 rad
half circle	180°	π rad = 3.14 rad
	114.6°	2.0 rad
quarter circle	90°	$\pi/2$ rad = 1.57 rad
	60°	$\pi/3$ rad = 1.05 rad
	57.3°	1.0 rad
	45°	$\pi/4$ rad = 0.79 rad
	30°	$\pi/6$ rad = 0.52 rad
	28.6°	0.5 rad

Table 7 Some conversions between radians and degrees

Note that π is a *number* (approximately 3.14) that frequently, but not always, appears in angles measured in radians.

6.2 Sine, cosine and tangent of an angle

Figure 6 shows a right angled triangle. The sides of the triangle are related by Pythagoras's theorem:

$$c^2 = b^2 + a^2$$

$$c = \sqrt{(a^2 + b^2)}$$

(Care! You can't 'cancel' the squares inside the bracket.)

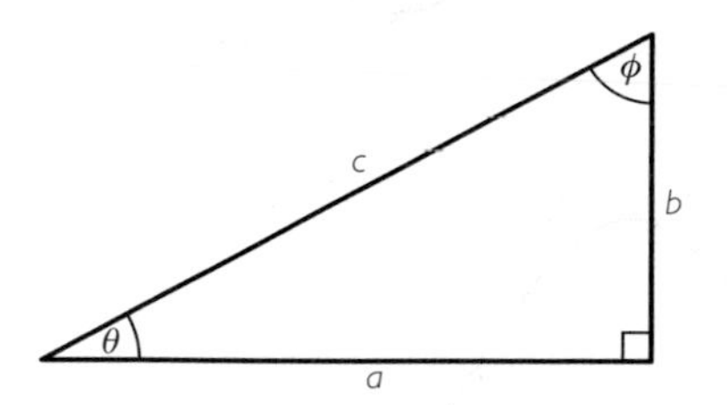

Figure 6 A right angled triangle

All similar triangles, i.e. those with the same angle θ, will have sides in the same proportion to one another. The ratios of the sides depend only on the angle θ.

The sine, cosine and tangent of the angle θ are known as **trigonometric ratios**.

- Sine of angle θ, $\sin\theta = \dfrac{\text{opposite side}}{\text{hypotenuse}} = \dfrac{b}{c}$
- Cosine of θ, $\cos\theta = \dfrac{\text{adjacent side}}{\text{hypotenuse}} = \dfrac{a}{c}$
- Tangent of θ, $\tan\theta = \dfrac{\text{opposite side}}{\text{adjacent side}} = \dfrac{b}{a}$

We can combine these to give another useful relationship. Since

$$\frac{b}{a} = \frac{b}{c} \div \frac{a}{c} \quad (c \text{ cancels}),$$

we can write

$$\tan\theta = \frac{\sin\theta}{\cos\theta}$$

Also

$$\sin\phi = \frac{a}{c} = \cos\theta \qquad \text{and} \qquad \cos\phi = \frac{b}{c} = \sin\theta$$

i.e. if two angles add up to 90°, then the cosine of one is equal to the sine of the other.

Using Pythagoras's theorem leads to another useful result. Dividing $c^2 = a^2 + b^2$ by c^2:

$$1 = \frac{a^2}{c^2} + \frac{b^2}{c^2} = \left(\frac{a}{c}\right)^2 + \left(\frac{b}{c}\right)^2$$

$$1 = (\cos\theta)^2 + (\sin\theta)^2,$$

which is true for any angle and is usually written as

$$\cos^2\theta + \sin^2\theta = 1$$

6.3 Graphs of trigonometric functions

For angles greater than 90°, Figure 7 shows how sin, cos and tan are defined. For some angles, negative numbers are involved. Figure 8 shows how the sin, cos and tan vary with angle θ. Note that we have labelled the axis in degrees and in radians.

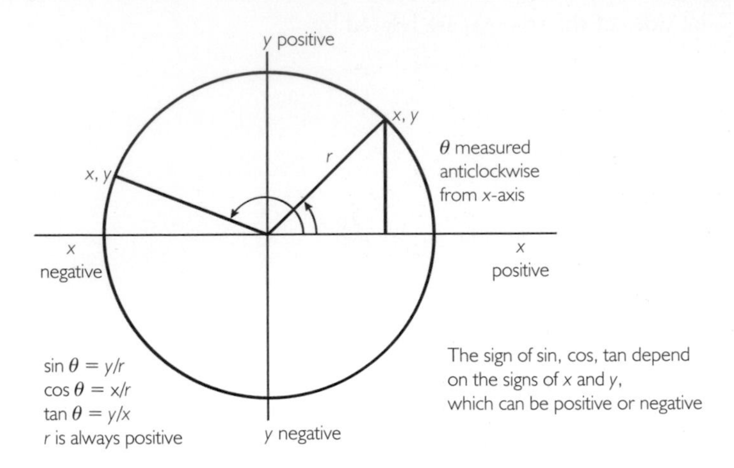

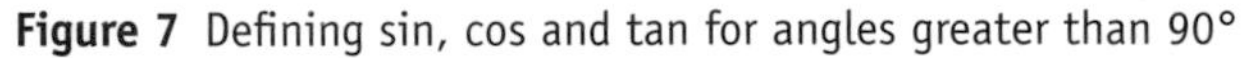

Figure 7 Defining sin, cos and tan for angles greater than 90°

Notice that $\sin\theta$ and $\cos\theta$ are always between +1 and −1, but $\tan\theta$ is infinite for some angles (notice the different scale in Figure 8(c)).

Also notice some useful values, e.g. $\sin 30° = \cos 60° = 0.5$. Look at the values of $\sin\theta$ and $\cos\theta$ when θ is a multiple of 90°.

Maths reference

Degrees and radians

See Maths note 6.1

6.4 Inverse sin, etc.

The angle whose sin is x is written $\sin^{-1} x$. We can write the relationships from Figure 6 as

$$\theta = \sin^{-1}\frac{b}{c} \qquad \theta = \cos^{-1}\frac{a}{c} \qquad \theta = \tan^{-1}\frac{b}{a} \qquad \phi = \sin^{-1}\frac{a}{c}$$

Beware! The index −1 here does *not* indicate a reciprocal:

$\sin^{-1} x$ is *not* the same as $\dfrac{1}{\sin x}$

6.5 Trigonometry on a calculator

You can find the sine, cosine and tangent of an angle on a calculator. For example, to find sin 30°, type 30 and press sin.

Many scientific calculators can be switched between 'degree' and 'radian' modes. The display will indicate which one you are in.

If you switch your calculator to 'radian' mode, you can find sin, etc., of angles in radians without having to convert to degrees. Check that you know how to do this.

With your calculator in radian mode, type π, ÷, 2 (you may need to press = as well) and then press sin or cos. You should get $\sin(\pi/2) = 1$, $\cos(\pi/2) = 0$. If you have your calculator in degree mode by mistake, you will find the sin or cos of 1.57° (3.14° ÷ 2).

Try finding the sin, cos and tan of some angles in degrees and in radians. Check that you get the same values as shown in Figure 8.

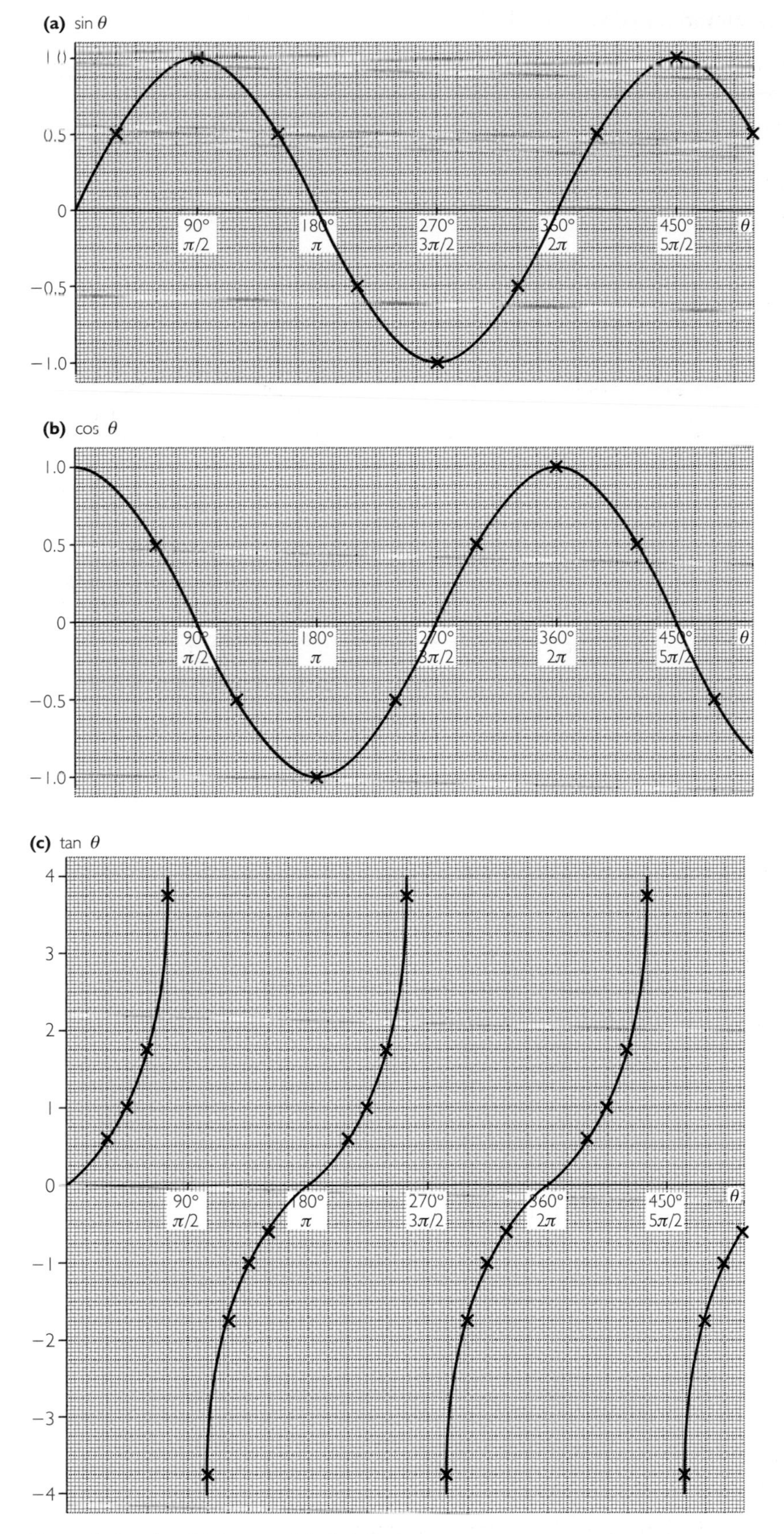

Figure 8 Graphs of trigonometric functions

If you know the sin, cos or tan of an angle and wish to determine the size of the angle, use the 'inv' key.

For example, to find the angle whose sin is 0.5, type 0.5, press inv and then press sin. You should get 30 if you have your calculator in degree mode. If you do this with your calculator in radian mode, you will get 0.5236 ($\approx \pi/6$).

6.6 The small angle approximations

There are some useful approximations involving the trigonometric ratios of small angles. These become evident when we express the sine and tangent of an angle θ in terms of the right angled triangles shown in Figure 9.

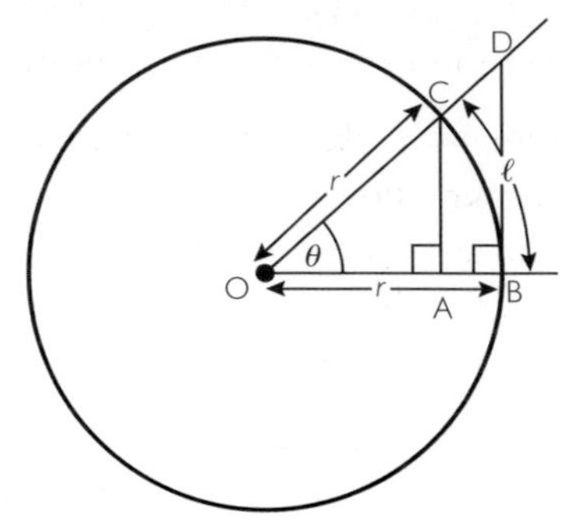

Figure 9 The sine and tangent of an angle

From the triangle OAC

$$\sin\theta = \frac{AC}{OC} = \frac{AC}{r}$$

and

$$\cos\theta = \frac{OA}{OC} = \frac{OA}{r}$$

From the triangle OBD

$$\tan\theta = \frac{BD}{OB} = \frac{BD}{r}$$

and

$$\cos\theta = \frac{OB}{OD} = \frac{r}{OB}$$

Figure 9 shows that $\tan\theta$ is always greater than $\sin\theta$ because BD is greater than AC.

As θ is made smaller, the lines AC and BD become closer together and more equal in length, and the lines OA and OD become closer to r, so *for small angles*:

$$\sin\theta \approx \tan\theta$$

and

$$\cos\theta \approx 1$$

With your calculator in degree mode, try finding the sin, cos and tan of the angles listed in Table 7, and some smaller angles. Notice that the approximations get better as the angles get smaller.

Small angles in radians

Comparison with Figure 5 shows that the size of θ *measured in radians* lies between $\sin\theta$ and $\tan\theta$ (the arc length ℓ is longer than AC and shorter than OD):

$$\sin\theta < \theta < \tan\theta$$

When θ is small,

$$AC \approx \ell \approx BD$$

and so *for small angles measured in radians* we have some additional approximations:

$$\sin\theta \approx \theta$$

and

$$\tan\theta \approx \theta$$

Switch your calculator into radian mode, and again try finding the sin, cos and tan of various angles. Notice that the approximation gets better at small angles.

7 Size and precision

7.1 Precision in measurements; experimental uncertainty

In any measurement, there is a limit to the precision of your result. Sometimes this **experimental uncertainty** arises because you get different answers when you repeat the measurement. For example, if you time an athlete running 100 metres, the same athlete will probably record different times on different occasions. The uncertainty in the measurements is indicated by the 'scatter' in the results.

For example, suppose a certain athlete records times of 12.5 s, 12.1 s, 12.6 s, 12.5 s and 12.3 s. The average time is t = 12.4 s. The difference between the average and the biggest or smallest value indicates the uncertainty Δt – in this case, $\Delta t \approx$ 0.3 s.

Sometimes the uncertainty arises because it is difficult to judge exactly what to measure. For example, if you are using a signal generator to produce a sound from a speaker and ajusting it to give the same pitch as a note from a guitar, it might be hard to judge the frequency exactly. If you think the best frequency f is 260 Hz, but are unsure by about 10 Hz either way, then the uncertaintly would be $\Delta f \approx$ 10 Hz.

Even if there is no problem deciding exactly what to measure, and you get the same answer each time you repeat the measurement, there is still an uncertainty because the measurement is limited by the instrument you are using. For example, if you use a digital ammeter to measure a current I, and you get 0.357 A each time, you can only be sure that the current is closer to 0.357 A than it is to either 0.356 A or 0.358 A – it could lie anywhere between 0.3565 A and 0.3575 A. So the uncertainty is $\Delta I \approx$ 0.0005 A.

Some books refer to **experimental error** rather than uncertainty. Don't be misled into thinking that they mean a mistake. However carefully and correctly you carry out a measurement, there will always be an uncertainty.

Experimental uncertainties apply to *all* measured quantities – including those you look up in a data book, though these values have usually been measured with much greater precision than you can achieve in a school or college laboratory.

7.2 Calculations with uncertainties; significant figures

If you carry out a calculation using a measured value, there will always be an uncertainty in your answer. You can use the uncertainties in the measurements to work out the uncertainty in the calculated value.

For example, suppose you measure a current of I = 0.24 A ± 0.01 A and a corresponding pd of V = 0.67 V ± 0.02 V.

On a calculator, the resistance found using the 'best' values is

$$R_{\text{best}} = \frac{V}{I} = \frac{0.67\text{V}}{0.24\text{A}} = 2.791\,6667\ \Omega$$

But, using the largest possible V (0.39 V) and the smallest possible I (0.23 A), the calculated resistance could be as large as

$$R_{\text{max}} = \frac{0.69\text{V}}{0.23\text{A}} = 3\ \Omega$$

Or, using the smallest V and the largest I, it could be as small as

$$R_{\text{min}} = \frac{0.65\text{V}}{0.25\text{A}} = 2.6\ \Omega$$

There are several things to notice! First, there are quite large differences between the three values. Second, the first value extends to the full length of the calculator display, whereas the others do not.

The large differences show that you cannot *possibly* say that the resistance is precisely 2.791 6667 Ω. This value is close to 2.8 Ω, and the other two differ by 0.2 Ω in either direction, i.e. the uncertainty in R is $\Delta R \approx 0.2\ \Omega$. The resistance can therefore be written as

$$R = 2.8\ \Omega \pm 0.2\ \Omega.$$

The second figure in this answer (the 8 after the decimal point) is uncertain, and so any further figures are meaningless.

Another way of putting this is to say that the answer has (only) two **significant figures** – the one before the decimal point and the first one after it. The rest of the figures in the original 'best' answer are meaningless. They are *not* significant.

7.3 A useful rule of thumb

In a calculation, the answer cannot be known any more precisely than the values used to calculate it. As a useful rule of thumb, the final answer has no more **significant figures** than the *least* precise value used in the calculation. (The example in Maths note 7.2 illustrates this.)

Suppose you did a calculation to find the frequency f of light whose wavelength is 468 nm (4.68×10^{-7} m). The speed of light is known very precisely: $2.997\,925 \times 10^{8}$ m s^{-1}.

Using speed ÷ wavelength

$$f = \frac{2.997\,925 \times 10^{8}\ \text{m s}^{-1}}{4.68 \times 10^{-7}\ \text{m}}$$

$$= 6.4058 \times 10^{14}\ \text{Hz}$$

However, we only knew the wavelength to three significant figures, so we cannot quote the frequency of this precisely. We must stick to the three significant figures and write

$$f = 6.41 \times 10^{14}\ \text{Hz}$$

There was in fact no point in using the very precise value for the speed of light. Values listed in data books are often rounded to, say, three significant figures if they are likely to be used only in calculations requiring this precision or less.

7.4 Significant figures and orders of magnitude

The speed of light to seven significant figures is $2.997\,925 \times 10^{8}$ m s^{-1}; the significant figures are 2997925.

Zeros in front of a number are not significant. The speed of light could be written (rather oddly) as $002.997\,925 \times 10^{8}$ m s^{-1} or $0.000\,299\,7925 \times 10^{12}$ m s^{-1} without making any difference to its value.

However, zeros at the end of a number are (or at least can be!) significant. If you wrote the speed of light as 299 792 500 m s^{-1}, that would imply that you knew that the last two figures were definitely zeros and not some other numbers. If they are, in fact, not known, it is better to use standard form so that the meaningless zeros can be dropped.

To five significant figures, the speed of light would be 2.9979×10^8 m s^{-1}. To three significant figures, it would be 3.00×10^8 m s^{-1}. Here the zeros *are* significant and should be written down, because 2.997… rounds to 3.00.

To one significant figure the speed of light would be 3×10^8 m s^{-1}.

If a value is rounded to just the nearest power of 10, then we say we are giving just the **order of magnitude**. Two values are said to have the same order of magnitude if one is between 1 and 10 times the other. For example, the wavelengths of red and blue light (about 400 nm and 700 nm) are within the same order of magnitude. But the wavelengths of infrared radiation range from about 10^{-6} m to about 10^{-3} m – they cover three orders of magnitude.

7.5 Error bars and error boxes

When plotting a graph of experimental data, you should take account of the uncertainties. Rather than representing each measurement by a point, you should draw an **error bar** to represent the range of possible values. Then use the vertical and horizontal error bars to draw an **error box** around each plotted point. Once you have plotted the error boxes, you can then draw a trend line on your graph. It might be possible to draw a straight line passing through all the boxes, even if you could not draw one through all the points.

8 Logarithms

8.1 Logs and powers of 10

If a number can be written as *just* a 'power of 10', then the power is the **logarithm** of that number; strictly speaking, it is the **logarithm to base 10**, or **common logarithm**, of the number, but it is often simply called the **log**.

Table 8 lists some examples using whole-number powers.

Number x	$\log_{10}(x)$
$100\,000 = 10^5$	5
$10\,000 = 10^4$	4
$1000 = 10^3$	3
$100 = 10^2$	2
$10 = 10^1$	1
$1 = 10^0$	0
$0.1 = 10^{-1}$	−1
$0.001 = 10^{-2}$	−2

Table 8 Some numbers and their common logarithms

In fact *any positive number* can be expressed as a power of 10, using powers that are not whole numbers. Most whole numbers have logs that are not themselves whole numbers or simple fractions. For example:

$$10^{0.6021} = 4.000$$

so

$$\log_{10}(4.000) = 0.6021$$

All numbers between 1 and 10 have base 10 logs that lie between 0 and 1. For example:

$$10^{0.333} = 2.153$$

so

$$\log_{10}(2.513) = 0.333$$

Similarly, all numbers between 10 and 100 have base 10 logs that lie between 1 and 2; all numbers between 100 and 1000 have base 10 logs between 2 and 3, and so on.

All numbers less than 1 have negative logs. For example:

$$\log_{10}(0.5) = -0.3010$$
$$\log_{10}(0.1) = -1.000$$

Maths reference

Powers that are not whole numbers.

See Maths note 1.5

8.2 Logs on a calculator

To find the common log of a number using a calculator, type in the number and then press the key marked log or lg.

This process can be reversed to find the **antilog** of a number. Type in the log whose number you want to find, then press the keys marked INV and log (or lg). By doing this, you can show that 4.000 is the antilog of 0.6021, and 2.153 is the antilog of 0.333.

Notice that using the INV and log keys to find the antilog of a number x gives exactly the same result as using the y^x key to find 10^x.

8.3 Logs; multiplication and division

There is a useful relationship between the logs of x and y and the log of their product xy.

Using the definition of a base 10 log:

if $\log_{10}(x) = a$, then $x = 10^a$

and if $\log_{10}(y) = b$, then $y = 10^b$

So we can write

$$xy = 10^{(a+b)}$$

In other words

$$\log_{10}(xy) = a + b$$

so

$$\log_{10}(xy) = \log_{10}(x) + \log_{10}(y)$$

You can illustrate this relationship using a calculator to look up the logs of various numbers and see how they relate to one another. For example: $\log_{10}(5) = 0.69897$, $\log_{10}(4) = 0.60206$,

$$\log_{10}(5 \times 4) = \log_{10}(20) = 1.30103$$
$$= 0.69897 + 0.60206$$
$$= \log_{10}(5) + \log_{10}(4)$$

There is a similar relationship for division:

$$\frac{x}{y} = 10^{(a-b)}$$

$$\log_{10}\left(\frac{x}{y}\right) = a - b$$

$$\log_{10}\left(\frac{x}{y}\right) = \log_{10}(x) - \log_{10}(y)$$

Maths reference

Combining powers
See Maths note 1.3

Logs and powers of 10
See Maths note 5.1

8.4 Logs and powers

We can extend Maths note 8.3 by considering powers. For example:

$$\log_{10}(x^3) = \log_{10}(x) + \log_{10}(x) + \log_{10}(x) = 3\log_{10}(x)$$

A similar line of reasoning says that for any whole number, n,

$$\log_{10}(x^n) = n\log_{10}(x)$$

This rule also works for powers that are not whole numbers, and for negative powers. For *any* value of y we can write

$$\log_{10}(x^y) = y\log_{10}(x)$$

For example

$$\log_{10}(9^{1/2}) = \log_{10}(\sqrt{9}) = \log_{10}(3)$$
$$\log_{10}(9) = 0.95424$$
$$\log_{10}(3) = 0.47712 = 0.95424 \div 2$$

and

$$\log_{10}(5^{-1}) = \log_{10}\left(\tfrac{1}{5}\right)\log_{10}(0.2)$$
$$\log_{10}(5) = 0.69897$$
$$\log_{10}(0.2) = -0.69897$$

8.5 Logs to other bases; natural logs

We can define logarithms to base 10 using numbers expressed as power of 10. But there is nothing special about the number 10. We can use any number as the base of logarithms.

Maths reference

Logs and powers of 10
See Maths note 8.1

For example, logarithms to base 2 use the fact that any number can be expressed as a power of 2, as shown in Table 10 (overleaf). Just as with base 10 logs, we are not restricted to whole-number powers.

By trying some examples using numbers from Table 10, you can demonstrate that the relationships set out in Maths notes 8.3 and 8.4 for base 10 logs also work with logs to base 2. (Notice that in *any* base, the log of 1 is zero.)

Number x	$\log_2(x)$
$16 = 2^4$	4
$8 = 2^3$	3
$4 = 2^2$	2
$2 = 2^1$	1
$1 = 2^0$	0
$0.5 = 2^{-1}$	−1
$0.25 = 2^{-2}$	−2
$1.4142 = 2^{1/2}$	0.5
$5 = 2^{2.32193}$	2.32193
$2.567 = 2^{1.36}$	1.36

Table 10 Some numbers and their logs to base 2

We can go further. The number at the base of a system of logs does not itself have to be a whole number. It can be *any* positive number. For *any* system of logs, it is always true that

$$\log(xy) = \log(x) + \log(y)$$

$$\log_{10}\left(\frac{x}{y}\right) = \log(x) - \log(y)$$

$$\log(x^y) = y\log(x)$$

Natural logs

Apart from common, base 10, logs, the system of logs most widely used in physics is that of **natural logarithms**. This system uses the number e as its base, where e is the **exponential number** (e ≈ 2.718) which arises in the mathematical description of many naturally occurring changes.

The natural logarithm of a number x is written as $\log_e(x)$ or $\ln(x)$. Most calculators have a natural log key, which is usually labelled $\ln x$.

The inverse of a natural log (its antilog) can be found on a calculator using the INV key and the $\ln x$ key. Finding the natural antilog of a number x is the same as finding e^x.

Table 11 lists some examples of natural logs. By picking suitable numbers from Table 11, you can demonstrate that the relationships for multiplication and division and powers also work for natural logs. Notice that $\log_e(e) = 1$, and that $\log_e(1) = 0$.

Maths reference

Exponential changes
See Maths note 9.1

Exponential functions
See Maths note 9.2

Number x	$\log_e(x)$
0.2	−1.69094
0.5	−069315
1	0.000000
2	0.69315
2.7183	1.0000
3	1.09861
4	1.38629
5	1.60943
9	2.19722
20	2.99573

Table 11 Some numbers and their natural logarithms

8.6 Using log scales

When we want to plot a graph of numbers that cover a large range of values, we often use a so-called **logarithmic scale** in order to fit the largest numbers on the graph paper while still being able to distinguish between the smallest numbers. Figure 10 shows such a scale. Notice that the powers of 10 are equally spaced – the scale might be better described as a 'powers scale'. The name 'logarithmic' becomes more obvious when we write the (base 10) logs next to the values of x. The logs of x go up in equal steps just like an ordinary graph scale.

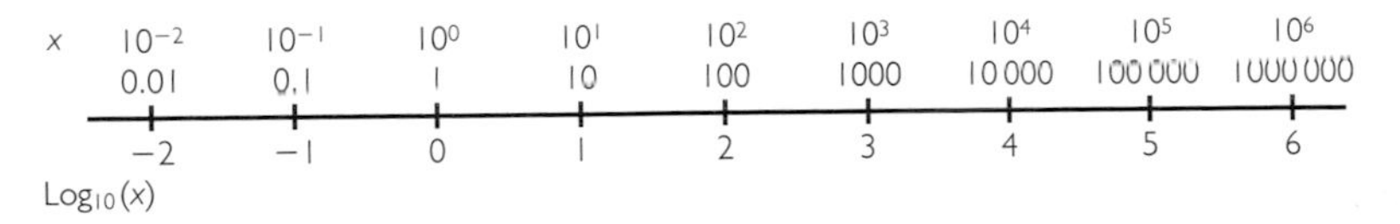

Figure 10 A logarithmic scale

When dealing with numbers other than whole-number powers of 10, there are three ways of plotting and reading a log scale — which all amount to the same thing really.

Using logs to plot and read a scale

If you are using ordinary linear graph paper, you need to use logs in order to plot and read a log scale accurately. It can be helpful to write the base 10 logs next to the numbers to be plotted, as was done in Figure 10. Figure 11 shows an expanded view of the part of Figure 10 between $x = 100$ and $x = 1000$, plotted on ordinary graph paper. Table 12 lists some numbers in this range and their base 10 logs.

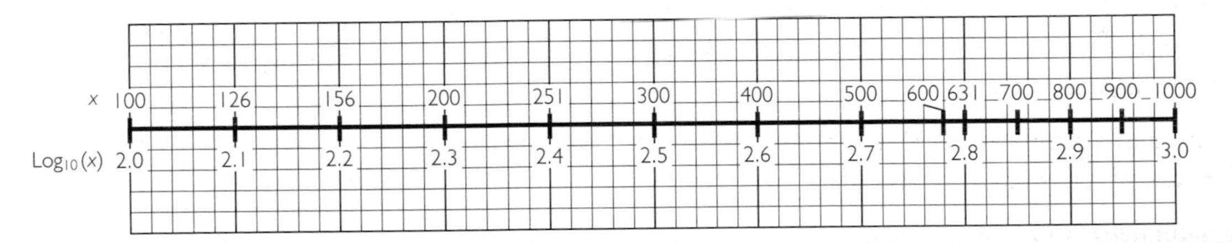

Figure 11 A log scale covering the range $x = 100$ to $x = 1000$

To plot a number x on the scale shown in Figure 11:

use a calculator to find $\log_{10}(x)$

plot $\log_{10}(x)$ on the side of the scale labelled $\log_{10}(x)$.

To read a number from the scale shown in Figure 11:

read $\log_{10}(x)$ from the side of the scale labelled $\log_{10}(x)$

find its antilog using a calculator.

x	$\log_{10}(x)$
100	2.000
126	2.100
158	2.200
200	2.300
251	2.400
300	2.477
400	2.602
500	2.698
600	2.778
631	2.800
700	2.845
800	2.903
900	2.954
1000	3.000

Table 12 The numbers plotted on Figure 11 and their logs

A short cut

If you do not require values to be plotted or read on a log scale very precisely, there is a useful short cut that avoids having to look up logs and antilogs.

From Figure 11, notice that on the side of the scale labelled *x*:

> 300 lies roughly halfway between 100 and 1000
>
> 200 lies roughly one-third of the way along from 100 to 1000
>
> the distance from 100 to 200 is exactly the same as that from 200 to 400 and from 400 to 800

This pattern is repeated between *any* two adjacent whole-number powers (e.g. 30 lies mid-way between 10 and 100 on a log scale, 0.3 lies mid-way between 0.1 and 1.0, etc.).

Bearing these points in mind, it is possible to estimate the value of *x* just by looking at where it lies on the scale between two whole-number powers.

Log graph paper

Figure 12 shows a piece of logarithmic graph paper. It covers the range of numbers between two adjacent powers of 10. The grid lines are labelled 1, 2, 3 and so on, corresponding to the positions of 100, 200, 300 on the top scale of Figure 11 – or to 1 ×, 2 ×, 3 × ... any power of 10 that you care to choose.

Plotting and reading values using such a piece of graph paper is just like using ordinary graph paper – the only difference is that the grid spacing varies across the page. There is no need to calculate logs or antilogs, as the paper is already labelled with values of *x*.

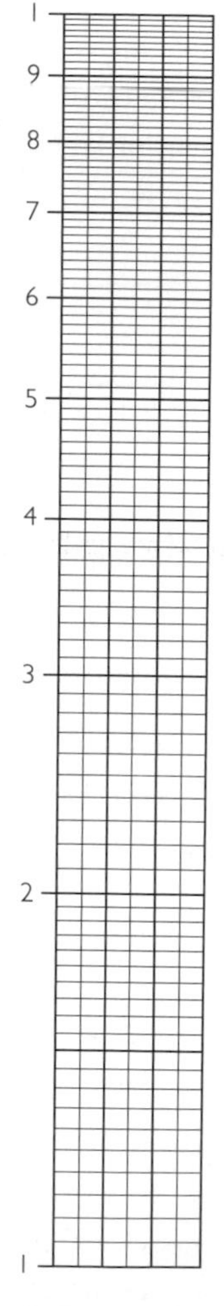

Figure 12 A piece of logarithmic paper

8.7 Using log graphs

There are essentially three reasons for plotting graphs using logarithmic scales:

- to represent a large range of values on a compact scale;
- to see whether data are described by an exponential relationship;
- to see whether data are described by a power law.

We will deal with each of the last two in turn.

Exponential relationships

An exponential relationship between two variables *x* and *y* is one that is described as an equation of the form

$$y = Ae^{kx}$$

where *A* and *k* are constants; *k* is sometimes called the **decay constant** (if negative) or the **growth constant**, but often has a particular name according to the situation being described. For example, the attenuation of a signal in a cable or optical fibre is described by the equation

$$I = I_0 e^{-\mu x}$$

where I_0 is the initial intensity of the signal, *I* is its intensity after travelling through a distance *x*, and μ the attenuation coefficient.

A plot of *y* against *x* gives a curve, in which *y* changes by equal fractions for equal steps in *x*.

A direct way to determine whether two variables are described by an exponential relationship is to plot a **log-linear graph**. For example, taking logs of the attenuation equation gives

$$\log(I) = \log(I_0) - \mu x \log(e)$$

This works whichever base of logs we choose. A graph of log (I) on the y-axis against x on the x-axis has the form $y = mx + c$; in other words, it is a straight line.

If we choose to take natural logs, the equation becomes simpler

$$\log(I) = \log_e(I_0) - \mu x \log_e(e)$$

i.e.

$$\log(I) = \log(I_0) - \mu x$$

A graph of log (I) against x has a gradient equal to $-\mu$.

This is an example of a general rule: a log-linear graph of an exponential relationship is a straight line whose gradient is equal to the growth or decay constant (see Figure 13).

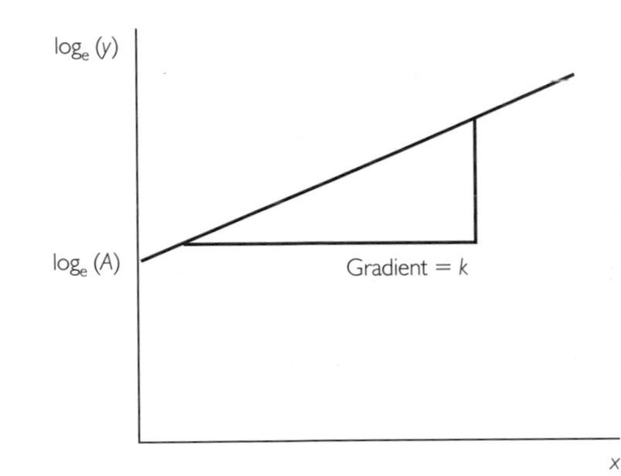

Figure 13 A log-linear graph of the relationship $y = Ae^{kx}$

Maths reference

Gradient of a linear graph
See Maths note 5.3

Logs; multiplication and division
See Maths note 8.3

Logs and powers
See Maths note 8.4

Logs to other bases; natural logs
See Maths 8.5

Power-law relationships

Two variables x and y are related by a **power law** if they obey an equation of the form

$$y = Ax^p$$

where A and p and constants; p is called the **exponent** (not to be confused with the exponential number e!).

An example of such relationship is Coulomb's law

$$F = \frac{kqQ}{r^2}$$

which describes how the magnitude of the electrostatic force, F, between two charges, q and Q, depends on their separation r.

A graph of x against y gives a curve whose shape depends on the exponent p.
A power-law curve can superficially look like an exponential growth or decay curve, but it does *not* follow the 'equal fractions in equal steps' pattern that characterises exponential curves.

One direct way to determine whether variables obey a power law is to plot a **log-log graph**. For example, taking logs (to *any* base) of the Coulomb's law equation gives

$$\log(F) = \log(kqQ) - 2\log(r)$$

A graph of log (F) against log (r) has the form $y = mx + c$. It is a straight line with gradient -2.

This is an example of a general rule. A log-log graph of a power-law relationship is a straight line whose gradient is equal to the exponent (see Figure 14).

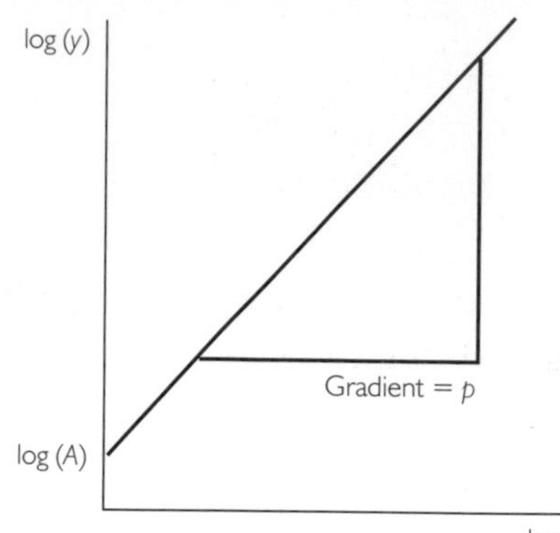

Figure 14 A log-log plot of the relationship $y = Ax^p$

Maths reference

Gradient of a linear graph
See Maths note 5.3

Logs; multiplication and division
See Maths note 8.3

Logs and powers
See Maths note 8.4

Logs to other bases; natural logs
See Maths note 8.5

9 Exponentials

9.1 Exponential changes

Many naturally occurring changes follow a pattern in which the rate of change is proportional to the amount, or number, of something present. Such a change is called an **exponential change**. Such changes are represented by equations of the form

$$\frac{dy}{dt} = kt \quad \text{or} \quad \frac{dy}{dx} = kx$$

depending on whether the change takes place over time, t, or over distance, x.

One example of exponential change is radioactive decay, where the number of radioactive disintegrations per second is proportional to the number of unstable nuclei present in a sample. Mathematically, this decay can be represented by an equation

$$\frac{dN}{dt} = -\lambda N$$

where N is the number of unstable nuclei and λ the **decay constant** for the particular isotope involved.

Another example is population growth

$$\frac{dN}{dt} = kN$$

where now N is the number of organisms in a population and k the **growth constant**.

Figure 15 shows graphs of exponential decay and growth. These graphs have certain characteristics that are unique to exponential changes (and which are related to one another):

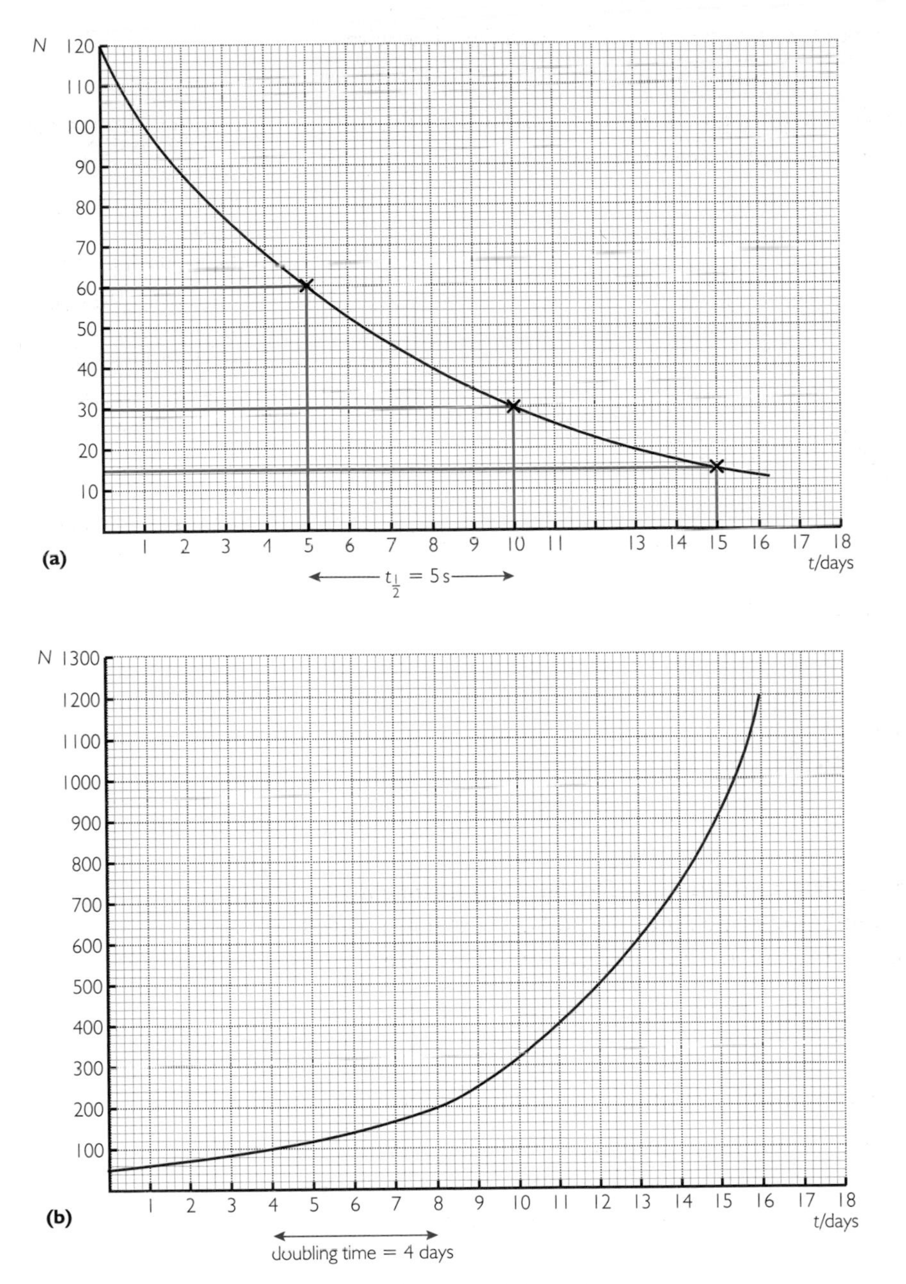

Figure 15 Graphs of exponential change; (a) decay, (b) growth

- the gradient at any point is proportional to the value of y at that point (this is described by the equations stated above)
- equal intervals of time or distance result in equal fractional changes in y
- the time or distance taken for y to halve, or double, is independent of the initial value of y; when dealing with decay over time, this time is called the **half-life** ($t_{1/2}$) of the decay.

9.2 Exponential functions

Exponential changes can always be described by an equation of the form

$$y = y_0 a^{bt} \quad \text{or} \quad y = y_0 a^{bx}$$

where a and b are constants and y_0 is the value of y when $x = 0$ or $t = 0$. For growth over time or distance b is positive, for decay b is negative.

For example:

- if $N = 2^t$, then when t increases by *1*, N doubles.
- if $y = 3 \times 10^{-x}$, then when x increases by 1, y is divided by 10.
- radioactive decay can be described by the equation

$$N = N_0 \times 2^{-(t/t_{1/2})}$$

where $t_{1/2}$ is the half-life of the decay.

All mathematical functions of the type $y = y_0 a^{bt}$ are called **exponential functions**.

One very commonly used exponential function uses the **exponential number** e (e ≈ 2.718). This number arises from the mathematical description of situations where the constant k in the growth or decay equation is numerically equal to 1.

If

$$\frac{dy}{dx} = y$$

then the change is described by

$$y = y_0 e^x$$

Similarly, if

$$\frac{dy}{dt} = y$$

then

$$y = y_0 e^t$$

The number e can be used to describe *all* exponential changes – and this is the most common way that we write them. If

$$\frac{dy}{dx} = ky$$

then

$$y = y_0 e^{kx}$$

and, similarly, if

$$\frac{dy}{dt} = ky$$

then

$$y = y_0 e^{kt}$$

For example, if a radioactive decay is described by the equation

$$\frac{dN}{dt} = -\lambda N$$

then it is also described by the equation

$$N = N_0 e^{-\lambda t}$$

Some texts write e^x as exp (x), and so on, so our last example would be written

$$N = N_0 \exp(-\lambda t)$$

9.3 Exponentials and logs

The exponential number e (e ≈ 2.718) is used as the base of so-called natural logarithms. By taking natural logs of exponential growth and decay equations, we arrive at some useful relationships. If

$$y = y_0 e^{kt}$$

then

$$\log_e (y) = \log_e (y_0) + kt$$

This form of the equation is useful as it shows how we can use a log-linear graph to determine whether a change is exponential and, if it is, to determine the value of k.

It also enables us to find the value of k if we know y_0 and y at a given time, t:

$$kt \log_e (y) - \log_e (y_0) = \log_e \left(\frac{y}{y_0}\right)$$

Similarly, in radioactive decay

$$N = N_0 \exp (-\lambda t)$$

so

$$\log_e(N) = \log_e(N_0) - \lambda t$$

$$\lambda t = \log_e \left(\frac{N_0}{N}\right)$$

This expression leads to a useful relationship between λ and the half-life, $t_{1/2}$, of the decay. If $t = t_{1/2}$, then by definition $N = N_0/2$,

so

$$\lambda t_{1/2} = \log_e (2)$$

$$t_{1/2} = \frac{\log_e (2)}{\lambda}$$

Maths reference

Logs to other bases; natural logs
See Maths note 8.5

Maths reference

Using log graphs
See Maths note 8.7

Maths reference

Logs; multiplication and division
See Maths note 8.3

Index